Spatial and Spatio-Temporal Geostatistical Modeling and Kriging

WILEY SERIES IN PROBABILITY AND STATISTICS

Established by WALTER A. SHEWHART and SAMUEL S. WILKS

A complete list of the titles in this series appears at the end of this volume.

Spatial and Spatio-Temporal Geostatistical Modeling and Kriging

José-María Montero and Gema Fernández-Avilés

University of Castilla-La Mancha, Spain

Jorge Mateu

University Jaume I of Castellón, Spain

WILEY

This edition first published 2015

Registered office
John Wiley & Sons Ltd, The Atrium, Southern Gate, Chichester, West Sussex, PO19 8SQ, United Kingdom

For details of our global editorial offices, for customer services and for information about how to apply for permission to reuse the copyright material in this book please see our website at www.wiley.com.

Library of Congress Cataloging-in-Publication Data

Montero, José María.

Spatial and spatio-temporal geostatistical modeling and kriging / José-María Montero, Department of Statistics, University of Castilla-La Mancha, Spain, Gema Fernández-Avilés, Department of Statistics, University of Castilla-La Mancha, Spain, Jorge Mateu, Department of Mathematics, University Jaume I of Castellon, Spain.

pages cm

Includes bibliographical references and index.

ISBN 978-1-118-41318-0 (pbk.)

1. Geology–Statistical methods. 2. Kriging. I. Fernández-Avilés, Gema. II. Mateu, Jorge. III. Title.

QE33.2.S82M66 2015

551.01'5195–dc23

2015015489

A catalogue record for this book is available from the British Library.

Typeset in 10/12pt TimesLTStd by SPi Global, Chennai, India

1 2015

To my parents Pepe and Maruja, my wife Gema, my children Beatriz and José, my sister Nines and my niece Eva … and also to Fito and Atlético de Madrid.
They never let me walk alone.

José-María Montero

To my parents, Jose and Juli, and to my husband, Chema. Thanks for loving me.

Gema Fernández-Avilés

To my sons and daughters, you are my life motif

Jorge Mateu

Contents

Foreword by Abdel H. El-Shaarawi

It is a great pleasure to write the Foreword for this important book on space-time modeling. I have known the authors for many years and am very familiar with their significant contributions to the subject. I consider the book to be an important addition to the current knowledge about space-time models.

The authors have written an excellent book on a topic of immense interest to a broad class of researchers, students, scientists and decision-makers. The book presents the current state of the art on modeling spatial/temporal complexity in a clear and accurate manner with illustrations using both artificial and real data examples that will help the readers to understand the steps required to build and assess spatial/temporal models. Although many of the examples discussed are related to pollution and the environment, the techniques presented are sufficiently general and can be easily used in other areas of investigation.

Clearly most phenomena encountered in real life include some aspects of space and time that need to be taken into account in the modeling process. The period since 1970 has been a time of spectacular growth in our understanding and ability to develop "valid" sophisticated spatial/temporal models to understand the changes in large systems; examples are climate change and long-range transport of air pollution and acid rain in the eastern part of North America and the Scandinavian countries. This growth can in large part be attributed to one factor: the exponential growth in our ability to both compute and measure data sets. We now have more powerful computers that allow us to store and analyze large data sets with relative ease and advanced instruments for measuring. The book shows us how to use these facilities effectively to develop, fit and assess models for spatio-temporal processes. The primary reason for working in this area is to be able to predict unknown values at unmeasured locations and at future times. Such models can thus be used to produce maps and to identify regions (problem areas) in the domain of interest where, for example, the level of pollution exceeds the permissible level and thus could be of importance to human or ecosystem health. The book describes how to include in the modeling processes the two components: a systematic component with available explanatory variables and the spatial and temporal correlation component, and how the two components interact to produce reliable forecasts.

The book consists of nine chapters which can be grouped into four parts. The first part consists of the first two chapters and is devoted to providing an excellent introduction to the nature and type of spatial data, along with the associated probabilistic and statistical terminology about random functions, stationarity, etc. The second part, consisting of Chapters 3 and 4, involves the development of valid covariance and semivariogram models in two-dimensional and three-dimensional space and explains the ways to fit these models to the corresponding empirical counterparts. This part covers the use of kriging equations to predict the values of the random function at non-observed points and blocks. A survey of different types of kriging is presented such as: simple kriging, ordinary kriging, universal kriging, direct residual kriging, iterative residual kriging, modified iterative residual kriging, median-polish kriging, and non-linear kriging (disjunctive kriging and indicator kriging). In addition, plenty of numerical examples are used to illustrate the step-by-step process of implementing the techniques covered.

The third section, Chapters 5–8, provides a natural extension of the methods in the second part to the case when the process evolves over time. Again the approach used starts with the empirical estimation of the space-time variogram and covariogram, then appropriate valid covariance models are fitted to the empirical estimates introduced. The final outcome is an adequate model that will be used for predictions. One important aspect of this part is the presentation of an extensive survey of available valid covariance functions. This will be of interest to researchers in the area of spatial temporal models because of the particularly detailed proofs of the theoretical work given in Appendix D.

The final part, Chapter 9, provides a brief discussion of functional geostatistics. This is an area of immense interest and is currently under extensive development. This chapter provides good starting point for research in this area.

The material in the book can be used for teaching a course on spatial statistics to graduate students or senior undergradute students. It is also an excellent reference book for researchers.

Abdel H. El-Shaarawi
National Water Research Institute
Burlington, Ontario, Canada

The American University in Cairo
Cairo, Egypt

Foreword by Hao Zhang

One of the areas that constantly presents stimulating research problems is spatial and spatio-temporal statistics. Most of the studies in environmental sciences, agriculture, natural resources and ecology involve data collected at various spatial locations and over a certain period of time. Technological advances make it possible to collect large-scale and massive amounts of data. Societal concerns about climate change and environmental issues are also contributory factors to explain the increased amount of interest in spatio-temporal statistics.

One distinctive feature of this kind of data is that they are correlated or dependent. This dependence should be included in the statistical modeling, and it is modeled through the covariance function or covariogram of a Gaussian stochastic process.

When I explain what spatial or spatio-temporal correlation means in my spatial statistics course, students often ask why there is spatial or temporal dependence. I often give a simplistic answer that the exact reasons are not known but it could be due to the unknown or unobserved variables. For example, despite our best effort in modeling the mean by incorporating all *available* observed explanatory variables, the residuals may still show spatial correlation. In this case, modeling the covariance appropriately may improve the efficiency of the estimation of the mean, and offset the effects of the unobserved variables that may affect the mean. In addition, the covariance function is essential for the prediction of a value at an unobserved location or time. Indeed, when making a prediction, there is an advantage in keeping the mean dependent on as few variables as possible.

This book provides a comprehensive coverage of covariance functions, some of which are relatively new. It includes details and some new materials that are hard to find in any other existing book. It is a very welcome addition to the current literature in spatio-temporal statistics. I trust readers will find it a very valuable resource.

Hao Zhang
Purdue University
West Lafayette, IN 47906
USA

List of figures

List of tables

About the companion website

This book's companion website: www.wiley.com/go/montero/spatial provides you with:

- Use of the widely spread and used free software R to show the open source code used throughout the book, and provide practical guidance to the practitioner.
- Use of extensively used libraries geoR, RandomFields, fields, gstat, CompRandFld, scatterplot3d, fda, and animation.
- Worked out examples with the code used for each chapter of the book.
- Descriptive coded examples, including a number of attractive graphical outputs.
- Developed spatio-temporal modelling code to run variogram and covariance fitting in space and in space-time, simulations and predictions.

1

From classical statistics to geostatistics

1.1 Not all spatial data are geostatistical data

As pointed out in Schabenberger and Gotway (2005, p. 6), because spatial data arise in a myriad of fields and applications, there is also a myriad of spatial data types, structures and scenarios. Thus, an exhaustive classification of spatial data would be a very difficult challenge and this is why we have opted for embracing the general, simple and useful classification of spatial data provided by Cressie (1993, pp. 8–13). Cressie's classification of spatial data is based on the nature of the spatial domain under study. Depending on this, we can have: geostatistical data, lattice data and point patterns.

Following Cressie (1993), let $\mathbf{s} \in \mathbb{R}^d$ be a generic location in a d-dimensional Euclidean space and $\{Z(\mathbf{s}): \mathbf{s} \in \mathbb{R}^d\}$ be a spatial random function, Z denoting the attribute we are interested in.

Geostatistical data arise when the domain under study is a fixed set D that is continuous. That is: (i) $Z(\mathbf{s})$ can be observed at any point of the domain (continuous); and (ii) the points in D are non-stochastic (fixed, D is the same for all the realizations of the spatial random function). From (i) it can be easily seen that geostatistical data are identified with spatial data with a continuous variation (the spatial process is indexed over a continuous space).

Some examples of geostatistical data are the level of a pollutant in a city, the precipitation or air temperature values in a country, the concentrations of heavy metals in the top soil of a region, etc. It is obvious that, at least theoretically, the level of a specific pollutant could be measured at any location of the city; the same can be said for measurements of precipitations or air temperatures across a

---

Spatial and Spatio-Temporal Geostatistical Modeling and Kriging, First Edition.
José-María Montero, Gema Fernández-Avilés, and Jorge Mateu.

Companion Website: www.wiley.com/go/montero/spatial

country or concentrations of a heavy metal across a region. However, in practice, an exhaustive observation of the spatial process is not possible. Usually, the spatial process is observed at a set of locations (for example, the level of a specific pollutant in a city is observed at the points where the monitoring stations are located) and, based on such observed values, geostatistical analysis reproduces the behavior of the spatial process across the entire domain of interest. Sometimes the goal is not so ambitious and the aim is the prediction at one or some few non-observed points or the estimation of an average value over small areas, or over the whole area under study. In geostatistical analysis the most important thing is to quantify the spatial correlation between observations (through the basic tool in geostatistics, the semivariogram) and use this information to achieve the above goals.

Figure 1.1 depicts the locations where the main pollutants are measured in Madrid, Spain (the location of the monitoring stations), along with the mapping of the level of nitrogen oxide (NOx) for the whole city (average of the NOx levels at 10 pm in the week days of the 50th week of 2008).

The fact that the attribute of interest is continuous or discrete has nothing to do with the data being geostatistical or not. Also, how observation points are selected (according to our convenience, using a monitoring network, using a probabilistic sampling scheme ...) has nothing to do with the data being geostatistical or not.

Lattice data arise when: (i) the domain under study D is discrete, that is, $Z(\mathbf{s})$ can be observed in a number of fixed locations that can be enumerated. These locations can be points or regions, but they are usually ZIP codes, census tracks, neighborhoods, provinces, etc., and the data in most of cases are spatially aggregated data over these areal regions. Although these regions can be regularly shaped, usually the shape they exhibit is irregular, and this, together with the spatially aggregated character of the data, is why lattice data are also called regional data. And (ii) the locations in D are

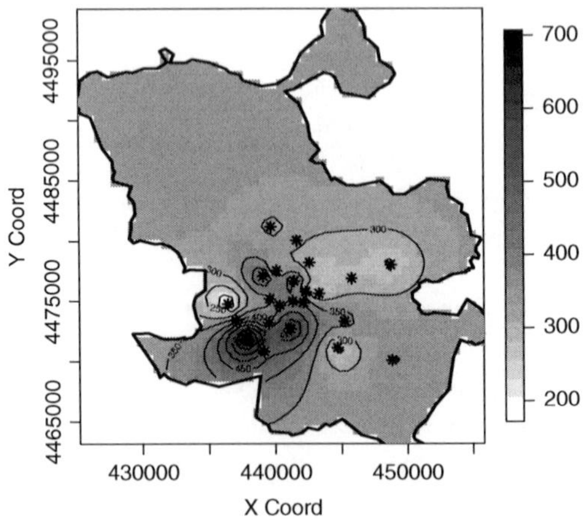

Figure 1.1 Location of the pollution monitoring stations in Madrid and map of predicted NOx levels (10 pm; average of the week days; 50th week of 2008) using geostatistical techniques.

non-stochastic. Of course, a core concept in lattice data analysis is the neighborhood. Some examples of lattice data include the unemployment rate by states, crime data by counties, agricultural yields in plots, average housing prices by provinces, etc. Unlike geostatistical data, lattice data can be exhaustively observed and in this case prediction makes no sense. However, smoothing and clustering acquire special importance when dealing with this type of spatial data. Similar to geostatistical data, the response measured can be discrete or continuous, and this has nothing to do with the data being lattice data or not.

Figure 1.2 depicts the percentage of households with problems of pollution and odors in each census tract of Madrid, Spain, in 2001. As can be observed, the attribute under study is aggregated over each census tract, the domain is the set of the 128 census tracts (discrete) and sites in the domain (the census tracts) are fixed (non-stochastic).

While in both geostatistical and lattice data the domain D is fixed, in point pattern data it is discrete or continuous, but random. Point patterns arise when the attribute under study is the location of events (observations). That is, the interest lies in where events of interest occur. Some examples of point patterns are the location of fires in Castilla-La Mancha, a Spanish region (see Figure 1.3), the location of trees in a forest or the location of nests in a breeding colony of birds, among many others. In these cases, it is obvious that D is random and the observation points do not depend on the

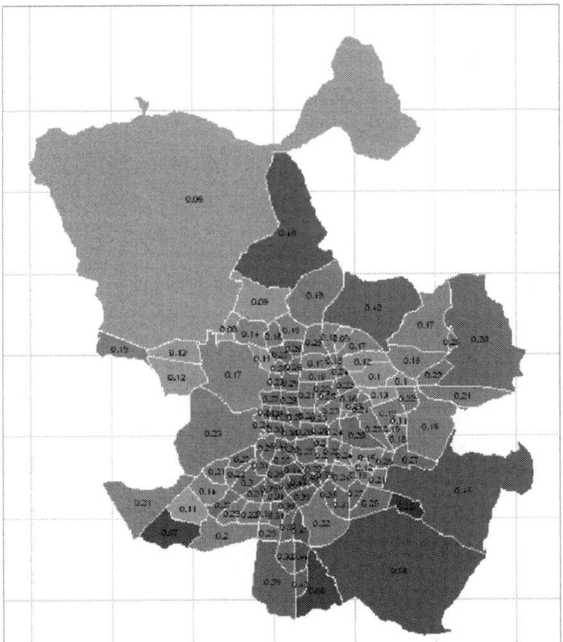

Figure 1.2 Percentage of households with problems of pollution and odors in Madrid, Spain, 2001 *(census tracts).*

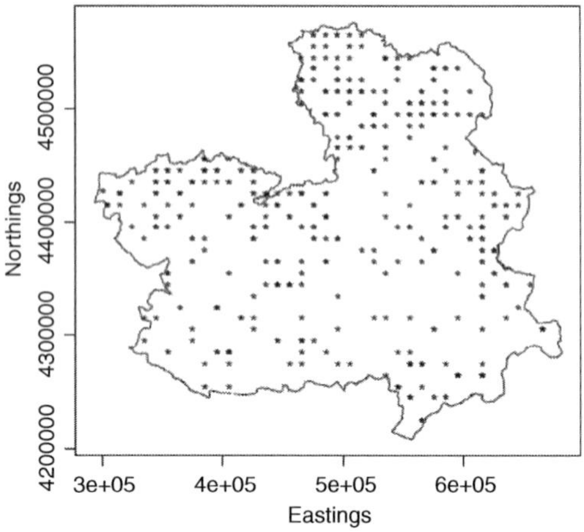

Figure 1.3 Fires in Castilla-La Mancha, Spain, 1998.

researcher. The realizations of spatial point processes are arrangements or patterns of points and we can observe all the points of such patterns or a sample. The main goal of point pattern analysis is to determine if the location of events tends to exhibit a systematic pattern over the area under study or, on the contrary, they are randomly distributed. More specifically, we are interested in analyzing if the location of events is completely random spatially (the location where events occur is not affected by the location of other events), uniform or regular (every point is as far from all of its neighbors as possible) or clustered or aggregated (the location of events is concentrated in clusters). Some other interesting questions in point pattern analysis include: How does the intensity of a point pattern vary over an area? Over what spatial scales do patterns exist? If along with the location of events a stochastic attribute is observed, the pattern is called a marked pattern; otherwise it will be named an unmarked pattern. Obviously, marked patterns extend the possibilities of spatial analysis.

The above refers to merely spatial data, but in recent years the spatio-temporal data analysis has become a core research area in a large variety of scientific disciplines. In the spatio-temporal context, the observed data are viewed as partial realizations of a spatio-temporal random function which spreads out in space and evolves in time. Thus, spatio-temporal data simultaneously capture spatial and temporal aspects of data.

Recalling some of the examples used to illustrate the types of spatial data, if we observe every hour the level of a specific pollutant in a city at the points where the monitoring stations are located, we have a spatio-temporal geostatistical dataset. Now, based on the spatio-temporal observations, we aim to reproduce the behavior of the spatio-temporal pollution process, or simply predict its value at a space-time point. Geostatistics takes advantage of the spatio-temporal correlations existing in the spatio-temporal data (the interaction of space-time is crucial) to make predictions at unobserved space-time locations. If we annually record the percentage of households with problems of pollution and odors in the census tracts of Madrid, we have a collection of spatio-temporal lattice data. Now we can study how the spatial percentage pattern evolves in time. If we observe the location of fires in Castilla-La Mancha for the last ten years, we have a spatio-temporal point pattern database. Now we can express the relationship of points not only by distance but by time lag, and can study whether there is complete spatio-temporal randomness in the disposition of the space-time events, or they exhibit a spatio-temporal aggregation, or their spatio-temporal disposition is regular or uniform.

1.2 The limits of classical statistics

As is well known, classic statistics is based on the independence of the observed values. These observed values are considered as independent realizations of the same random variable. However, when the observed values are anchored in space, the hypothesis of independence is no longer acceptable. As stated in the First Law of Geography: "Everything is related to everything else, but near things are more related than distant things" (Tobler 1970).

In this section we illustrate some of the consequences of ignoring the spatial dependencies in the data and using classical statistical methods with spatial data. For the sake of simplicity, we will focus on the spatial case but the consequences of using classical statistics with spatio-temporal data are the same.

Suppose that $\{Z(\mathbf{s}_1), Z(\mathbf{s}_2), \ldots, Z(\mathbf{s}_n)\}$ are n identically distributed observations recorded at spatial points $\{\mathbf{s}_1, \mathbf{s}_2, \ldots, \mathbf{s}_n\}$. More specifically, suppose that they follow a Gaussian distribution with unknown mean, μ, and known variance, σ_0^2, and that covariances between the observed points are positive and diminish with the distance between them, h, according to the expression: $C(h) = \sigma_0^2 \left(1 - \left(\frac{3}{2}\frac{h}{110} - \frac{1}{2}\frac{h^3}{110^3}\right)\right)$.

If our goal is the estimation of the unknown mean and we ignore the existing spatial correlations, the sample mean, $\bar{Z}$, would undoubtedly be the estimator proposed for μ. Ignoring the spatial correlations, it is well known that $E(\bar{Z}) = \mu$ and $V(\bar{Z}) = \sigma_0^2/n$. But, if we consider such correlations, the sample mean continues to be an unbiased estimator of μ but now its variance is:

$$\begin{aligned}
V(\bar{Z}) &= \frac{1}{n^2}\sum_{i=1}^{n}\sum_{j=1}^{n} C(Z(\mathbf{s}_i), Z(\mathbf{s}_j)) \\
&= \frac{1}{n^2}\left(n\sigma_0^2 + 2\sum_{i<j}^{n} C\left(Z(\mathbf{s}_i), Z(\mathbf{s}_j)\right)\right) \\
&= \frac{1}{n^2}\left(n\sigma_0^2 + 2\sum_{i<j}^{n} \sigma_0^2\left(1 - \left(\frac{3}{2}\frac{h}{110} - \frac{1}{2}\frac{h^3}{110^3}\right)\right)\right) \\
&= \frac{\sigma_0^2}{n}\left(1 + \frac{2}{n}\sum_{i<j}^{n}\left(1 - \left(\frac{3}{2}\frac{h}{110} - \frac{1}{2}\frac{h^3}{110^3}\right)\right)\right),
\end{aligned} \tag{1.1}$$

that is, the variance of the sample mean is larger than in the random sample case.

If we ignore the existing correlations in the data and use $\bar{Z}$ as an estimator of μ, the under-estimation of $V(\bar{Z})$ brings unfortunate consequences for inference about μ. The classical confidence intervals for a specific confidence level will be narrower than they really are, or, in other words, the confidence level of classical intervals is larger than it really is. When testing, if spatial correlations are not taken into account, the p-values will be larger than they really are and this will lead to undesirable rejections of the null hypothesis. In addition, the power of the tests will be overstated.

If the number of observations is 16 and they are recorded over a regularly spaced 4×4 grid ($75m \times 75m$) we have, ignoring the spatial correlations between observations, $V(\bar{Z}) = \sigma_0^2$, so that the 95% confidence interval for μ is $\left(\bar{Z} - 1.96\frac{\sigma_0}{\sqrt{16}}; \bar{Z} + 1.96\frac{\sigma_0}{\sqrt{16}}\right)$ and the test-statistic for $\mu = 0$ is $t = \frac{\bar{Z}}{\sigma_0\sqrt{1/16}}$.

When we take into account the spatial correlations between the observations, we obtain:

(i) $V(\bar{Z}) = \frac{\sigma_0^2}{n}\left(1 + \frac{2}{n}\sum_{i<j}^{n}\left(1 - \left(\frac{3}{2}\frac{h}{110} - \frac{1}{2}\frac{h^3}{110^3}\right)\right)\right) = 0.3969\sigma_0^2$.

(ii) 95% confidence interval:

$$\left(\bar{Z} - 1.96 \times 0.630\sigma_0; \bar{Z} + 1.96 \times 0.630\sigma_0\right) = \left(\bar{Z} - 1.235\sigma_0; \bar{Z} + 1.235\sigma_0\right).$$

(iii) Test-statistic for testing $\mu = 0 : t = \frac{\bar{Z}}{0.630\sigma_0}$.

As can be seen, when we take into account the spatial correlations, the variance of $\bar{Z}$ lengthens by more than six, the width of the 95% confidence interval increases by 2.52 and the value of the test-statistic decreases by 2.52.

When estimating the mean, the classical estimator is unbiased in the presence of spatial correlations. However, if we are interested in the estimation of σ^2, the quasi-variance S^{*2} is not. In effect,

$$E(S^{*2}) = \sigma^2 - \left(\frac{2}{n(n-1)}\sum_{i<j}^{n} C\left(Z(\mathbf{s}_i), Z(\mathbf{s}_j)\right)\right). \tag{1.2}$$

The lesson taken from the above example is that ignoring spatial correlations and continuing to rely on the best estimators for the case of independent observations is not a good idea. These are not the appropriate estimators when data are correlated and using them usually leads to wrong decisions. A better idea would be to obtain the best estimators for the case of spatial correlations.

The perverse effects of not taking into account the spatial correlation and using classic estimators also appear in the field of prediction. Some examples can be found in Cressie (1993, pp. 15–17) and Schabenberger and Gotway (2005, pp. 32–4), among others.

1.3 A real geostatistical dataset: data on carbon monoxide in Madrid, Spain

In order to illustrate the main concepts presented in the book, we use toy examples, theoretical examples, classical examples, and examples based on simulated and real data. Pretend, theoretical and classical examples as well as simulated data-based examples are useful when illustrating and helping to better understand the core geostatistical concepts. However, the use of real data reveals some difficulties that the researcher may encounter during the process of applying geostatistical procedures, which are not usually encountered when data are simulated or simply any set of numbers. This is why we find it extremely exciting to deal with real data.

In this book real data has been used to illustrate specific geostatistical questions. For example, in Section 1.1 data on the percentage of households with problems of pollution and odors in each census tract of Madrid, Spain, 2001, and on the number of fires in Castilla-La Mancha, Spain, 1998, were used to illustrate lattice data and point pattern data, respectively; in Section 4.8 real data on coal-ash for the Robena Mine Property in Greene County, Pennsylvania, are used to illustrate the median-polish kriging procedure; in Section 9.1 daily temperature data for 35 Canadian monitoring stations are used to convert raw data into functional data.

But the real database we use in this book to illustrate the spatial, spatio-temporal, and functional kriging procedures focuses on air pollution in Madrid, Spain, and more specifically on carbon monoxide (CO). The reasons for choosing a database on air pollution (and specifically on CO) for this purpose are the following:

(i) The health effects of air pollution. Air pollution is one of the most important pollution problems in the world. Many health problems (e.g., respiratory and cardiovascular) can be caused or worsened by exposure to air pollution on a daily basis. Therefore, it is not surprising that the World Health Organisation has ranked urban air pollution the 13th contributor to global deaths in its 2012 World Health Report.

(ii) The economic costs (explicit or accounting costs and implicit or social costs) of pollution at the local, provincial, regional, national, or global scale are considerable, and the cost of ignoring them would be even higher.

(iii) The urban environment is presently an issue of the utmost concern, not only for the citizens themselves, increasingly more aware of the conditions of the environment they live in, but also for health authorities and political leaders, who have also become aware of these problems.

The reason for choosing the city of Madrid as the study area is that it is one of the European largest cities where air pollution remains a serious problem. It is the third largest city in the European Union in terms of population and is undergoing

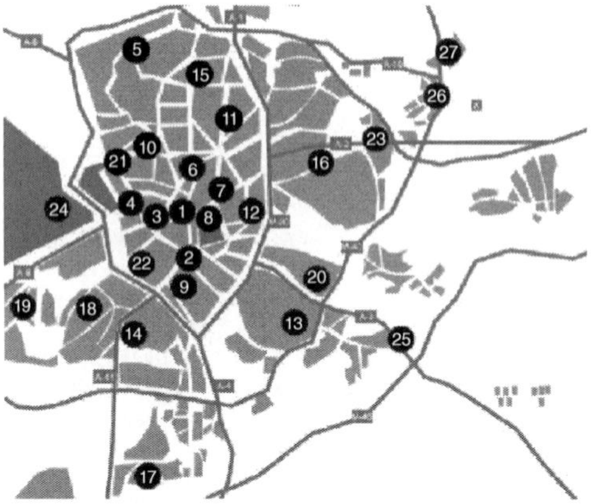

Figure 1.4 Location of the monitoring stations in the city of Madrid.

a significant suburbanization process that is resulting in both people and workplaces being concentrated not only in the heart of the city, but also in surrounding areas. This process is resulting in an increase in Madrid dwellers being much more dependent on cars than in the past, as many of the jobs that have been created are located in areas up to 20km from the center of the city, which means more vehicle exhaust emissions than desired (the main source of CO in the city of Madrid). Although Madrid does have a remarkable urban transport system, the truth is that in autumn, winter, and the month of July, there is a massive tendency to use cars, as well as the urban trips that are so necessary for everyday economic activity.

In the city of Madrid, as in all large cities in the world, there is a monitoring network that measures the level of the pollutants particularly harmful to human health (CO, NO_2, NOx, O_3, PM_{10}, and SO_2, among others) on an hourly basis. The Atmosphere Pollution Monitoring Network (APMN) of the city of Madrid (http://www.mambiente.munimadrid.es/opencms/opencms/calaire) is made up of 27 fixed monitoring stations (see Figure 1.4), although only 23 have been continuously operative (monitoring stations 2, 17, 26, and 27 have experienced in the last few years a number of interruptions in their operation that led us to remove them from the database).

As a consequence of the fact that the main source of CO emissions in the city of Madrid is vehicle exhaust emissions, it is no surprise that (i) levels of CO are different on weekdays and weekends and holidays, and controlling such levels is only of interest in the week (or work) days; and that (ii) there are no significant differences in the levels of CO for the five weekdays. This is the reason why a "typical working day" was created for each week (that is, weekends and bank holidays were eliminated) and the hourly average for working days was computed.

Thus, taking as a starting point the hourly data (in mg/cm^3) provided by APMN in 2008, we constructed a new database as follows:

1. We removed the weekdays and holidays from the initial database and considered only the weekdays as a single entity: "typical working day."

2. For each of the 52 weeks of the year 2008 we averaged the log CO data registered on the five weekdays at each of the 23 monitoring stations operating in Madrid in 2008 on an hourly basis. The reason for making a logarithmic transformation of the raw data and working with the log of CO is that CO hourly measurements do not follow a Gaussian distribution. However, after the logarithmic transformation of the data, a significant departure from a Gaussian distribution in any of the stations cannot be found.

3. As a consequence, for each of the 23 monitoring stations, we have $52 \times 24 = 1248$ spatio-temporal data $\bar{x}^k_{i,j}$, where $\bar{x}$ represents the averaged log of CO, $i = 1, \ldots, 24$ indicates the hour, $j = 1, \ldots, 52$, the week, and $k = 1, \ldots, 23$, the monitoring station. For example, $\bar{x}^9_{13,40}$ indicates the average of log CO measurements registered in station 9 at 13 hours on the five weekdays of week 40. Notwithstanding, in order to directly refer to the pollutant under study, in what follows we will refer to such a mean as $logCO^*$, the asterisk indicating the arithmetic mean.

4. Therefore, the final database contains a total of $23 \times 1248 = 28704$ $logCO^*$ data available from the book's website: www.wiley.com/go/montero/spatial.

Although the monitoring stations that make up the APMN are located in accordance with their ultimate purpose, it is also true that on numerous occasions the size of such stations and/or local government regulations in force at the time are significant restrictions when it comes to locating them in the optimal places. For this reason it is of vital importance to predict the level of the main pollutants in places that are especially affected by traffic congestion and, particularly, at the main intersections in the center of the city, because this is where the health of a large number of Madrid residents and people visiting the city suffers. We will use spatial, spatio-temporal, and functional kriging procedures to, among other things, predict the level of CO at such non-observed sites.

2

Geostatistics: preliminaries

2.1 Regionalized variables

As mentioned in Chapter 1, geostatistics can be defined as the study of regionalized phenomena, that is, phenomena that stretch across space and which have a certain spatial organization or structure.

However, geostatistics is not applied to the regionalized phenomenon as such, which is a physical reality, but to a mathematical description of that reality, that is, a numerical function called the *regionalized variable* or *regionalization*, defined in a geographical space, which is supposed to correctly represent and measure that phenomenon.

In order to delve deeper into the concept of a regionalized variable or regionalization, let us imagine we are interested in a feature of a given phenomenon that spans across space and that several measurements are taken in a domain D at a given moment in time. If the measurements are taken on objects or the like, the objects sampled can be considered a fragment of a larger collection of objects, as many more measurements could have been taken, but were not because the measurement process is very time-consuming and/or very costly, because a great deal of effort is required to collect them, or because of other reasons. If the observations were made at certain points in the domain, infinite measurements could be taken.

Definition 2.1.1 *When* $\mathbf{s}$ *spans the domain under study, D, the set* $\{z(\mathbf{s}), \mathbf{s} \in D\}$, *is called a regionalized variable or regionalization,* $\{z(\mathbf{s}_i),\ i = 1, 2, 3...\}$ *being a collection of values of the regionalized variable and each value of that collection being a regionalized value.*

It is true that a deterministic approach can be used to describe or model a regionalized phenomenon and obtain an accurate assessment of the values of the regionalization

Spatial and Spatio-Temporal Geostatistical Modeling and Kriging, First Edition.
José-María Montero, Gema Fernández-Avilés, and Jorge Mateu.

Companion Website: www.wiley.com/go/montero/spatial

on the basis of a limited number of observations. However, this requires in-depth knowledge of the origin of the phenomenon and the physical or mathematical laws that govern the evolution of the regionalized variable. Furthermore, many of the regionalized phenomena that are usually studied are so complex that a deterministic approach can only partially portray them. That is why the deterministic approach is rejected and the probabilistic approach, which permits modeling both the knowledge of and also the uncertainty surrounding the regionalized random phenomenon, is adopted.

2.2 Random functions

From the probabilistic perspective, the regionalized value can be seen as the result of a random mechanism, a mechanism that is called the *random variable* (rv). If the regionalized values at all the points in the domain D are considered, that is, if we consider the regionalized variable, it can be seen as a reality of an infinitely large set of rv's one at each point in the domain, which is known as a spatial *random function*, (synonyms for this are: stochastic process, random field).

Definition 2.2.1 *When* $\mathbf{s}$ *spans the domain under study, D, we have a family of rv's* $\{Z(\mathbf{s}),\ \mathbf{s} \in D\}$, *which constitutes a spatial random field.*

This methodological decision is one of the cornerstones of geostatistics: *the regionalized variable is interpreted as a realization of a spatial rf.* At this point, we must state that the regionalized variable is often highly locally irregular (which makes it impossible to represent using a deterministic mathematical function) and has a certain spatial organization or structure. The probabilistic approach, or probabilistic geostatistics, which interprets the regionalized variable as a realization of a rf, can take into account all the aspects of regionalization mentioned above, because, as stated in Emery (2000, p. 55):

(i) At each location $\mathbf{s}$, $Z(\mathbf{s})$ is a rv (hence the erratic aspect).

(ii) For any given set of points $\mathbf{s}_1, \mathbf{s}_2, \ldots, \mathbf{s}_k$, the rv's $Z(\mathbf{s}_1), Z(\mathbf{s}_2), \ldots, Z(\mathbf{s}_k)$ are linked by a network of spatial correlations responsible for the similarity of the values they take (hence the structured aspect).

In Figure 2.1 both the local erratic aspect and the global structure (or organization) of the regionalized variable can be appreciated.

Let $Z(\mathbf{s})$ be a rf and let us consider the set of points $(\mathbf{s}_1, \ldots, \mathbf{s}_k)$. Then, the rf $Z(\mathbf{s})$ is characterized by its k-dimensional distribution function. The set of k-dimensional distribution functions for all values of k and all possible choices of $(\mathbf{s}_1, \ldots, \mathbf{s}_k)$ in the domain is called the *spatial law of probability.*

Definition 2.2.2 *For a given rf,* $Z(\mathbf{s})$, *the k-dimensional distribution function*

$$F(z(\mathbf{s}_1), \ldots, z(\mathbf{s}_k)) : \mathbb{R}^d \rightarrow [0, 1]$$

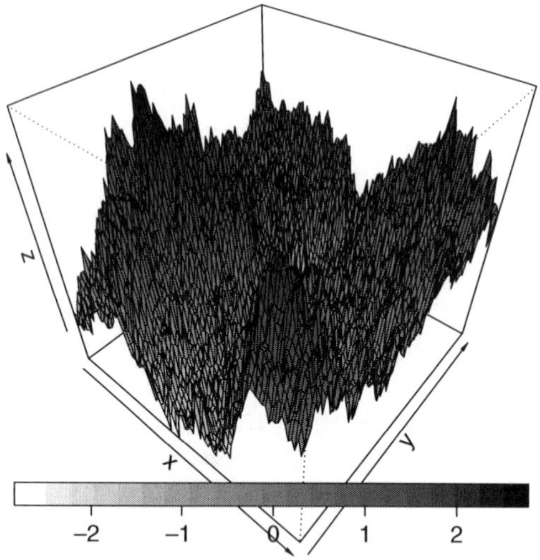

Figure 2.1 Simulation of a regionalized variable.

is defined as:

$$F_{Z_1,\ldots,Z_k}(z(\mathbf{s}_1),\ldots,z(\mathbf{s}_k)) = P[Z(\mathbf{s}_1) \leq z(\mathbf{s}_1),\ldots,Z(\mathbf{s}_k) \leq z(\mathbf{s}_k)]. \tag{2.1}$$

In linear geostatistics it is enough to know the first two moments of the distribution of $Z(\mathbf{s})$. What is more, in most practical applications the available information does not allow higher-order moments to be inferred.

Definition 2.2.3 *The expectation, expected value or first-order moment of a rf is defined as a non-random function of* **s** *that coincides at each point with the expectation of the rv at that point:*

$$\mu(\mathbf{s}) = E(Z(\mathbf{s})) \text{ where } \mu(\mathbf{s}_i) = E(Z(\mathbf{s}_i)), \forall i \in \mathbb{N}. \tag{2.2}$$

It is also called the drift of the rf, especially when it varies with location.

The three second-order moments in geostatistics are: variance, covariance and variogram.

Definition 2.2.4 *The variance of a rf is defined as a non-random function of* **s** *that coincides at each point with the variance of the rv at that point:*

$$V(\mathbf{s}) = V(Z(\mathbf{s})) \text{ where } V(\mathbf{s}_i) = V(Z(\mathbf{s}_i)), \forall i \in \mathbb{N}. \tag{2.3}$$

Definition 2.2.5 *The covariance function of a rf is defined as a non-random function of* $\mathbf{s}_i$ *and* $\mathbf{s}_j$*, such that for any pair of values* $(\mathbf{s}_i, \mathbf{s}_j)$ *coincides with the covariance between the rv at those two points:*

$$C(\mathbf{s}_i,\mathbf{s}_j) = C(Z(\mathbf{s}_i),Z(\mathbf{s}_j)) = E\left((Z(\mathbf{s}_i)-\mu(\mathbf{s}_i))(Z(\mathbf{s}_j)-\mu(\mathbf{s}_j))\right), \forall \mathbf{s}_i,\mathbf{s}_j \in D. \tag{2.4}$$

Definition 2.2.6 *The variogram of the rf is defined as the variance of the first differences of the rf:*

$$2\gamma(\mathbf{s}_i - \mathbf{s}_j) = V\left((\mathbf{s}_i) - Z(\mathbf{s}_j)\right), \forall \mathbf{s}_i, \mathbf{s}_j \in D. \tag{2.5}$$

The function $\gamma(\mathbf{s}_i, \mathbf{s}_j)$ *is called the semivariogram.*

Definition 2.2.7 $Z(\mathbf{s})$ *is a Gaussian rf if for all k and any given set of points* $\mathbf{s}_1, \ldots, \mathbf{s}_k$, *the joint distribution of* $Z(\mathbf{s}_1), \ldots, Z(\mathbf{s}_k)$ *is a multivariate Gaussian distribution.*

A multivariate Gaussian distribution is characterized by a mean vector and a variance-covariance matrix, such that the two first moments of a Gaussian rf completely determine its probability structure. The Gaussian distribution of the rf is a common assumption in geostatistics.

2.3 Stationary and intrinsic hypotheses

2.3.1 Stationarity

As stated by Journel and Huijbregts (1978, p. 30) among others, interpreting a regionalized variable in probabilistic terms as a particular realization of a given rf $\{Z(\mathbf{s}), \mathbf{s} \in D\}$ makes operational sense when it is possible to infer part or all of the law of probability which defines that rf. In this sense, as we will see below, stationarity, which indicates a certain degree of homogeneity in the regionalization across space, is a desirable quality.

Indeed, it would be impossible to infer the probability law of a rf if there were only one realization of the rf, one sole regionalization (it would be like having a sample size of one). In order to make inferences consistently, many realizations are necessary. However, in reality there is only one. Moreover, even though the main problem when making inferences is that referred to above, only part of the realization available is known: at the sampled locations, a set we will call the *observed realization* or the *observed regionalization*. The solution to this second problem is to adopt the *hypothesis of stationarity* or *spatial homogeneity*. The idea behind the hypothesis of stationarity is to substitute repetitions of the (inaccessible) realizations of the rf with repetitions in space, that is, the values observed at different locations in the domain under study have the same characteristics and can be considered as realizations of the same rf in mathematical terms but, in reality, these realizations are not independent, and an additional hypothesis, ergodicity, is normally assumed; see Chilès and Delfiner (1999, pp. 19–22) for details. The hypothesis of stationarity means that the spatial law of probability of the rf, or part of it, is translation invariant. That is, *the probabilistic properties of a set of observations do not depend on the specific locations where they have been measured, but only on their separations.*

Therefore, in mathematical and probabilistic terms, the hypothesis of stationarity refers to the regular behavior in space of the moments of the rf, or the function itself and, as we will see later, there are different degrees of stationarity. This hypothesis will allow us to act as if all the variables that make up the rf had the same probability distribution (or the same moments; we can even relax this assumption) and, as a consequence, we are able to make inferences.

Using the assumed level of spatial homogeneity of the rf that (supposedly) generates the observed realization as a basis, we have the following casuistry: Stationary random function in the strict sense, second-order stationary random function, and intrinsically stationary random function or random function of stationary increments.

2.3.2 Stationary random functions in the strict sense

Definition 2.3.1 *The rf* $\{Z(\mathbf{s}),\ \mathbf{s} \in D\}$ *is said to be stationary in the strict sense, or strictly stationary, if the families of rv's* $Z(\mathbf{s}_1),\ Z(\mathbf{s}_2), \ldots,\ Z(\mathbf{s}_k)$ *and* $Z(\mathbf{s}_1 + \mathbf{h}), Z(\mathbf{s}_2 + \mathbf{h}), \ldots, Z(\mathbf{s}_k + \mathbf{h})$ *have the same joint distribution function for all* k, *and for any given spatial points* $\mathbf{s}_1, \mathbf{s}_2, \ldots, \mathbf{s}_k$ *and any translation vector* $\mathbf{h} \in \mathbb{R}^d$, *providing the values* $\mathbf{s}_1,\ \mathbf{s}_2, \ldots,\ \mathbf{s}_k,\ \mathbf{s}_1 + \mathbf{h},\ \mathbf{s}_2 + \mathbf{h}, \ldots,\ \mathbf{s}_k + \mathbf{h}$, *are contained in the domain D.*

In other words, the joint distribution function of $\{Z(\mathbf{s}_1), Z(\mathbf{s}_2), \ldots, Z(\mathbf{s}_k)\}$ is unaffected by the translation of an arbitrary quantity $\mathbf{h}$. As a result, density functions with a dimension lower than k do not depend on location either. Figure 2.2 depicts the location of four pairs of points (in the 2D case). In the case of strict stationarity, $\{Z(\mathbf{s}_1),\ Z(\mathbf{s}_2)\}$, $\{Z(\mathbf{s}_3),\ Z(\mathbf{s}_4)\}$, and $\{Z(\mathbf{s}_5),\ Z(\mathbf{s}_6)\}$ have the same bivariate distribution function, because the distance between the pairs $\{\mathbf{s}_1, \mathbf{s}_2\}$, $\{\mathbf{s}_3, \mathbf{s}_4\}$, and $\{\mathbf{s}_5, \mathbf{s}_6\}$ is the same.

Generally speaking, this is a tremendously strict condition, which is why this hypothesis is normally relaxed to the so-called assumption of second-order stationarity, which limits the stationarity hypothesis to the first two moments of the rf (recall that in linear geostatistics we are only interested in the two first moments of the rf).

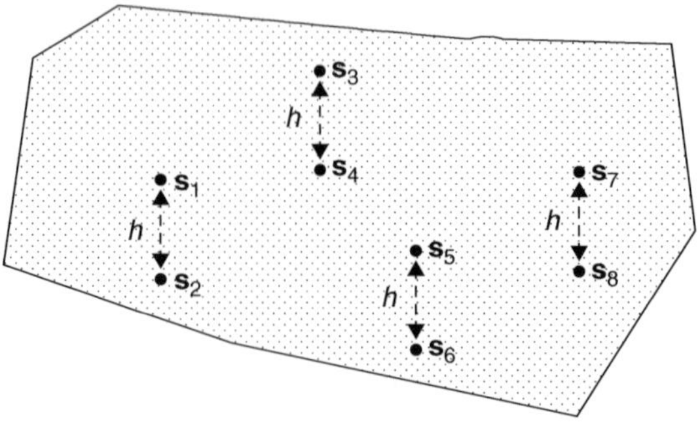

Figure 2.2 Four pairs of points separated by a distance h in a 2D domain.

2.3.3 Second-order stationary random functions

Definition 2.3.2 *The rf* $\{Z(\mathbf{s}),\ \mathbf{s} \in D\}$ *is said to be second-order stationary, weakly stationary or stationary in the broad sense, if it has finite second-order moments (that is, the covariance exists) and verifies that:*

- *The expectation exists and is constant, and therefore does not depend on the location* $\mathbf{s}$*:*

$$E\left(Z(\mathbf{s})\right) = \mu(\mathbf{s}) = \mu. \tag{2.6}$$

- *The covariance exists for every pair of rv's,* $Z(\mathbf{s})$ *and* $Z(\mathbf{s}+\mathbf{h})$*, and only depends on the vector* $\mathbf{h}$ *that joins the locations* $\mathbf{s}$ *and* $(\mathbf{s}+\mathbf{h})$*, but not specifically on them:*

$$C\left(Z(\mathbf{s}), Z(\mathbf{s}+\mathbf{h})\right) = C(\mathbf{h}),\ \forall \mathbf{s} \in D \ \textit{ and } \ \mathbf{h}. \tag{2.7}$$

As the covariance function $C(\mathbf{h})$ of a second-order stationary rf is only a function of $\mathbf{h}$, the variance of the rf exists and is finite, and constant:

$$V(Z(\mathbf{s})) = C(\mathbf{0}) = \sigma^2. \tag{2.8}$$

In light of Equations (2.6) and (2.8), the second-order stationarity hypotheses can be interpreted as if the regionalized variable takes values that fluctuate around a constant value (the mean) and the variation of these fluctuations is the same everywhere in the domain.

Under the second-order stationary hypothesis, the rv's corresponding to the eight points depicted in Figure 2.2 have the same mean and variance, and the covariance between $Z(\mathbf{s}_1)$ and $Z(\mathbf{s}_2)$, $Z(\mathbf{s}_3)$ and $Z(\mathbf{s}_4)$, and $Z(\mathbf{s}_5)$ and $Z(\mathbf{s}_6)$ is also the same, because the four pairs of points are separated by the same distance.

In some cases, in order to model the spatial dependence of second-order stationary rf's, the *correlogram*, or *correlation function*, is used instead of the covariogram and is defined by:

$$Corr\left(Z(\mathbf{s}), Z(\mathbf{s}+\mathbf{h})\right) = \frac{C(\mathbf{h})}{C(\mathbf{0})} = \rho(\mathbf{h}). \tag{2.9}$$

In the case of second-order stationarity, the covariance function and the semivariogram are equivalent when it comes to defining the structure of spatial dependence displayed by the phenomenon, as they verify the following mutual relationship:

$$\begin{aligned}
\gamma(\mathbf{h}) &= \frac{1}{2} V\left(Z(\mathbf{s}+\mathbf{h}) - Z(\mathbf{s})\right) \\
&= \frac{1}{2}\left(V(Z(\mathbf{s}+\mathbf{h})) + V(Z(\mathbf{s})) + 2C(Z(\mathbf{s}+\mathbf{h}), Z(\mathbf{s}))\right) \\
&= \frac{1}{2}\left(C(\mathbf{0}) + C(\mathbf{0}) + 2C(\mathbf{h})\right) \\
&= C(\mathbf{0}) - C(\mathbf{h}).
\end{aligned}$$

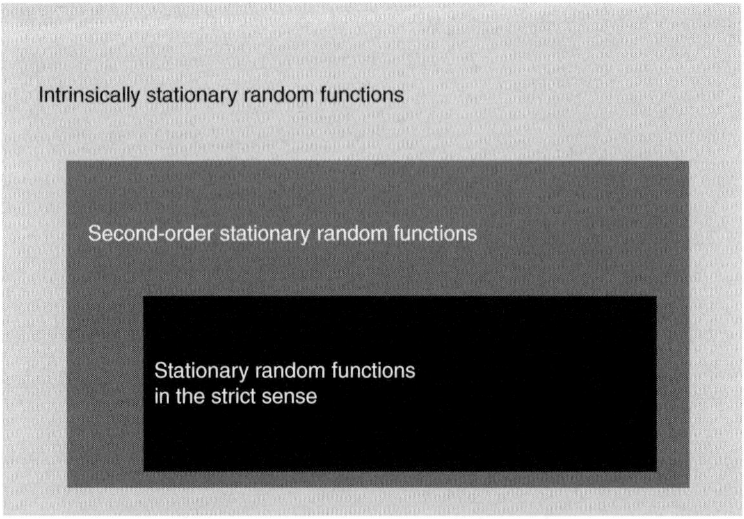

Figure 2.3 Stationary and intrinsic hypotheses.

Note that if a rf is strictly stationary, then it is also stationary in the broad sense. The converse, however, is generally not true (Figure 2.3).

Obviously, for Gaussian rf's, second-order stationarity is equivalent to strict stationarity.

A rf is said to be quasi-stationary when the corresponding stationary hypothesis (usually the hypothesis of second-order stationarity) is valid only for distances $|\mathbf{h}| < d$, where d is a limit distance. That is, in the second-order quasi-stationary case (usually referred to as the quasi-stationary case) $\mu(\mathbf{s}+\mathbf{h}) \approx \mu(\mathbf{s})$ if $|\mathbf{h}| < d$ and $C(Z(\mathbf{s}+\mathbf{h}) - Z(\mathbf{s})) = C(\mathbf{h})$ if $|\mathbf{h}| < d$.

Second-order stationarity can also be considered a strict assumption on many occasions, as it implies the existence of the variance in the rf. A phenomenon may have infinite variation capacity and be impossible to model using a rf with finite variance. However, there are cases in which the increments or differences $Z(\mathbf{s}+\mathbf{h}) - Z(\mathbf{s})$ do have finite variance and, therefore, are second-order stationary. This type of rf is described as being intrinsically stationary.

In this book, unless specified otherwise, we will always consider the second-order stationary context.

2.3.4 Intrinsically stationary random functions

Definition 2.3.3 *The rf $\{Z(\mathbf{s}),\ \mathbf{s} \in D\}$ is said to be intrinsically stationary (or simply intrinsic) if, for any given translation vector $\mathbf{h}$, the first-order increments $Z(\mathbf{s}+\mathbf{h}) - Z(\mathbf{s})$ are second-order stationary, despite the rf itself not requiring this condition, that is:*

$$E(Z(\mathbf{s}+\mathbf{h}) - Z(\mathbf{s})) = \mu(\mathbf{s}), \tag{2.10}$$

where $\mu(\mathbf{s})$, the drift, is necessarily linear in $\mathbf{h}$, *and*

$$C\left((Z(\mathbf{s}+\mathbf{h})-Z(\mathbf{s})),(Z(\mathbf{s}+\mathbf{h}+\mathbf{h}')-Z(\mathbf{s}+\mathbf{h}'))\right)=C(\mathbf{h},\mathbf{h}'), \tag{2.11}$$

which is equivalent to[1]*:*

$$\frac{1}{2}V\left(Z(\mathbf{s}+\mathbf{h})-Z(\mathbf{s})\right)=\gamma(\mathbf{h}), \tag{2.12}$$

which is only a function of $\mathbf{h}$.

Obviously, if the linear drift is zero:

$$E\left(Z(\mathbf{s}+\mathbf{h})-Z(\mathbf{s})\right)=0$$

and

$$E\left(Z(\mathbf{s}+\mathbf{h})-Z(\mathbf{s})\right)^2=\gamma(\mathbf{h}),$$

Unless specified otherwise, when referring to the intrinsic hypotheses we will consider this simplified form.

If a rf is second-order stationary, then it is also intrinsically stationary. However, the converse is not necessarily true (see Figure 2.3). Intrinsic rf's that are not second-order stationary are called *strictly intrinsic* rf's.

In particular, a Gaussian intrinsic rf is an intrinsic rf whose increments follow a multivariate Gaussian distribution.

A rf is said to be quasi-intrinsic when the intrinsic hypotheses is valid only for distances $|\mathbf{h}|<d$, where d is a limit distance.

Example 2.1 (The Wiener-Levy process) *Let us consider the Wiener-Levy discrete process, $Z(\mathbf{s}_{k+1})=Z(\mathbf{s}_k)+\epsilon_k$, with ϵ_k independent $N(0,1)$ rv's, where values are taken at the univariate points $\mathbf{s}_i$, $i=1,2,\ldots,k,k+1,\ldots$.*

Then, $V\left(Z(\mathbf{s}_{k+h})\right)=V\left(Z(\mathbf{s}_k)+\epsilon_k+\epsilon_{k+1}+\ldots+\epsilon_{k+h-1}\right)=V\left(Z(\mathbf{s}_k)\right)+h$, that is, the variance increases indefinitely when h increases (it is not finite) and depends on k (see Figure 2.4, top panel).

If we consider the increments $Z(\mathbf{s}_{k+h})-Z(\mathbf{s}_k)$ (Figure 2.4 bottom panel), their mean is null and, what is the most important, their variance exists and is finite, and does not depend on k:

$$E\left(Z(\mathbf{s}_{k+h})-Z(\mathbf{s}_k)\right)=E\left(\sum_{i=k}^{k+h-1}\epsilon_i\right)=0$$

$$V\left(Z(\mathbf{s}_{k+h})-Z(\mathbf{s}_k)\right)=V\left(\sum_{i=k}^{k+h-1}\epsilon_i\right)=h$$

which implies $\gamma(h)=\frac{h}{2}$.

[1] It is sufficient to set $\mathbf{h}'=\mathbf{0}$.

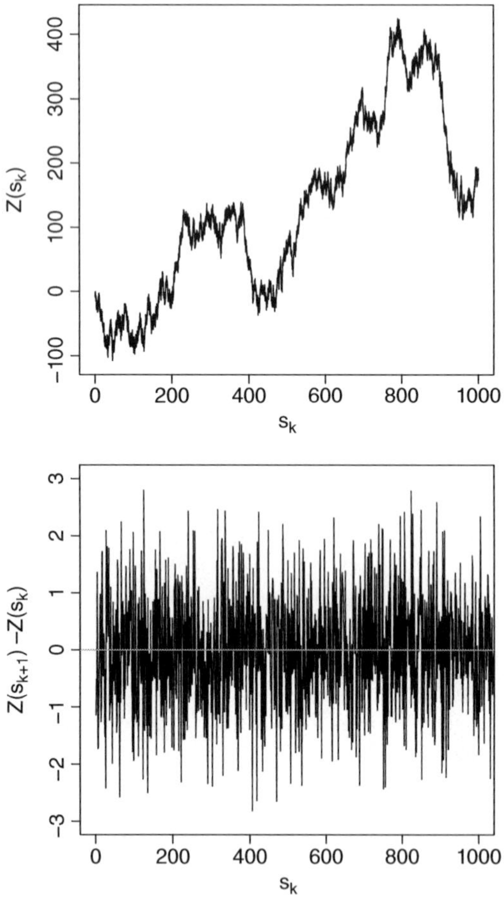

Figure 2.4 Top panel: Realization of a Wiener-Levy process. Bottom panel: First-order increments of the realization of the above Wiener-Levy process.

2.3.5 Non-stationary random functions

Definition 2.3.4 *A rf* $\{Z(\mathbf{s}), \mathbf{s} \in D\}$ *for which the mean and/or the covariance function depend on the location (are not translation invariant) is said to be a non-stationary rf.*

When a rf $\{Z(\mathbf{s}), \mathbf{s} \in D\}$ has a drift, i.e. its mean is non-constant and varies with location, and its first-order increments $Z(\mathbf{s}+\mathbf{h}) - Z(\mathbf{s})$ are non-stationary, it is said that the rf is a *non-intrinsic* rf (some authors call them *intrinsic random functions* of order $k > 0$).

2.4 Support

The support V of a rf refers to the geometry of the characteristic represented by such a rf. Such a geometry can be a point, a segment, a surface, or a volume. In the first case, the rf represents a punctual characteristic. In the second, third, and fourth cases it represents the mean value of the characteristic along a line, across a surface, and in a volume, respectively. In many practical applications the regionalized variable is measured as the average over a certain volume or surface (even over a line). In some cases punctual measurements of the rf simply makes no sense. An example would be the porosity of a material (the relationship between the volume of gaps and the total volume), as the random function would only take the values of 0 or 1 depending on whether the point observed is solid or not, respectively.

Therefore, we will distinguish two types of support: punctual support and non-punctual or block support (we will refer to volumes and surfaces, in general terms, as blocks, following the tradition in mining).

For practical considerations, block supports will be centered on a point $\mathbf{s}$. The average over the block will be denoted as $Z_V(\mathbf{s}) = \frac{1}{|V|}\int_V Z(\mathbf{s})ds$, with $|V| = \int_V ds$, the stochastic integral being defined as a limit (in mean square) of approximating sums.

Obviously, the size of the observation support implies a series of statistical consequences, the most important one being that measurements over blocks have less variability than measurements taken at points. This close relation between the support of the observations and the statistical distribution of their values has relevant consequences in practice. The analysis of these consequences when changing the size of support is known as the change of support problem.

We will also use the distinction point support-block support to indicate if the prediction is made at a non-observed point or in a block, respectively.

3

Structural analysis

3.1 Introduction

The key question this chapter raises is the following: How do we express in a function the structure of the spatial dependence or correlation present in the realization observed? This question, known in the geostatistics literature as the structural analysis of the spatial dependence or, in short, the structural analysis, is a key issue in the subsequent process of optimal prediction (kriging), as the success of the kriging methods is based on the functions yielding information about the spatial dependence detected.

The functions referred to above are covariance functions (also called covariograms) and semivariograms, but they must meet a series of requisites. As we only have the observed realization, in practice, the covariance functions and semivariograms derived from it may not satisfy such requisites. For this reason, one of the theoretical models (also called the valid models) that do comply must be fitted to it.

This is why in this chapter we deal with the main valid covariance and semivariogram models, the construction of the empirical or experimental counterparts, and how to fit a valid model to its empirical counterpart. We will focus mainly on semivariograms, as they cover a broader spectrum of regionalized variables than covariance models.

At this stage, as we will see later, all physical knowledge or any other regarding the regionalized phenomenon being studied is very welcome.

Some authors include the exploratory data analysis in the topic of structural analysis, but we only focus on the analysis of the spatial dependence detected in the observed realization and how to represent it through a valid covariance function or a semivariogram (depending on the stationary hypothesis assumed).

Spatial and Spatio-Temporal Geostatistical Modeling and Kriging, First Edition.
José-María Montero, Gema Fernández-Avilés, and Jorge Mateu.

Companion Website: www.wiley.com/go/montero/spatial

3.2 Covariance function

3.2.1 Definition and properties

As mentioned in Definition 2.2.5, the covariance function of a spatial rf is given by:

$$\begin{aligned} C(\mathbf{s}_i, \mathbf{s}_j) &= C\left(Z(\mathbf{s}_i), Z(\mathbf{s}_j)\right) \\ &= E\left(\left(Z(\mathbf{s}_i) - \mu(\mathbf{s}_i)\right)\left(Z(\mathbf{s}_j) - \mu(\mathbf{s}_j)\right)\right), \quad \forall \mathbf{s}_i, \mathbf{s}_j \in D. \end{aligned} \tag{3.1}$$

Under the hypothesis of second-order stationarity, the covariance function verifies the following theoretical properties:

(i) It only depends on the vector **h** to join the locations, but not on the locations themselves:

$$C(\mathbf{h}) = E\left((Z(\mathbf{s}+\mathbf{h}) - \mu)(Z(\mathbf{s}) - \mu)\right), \quad \forall \mathbf{s}, \mathbf{s}+\mathbf{h} \in D \subset \mathbb{R}^d, \tag{3.2}$$

μ being the constant mean of the rf. Hence, a covariogram indicates how the correlation between $Z(\mathbf{s})$ and $Z(\mathbf{s}+\mathbf{h})$ evolves with the separation given by the vector **h**. These functions depend on both the distance between the locations **s** and $\mathbf{s}+\mathbf{h}$ and also on the direction of the vector that joins them.

Definition 3.2.1 *When the covariance function only depends on the distance between the locations* **s** *and* $\mathbf{s}+\mathbf{h}$, *it is called isotropic. When it depends on both the distance and direction of vector* **h**, *it is described as* anisotropic.

Definition 3.2.2 *A rf that is intrinsically stationary and isotropic is said to be homogenous.*

(ii) It is bounded by its value at the origin, that is, by the variance of the rf:

$$|C(\mathbf{h})| \leq C(\mathbf{0}) = V(Z(\mathbf{s})). \tag{3.3}$$

Proof.
On the one hand,

$$\begin{aligned} 0 &\leq E\left(\left((Z(\mathbf{s}+\mathbf{h}) - \mu) - (Z(\mathbf{s}) - \mu)\right)^2\right) \\ &= E\left((Z(\mathbf{s}+\mathbf{h}) - \mu)^2 + (Z(\mathbf{s}) - \mu)^2 - 2\,(Z(\mathbf{s}+\mathbf{h}) - \mu)\,(Z(\mathbf{s}) - \mu)\right) \\ &= 2C(\mathbf{0}) - 2C(\mathbf{h}). \end{aligned}$$

On the other hand,

$$\begin{aligned} 0 &\leq E\left(\left((Z(\mathbf{s}+\mathbf{h}) - \mu) + (Z(\mathbf{s}) - \mu)\right)^2\right) \\ &= E\left((Z(\mathbf{s}+\mathbf{h}) - \mu)^2 + (Z(\mathbf{s}) - \mu)^2 + 2\,(Z(\mathbf{s}+\mathbf{h}) - \mu)\,(Z(\mathbf{s}) - \mu)\right) \\ &= 2C(\mathbf{0}) + 2C(\mathbf{h}). \end{aligned}$$

Therefore, $C(\mathbf{h}) \leq C(\mathbf{0})$ and $-C(\mathbf{h}) \leq C(\mathbf{0})$, that is, $|C(\mathbf{h})| \leq C(\mathbf{0})$.

(iii) It is an even function, that is, $C(\mathbf{h}) = C(-\mathbf{h})$.

Proof.
By definition:

$$C(\mathbf{h}) = E\big((Z(\mathbf{s}+\mathbf{h}) - \mu)(Z(\mathbf{s} - \mu))\big),$$

and setting $\mathbf{r} = \mathbf{s} + \mathbf{h}$,

$$C(\mathbf{h}) = E\left((Z(\mathbf{r}) - \mu)(Z(\mathbf{r} - \mathbf{h}) - \mu)\right) = C(-\mathbf{h}).$$

(iv) It is a positive-definite function, that is, in terms of the coordinate difference $\mathbf{s}_i - \mathbf{s}_j = \mathbf{h}$,

$$\sum_{i=1}^{n}\sum_{j=1}^{n} \lambda_i\lambda_j C(\mathbf{s}_i - \mathbf{s}_j) \geq 0\,, \forall n \in \mathbb{N}^*, \forall \lambda_1, \ldots, \lambda_n \in \mathbb{R},$$

$$\forall \mathbf{s}_1, \ldots, \mathbf{s}_n \in D \subset \mathbb{R}^d. \tag{3.4}$$

Proof.
Consider the linear combination $\sum_{i=1}^{n} \lambda_i Z(\mathbf{s}_i)$. Its variance is given by:

$$V\left(\sum_{i=1}^{n} \lambda_i Z(\mathbf{s}_i)\right) = \sum_{i=1}^{n}\sum_{j=1}^{n} \lambda_i\lambda_j C(\mathbf{s}_i - \mathbf{s}_j).$$

Since $V\left(\sum_{i=1}^{n} \lambda_i Z(\mathbf{s}_i)\right) \geq 0$, for any choice of n, λ_i and $\mathbf{s}_i$, the function $C(\mathbf{h})$ is a positive-definite function on $\mathbb{R}^d$.

This is an important property. Indeed, it is the only really necessary condition for a function to be a covariance. Take into account that the ultimate objective of kriging is to construct linear predictors based on the observed variables $Z(\mathbf{s}_i)$ that are unbiased and of minimum variance. Obviously, their expectation and variance must be finite, and the latter non-negative.
Other properties that the covariance functions on $\mathbb{R}^d$ satisfy are what Chilès and Delfiner (1999, p. 60) call *stability properties*:

(v) If $C_k(\mathbf{h}), \forall k \in \mathbb{N}$ are covariance functions on $\mathbb{R}^d$ and $lim_{k\to\infty} C_k(\mathbf{h}) = C(\mathbf{h}), \forall \mathbf{h} \in \mathbb{R}^d$, then $C(\mathbf{h})$ is a covariogram on $\mathbb{R}^d$, provided that this limit exists $\forall \mathbf{h}$. This property results from the definition of positive-definiteness.

(vi) Every linear combination of covariance functions with positive coefficients is also a covariance function. To demonstrate this property, it suffices to consider the rf $Z(\mathbf{s})$ as a linear combination of n independent rf's $Y_i(\mathbf{s})$.

In more general terms, if $C(\mathbf{h}, \theta)$ is a valid covariogram on $\mathbb{R}^d, \forall\theta \in A \subset \mathbb{R}$, and if $\mu(d\theta)$ is a positive measure on A, then:

$$\int_A C(\mathbf{h}, \theta)\mu(d\theta) \tag{3.5}$$

is a covariogram in $\mathbb{R}^d$, provided that the integral exists $\forall\mathbf{h} \in \mathbb{R}^d$.

(vii) The product of covariance functions is also a covariance function. This can be demonstrated considering $Z(\mathbf{s})$ as the product of n independent rf's $Y_i(\mathbf{s})$.

3.2.2 Some theoretical isotropic covariance functions

We will focus on isotropic covariance functions because, as can be seen in Section 3.7, anisotropic models can be reduced to isotropic ones.

We denote the module of the vector $\mathbf{h}$ as $|\mathbf{h}|$ (as the norm of the vector $\mathbf{h}$ in the Euclidean space coincides with its module, $\|\mathbf{h}\| = |\mathbf{h}|$).

First, let us say a word of caution: a positive-definite isotropic function on $\mathbb{R}^{d_1}$ is also a positive-definite isotropic function on $\mathbb{R}^{d_2}, d_2 \leq d_1$ (the opposite is generally not true). Thus, an isotropic covariogram on $\mathbb{R}^{d_1}$ is also isotropic on $\mathbb{R}^{d_2}, d_2 \leq d_1$, but not necessarily on $\mathbb{R}^{d_2}, d_2 > d_1$.

Example 3.1 (Tent covariogram) *Consider the following covariogram:*

$$C(h) = \begin{cases} \sigma^2\left(1 - \frac{|h|}{a}\right) & \text{if } 0 < |h| \leq a \\ 0 & \text{if } |h| > a \end{cases},$$

called a tent covariogram and valid on $\mathbb{R}^1$. Take a $k \times k$ square grid of spacing $\frac{a}{2^{1/2}}$ and, in Equation (3.4), set the coefficients $\lambda_i\lambda_j = 1$ for $i = j$ and $\lambda_i\lambda_j = -1$ for $i \neq j$. It can easily be checked that:

$$V\left(\sum_{i=1}^{k\times k} \lambda_i Z(\mathbf{s}_i)\right) = \sum_{i=1}^{k\times k}\sum_{j=1}^{k\times k} \lambda_i\lambda_j C(\mathbf{s}_i - \mathbf{s}_j) = \sigma^2\left(k^2 - [4k(k-1)(1-2^{-1/2})]\right),$$

which for k sufficiently large is negative. For example, for $k = 8$, as proposed by Cressie (1993, pp. 84–5), $V\left(\sum_{i=1}^{k\times k} \lambda_i Z(\mathbf{s}_i)\right) = -1.608081\sigma^2$. Therefore, the tent covariogram is not positive-definite on $\mathbb{R}^2$.

Second, in a covariance function, we have to pay special attention to:

- the behavior near the origin, which is related to the continuity and the degree of spatial regularity of the rf. Specifically, the *value at the origin* (m) coincides with the variance of the rf, $C(\mathbf{0})$.

- the behavior at large distances, because the rate of the covariogram decrease indicates the degree of dissimilarity of values in ever more distant pairs of locations. We call *range* the distance beyond which it is considered the covariance cancels out.

The best-known (and used in practical applications) isotropic covariance models that verify the properties detailed in Section 3.2.1 are shown below.

3.2.2.1 The spherical model

$$C(|\mathbf{h}|) = \begin{cases} m\left(1 - \left(\frac{3}{2}\frac{|\mathbf{h}|}{a} - \frac{1}{2}\frac{|\mathbf{h}|^3}{a^3}\right)\right) & \text{if } 0 \leq |\mathbf{h}| \leq a \\ 0 & \text{if } |\mathbf{h}| > a \end{cases}. \tag{3.6}$$

This model, valid on $\mathbb{R}^1$, $\mathbb{R}^2$ and $\mathbb{R}^3$, is the most used covariogram in practical applications. The covariance function decreases almost constantly with distance, starting at its value at the origin, m, and canceling out at a distance a (the scale parameter) range of the function.

This behavior often matches reasonably well the observed data. The spherical model represents a continuous phenomenon, albeit non-differentiable, that is, a graph of the phenomenon may display abrupt changes in slope.

3.2.2.2 The exponential model

$$C(|\mathbf{h}|) = m \exp\left(-\frac{|\mathbf{h}|}{a}\right), \quad a > 0. \tag{3.7}$$

The exponential covariance function is valid on $\mathbb{R}^d, d \geq 1$, and decreases exponentially as the distance between points increases. This model is continuous but not differentiable at the origin, and reaches a value of zero asymptotically, considering $a' = 3a$ as the distance above which the covariance is practically non-existent, as it would be worth $0.05m$. That is, $3a$ is the distance above which the covariance is only 5% of the value recorded at the origin and is termed the "practical range." The scale parameter a determines how quickly the covariance decreases with distance.

3.2.2.3 The Gaussian model

$$C(|\mathbf{h}|) = m \exp\left(-\frac{|\mathbf{h}|^2}{a^2}\right), \quad a > 0. \tag{3.8}$$

The Gaussian model is valid on $\mathbb{R}^d, d \geq 1$, and the covariance reaches a value of zero asymptotically, $a' = a\sqrt{3}$ being the distance above which the covariance is 5% of its value at the origin, that is, the practical range. The scale parameter a indicates the rate at which the function decreases with distance. The Gaussian model corresponds to infinitely differentiable (and, as a consequence, very soft) rf's. This kind of rf's are not common in practice and thus the Gaussian model is unusual in practical applications.

Both the exponential and the Gaussian covariance models belong to the *Stable family*:

$$C(|\mathbf{h}|) = \exp\left(-\frac{|\mathbf{h}|}{a}\right)^{\alpha}, \qquad a > 0, \quad 0 < \alpha \leq 2. \tag{3.9}$$

Specifically, setting $\alpha = 1$ and $\alpha = 2$ leads to the exponential and Gaussian models, respectively.

Figure 3.1 depicts spherical, exponential and Gaussian covariograms with $m = 1$ and different values of the scale parameter a. It can be seen that: (i) both the spherical and exponential models exhibit a linear behavior near the origin; however, the behavior near the origin that the Gaussian model exhibits is parabolic. (ii) The smaller the scale parameter, a, the more quickly the covariance decreases. (iii) The exponential and Gaussian models drop asymptotically towards 0 for $|\mathbf{h}| \to \infty$.

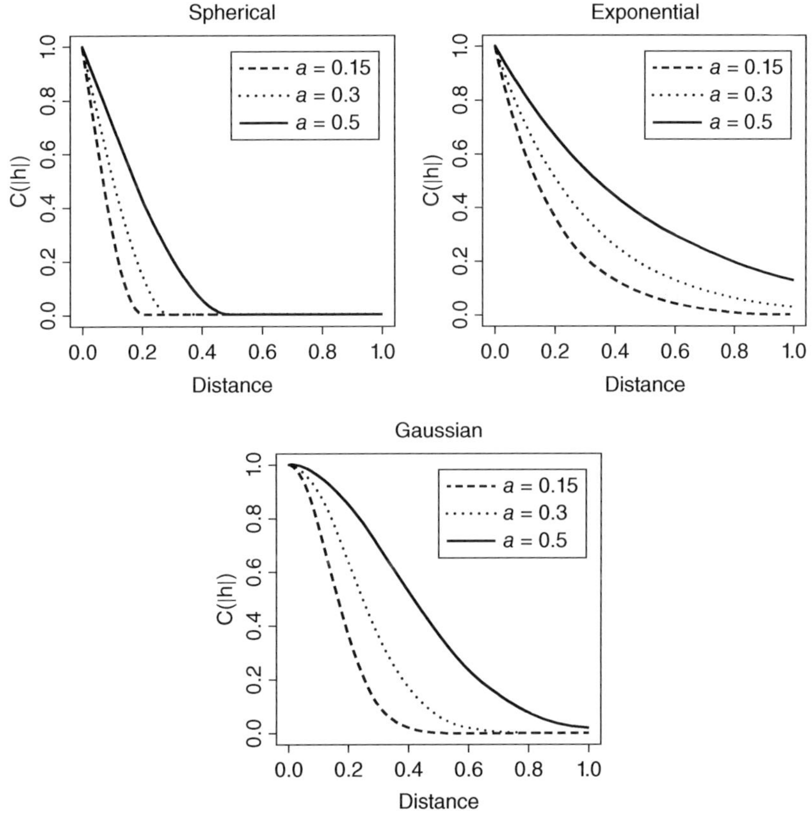

Figure 3.1 Spherical, exponential, and Gaussian covariance models with $m = 1$ and different values of the scale parameter.

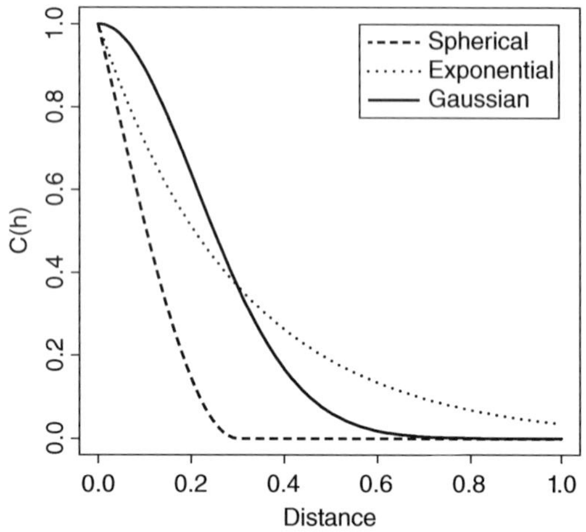

Figure 3.2 Spherical, exponential, and Gaussian models with $m = 1$ and $a = 1/3$.

Figure 3.2 shows the different behavior of the spherical, exponential, and Gaussian models with both the same maximal value of the covariance ($m = 1$) and also the same scale parameter ($a = 1/3$).

3.3 Empirical covariogram

In practice, what we have is the realization observed (and therefore sketchy) of the rf under study, that is, a set of georeferenced data in a given domain. Using these data as a basis, we must infer the spatial dependence structure of the phenomenon.

In the second-order stationary framework, in order to estimate the covariance function, the Method-of-Moments estimator (MoM) is generally used:

$$\hat{C}(\mathbf{h}) = \frac{1}{\#N(\mathbf{h})} \sum_{N(\mathbf{h})} (Z(\mathbf{s}_i + \mathbf{h}) - \bar{Z})(Z(\mathbf{s}_i) - \bar{Z}), \tag{3.10}$$

with $\bar{Z} = \sum_{i=1}^{n} Z(\mathbf{s}_i)$ being an estimator of the constant mean μ, and $\#N(\mathbf{h})$ the number of pairs of locations that are found at a distance $\mathbf{h}$. This covariogram is called *empirical* or *experimental* because it is calculated using the information available.

Note that the empirical covariogram uses the sample mean, $\bar{Z}$, as an estimator of the mean of the rf, μ. However, in the case of spatial dependence, $\bar{Z}$, despite still being an unbiased estimator of μ, is not that of minimum variance. A solution to this problem would involve substituting the arithmetic mean with another mean that would take into account such spatial dependence, but we would have to know the spatial covariance function to be able to calculate it, which leads to a vicious circle.

As a consequence of using $\bar{Z}$ as an estimator of μ, the MoM estimator $\hat{C}(\mathbf{h})$ is a biased estimator of $C(\mathbf{h})$: $E\left(\hat{C}(\mathbf{h})\right) \neq C(\mathbf{h})$. The smaller $\#N(\mathbf{h})$, the larger the bias.

The empirical covariogram should not be used for prediction because it could not verify the most important condition of a covariogram to be considered valid: positive-definiteness. In addition, it is defined only at a few distances, and usually has no physical meaning.

This is why a theoretical model, such as that detailed previously, must be fitted to it. Details about model fitting are given in Section 3.8.

3.4 Semivariogram

3.4.1 Definition and properties

As introduced in Chapter 2, the semivariogram is given by:

$$\gamma(\mathbf{s}_i - \mathbf{s}_j) = \frac{1}{2}V(Z(\mathbf{s}_i) - Z(\mathbf{s}_j)), \forall \mathbf{s}_i, \mathbf{s}_j \in D. \tag{3.11}$$

Under the second-order stationary hypothesis and the intrinsic hypothesis (with no drift), it can be written as:

$$\gamma(\mathbf{h}) = \frac{1}{2}V\left(Z(\mathbf{s}+\mathbf{h}) - Z(\mathbf{s})\right) = \frac{1}{2}E\left((Z(\mathbf{s}+\mathbf{h}) - Z(\mathbf{s}))^2\right), \tag{3.12}$$

showing how the dissimilarity between $Z(\mathbf{s}+\mathbf{h})$ and $Z(\mathbf{s})$ develops with distance $\mathbf{h}$.

The semivariogram is the instrument used *par excellence* to describe the spatial dependence in the regionalized variable. The reason is that it covers a broader spectrum of regionalized variables than the covariance function, which is confined to second-order stationary rf's. This spectrum includes intrinsically stationary rf's, in which the covariance cannot be defined.

In the second-order stationary framework, both the semivariogram and the covariogram are theoretically equivalent. Indeed, as seen in Section 2.3.3,

$$C(\mathbf{h}) = C(\mathbf{0}) - \gamma(\mathbf{h}). \tag{3.13}$$

But even in this case, the semivariogram is preferred to the covariogram because it does not require the knowledge of the mean of the rf. In practice, this mean is unknown and it must be estimated from the data, which introduces a bias (see Section 3.3).

The semivariogram of an intrinsically stationary rf depends on the vector $\mathbf{h}$ that joins the locations (both on the distance between $\mathbf{s}$ and $\mathbf{s}+\mathbf{h}$ and also on the direction, but not on the locations themselves). Thus, in general terms, it is anisotropic. When the semivariogram depends only on distance, it is called isotropic.

As is the case with the covariogram, a semivariogram cannot be an arbitrary function and must verify the following theoretical properties:

(i) It is identically null at the origin. By definition, $\gamma(\mathbf{0}) = 0$, although in practice, it normally displays a discontinuity at the origin. The discontinuity at the origin is called the nugget effect.

(ii) It is an even function, that is, $\gamma(\mathbf{h}) = \gamma(-\mathbf{h})$. In effect, setting $\mathbf{s} + \mathbf{h} = \mathbf{r}$:

$$\gamma(\mathbf{h}) = \gamma(Z(\mathbf{r}) - Z(\mathbf{r} - \mathbf{h})) = \gamma(-\mathbf{h}). \tag{3.14}$$

(iii) It always takes values greater than or equal to zero

$$\gamma(\mathbf{h}) \geq 0, \tag{3.15}$$

whereas the covariogram can take negative values.

(iv) In the context of intrinsic stationarity, it is a conditionally negative-definite function, that is,

$$\sum_{i=1}^{n} \sum_{j=1}^{n} \lambda_i \lambda_j \gamma(\mathbf{s}_i - \mathbf{s}_j) \leq 0,$$

for any given set $\mathbf{s}_1, \ldots, \mathbf{s}_n$ of arbitrary locations and for any given set of real numbers $\lambda_1, \ldots, \lambda_n$ satisfying the condition $\sum_{i=1}^{n} \lambda_i = 0$.

Proof.
In the second-order stationary case we have:

$$\begin{aligned} 0 \leq V\left(\sum_{i=1}^{n} \lambda_i Z(\mathbf{s}_i)\right) &= \sum_{i=1}^{n} \sum_{j=1}^{n} \lambda_i \lambda_j C(\mathbf{s}_i - \mathbf{s}_j) \\ &= \sum_{i=1}^{n} \sum_{j=1}^{n} \lambda_i \lambda_j \left(C(\mathbf{0}) - \gamma(\mathbf{s}_i - \mathbf{s}_j)\right) \\ &= C(\mathbf{0}) \sum_{i=1}^{n} \lambda_i \sum_{j=1}^{n} \lambda_j - \sum_{i=1}^{n} \sum_{j=1}^{n} \lambda_i \lambda_j \gamma(\mathbf{s}_i - \mathbf{s}_j). \end{aligned} \tag{3.16}$$

But in the intrinsic case $C(\mathbf{0})$ may not exist and to avoid the cases when the variance of the rf is not definite, the condition $\sum_{i=1}^{n} \lambda_i = 0$ is imposed. Thus, in the intrinsic case we have that:

$$\sum_{i=1}^{n} \sum_{j=1}^{n} \lambda_i \lambda_j \gamma(\mathbf{s}_i - \mathbf{s}_j) \leq 0,$$

that is, the semivariogram must be conditionally negative-definite.
Clearly, any linear combination of increments satisfies the condition $\sum_{i=1}^{n} \lambda_i = 0$. Conversely, any combination verifying this condition can be expressed as a linear combination of increments.

In effect, if we choose an arbitrary point $\mathbf{s}_0$, we can define the increments $Z(\mathbf{s}_i) - Z(\mathbf{s}_0)$ for all i. Now, under the condition $\sum_{i=1}^{n} \lambda_i = 0$,

$$\sum_{i=1}^{n} \lambda_i Z(\mathbf{s}_i) = \sum_{i=1}^{n} \lambda_i \left(Z(\mathbf{s}_i) - Z(\mathbf{s}_0) \right).$$

The variance of this linear combination exists and is given by:

$$\begin{aligned} V\left(\sum_{i=1}^{n} \lambda_i Z(\mathbf{s}_i) \right) &= V\left(\sum_{i=1}^{n} \lambda_i \left(Z(\mathbf{s}_i) - Z(\mathbf{s}_0) \right) \right) = \\ &= \sum_{i=1}^{n} \sum_{j=1}^{n} \lambda_i \lambda_j E\left(\left(Z(\mathbf{s}_i) - Z(\mathbf{s}_0) \right) \left(Z(\mathbf{s}_j) - Z(\mathbf{s}_0) \right) \right) \\ &= \sum_{i=1}^{n} \sum_{j=1}^{n} \lambda_i \lambda_j C'(\mathbf{s}_i, \mathbf{s}_j), \end{aligned} \tag{3.17}$$

$C'(\mathbf{s}_i, \mathbf{s}_j)$ being the covariance of the increments regarding $Z(\mathbf{s}_0)$. Furthermore,

$$\begin{aligned} \gamma(\mathbf{s}_i - \mathbf{s}_j) &= \frac{1}{2} V\left(\left(Z(\mathbf{s}_i) - Z(\mathbf{s}_0) \right) - \left(Z(\mathbf{s}_j) - Z(\mathbf{s}_0) \right) \right) \\ &= \frac{1}{2} \left(C'(\mathbf{s}_i, \mathbf{s}_i) + C'(\mathbf{s}_j, \mathbf{s}_j) - 2C'(\mathbf{s}_i, \mathbf{s}_j) \right), \end{aligned} \tag{3.18}$$

from which we obtain:

$$C'(\mathbf{s}_i, \mathbf{s}_j) = \frac{1}{2} C'(\mathbf{s}_i, \mathbf{s}_i) + \frac{1}{2} C'(\mathbf{s}_j, \mathbf{s}_j) - \gamma(\mathbf{s}_i - \mathbf{s}_j). \tag{3.19}$$

Now, by substituting this expression in (3.17), we obtain:

$$\begin{aligned} V\left(\sum_{i=1}^{n} \lambda_i Z(\mathbf{s}_i) \right) &= \sum_{i=1}^{n} \sum_{j=1}^{n} \lambda_i \lambda_j C'(\mathbf{s}_i, \mathbf{s}_j) \\ &= \frac{1}{2} \sum_{j=1}^{n} \lambda_j \sum_{i=1}^{n} \lambda_i C'(\mathbf{s}_i, \mathbf{s}_i) \\ &\quad + \frac{1}{2} \sum_{i=1}^{n} \lambda_i \sum_{j=1}^{n} \lambda_j C'(\mathbf{s}_j, \mathbf{s}_j) \\ &\quad - \sum_{i=1}^{n} \sum_{j=1}^{n} \lambda_i \lambda_j \gamma(\mathbf{s}_i - \mathbf{s}_j) \\ &= - \sum_{i=1}^{n} \sum_{j=1}^{n} \lambda_i \lambda_j \gamma(\mathbf{s}_i - \mathbf{s}_j), \end{aligned} \tag{3.20}$$

requiring $\gamma(\mathbf{h})$ to be conditionally negative-definite.

A linear combination $\sum_{i=1}^{n} \lambda_i Z(\mathbf{s}_i)$ with both finite expectation and variance is said to be "admissible" or "allowable." Under the second-order stationary hypothesis all linear combinations are allowable. However, under the intrinsic hypothesis only the linear combinations of increments (which identify with those where the sum of the coefficients is null) are allowed.
As a consequence, the class of semivariogram models is richer than the class of covariance functions, because it includes both bounded (with covariance functions counterpart) and also unbounded (without covariance functions counterpart). Therefore, under the intrinsic hypothesis, the spectrum of semivariograms is larger than under the second-order stationary one, but the coefficients of the linear combination are restricted to sum to zero. Under the second-order stationarity hypothesis, the spectrum of semivariogram models is more limited, but no condition is imposed on the coefficients of the linear combination (all of them are allowed).

(v) The semivariogram of a second-order stationary rf is finite. That of an intrinsically stationary rf with no drift could grow to infinity, but not uncontrollably, as it grows more slowly than $|\mathbf{h}|^2$ when $|\mathbf{h}| \to \infty$.[1] That is,

$$\lim_{|\mathbf{h}|\to\infty} \frac{\gamma(\mathbf{h})}{|\mathbf{h}|^2} = 0. \tag{3.21}$$

3.4.2 Behavior at intermediate and large distances

The semivariogram is, generally, a non-decreasing monotone function, so that the variability of the first increments of the rf increases with distance.

The semivariograms corresponding to second-order stationary rf's exhibit a typical behavior at intermediate and large distances: They rise from the origin and increase monotonically with distance (if the corresponding covariogram decreases monotonically) until approaching its limiting value, the *a priori* variance of the rf, $C(\mathbf{0})$, either exactly or asymptotically (see Figure 3.3).

The slope of the semivariogram indicates the change in the dissimilarity of the values of the regionalized variable with distance. The above-mentioned limiting value of the semivariogram is called the variance sill, or simply the sill (m), and the distance at which the sill is reached is termed the range, which defines the threshold of spatial dependence, that is, the zone of influence of the rf. In other words, the range is the distance beyond which the values of the regionalized variable have no spatial dependence. Figure 3.4 illustrates that the larger the range, the larger the zone of influence of the rf.

When the sill is reached asymptotically there is not a well-marked range, but a practical range (the distance at which the semivariogram takes the value 0.95 m). This practical range is closely related to the scale parameter, a, of the semivariogram

[1] For a demonstration of this property, see Montero and Larraz (2008).

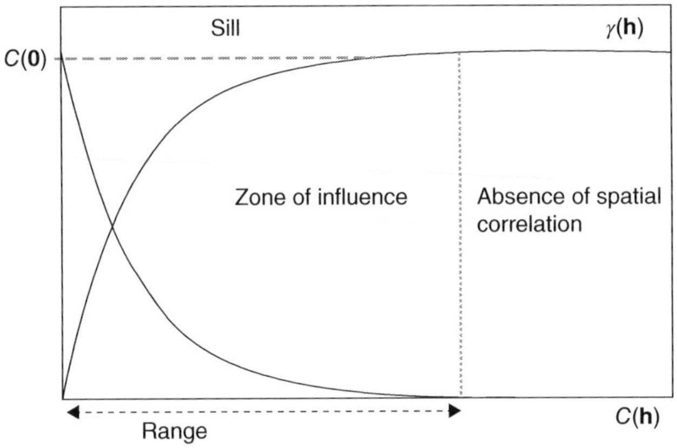

Figure 3.3 Bounded semivariogram and its covariogram counterpart.

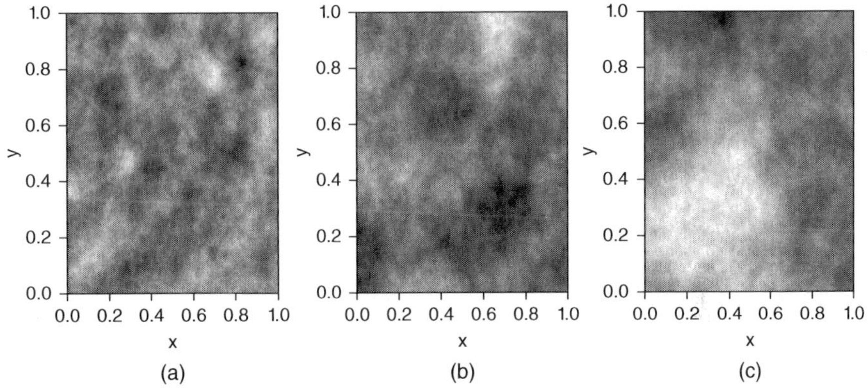

Figure 3.4 Simulations of a rf having semivariograms that only differ in the range: (a) range $= 0.15$*; (b) range* $= 0.33$*; (c) range* $= 0.66$.

(if the sill is reached exactly, that is, in the case of a flat sill, a coincides with the range).

Semivariograms that do not reach a sill are very common, and in particular this fact can occur when dealing with non-stationary rf's (e.g. the existence of a drift), intrinsically stationary rf's (see point (v) in Section 3.4.1), or even second-order stationary rf's (when the range exceeds the largest distance for which the semivariogram can be estimated).

3.4.3 Behavior near the origin

The behavior of the semivariogram near the origin, and particularly at the origin, is closely linked to the continuity and degree of regularity of the rf. The more continuous and regular across space the rf, the smoother and the more spatially structured the realizations and the more regular the behavior of the semivariogram near the origin, that is, for very short distances **h**. Thus, it is not surprising that the partial derivatives of the rf contain very important information about its nature, and specifically about its continuity.

Why do we focus on inferring the degree of continuity of the rf from the behavior of the semivariogram near the origin? The intuitive answer is that this behavior gives information on the interval of distances for which the spatial dependence is high. The theoretical reason involves the concepts of mean-square continuity and differentiability of the rf. In the second-order stationary case, mean-square continuity at point **s** means that:

$$\lim_{\mathbf{h}\to 0} E\left((Z(\mathbf{s}+\mathbf{h}) - Z(\mathbf{s}))^2\right) = 0,$$

which implies that:

$$\lim_{\mathbf{h}\to 0} 2V(Z(\mathbf{s})) - 2C(\mathbf{h}) = \lim_{\mathbf{h}\to 0} 2(C(\mathbf{0}) - 2C(\mathbf{h})) = \lim_{\mathbf{h}\to 0} \gamma(\mathbf{h}) = 0.$$

That is, unless $C(\mathbf{h}) \to C(\mathbf{0})$, or alternatively $\gamma(\mathbf{h}) \to 0$, as $\mathbf{h} \to 0$, the rf cannot be continuous at location **s**.

From the above, it follows that an intrinsic rf is mean-square continuous if and only if the semivariogram is continuous at $\mathbf{h} = 0$, because in such a case it is continuous at all $\mathbf{h} \in \mathbb{R}^d, d \geq 1$.

It must be noted that mean-square convergence does not imply almost certain convergence. As a consequence, the mean-square continuity of a rf does not imply the continuity of its realizations.

The smoothness of the rf is related to the number of times it is mean-square differentiable. More specifically, the more times it is mean-square differentiable, the greater is the degree of smoothness of the rf (details can be seen in Chilès and Delfiner 1999, pp. 58–9 and Cressie 1993, p. 60).

As in the case of continuity, the mean-square differentiability of a rf does not imply differentiability of its realizations.

The degree of smoothness of the rf is given by the order of the semivariogram at the origin, which is defined as the real number α so that:

$$\lim_{|\mathbf{h}|\to 0} \frac{\gamma(\mathbf{h})}{|\mathbf{h}|^\alpha} = c, \tag{3.22}$$

c being a real number.

The greater α, the smoother the rf. If the tangent at the origin is null (indicative of parabolic behavior), then $\alpha > 1$; otherwise, $\alpha \leq 1$. The most frequent case is $\alpha = 1$; in this case, the behavior of the semivariogram near the origin is linear.

When dealing with realizations, we can have the following typical behavior near the origin:

- A linear behavior, that is, $\gamma(\mathbf{h})_{|\mathbf{h}|\to 0} \to f(|\mathbf{h}|)$, which is typical of continuous regionalized variables, at least piecewise, but not differentiable. A graphical representation of one of these regionalized variables will be full of peaks. The amplitude of fluctuations increases with the inter-point distance and is proportional to the slope of the tangent at the origin. Thus, the degree of regularity of these regionalized variables is much less than in the case where the semivariogram shows parabolic behavior near the origin.

- A parabolic behavior, that is, $\gamma(\mathbf{h})_{|\mathbf{h}|\to 0} \to f(|\mathbf{h}|^2)$, which is typical of highly regular regionalized variables (the graphical representation of this kind of rf's does not present peaks). If such a behavior persists over large distances, it can also be associated with the existence of a strong drift.
- A discontinuity at the origin. This case will be addressed in Section 3.4.4.

Figure 3.5 shows that a simulation coming from a rf with a semivariogram exhibiting a parabolic behavior near the origin is "smoother" or less "fuzzy," that is, more structured, than another coming from a rf with a semivariogram with the same characteristics except that it shows a linear behavior near the origin.

3.4.4 A discontinuity at the origin

Although the semivariogram must be identically null at the origin, sometimes in practice this does not occur. This discontinuity at 0 is called the *nugget effect*, and usually indicates that the regionalized variable is very irregular, maybe discontinuous.

In order to model a semivariogram with a discontinuity at the origin, it suffices to add it to whichever valid semivariogram is found. Figure 3.7, right panel, depicts a semivariogram with a nugget effect. Of note is that now the sill is divided into two components: the partial sill and the nugget effect. The ratio of the nugget effect to the sill is often referred as the *relative nugget effect*. The ratio of the partial sill to the sill is termed by Schabenberger and Gotway (2005, p. 140) the *relative structured*

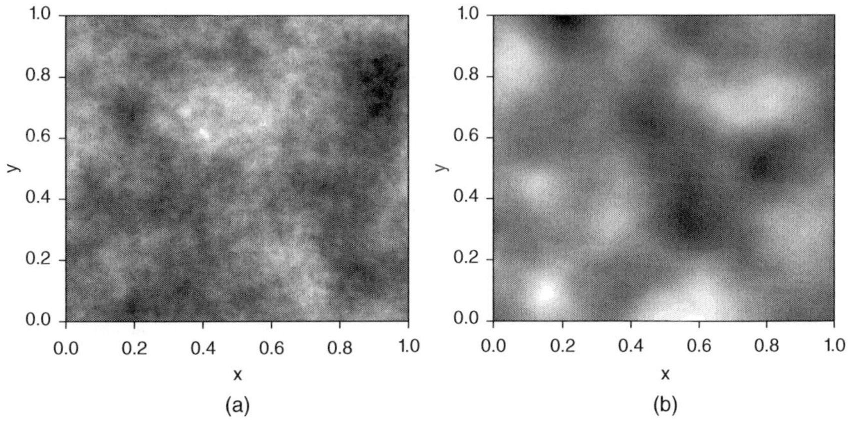

Figure 3.5 2D representation of simulations of two rf's with a semivariogram that only differs in the behavior near the origin: (a) linear, (b) parabolic.

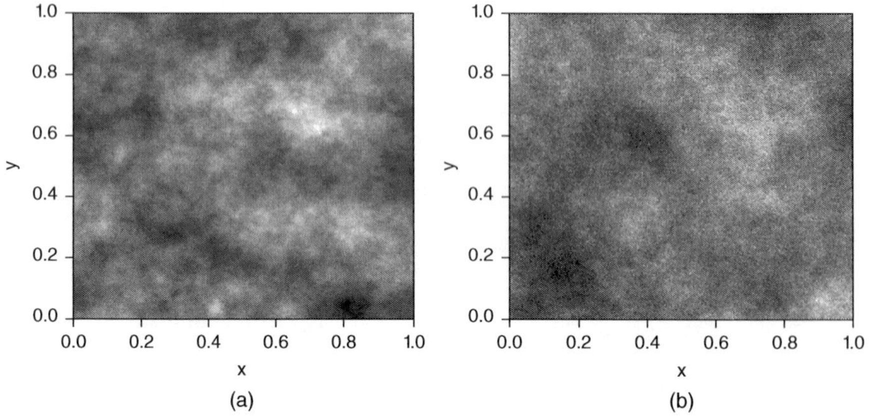

Figure 3.6 Simulated fields of values using a semivariogram with scale parameter $a = 0.33$*: (a) nugget effect*$=0$*; (b) nugget effect*$=0.25$.

variability, and since the nugget effect reduces the smoothness of the rf indicates the degree of spatial structure existing in the rf.

Figure 3.6 shows the 2D representation of two simulations of a rf: with no nugget effect (Figure 3.6(a)) and with a nugget effect (Figure 3.6(b)). It can be noted that the simulation with nugget effect is much less structured.

According to Chilès and Delfiner (1999, p. 52), the possible causes of a nugget effect include a microstructure with a range shorter than the sampling support (this case is known as the true nugget effect), a structure with range shorter than the smallest inter-point distance, and measurement or positioning errors. Usually, the origin of the nugget effect is not unique and can be attributable to various causes. Thus, all kind of knowledge about the phenomenon under study is very welcome when it comes to modeling the semivariogram.

Let us comment on the cases of a structure with range shorter than the smallest inter-point distance and measurements errors.

In the first case, Figure 3.7, we have considered a nested semivariogram composed of two structures. The first represents the micro-scale structure and has a sill m which is reached at a lag distance of some few centimeters (Figure 3.7(a)). The second represents the macro-scale (in kilometers). The nested semivariogram (Figure 3.7(b)) will reach the sill of the micro-scale structure in only a few centimeters, so that this sill will be confounded with a discontinuity at the origin of size m.

In the case of measurement errors, a situation often found in practice, we can consider $Z^*(\mathbf{s}) = Z(\mathbf{s}) + \epsilon(\mathbf{s})$, where the asterisk indicates measurement and $Z(\mathbf{s})$ and $\epsilon(\mathbf{s})$ are independent. Then,

$$\gamma_{Z^*(\mathbf{s})}(\mathbf{h}) = \gamma_{Z(\mathbf{s})}(\mathbf{h}) + \gamma_{\epsilon(\mathbf{s})}(\mathbf{h}), \tag{3.23}$$

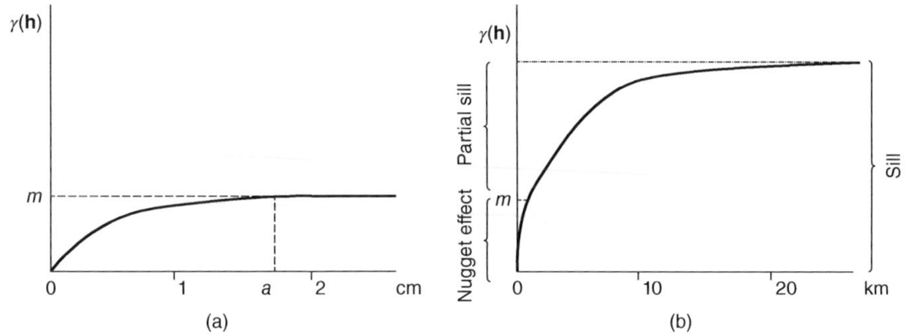

Figure 3.7 Nested semivariogram.

and assuming that the measurement errors are spatially uncorrelated and have a constant variance σ_ϵ^2,

$$\gamma_{Z^*(\mathbf{s})}(\mathbf{h}) = \gamma_{Z(\mathbf{s})}(\mathbf{h}) + \sigma_\epsilon^2(\mathbf{h}), \quad \mathbf{h} \neq 0, \tag{3.24}$$

that is, the data have a nugget effect, σ_ϵ^2, in addition to the nugget effect that could have the rf.

The limiting case of the nugget effect is the pure nugget effect. In this case, the semivariogram is constant for any given distance, indicating the absence of spatial dependence.

3.5 Theoretical semivariogram models

The first thing to point out, as noted in Montero and Larraz (2008), is that the adjective "theoretical" can lead to confusion. Theoretical semivariograms are nothing more than functions with a simple analytical expression that fulfill the properties detailed in Section 3.2.1, especially the conditionally negative-definiteness condition. These valid semivariograms are used to represent empirical semivariograms that are obtained from the available regionalization, as they could not fulfill the above referred properties, which would give rise to various problems when undertaking kriging procedures. However, it is important to take into account that their expressions have not been deduced from any special hypothesis, nor are they intended to represent specific rf's.

In what follows we present the most interesting theoretical or valid isotropic semivariograms. We focus on isotropic models because, as we will see in Section 3.7, anisotropies can be represented by reducing them to the isotropic case by a linear transformation of the coordinates, or by representing separately each of the directional variabilities, depending on the type of anisotropy. We classify them following the approach suggested by Journel and Huijbregts (1978), although introducing some slight changes:

- Semivariograms with a sill or transition semivariograms.
- Semivariograms with both a sill and hole effect.
- Semivariograms without a sill.

3.5.1 Semivariograms with a sill

These semivariograms are associated with the second-order stationary hypothesis. Thus, they have a covariogram counterpart (see Section 3.2.2). They also received the name of *transition models* because the distance at which the sill is reached represents the transition from the state of existence of spatial correlation to the state of absence of such spatial correlation.

3.5.1.1 The spherical model

This model, only valid on $\mathbb{R}^1, \mathbb{R}^2$ and $\mathbb{R}^3$, is defined by:

$$\gamma(|\mathbf{h}|) = \begin{cases} m\left(1.5\frac{|\mathbf{h}|}{a} - 0.5\left(\frac{|\mathbf{h}|}{a}\right)^3\right) & \text{if } |\mathbf{h}| \le a \\ m & \text{if } |\mathbf{h}| > a \end{cases}, \tag{3.25}$$

where $m = C(\mathbf{0})$ is the value of the semivariogram when it reaches the sill, and a is the range.

This semivariogram model exhibits a linear behavior near the origin (its order at the origin is $\alpha = 1$), which indicates continuity but a certain degree of irregularity in the rf. As can easily be checked,

$$\frac{\partial\gamma(|\mathbf{h}|)}{\partial\mathbf{h}} = m\left(\frac{1.5}{a} - \frac{|\mathbf{h}|^2}{a^3}\right),$$

so that the slope of the semivariogram at the origin is $1.5\, m/a$. The tangent at the origin intersects the sill at $|\mathbf{h}| = 2/3a$.

As for its behavior at large distances, it reaches the sill at $|\mathbf{h}| = a$. This well-defined range, along with its simple polynomial expression and its validity on $\mathbb{R}^1$, $\mathbb{R}^2$ and $\mathbb{R}^3$, are some of the reasons for the wide use of the spherical semivariogram in practical applications. But the main reason is that an almost linear behavior up to a certain distance (the range) and then a stabilization matches a large variety of observed regionalizations.

As pointed out by Webster and Oliver (2001, p. 115), the spherical model represents transition features that have a common extent and which appear as patches, some with large values and others with small ones. Figure 3.8 displays the plot of the spherical model as well as the 2D and 3D representations of a simulation of a rf having such a semivariogram model with the same sill ($m = 1$) and different ranges: $a = 0.15$ (left) and $a = 0.5$ (right). If we pay attention to the 2D representations, we can note that the average diameter of the patches of small and large values is represented by

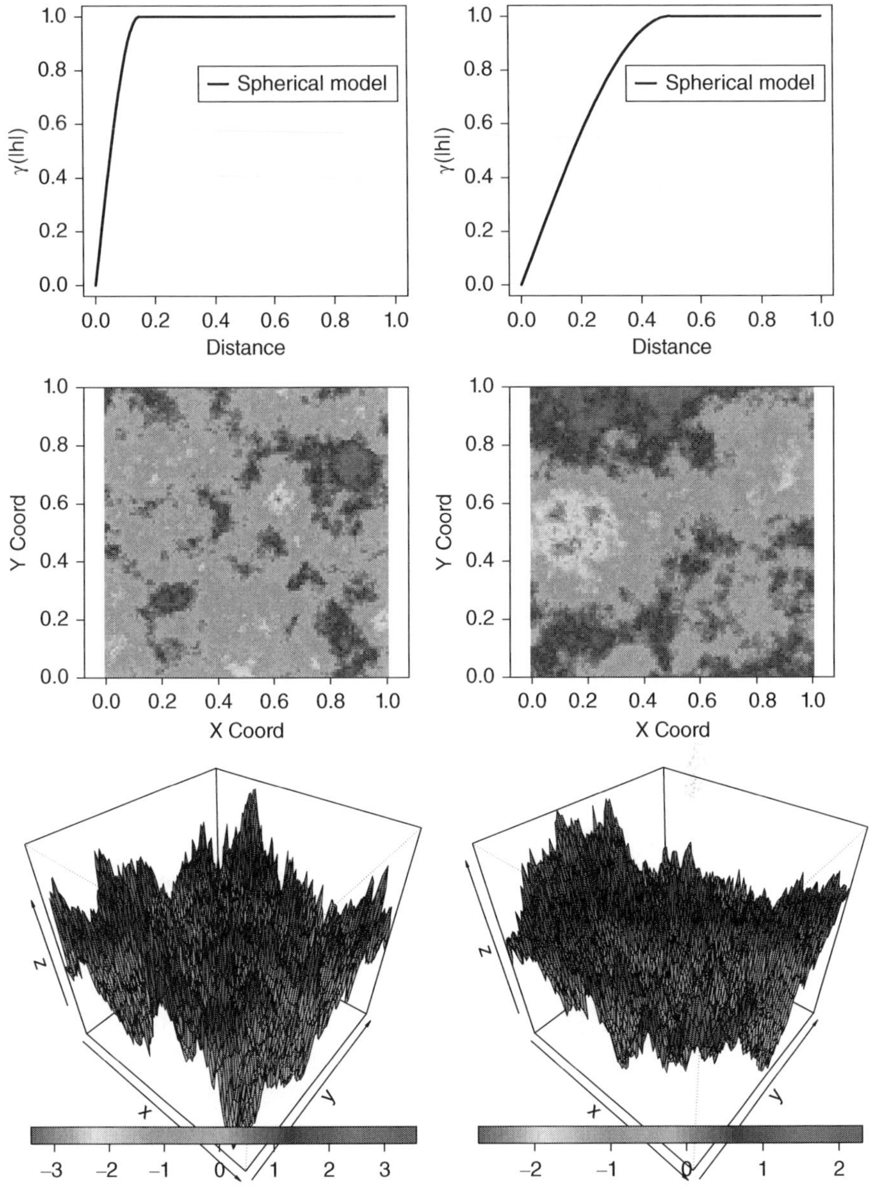

Figure 3.8 Upper panel: Spherical model. Left: $a = 0.15, m = 1$. Right: $a = 0.50$, $m = 1$. Middle panel: Simulation of a rf having a spherical semivariogram (2D representation). Left: $a = 0.15, m = 1$. Right: $a = 0.50, m = 1$. Bottom panel: Simulation of a rf having a spherical semivariogram (3D representation). Left: $a = 0.15, m = 1$. Right: $a = 0.50, m = 1$. (See color figure in color plate section.)

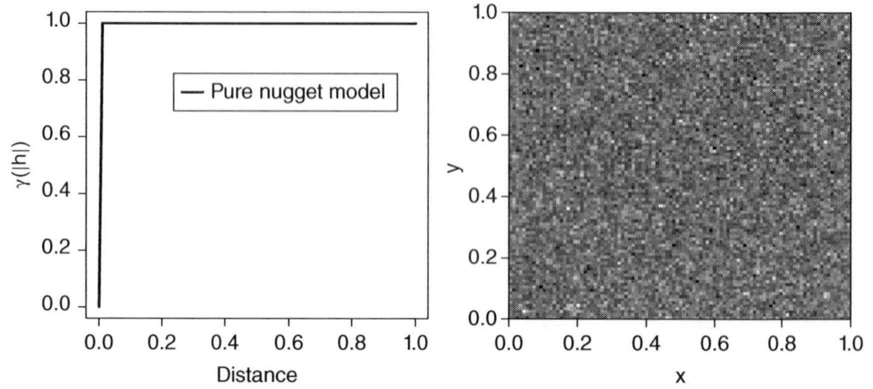

Figure 3.9 Left: Pure nugget semivariogram. Right: Simulation of a non-spatially correlated rf (2D representation).

the range of the model. This can also be noticed in the 3D representation of the simulated values.

The *pure nugget* semivariogram, which reflects the absence of spatial dependence in the rf, could be seen as a particular case of the spherical model when $a \to 0$:

$$\gamma(|\mathbf{h}|) = \begin{cases} m & \text{if } |\mathbf{h}| = 0 \\ 0 & \text{if } |\mathbf{h}| > 0 \end{cases}, \quad m > 0. \tag{3.26}$$

However, there is an important difference between both models: the spherical model corresponds to a continuous rf, whereas the pure nugget model corresponds to a discontinuous one.

Figure 3.9 shows the typical plot of a pure nugget effect semivariogram (left) and the 2D representation of a simulation of a non-spatially correlated rf (right). As can be seen in the right part of the Figure 3.9, there is no detectable pattern of spatial variation.

3.5.1.2 The exponential model

The exponential model is valid on $\mathbb{R}^d, d \geq 1$, and is given by the expression:

$$\gamma(|\mathbf{h}|) = m\left(1 - \exp\left(-\frac{|\mathbf{h}|}{a}\right)\right). \tag{3.27}$$

Like the spherical semivariogram, the exponential model exhibits a linear behavior near the origin (the order of the semivariogram at the origin is $\alpha = 1$), which indicates continuity but a certain degree of irregularity in the rf (Figure 3.10, middle and bottom panels). For exponential semivariograms,

$$\frac{\partial \gamma(|\mathbf{h}|)}{\partial \mathbf{h}} = m\left(\frac{1}{a}\exp\left(-\frac{|\mathbf{h}|}{a}\right)\right), \tag{3.28}$$

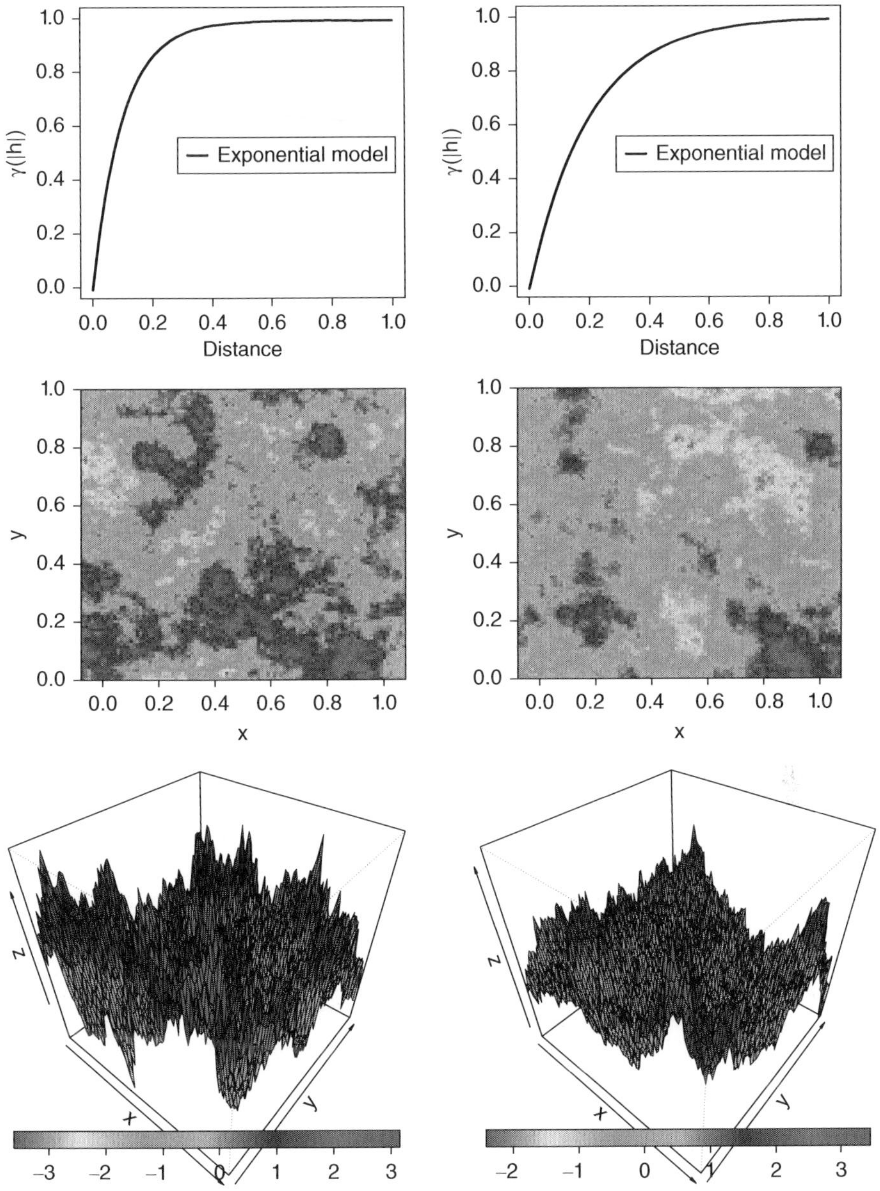

Figure 3.10 Upper panel: Exponential model. Left: $a = 0.10, m = 1$. Right: $a = 0.25, m = 1$. Middle panel: Simulation of a rf having an exponential semivariogram (2D representation). Left: $a = 0.10, m = 1$. Right: $a = 0.25, m = 1$. Bottom panel: Simulation of a rf having an exponential semivariogram (3D representation). Left: $a = 0.10, m = 1$. Right: $a = 0.25, m = 1$. (See color figure in color plate section.)

so that its slope at the origin is m/a, 1.5 times less than in the spherical case for the same range and sill.

As for the behavior at large distances, unlike the spherical model, exponential semivariograms reach the sill only asymptotically when $|\mathbf{h}| \to \infty$ (Figure 3.10, upper panel). For this model is defined the effective range or practical range, a', as the distance for which $\gamma(a') = m\left(1 - \exp\left(-\frac{a'}{a}\right)\right) = 0.95\,m$, that is, the distance above which the covariance is only 5% of the value recorded at the origin. This means that $a' \cong 3a$.

As a consequence, the distance at which the tangent at the origin of the exponential model intersects the sill is $|\mathbf{h}| = a = a'/3$. Thus, the spherical semivariogram reaches its sill faster than the exponential semivariogram. This is one of the more important differences between both models.

Figure 3.10 displays the plot of the exponential model and the 2D and 3D representations of a simulation of a rf with such a semivariogram model with the same sill and different scale parameter. It can be seen that the patches of small and large values are similarly irregular in both cases. This is because the exponential model is intimately related to transition features in which the structures have random extents. The average size of the patches indicates how large the scale parameter is and, therefore, the practical range.

3.5.1.3 The Gaussian model

This is a valid model on $\mathbb{R}^d, d \geq 1$, and is defined by:

$$\gamma(|\mathbf{h}|) = m\left(1 - \exp\left(-\frac{|\mathbf{h}|^2}{a^2}\right)\right). \tag{3.29}$$

Unlike the spherical and exponential models, the Gaussian model exhibits a parabolic behavior near the origin (the order of the Gaussian model at the origin is $\alpha = 2$). Thus, it can be associated with infinitely differentiable (and thus very regular) second-order stationary functions (Figure 3.11, middle and bottom panels). Such a large regularity hardly ever appears in practical applications and, as a consequence, the Gaussian model is considered unrealistic under normal circumstances.

For Gaussian semivariograms:

$$\frac{\partial\gamma(|\mathbf{h}|)}{\partial|\mathbf{h}|} = 2m\left(\frac{|\mathbf{h}|}{a^2}\exp\left(-\frac{|\mathbf{h}|}{a}\right)^2\right), \tag{3.30}$$

so that the slope at the origin is null.

Like the exponential model, the Gaussian model reaches the sill only asymptotically when $|\mathbf{h}| \to \infty$ (Figure 3.11, upper panel). The practical range is $a' = a\sqrt{3}$, the distance for which $\gamma(a') = m\left(1 - \exp\left(-\left(\frac{a'}{a}\right)^2\right)\right) = 0.95\,m$.

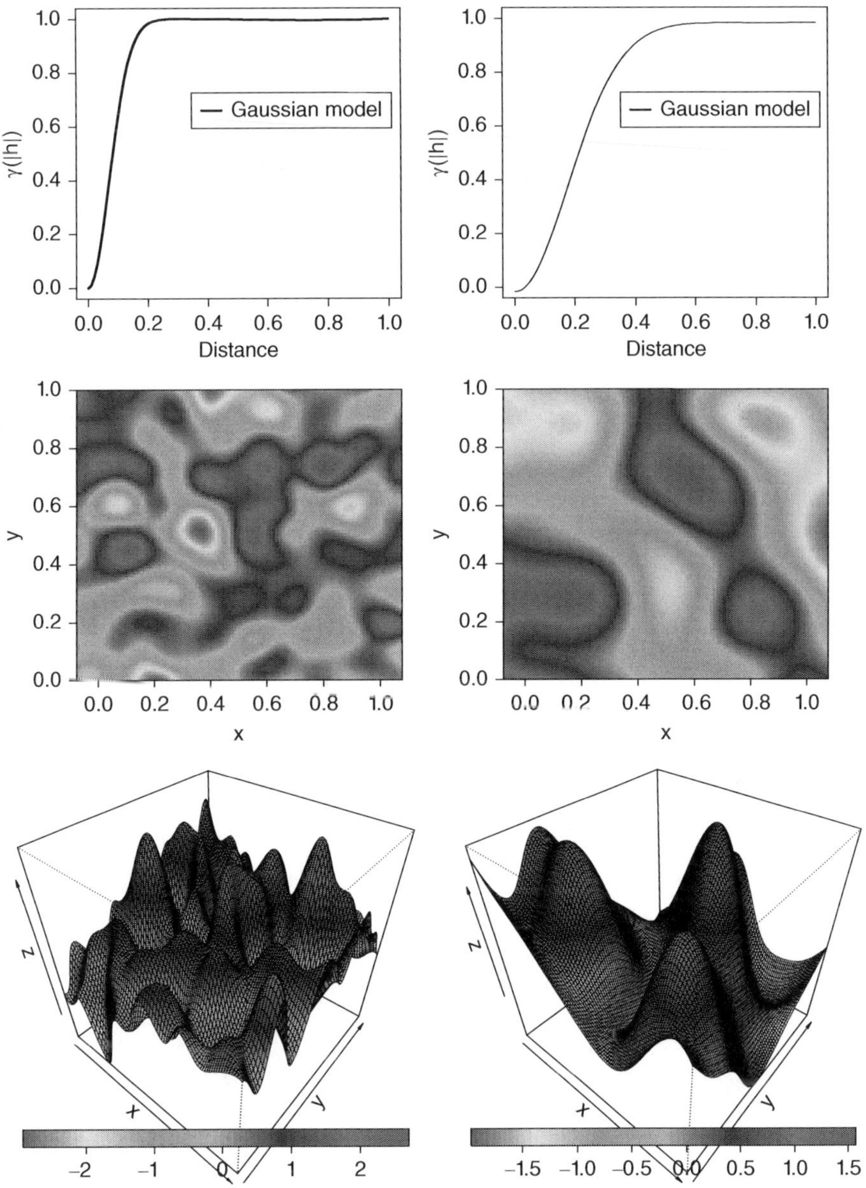

Figure 3.11 *Upper panel: Gaussian model. Left:* $a = 0.15, m = 1$. *Right:* $a = 0.25$, $m = 1$. *Middle panel: Simulation of a rf having a Gaussian semivariogram (2D representation). Left:* $a = 0.10, m = 1$. *Right:* $a = 0.25, m = 1$. *Bottom panel: Simulation of a rf having a Gaussian semivariogram (3D representation). Left:* $a = 0.10, m = 1$. *Right:* $a = 0.25, m = 1$. *(See color figure in color plate section.)*

Finally, as stated by Journel and Huijbregts (1978, p. 165), a parabolic drift effect does not stabilize around a sill for large distances and, thus, cannot be confused with the parabolic behavior at the origin exhibited by the Gaussian model. However, for small distances $|\mathbf{h}| < \frac{2}{3}a$, it makes little difference whether a very regular local variation is interpreted as a drift effect or as a stationary Gaussian structure.

Figure 3.11 (upper panel) shows the plot of two Gaussian models with the same asymptotic sill and different ranges. From the 2D and 3D representations of a simulation of a rf having a Gaussian semivariogram (middle and bottom panels, respectively) it can easily be seen why this model is not used in practice.

3.5.1.4 The cubic model

This is defined by:

$$\gamma(|\mathbf{h}|) = \begin{cases} m\left(7\frac{|\mathbf{h}|^2}{a^2} - \frac{35}{4}\frac{|\mathbf{h}|^3}{a^3} + \frac{7}{2}\frac{|\mathbf{h}|^5}{a^5} - \frac{3}{4}\frac{|\mathbf{h}|^7}{a^7}\right) & \text{if } 0 \leq |\mathbf{h}| \leq a \\ m \quad \text{if } |\mathbf{h}| \geq a \end{cases} . \tag{3.31}$$

This model, valid on $\mathbb{R}^1$, $\mathbb{R}^2$ and $\mathbb{R}^3$, is, in general, similar to the Gaussian model. It also has a parabolic behavior near the origin, but it is not infinitely differentiable (Figures 3.12 and 3.13). Another difference is that the cubic model reaches a flat sill at distance a.

In effect,

$$\frac{\partial\gamma(|\mathbf{h}|)}{\partial(|\mathbf{h}|)} = m\left(\frac{14|\mathbf{h}|}{a^2} - \frac{105}{4}\frac{|\mathbf{h}|^2}{a^3} + \frac{35}{2}\frac{|\mathbf{h}|^4}{a^5} - \frac{21}{4}\frac{|\mathbf{h}|^6}{a^7}\right),$$

which is null for $|\mathbf{h}| = 0$ and $|\mathbf{h}| = a$.

Figure 3.14 offers the comparison of the simulations of two rf's, one having a Gaussian semivariogram and the other having a cubic one (both of them with the same sill and scale parameter).

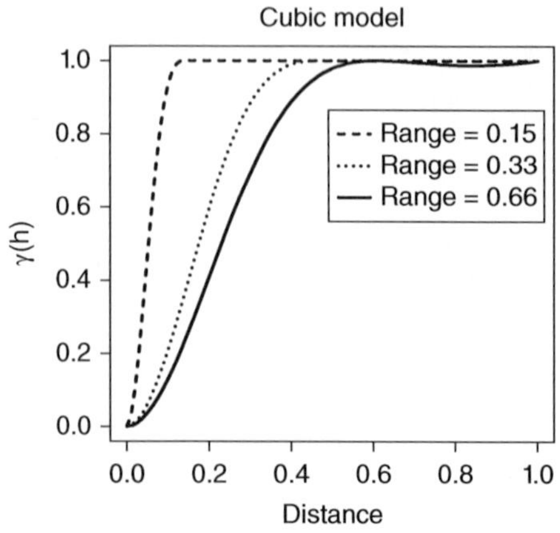

Figure 3.12 Cubic model with different ranges and the same sill ($m = 1$).

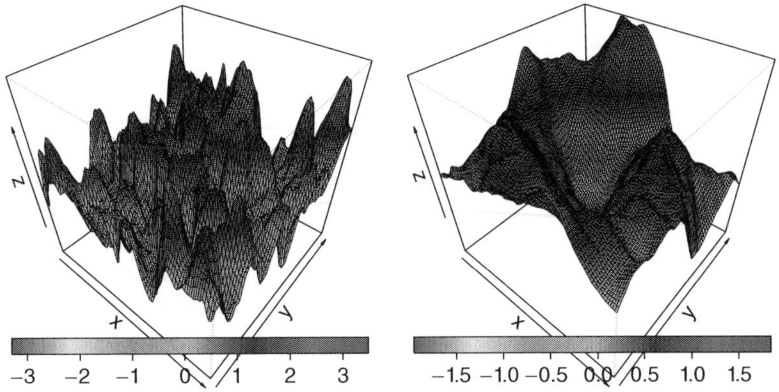

Figure 3.13 Simulation of a rf with cubic semivariogram model (3D representation). Left: $a = 0.15, m = 1$. Right: $a = 0.66, m = 1$.

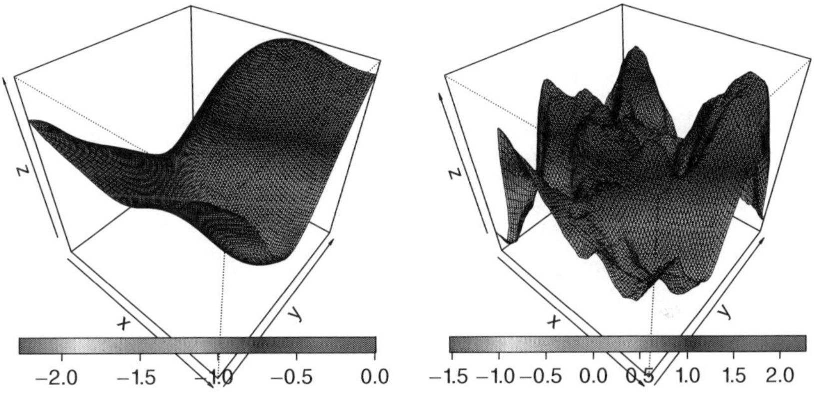

Figure 3.14 Left: 3D representation of a simulation of a rf having a Gaussian semivariogram with $a = 0.50, m = 1$. Right: 3D representation of a simulation of a rf having a cubic semivariogram with $a = 0.50, m = 1$.

Other interesting semivariograms with a sill are the stable model, the Cauchy model and the K-Bessel model. These models include not only the parameters m and a, but also an additional shape parameter, α.

3.5.1.5 The stable model

The stable model is defined by:

$$\gamma(|\mathbf{h}|) = m\left(1 - \exp\left(-\left(\frac{|\mathbf{h}|}{a}\right)^{\alpha}\right)\right), \quad 0 < \alpha \leq 2, \tag{3.32}$$

and is a semivariogram on $\mathbb{R}^d, d \geq 1$.

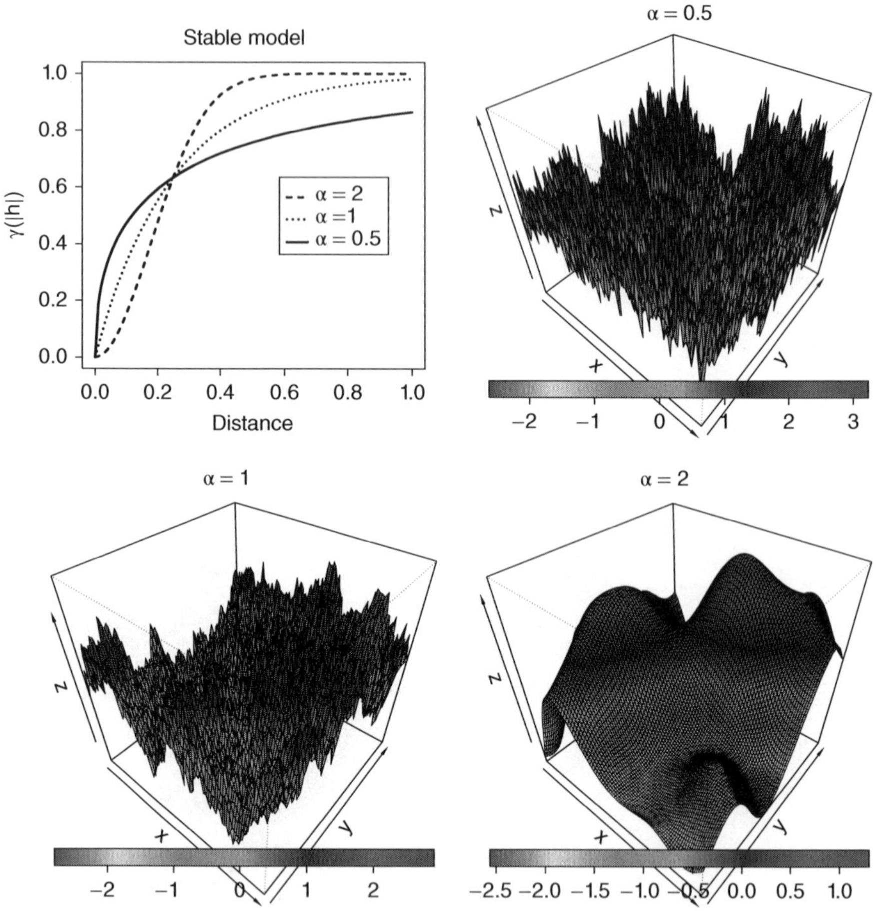

Figure 3.15 Upper panel, left: Stable model with the same sill ($m = 1$) and scale parameter ($a = 0.25$) but different shape parameter α. Upper panel, right: 3D representation of a simulation of a rf having a stable semivariogram ($a = 0.25, m = 1, \alpha = 0.5$). Bottom panel, left: 3D representation of a simulation of a rf having a stable semivariogram ($a = 0.25, m = 1, \alpha = 1$). Bottom panel, right: 3D representation of a simulation of a rf having a stable semivariogram ($a = 0.25, m = 1, \alpha = 2$). (See color figure in color plate section.)

Note that for $\alpha = 1$ we obtain the exponential model and that $\alpha = 2$ leads to the Gaussian semivariogram model (see Figure 3.15).

The practical range of the stable model is $a' = a\sqrt[\alpha]{3}$.

3.5.1.6 The generalized Cauchy model

This model is valid in $\mathbb{R}^d, d \geq 1$, and is defined by:

$$\gamma(|\mathbf{h}|) = m\left(1 - \frac{1}{\left(1+\left(\frac{|\mathbf{h}|}{a}\right)^2\right)^{\alpha}}\right). \quad (3.33)$$

For $\alpha = 1$ it is known as the Cauchy model.

The generalized Cauchy model exhibits a parabolic behavior near the origin, since its Taylor expression contains only even terms, and for small values of α, in particular $\alpha < 2$, reaches the sill really slowly (its practical range is $a' = a(\sqrt[\alpha]{20} - 1)^{1/2}$). Figure 3.16 shows the Cauchy model for different values of the shape parameter.

3.5.1.7 The K-Bessel model

This model is valid on $\mathbb{R}^d, d \geq 1$, and is given by:

$$\gamma(|\mathbf{h}|) = m\left(1 - \frac{1}{2^{\alpha-1}\Gamma(\alpha)}\left(\frac{|\mathbf{h}|}{a}\right)^{\alpha} K_\alpha\left(\frac{|\mathbf{h}|}{a}\right)\right), \quad \alpha > 0, \quad (3.34)$$

where K_α is the modified Bessel function of the second kind of order α, which is defined by:

$$K_\alpha(\nu) = \frac{\pi}{2\sin(\alpha\pi)}\left(\sum_{k=0}^{\infty}\frac{1}{k!\Gamma(-\alpha+k+1)}\left(\frac{\nu}{2}\right)^{2k-\alpha} - \sum_{k=0}^{\infty}\frac{1}{k!\Gamma(\alpha+k+1)}\left(\frac{\nu}{2}\right)^{2k+\alpha}\right).$$

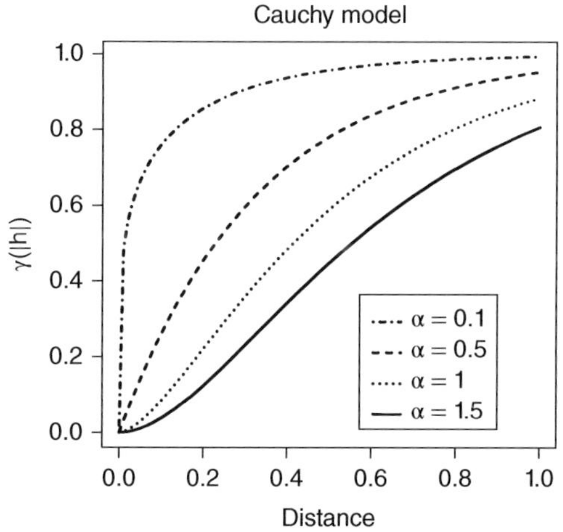

Figure 3.16 Cauchy models with the same sill ($m = 1$) and scale parameter ($a = 0.33$) but different shape parameter α.

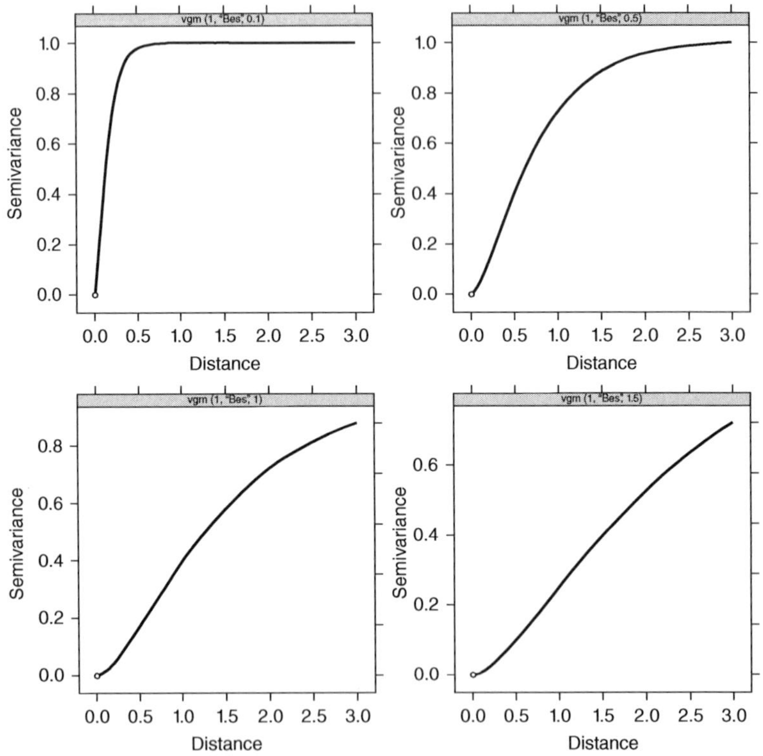

Figure 3.17 K-Bessel model with the same scale parameter ($a = 0.5$) and the same sill ($m = 0.5$) but different shape parameter α.

The K-Bessel model can have any type of behavior near the origin (see Figure 3.17). For $\alpha = 1/2$ the exponential model is obtained.

3.5.2 Semivariograms with a hole effect

The models presented in the previous section are non-decreasing functions of the separation between two points. Thus, they are only valid for positive spatial dependence. However, a semivariogram could not growth monotonically but would display a hole effect or waves. That is, a semivariogram can also allow for negative spatial dependence, and more specifically for the alternation of positive and negative spatial dependence.

These models are termed hole-effect models and are used to define underlying components or oscillations that generally have a physical meaning. They can have a linear or parabolic behavior near the origin, can have a sill or not, and can be periodicals (or pseudo-periodicals) or not.

In general, if the periodicity or pseudo-periodicity is well known, it is recommended filtering it. If the information available is not large enough to guarantee the existence of such periodicities, they should be ignored because the waves that usually appear in empirical semivariograms can be due to other multiple causes (for example, the correlation between the empirical values).

Next, we focus on the J-Bessel model (which has a sill) and one of its particularizations, the cardinal sine model.

3.5.2.1 The J-Bessel model

This model is given by:

$$\gamma(|\mathbf{h}|) = m\left(1 - \left(\frac{2a}{|\mathbf{h}|}\right)^{\alpha} \Gamma(\alpha+1)\, J_{\alpha}\left(\frac{|\mathbf{h}|}{a}\right)\right), \tag{3.35}$$

where α is a shape parameter, a is a scale parameter, Γ is the Euler function which interpolates the factorial, and J_{α} is the J-Bessel function of the first kind of order α given by:

$$J_{\alpha}(v) = \left(\frac{v}{2}\right)^{2} \sum_{k=0}^{\infty} \frac{-1^{k}}{k!\Gamma(\alpha+k+1)} \left(\frac{v}{2}\right)^{2k},$$

and is valid for $\mathbb{R}^d, d \leq 2(\alpha+1)$.

3.5.2.2 The cardinal sine model

This model is the particularization of the J-Bessel model for $\alpha = 1/2$ and is one of the few hole effect models valid on $\mathbb{R}^3$ (see Figure 3.18). Other important features of this model include being a pseudo-periodical model and its parabolic behavior near the origin and, thus, its association with very continuous structures.

According to Emery (2000), the practical range of the cardinal sine model is $20.37a$. The amplitude of the hole effect is 1.21, the maximum amplitude that can be obtained with an isotropic model on $\mathbb{R}^3$.

There are also non-periodical hole effect models with a linear behavior near the origin. They include the truncated polynomial models and the generalized stable model (see Emery 2000, p. 102, for details).

3.5.3 Semivariograms without a sill

These models go beyond the second-order stationary hypothesis. They are unbounded and correspond to rf's with unlimited capacity for spatial dispersion and, as a consequence, neither their variance nor their covariances can be defined. More specifically, they correspond to rf's that are intrinsically stationary, but not second-order stationary.

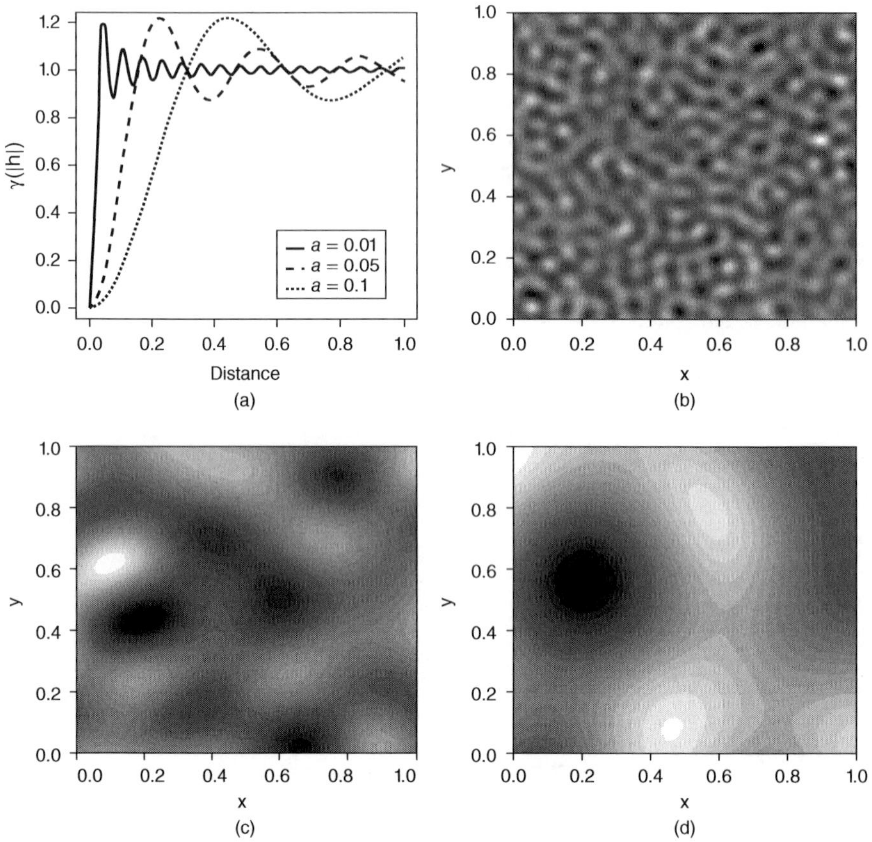

Figure 3.18 Cardinal sine models with the same sill ($m = 1$) and different values of the scale parameter: (a) plot of the models; (b), (c) and (d) simulation of a rf having a cardinal sine model with $a = 0.01, 0.05, 0.10$, respectively (2D representation).

3.5.3.1 The power model

The power model is defined by:

$$\gamma(|\mathbf{h}|) = (|\mathbf{h}|)^{\alpha}, \quad \text{with} \quad 0 < \alpha < 2, \tag{3.36}$$

and is valid on $\mathbb{R}^d, d \geq 1$.

Depending on the value of α, the power model exhibits a large variety of behaviors near the origin (see Figure 3.19).

For $\alpha \geq 2$ the condition $\lim_{h \to \infty} \frac{\gamma(|\mathbf{h}|)}{h^2} = 0$ is not satisfied. That is, the model does not satisfy the intrinsic hypothesis. The case $\alpha = 0$ corresponds to a pure nugget effect. The case $\alpha = 1$ results in the linear model.

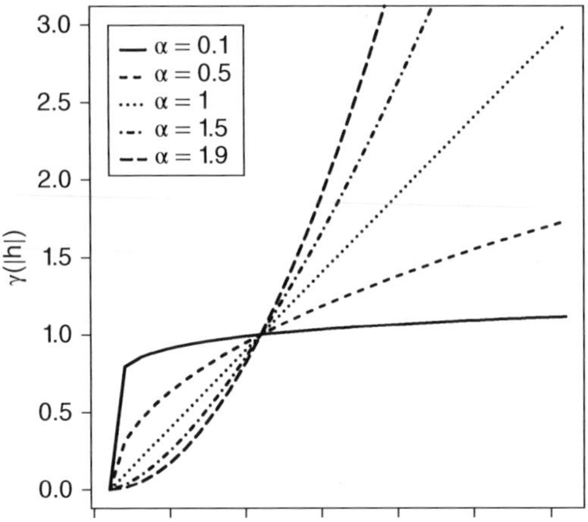

Figure 3.19 Power models.

3.5.3.2 The linear model

It is defined by:

$$\gamma(|\mathbf{h}|) = \begin{cases} 0 & \text{if } |\mathbf{h}| = 0 \\ |\mathbf{h}| & \text{if } |\mathbf{h}| > 0 \end{cases}. \tag{3.37}$$

The linear model is the particular case of the power model when $\alpha = 1$ (Figure 3.20). Since $0 < \alpha = 1 < 2$ the linear model can be associated with the intrinsic, but not

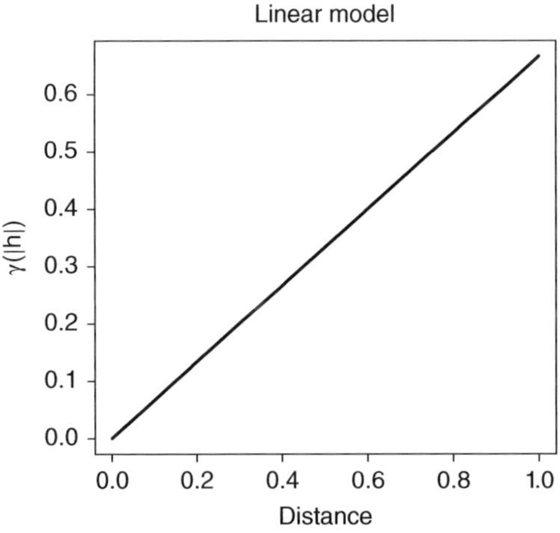

Figure 3.20 Linear model.

second-order, stationary hypothesis. Schabenberger and Gotway (2005) point out that it could also be indicative of a second-order stationary rf with a semivariogram exhibiting a linear behavior near the origin, but whose sill has not been reached across the observed lag distances.

3.5.3.3 The logarithmic model

The logarithmic model (Figure 3.21) is given by:

$$\gamma(|\mathbf{h}|) = b \log |\mathbf{h}| \quad \text{if} \quad |\mathbf{h}| \geq 0, \tag{3.38}$$

where b is a constant, and is a semivariogram on $\mathbb{R}^d, d \geq 1$. It is not defined at the origin ($\gamma\,(\mathbf{h}) \to -\infty$ when $\mathbf{h} \to 0$), thereby not complying with one of the theoretical requisites of semivariogram models, nor does it have a sill ($\gamma\,(\mathbf{h}) \to \infty$ when $\mathbf{h} \to \infty$). This is why this model is used when dealing with variables regularized by a sampling support (see Chilès and Delfiner 1999, pp. 90–1, for details).

3.5.4 Combining semivariogram models

In practice, especially in case of a large number of data, experimental semivariograms do not seem to have an appearance as simple as that of the theoretical models presented in Section 3.4. This is not a problem, since by combining (synonyms: superposing, nesting) theoretical semivariograms we can obtain new and more complex models.

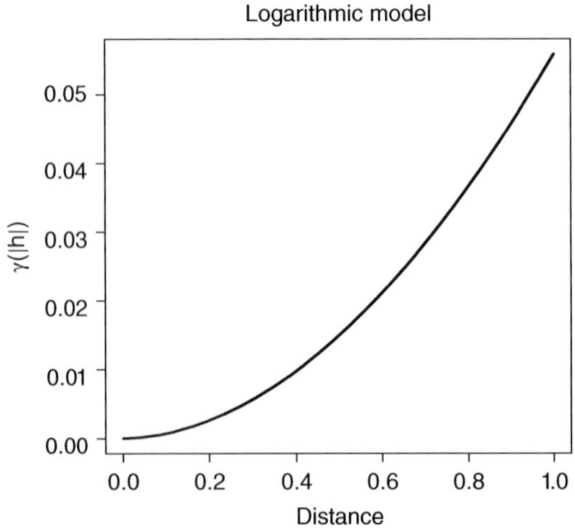

Figure 3.21 Logarithmic model ($b = 1$).

One of these cases was introduced in Section 3.4.3 to illustrate semivariograms with a discontinuity at the origin. Another typical case is when spatial dependence occurs at different scales. In this case, the semivariogram exhibits changes in its slope. As stated in Wackernagel (2003, p. 101), a regionalized phenomenon can be thought of as being the sum of several independent sub-phenomena acting at different characteristic scales.[2] Thus, a rf associated with a nested semivariogram is the sum of spatial components characterizing different spatial scales. More specifically, in the intrinsic case, and for a rf composed of two spatial components ($Z(\mathbf{s}) = Z(\mathbf{s}_1) + Z(\mathbf{s}_2)$) and whose first increments are uncorrelated, we have:

$$\begin{aligned}\gamma(\mathbf{h}) &= \frac{1}{2}E\big(Z(\mathbf{s}_i+\mathbf{h}) - Z(\mathbf{s}_i)\big)^2\\ &= \frac{1}{2}E\big(\big(Z_1(\mathbf{s}_i+\mathbf{h}) + Z_2(\mathbf{s}_i+\mathbf{h})\big) - \big(Z_1(\mathbf{s}_i) - Z_2(\mathbf{s}_i)\big)\big)^2\\ &= \frac{1}{2}E\big(\big(Z_1(\mathbf{s}_i+\mathbf{h}) - Z_1(\mathbf{s}_i)\big) + \big(Z_2(\mathbf{s}_i+\mathbf{h}) - Z_2(\mathbf{s}_i)\big)\big)^2\\ &= \gamma_1(\mathbf{h}) + \gamma_2(\mathbf{h}).\end{aligned} \tag{3.39}$$

Example 3.2 (Combining semivariogram models) *Let us combine a pure nugget model, $\gamma_1(\mathbf{h})$, and two spherical semivariograms, $\gamma_2(\mathbf{h})$ and $\gamma_3(\mathbf{h})$:*

$$\begin{aligned}\gamma_1(\mathbf{h}) &= m_1\\ \gamma_2(\mathbf{h}) &= m_2\left(1.5\left(\frac{|\mathbf{h}|}{a_2}\right) + 0.5\left(\frac{|\mathbf{h}|}{a_2}\right)^3\right)\\ \gamma_3(\mathbf{h}) &= m_3\left(1.5\left(\frac{|\mathbf{h}|}{a_3}\right) + 0.5\left(\frac{|\mathbf{h}|}{a_3}\right)^3\right),\end{aligned}$$

where $a_2 < a_3$.

The combined semivariogram is a double spherical model with nugget effect:

$$\gamma(|\mathbf{h}|) = \begin{cases} m_1 + m_2\left(1.5\left(\frac{|\mathbf{h}|}{a_2}\right) + 0.5\left(\frac{|\mathbf{h}|}{a_2}\right)^3\right) + m_3\left(1.5\left(\frac{|\mathbf{h}|}{a_3}\right) + 0.5\left(\frac{|\mathbf{h}|}{a_3}\right)^3\right), \\ \qquad |\mathbf{h}| \le a_2 \\ m_1 + m_2 + m_3\left(1.5\left(\frac{|\mathbf{h}|}{a_3}\right) + 0.5\left(\frac{|\mathbf{h}|}{a_3}\right)^3\right), \quad a_2 < |\mathbf{h}| \le a_3 \\ m_1 + m_2 + m_3, \quad |\mathbf{h}| > a_3. \end{cases} \tag{3.40}$$

In Figure 3.22 it can be seen how each scale of variation integrates the variability of all the smaller scales.

[2] In reality, the independence of the sub-phenomena could be a questionable assumption.

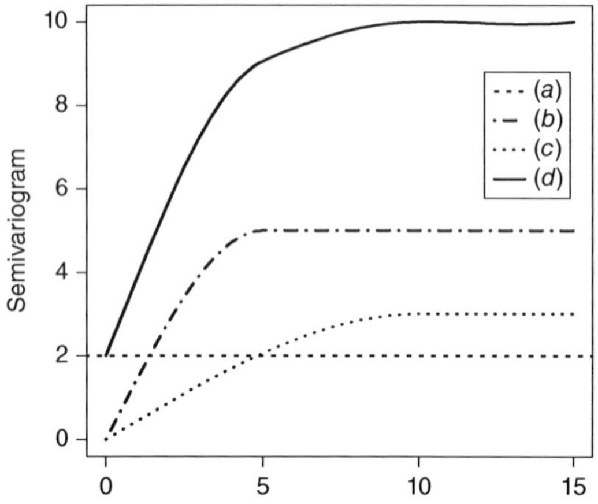

Figure 3.22 Nested model composed of: (a) Pure nugget semivariogram ($m = 2$); (b) Spherical semivariogram ($a_2 = 5$, $m_2 = 5$); (c) Spherical semivariogram ($a_3 = 10$, $m_3 = 3$); (d) Nested semivariogram (3.40).

3.6 Empirical semivariogram

As mentioned in Section 3.3, in practice, we only have the observed regionalization, that is, a fragmentary set of geo-referenced data in a given domain, and we must use these data to infer the spatial dependence structure of the phenomenon being studied.

Bearing in mind that the semivariogram is the instrument used *par excellence* to describe the spatial dependence in the regionalized variable (it covers a broader spectrum of regionalized variables than the covariogram), in practice, the empirical (or experimental) semivariogram is the instrument used to estimate the structure of spatial variability existing in the phenomenon of interest. Obviously, when performing this estimation process, it is desirable to have some theoretical knowledge of such a phenomenon, and the more the better.

Remark 3.6.1 *The empirical semivariogram is the instrument used to estimate the spatial variability of the phenomenon being studied.*

In the context of intrinsic stationarity, the estimator of the semivariogram is obtained by the MoM (as proposed by Matheron 1962; in this sense the MoM estimator of the semivariogram is also termed the classical estimator) and is defined by:

$$\hat{\gamma}(\mathbf{h}) = \frac{1}{2\#N(\mathbf{h})} \sum_{N(\mathbf{h})} \left(Z(\mathbf{s}_i + \mathbf{h}) - Z(\mathbf{s}_i)\right)^2, \tag{3.41}$$

where $Z(\mathbf{s}_i)$ are the values of the characteristic of interest at the points $\mathbf{s}_i$ when we also know the value of this characteristic at $\mathbf{s}_i + \mathbf{h}$, and $\#N(\mathbf{h})$ is the number of pairs of locations separated by a vector $\mathbf{h}$.

The plot $\hat{\gamma}(\mathbf{h})$ versus $|\mathbf{h}|$ is known as the *empirical semivariogram*.

The MoM estimator is unbiased. In effect,

$$\begin{aligned}
E\left(\hat{\gamma}(\mathbf{h})\right) &= E\left(\frac{1}{2\#N(\mathbf{h})}\sum_{N(\mathbf{h})}\left(Z(\mathbf{s}_i+\mathbf{h})-Z(\mathbf{s}_i)\right)^2\right)\\
&= \frac{1}{\#N(\mathbf{h})}\sum_{N(\mathbf{h})}\frac{1}{2}E\left(Z(\mathbf{s}_i+\mathbf{h})-Z(\mathbf{s}_i)\right)^2\\
&= \frac{1}{\#N(\mathbf{h})}\sum_{N(\mathbf{h})}\frac{1}{2}V\left(Z(\mathbf{s}_i+\mathbf{h})-Z(\mathbf{s}_i)\right)\\
&= \frac{1}{\#N(\mathbf{h})}\sum_{N(\mathbf{h})}\gamma(\mathbf{h})\\
&= \gamma(\mathbf{h}).
\end{aligned}$$

In practice, the empirical semivariogram is usually computed for lag distances inferior to half the diameter of the domain. The main reason for this is that the number of pairs diminishes with distance, and at large distances the number of pairs is not large enough to yield a credible estimate.

It is of note that while the empirical covariogram (under the second-order stationary hypothesis) is a biased estimator of the covariance function(see Section 3.3), the empirical semivariogram is an unbiased estimator of the semivariogram (under the intrinsic hypothesis), the reason being that the unknown mean of the rf is not involved in the expression of the semivariogram. It is also important to highlight that, in the case of second-order stationarity,

$$\hat{\gamma}(\mathbf{h}) \neq \hat{C}(\mathbf{0}) - \hat{C}(\mathbf{h}). \tag{3.42}$$

In effect, let $\mathbf{s}_i - \mathbf{s}_j = \mathbf{h}$, then

$$\begin{aligned}
\hat{\gamma}(\mathbf{h}) &= \frac{1}{2\#N(\mathbf{h})}\sum_{N(\mathbf{h})}\left(\left(Z(\mathbf{s}_i)-\bar{Z}\right)-\left(Z(\mathbf{s}_j)-\bar{Z}\right)\right)^2\\
&= \frac{1}{\#N(\mathbf{h})}\sum_{N(\mathbf{h})}\left(Z(\mathbf{s}_i)-\bar{Z}\right)^2-\hat{C}(\mathbf{h}),
\end{aligned}$$

and taking into account that $\frac{1}{\#N(\mathbf{h})}\sum_{N(\mathbf{h})}\left(Z(\mathbf{s}_i)-\bar{Z}\right)^2 \neq \frac{1}{n}\sum_{i=1}^{n}\left(Z(\mathbf{s}_i)-\bar{Z}\right)^2 = \hat{C}(\mathbf{0})$ we have that:

$$\hat{\gamma}(\mathbf{h}) \neq \hat{C}(\mathbf{0}) - \hat{C}(\mathbf{h}).$$

Thus, a semivariogram constructed from the empirical covariogram will be biased (the bias tends to disappear as $\neq N(\mathbf{h}) \rightarrow n$).

If all the sampled points are regularly distributed in space, the distances between them will be pre-established. But if they are distributed irregularly, which is very

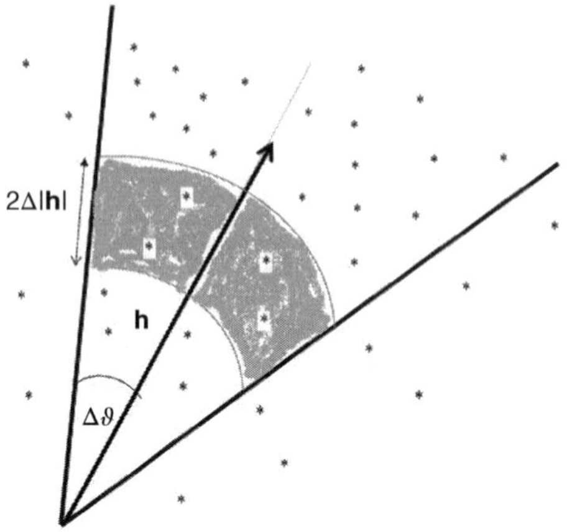

Figure 3.23 *Tolerance region on* $\mathbb{R}^2$.

common, the number of different distances between points sampled will increase and there will be few pairs for each distance, especially at large distances (there may not even be two pairs at the same distance). It is obvious that ideally we aim for a large number of pairs at each distance. For this reason, it is also common in practice to consider non-overlapping *tolerance regions* (the shaded area of Figure 3.23), based on distance intervals (normally of equal length) and a tolerance angle around the direction defined by the vector **h**. More specifically, we specify tolerance in both the module of **h** ($\pm\triangle|\mathbf{h}|$) and its direction ($\pm\triangle\vartheta$). The empirical semivariogram is calculated considering all pairs of points whose separation vector **h** falls inside the tolerance region; the mean of the squared semidifference of its values is assigned to a vector **h** that represents all the vectors included in the tolerance region.

In order to check for anisotropies it is advisable to construct the empirical semivariogram in several directions (for example, $\vartheta = 0, \pi/4, \pi/2, 3\pi/4$).

Obviously, as the length of the distance interval and the tolerance angle increase, the fluctuations in the empirical semivariogram decrease, albeit at the expense of increasing the discretization error committed (see Figure 3.24).

Example 3.3 (Empirical semivariogram) *Let us consider* 25 *data distributed on a regularly spaced* 100×100 *grid. Figure 3.25 indicates the location of such data in the grid (the bigger the point size, the larger the data value). More specifically, both the coordinates and also the data values in the observed points are listed in Table 3.1.*

The calculations needed to construct the MoM or classical empirical semivariogram can be found in Table 3.2. We used `geoR` *package.*

As expected, the number of pairs decreases with distance, which implies that the semivariogram values for large distances are not reliable (in the last column of Table 3.2 the variance of values $\frac{1}{2}(Z(\mathbf{s}+\mathbf{h}) - Z(\mathbf{s}))^2$ *increases with distance). This is why the usual practice is to compute the semivariogram values up to half the*

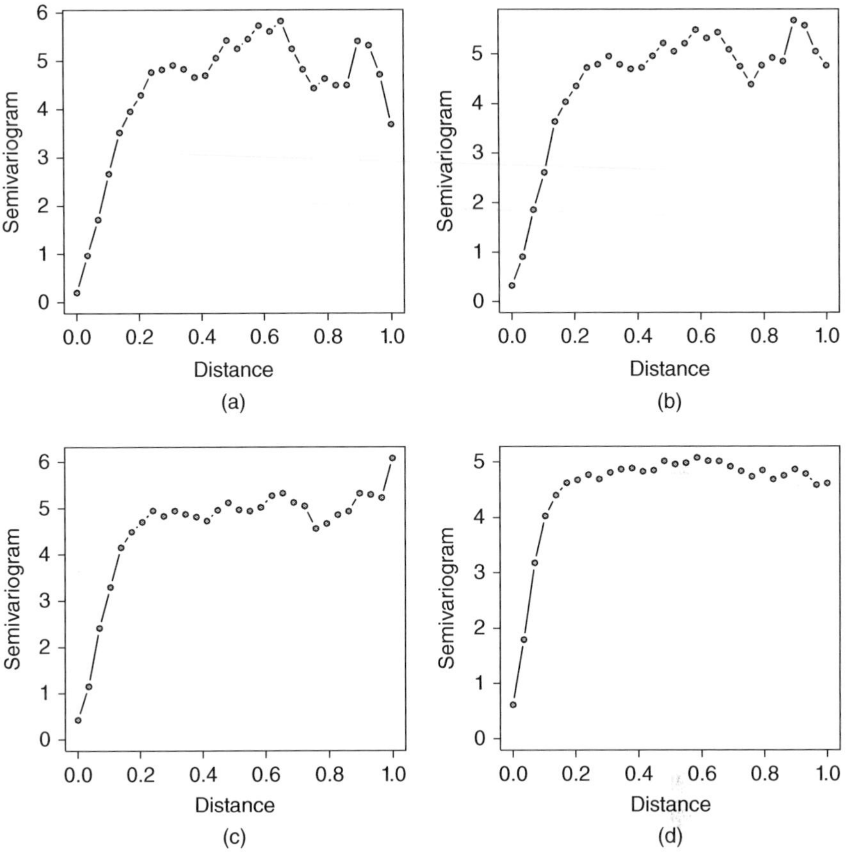

Figure 3.24 Effect of the tolerance angle on a North-South empirical semivariogram. Tolerance angle: (a) $\pi/27$, (b) $\pi/18$, (c) $\pi/6$, (d) $\pi/3$. The observed regionalization was simulated with an spherical model ($m = 5, a = 0.3$).

maximal distance. Also, as expected, the semivariogram values increase (albeit not monotonically) up to a certain distance and then stabilize. These oscillations are normal.

The empirical semivariogram is depicted in Figure 3.26. It can easily be seen that it stabilizes around a value of 30 *(the sill) for a lag distance of nearly* 75 *m (the range). The scatter diagram in the right panel of Figure 3.26 (the semivariogram cloud) plots the values* $\frac{1}{2}(Z(\mathbf{s}+\mathbf{h}) - Z(\mathbf{s}))^2$ *against the* 14 *distances considered. This semivariogram cloud allows the dispersion of the above squared differences around the empirical semivariogram value for every distance* $\mathbf{h}$ *to be shown. In the semivariogram cloud some atypical points can be observed, but since the data were obtained from a Gaussian rf, this atypical behavior could not be really atypical but a result of the high skewness of the variable* $(Z(\mathbf{s}+\mathbf{h}) - Z(\mathbf{s}))^2$*. This detail, which can be of*

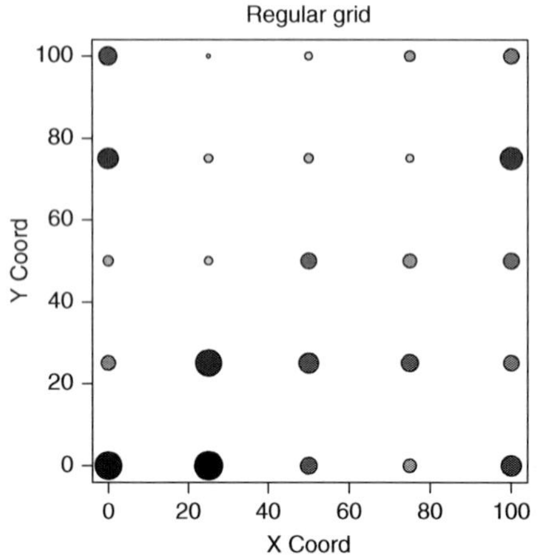

Figure 3.25 Twenty-five observed values in a 100×100 *grid.*

Table 3.1 Coordinates and value of the observed data.

X coord	Y coord	Value
0	0	19.64
25	0	20.67
50	0	11.40
75	0	8.72
100	0	14.19
0	25	9.76
25	25	18.98
50	25	14.05
75	25	11.76
100	25	10.43
0	50	6.56
25	50	4.88
50	50	10.76
75	50	9.10
100	50	10.69
0	75	14.88
25	75	5.37
50	75	5.71
75	75	4.43
100	75	15.69
0	100	12.77
25	100	1.37
50	100	4.36
75	100	6.61
100	100	10.26

Table 3.2 Empirical semivariogram values together with the variance of half the squared semidifferences used to compute them.

Lag distances	Number of pairs	Semivariogram values	Variance
25.00	40	15.20	495.56
35.36	32	19.63	533.00
50.00	30	25.82	927.83
55.90	48	22.24	695.08
70.71	18	18.60	730.85
75.00	20	34.02	1542.93
79.06	32	34.03	1044.06
90.14	24	29.60	1110.00
100.00	10	27.19	2888.95
103.08	16	34.89	2161.79
106.06	8	33.81	1177.60
111.80	12	33.21	1348.54
125.00	8	30.07	1228.88
141.42	2	22.53	462.93

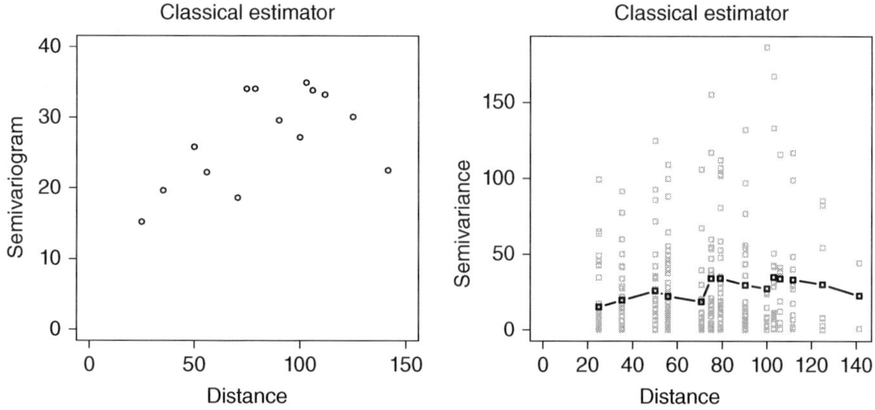

Figure 3.26 Left: Empirical semivariogram (classic estimator). Right: Semivariogram cloud.

great help in the interpretation and understanding of the result obtained, and also in the fitting of a valid model to the empirical semivariogram points, is lost in the standard plot (the plot in the left panel of Figure 3.26).

Example 3.4 (Constructing an empirical semivariogram: Data on carbon monoxide in Madrid, week 50, 10 pm) *Now we focus on the registered levels of carbon monoxide (CO) in Madrid, the real geostatistical database described on Section 1.3. The data have the format* $(x_i, y_i, logCO_i^*) : i = 1, \ldots, 23,$ *where*

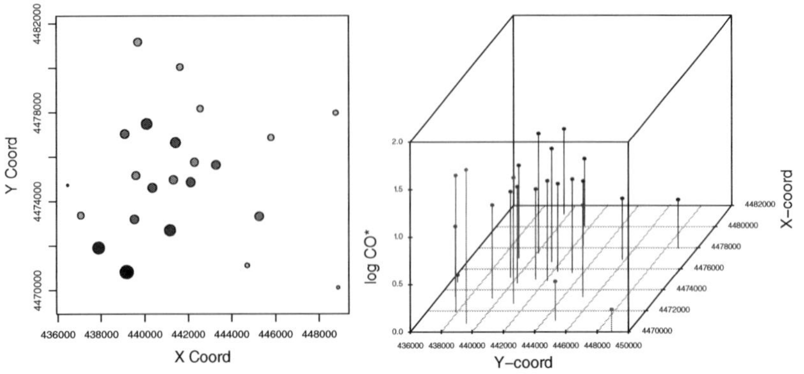

Figure 3.27 Observed values of logCO at the* 23 *monitoring stations operating in Madrid, week* 50, 10 *pm. Left panel: 2D representation (black: higher values; white: lower values). Right panel: 3D representation.*

(x_i, y_i) identifies the UTM coordinates where the 23 *operative monitoring stations are located and $logCO_i^*$ is the average of the logarithm of carbon monoxide levels (mg/cm^3) observed at the ith monitoring station in the labor days of week* 50 (2008) *at* 10 *pm. Figure 3.27 depicts the observed values of logCO*.*

Since the minimum and maximum distances between two monitoring stations are 820 *and* 14360 *m., respectively, following the usual rule we centered the first lag distance at* 800 *m and computed the empirical semivariogram values from such a distance to half the maximum distance (more specifically, we centered the last lag distance at* 7000 *m.). Since the monitoring stations are irregularly distributed across Madrid, we divided such an interval in* 10 *lag distances of length* 688.889 *m., and used a tolerance angle of $\pi/8$. This guarantees an acceptable number of pairs for each lag distance.*

The calculations needed to obtain the values of the empirical semivariogram are listed in Table 3.3. The plot of the classical empirical semivariogram (the MoM estimator) and the semivariogram cloud can be seen in Figure 3.28. Both calculations and plots were made using `geoR`.

In light of Figure 3.28, it can be guessed that there is a linear behavior near the origin (it is no surprise in light of Figure 3.27) and a well-defined behavior for distances less than 5000 *m. However, though at large distances the semivariogram seems to reach a sill (around* 0.14*), it is not completely clear that it could not be higher and reached asymptotically (the practical range being much larger than* 5000 *m). Thus, assuming a linear behavior near the origin our first candidate is a spherical model, though an exponential model could also be a good choice to represent the spatial dependencies observed in the data. Of course, the knowledge of the city of Madrid and how pollution develops in is of great help when it comes to choosing one of the two candidate models. In the case of CO in Madrid, that knowledge favors the spherical model.*

Table 3.3 Data on carbon monoxide in Madrid, week 50, 10 pm: Empirical semivariogram values for each lag distance together with the variance of half the squared semidifferences used to compute them.

Lag distances	Number of pairs	Semivariogram values	Variance
800.000	6	0.0108	0.0002
1488.889	12	0.0507	0.0074
2177.778	22	0.0459	0.0030
2866.667	20	0.0995	0.0443
3555.556	27	0.0936	0.0188
4244.444	25	0.1067	0.0259
4933.333	22	0.1852	0.0793
5622.222	24	0.1657	0.0383
6311.111	18	0.0903	0.0099
7000.000	13	0.1476	0.0295

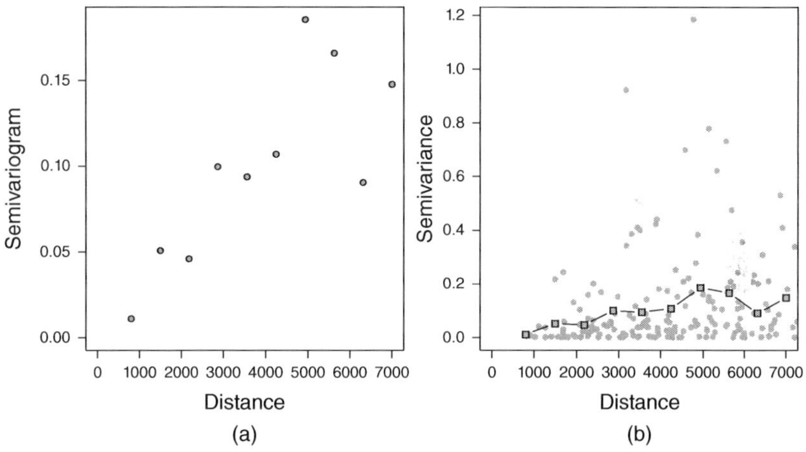

Figure 3.28 Data on carbon monoxide in Madrid, week 50, 10 *pm: (a) Classical empirical semivariogram; (b) Semivariogram cloud.*

One possible objection to the MoM estimator of the semivariogram is that it is not robust to extreme values.[3] As stated in Cressie (1993, p. 40), it is badly affected by atypical observations due to the quadratic terms in the summand of Equation (3.41).

[3] The distinction between the concepts of resistance and robustness is largely arbitrary. As stated in Chilès and Delfiner (1999, pp. 39–40), the concept of resistance appeared in exploratory data analysis and does not involve distributional assumptions; in this sense, an estimation procedure is resistant if its result is insensitive to a few erroneous or anomalous data. Robustness is related to a statistical model. An estimation procedure is robust if the data do not exactly conform to the true statistical model, it still gives good efficiency, and in the case of larger deviations, it does not give absurd results. In light of the above, and given that most texts in geostatistics refer to robustness, we will also use the term robustness.

Another objection arises from its highly skewed probability distribution: under the assumption that the rf is Gaussian, $\{Z(\mathbf{s}+\mathbf{h}) - Z(\mathbf{s})\}^2$ is distributed as $2\gamma(\mathbf{h})\chi_1^2$, where χ_1^2 is a Chi-square distribution with one degree of freedom, which displays a high degree of positive skewness. This implies that:

$$
\begin{aligned}
&E\left((Z(\mathbf{s}+\mathbf{h}) - Z(\mathbf{s}))^2\right) = 2\gamma(\mathbf{h})\\
&V\left((Z(\mathbf{s}+\mathbf{h}) - Z(\mathbf{s}))^2\right) = 8(\gamma(\mathbf{h}))^2\\
&C\left((Z(\mathbf{s}_1+\mathbf{h}_1) - Z(\mathbf{s}_1))^2, (Z(\mathbf{s}_2+\mathbf{h}_2) - Z(\mathbf{s}_2))^2\right)\\
&= C\big((Z(\mathbf{s}_1+\mathbf{h}_1) - Z(\mathbf{s}_1)), (Z(\mathbf{s}_2+\mathbf{h}_2) - Z(\mathbf{s}_2))\big)^2\\
&= 2\big(\gamma((\mathbf{s}_1+\mathbf{h}_1) - \mathbf{s}_2) + \gamma(\mathbf{s}_1 - (\mathbf{s}_2+\mathbf{h}_2)) - \gamma(\mathbf{s}_1+\mathbf{h}_1 - (\mathbf{s}_2+\mathbf{h}_2)) - \gamma(\mathbf{s}_1 - \mathbf{s}_2)\big)^2.
\end{aligned}
\tag{3.43}
$$

Thus, $2\gamma(\mathbf{h})$ is the expected value of a highly skewed rv, and unless the number of increments is large enough, the half of the mean of such increments (the empirical semivariogram) will not be a stable estimator of the theoretical semivariogram.

Finally, a third objection is the following: the estimates of the semivariogram at two different lag distances are usually correlated because the observations at these distances are spatially correlated and the same points are used in estimating the semivariogram at those two lag distances.

The statistical literature regarding the issue of the MoM estimator not being robust provides several solutions to the problem including that of Cressie and Hawkins (1980), Armstrong and Delfiner (1980) and Gunst and Hartfield (1997) among others. In what follows, we will focus on them, but it is a fact that in practice the classical estimator is still the most widely used.

In the Gaussian context, Cressie and Hawkins (1980) found, by using the set of Box and Cox power transformations, that the skewness and kurtosis coefficients of the fourth-root of χ_1^2 are 0.08 and 2.48, respectively. Then, estimates of location can be applied to the $\#N(\mathbf{h})$ transformed differences $|Z(\mathbf{s}+\mathbf{h}) - Z(\mathbf{s})|^{\frac{1}{2}}$ and eventually these estimates are raised to the fourth power to bring them back to the correct scale and adjusted for bias. The resulting robust estimator is:

$$
\hat{\gamma}_{CH}(\mathbf{h}) = \frac{1}{2}\left(0.457 + \frac{0.494}{\#N(\mathbf{h})}\right)^{-1}\left(\frac{1}{\#N(\mathbf{h})}\sum_{N(\mathbf{h})} |Z(\mathbf{s}_i+\mathbf{h}) - Z(\mathbf{s}_i)|^{1/2}\right)^4, \tag{3.44}
$$

which, when $\#N(\mathbf{h})$ is sufficiently large, is reduced to:

$$
\hat{\gamma}_{CH}(\mathbf{h}) = \frac{1}{2}(0.457)^{-1}\left(\sum_{N(\mathbf{h})} \frac{|Z(\mathbf{s}_i+\mathbf{h}) - Z(\mathbf{s}_i)|^{1/2}}{\#N(\mathbf{h})}\right)^4. \tag{3.45}
$$

The factor multiplying the fourth power of $\frac{1}{2\#N(\mathbf{h})}\sum_{N(\mathbf{h})}|Z(\mathbf{s}+\mathbf{h})-Z(\mathbf{s})|^{\frac{1}{2}}$ derives from the following result:

$$E\left(\frac{1}{2\#N(\mathbf{h})}\sum_{N(\mathbf{h})}|Z(\mathbf{s}+\mathbf{h})-Z(\mathbf{s})|^{\frac{1}{2}}\right)=2\gamma(\mathbf{h})\left(0.457+\frac{0.494}{\#N(\mathbf{h})}+\frac{0.045}{(\#N(\mathbf{h}))^2}\right),$$

where the term $\frac{0.045}{(\#N(\mathbf{h}))^2}$ has a negligible contribution to the bias correction, especially when $\#N(\mathbf{h})$ is large.

There is another advantage to using $|Z(\mathbf{s}_i)-Z(\mathbf{s}_j)|^{\frac{1}{2}}$ over $|Z(\mathbf{s}_i)-Z(\mathbf{s}_j)|^2$: The summands in each $\sum_{N(\mathbf{h})}|Z(\mathbf{s}_i)-Z(\mathbf{s}_j)|^2$ and $\sum_{N(\mathbf{h})}|Z(\mathbf{s}_i)-Z(\mathbf{s}_j)|^{\frac{1}{2}}$ are not independent, and the more dependent they are, the less efficient is the average in estimating the semivariogram. It can be shown (Cressie 1993, pp. 75–6) that, in general, the summands $\sum_{N(\mathbf{h})}|Z(\mathbf{s}_i)-Z(\mathbf{s}_j)|^{\frac{1}{2}}$ are less correlated than the summands $\sum_{N(\mathbf{h})}|Z(\mathbf{s}_i)-Z(\mathbf{s}_j)|^2$.

Armstrong and Delfiner (1980) defined quantile semivariograms as:

$$\hat{\gamma}_p(\mathbf{h}) = Q_p\left\{\frac{1}{2}\left(Z(\mathbf{s}_i+\mathbf{h})-Z(\mathbf{s}_i)\right)^2\right\}, \tag{3.46}$$

where Q_p denotes the quantile associated to the proportion p, $(0 < p < 1)$. If, for each value of $\mathbf{h}$, we arrange the values of $\frac{1}{2}(Z(\mathbf{s}+\mathbf{h})-Z(\mathbf{s}))^2$ in ascending order and construct the cumulative frequency curve, we can easily calculate any quantile semivariogram.

Obviously, the median semivariogram figures prominently among quantile semivariograms because the median is more robust than the mean.

As seen above, under the assumption that the rf is Gaussian, $(Z(\mathbf{s}+\mathbf{h})-Z(\mathbf{s}))^2$ is distributed as $2\gamma(\mathbf{h})\chi_1^2$, where χ_1^2 is a Chi-square with one degree of freedom. Therefore, quantile semivariograms are proportional to the semivariogram $\gamma(\mathbf{h})$, the proportionality factors being the quantiles of the Chi-square distribution with one degree of freedom. More specifically, the proportionality factors of the quartile semivariograms are:

$$Q_{0.25}(\chi_1^2)=0.101, \quad Q_{0.50}(\chi_1^2)=0.455, \quad Q_{0.75}(\chi_1^2)=1.324.$$

Gunst and Hartfield (1997) suggest, among other options, applying *trimmed fixed percentages*, which as the name indicates, consists of trimming a fixed percentage of values of the observed realization to later calculate the classical empirical semivariogram using the observations that remain in the set.

One alternative to trimming is the robust M-estimator of location. In order to calculate it, let $\#N(\mathbf{h})$ be the number of pairs of locations that can be joined by a vector of module h. Let $M_h = Med\left\{\left|Z(\mathbf{s}_i)-Z(\mathbf{s}_j)\right|^{1/2}\right\}$ be the median of the square root of the differences between observations located at distance h and let S_h be the corresponding median absolute deviation, $S_h = \frac{1}{0.6745} \; Med\left\{ \left| \left|Z(\mathbf{s}_i) - Z(\mathbf{s}_j)\right|^{1/2} - M_h \right| \right\}$. Then, the robust M-estimator of location is calculated as:

$$\hat{\gamma}_M(\mathbf{h}) = \frac{\left\{ M_h + S_h \; \frac{\sum \psi(y_{ij})}{\sum \dot{\psi}(y_{ij})} \right\}^4}{2\,(0.457 + 0.494/\#N(\mathbf{h}))}, \tag{3.47}$$

where $y_{ij} = \frac{1}{S_h}\left(\left|Z(\mathbf{s}_i) - Z(\mathbf{s}_j)\right|^{1/2} - M_h \right)$ and $\dot{\psi}$ is the derivative of Tukey's biweight ψ (see Hampel *et al.* 1986 and Staudte and Sheather 1990):

$$\psi(t) = \begin{cases} t(4^2 - t^2)^2 & |t| \leq 4 \\ 0 & |t| > 4 \end{cases}. \tag{3.48}$$

The bias correction in the denominator of (3.47) is the same as that in the robust estimator (3.44) since both are based on estimating the center of the distribution of $\{|Z(\mathbf{s}_i) - Z(\mathbf{s}_j)|\}^{1/2}$.

Another alternative is to perform a test to check if some of the observed data highly influences the shape of the empirical semivariogram. In such a case, those data should be removed from the dataset and the Cressie-Hawkins empirical semivariogram is constructed from the remaining data. An observed value $z(\mathbf{s}_i)$ is considered to highly influence the empirical semivariogram when:

$$\left| \frac{z(\mathbf{s}_i) - Med\{z(\mathbf{s})\}}{\frac{1}{0.6745} Med\{z(\mathbf{s}_i) - Med\{z(\mathbf{s})\}\}} \right| > 3, \tag{3.49}$$

$z(\mathbf{s})$ representing the observed realization.

Example 3.5 (Illustrating the impact of outliers) *Consider the simplified example of a spatial dataset with 9 locations. Let $Z(\{x, y\})$ denote the observed value at coordinates $\mathbf{s} = \{x, y\}$. Data are displayed in Figure 3.29. They include the outlier $Z(\{1, 0.5\})$. Calculations needed to obtain the classical and the Cressie-Hawkins robust estimators are shown in Tables 3.4 and 3.5, respectively. We used five distance classes and $Z(\{1, 0.5\})$ contributes to the estimates of the four first classes.*

As can be observed, the great impact of $Z(\{1, 0.5\})$ in the classical empirical semivariogram is largely mitigated in the Cressie-Hawkins estimator.

In fact, if we remove the observation $Z(\{1, 0.5\})$ from the data (Figure 3.30), the semivariogram values of both the classical and Cressie-Hawkins estimators, as well as their variances, approach greatly (see Tables 3.6 and 3.7).

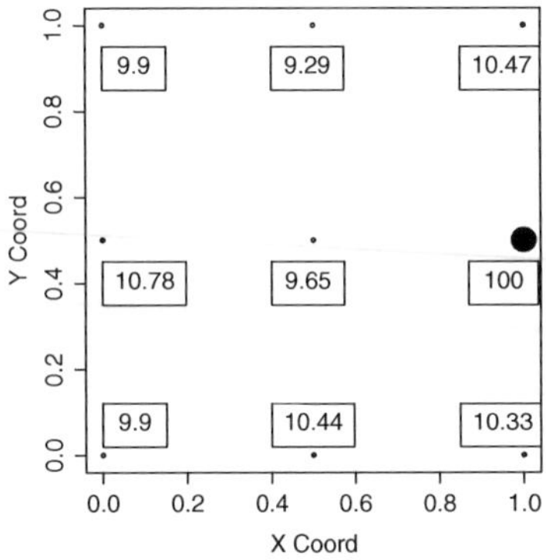

Figure 3.29 Observed points and data values.

Table 3.4 Empirical semivariogram values (MoM estimator).

Lag distances	Number of pairs	Semivariogram values	Variance
0.50	12	1009.241	3054075.453
0.71	8	1015.761	3094164.050
1.00	6	663.501	2199946.429
1.12	8	1014.714	3087897.376
1.41	2	0.126	0.001

Table 3.5 Empirical semivariogram values (Cressie-Hawkins estimator).

Lag distances	Number of pairs	Semivariogram values	Variance
0.50	12	2.982	14.108
0.71	8	2.939	14.371
1.00	6	2.058	11.018
1.12	8	2.866	14.680
1.41	2	0.704	0.003

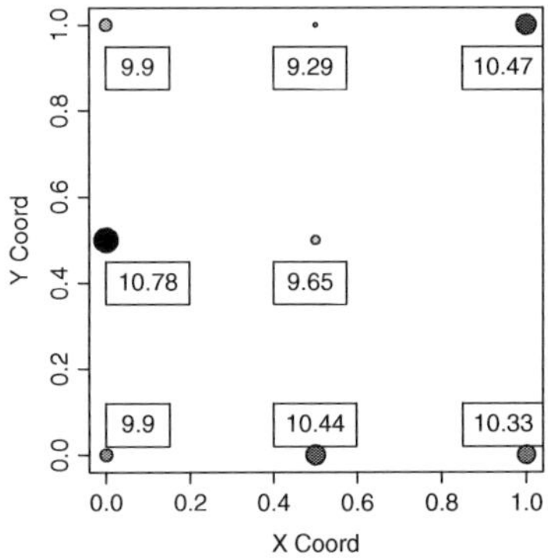

Figure 3.30 Observed points and data values (without the outlier).

Table 3.6 Empirical semivariogram values after removing the outlier (MoM estimator).

Lag distances	Number of pairs	Semivariogram values	Variance
0.50	9	0.311	0.051
0.71	6	0.298	0.143
1.00	5	0.184	0.060
1.12	6	0.170	0.031
1.41	2	0.126	0.001

Table 3.7 Empirical semivariogram values after removing the outlier (Cressie-Hawkins estimator).

Lag distances	Number of pairs	Semivariogram values	Variance
0.50	9	0.816	0.051
0.71	6	0.755	0.067
1.00	5	0.580	0.122
1.12	6	0.658	0.063
1.41	2	0.704	0.003

3.7 Anisotropy

As introduced in Section 3.4, a theoretical semivariogram model $\gamma(\mathbf{h})$ is anisotropic if it depends on the direction of the separation vector $\mathbf{h}$. Otherwise it is called isotropic. That is, if we consider three dimensions, in the anisotropic case the semivariogram depends on the separation vector $\mathbf{h} = (h_1, h_2, h_3)$ instead of $|\mathbf{h}| = \sqrt{h_1^2 + h_2^2 + h_3^2}$. Isotropy implies that the directional semivariograms (in the main directions of the space) coincide, and therefore the "global" semivariogram will be omnidirectional and will only depend on the length of the vector $\mathbf{h}$. Of course, the hypothesis of isotropy simplifies the modeling of spatial dependence remarkably. Notwithstanding, we can assume in many cases that dependence is equal in any given direction.

Empirical calculations can reveal different behaviors on behalf of the empirical semivariogram depending on direction (though this requires an observed realization that is broader than those normally available). There are two types of anisotropy: geometric or elliptical, and zonal.

In the stationary case, *geometric* or *elliptical* anisotropy implies that the sills of at least two directional semivariograms, despite being equal, are reached at different distances[4] (these semivariograms have different ranges). This is the case, for example, of airborne pollution: it is expected to exhibit geometrical anisotropy in the prevailing

[4] In the case of linear semivariograms, which go beyond the second-order stationarity, the slope will vary with direction.

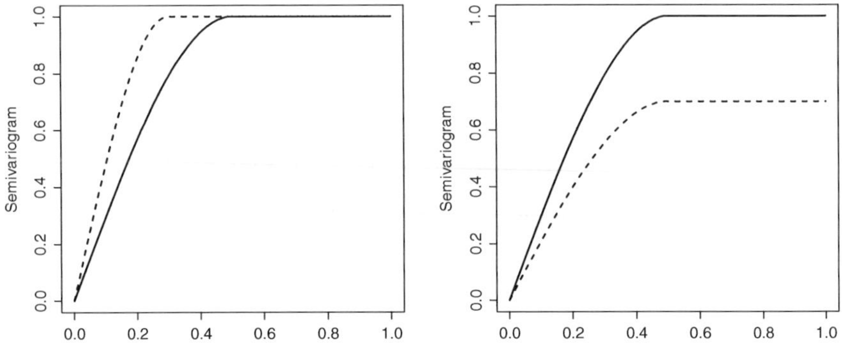

Figure 3.31 Left panel: Geometric anisotropy. Right panel: Zonal anisotropy.

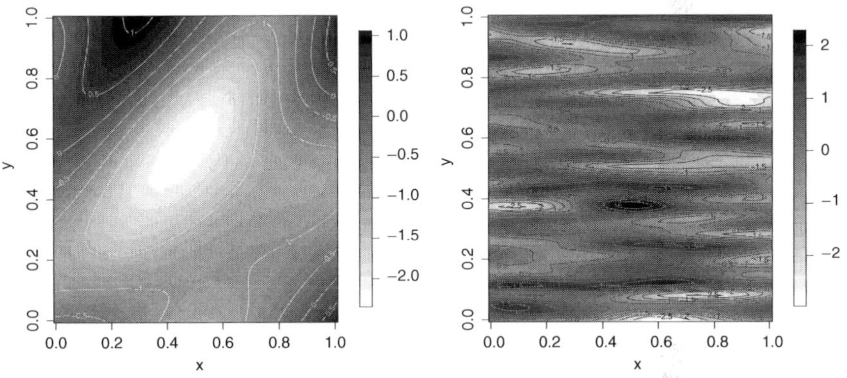

Figure 3.32 Simulation of two rf's. Left panel: The geometric anisotropy case; Right panel: The zonal anisotropy case.

wind direction and perpendicular to it. In the case of *zonal anisotropy*, the sill changes with direction while the range remains constant. Figure 3.31 illustrates both types of anisotropy. Figure 3.32 shows a simulated realization of a rf exhibiting geometric anisotropy and another of a rf exhibiting zonal anisotropy, in the two-dimensional case (we have included the iso-semivariogram lines). Figure 3.33 depicts the isotropic case. In practice, zonal anisotropy is seldom found alone; it is more frequent to find a mixture of both the geometric and zonal anisotropy (*hybrid* anisotropy).

In practice, if a semivariogram map or map of iso-semivariogram lines (the plot of iso-values of the empirical semivariogram depending on both the length and direction of vector $\mathbf{h}$) is used to detect the presence of anisotropy, when the iso-semivariogram lines are circular around the origin, the semivariogram will only depend on the length of the vector $\mathbf{h}$, indicating isotropy. In contrast, geometric anisotropy is indicated by elliptical concentric iso-semivariogram lines along the two main (perpendicular) axes of anisotropy (see Figure 3.34). The angle that the main axis forms with the axis of abscissas is known as the *anisotropy angle*, ϑ, and

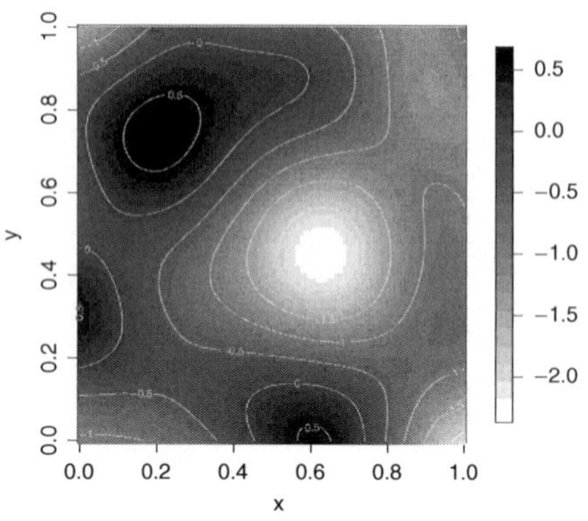

Figure 3.33 Simulation of a isotropic rf.

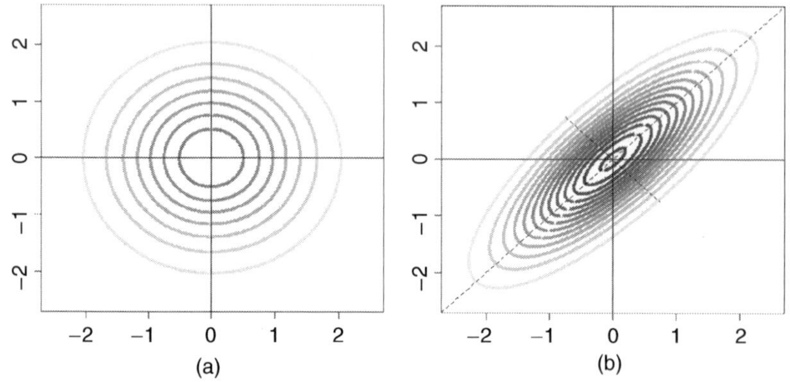

Figure 3.34 Semivariogram maps: (a) The isotropic case (circular contours); (b) The anisotropic case (elliptic contours, $\vartheta = \pi/4, \lambda = 2$). The axes depict lag distances in the corresponding coordinate system.

indicates its orientation. The quotient of the ranges in the major and minor axes of the ellipse is called the *anisotropy ratio* of the ellipse, λ. In the case of zonal anisotropy the iso-semivariogram lines cannot be fitted to a second-degree curve.

In order to correct the geometric anisotropy (in two dimensions),[5] we must substitute the elliptical coordinates for the others referred to a circle, since they are the respective ranges which are different in this type of anisotropy.

Let (x, y) be the coordinates of a point, a_{α_i} the range of the semivariogram in the direction α_i, ϑ the anisotropy angle, and λ the ratio of anisotropy of the ellipse, that is the quotient of the ranges in the mayor and minor axes of the ellipse (a_{α_2} and a_{α_1},

[5] The generalization to the three-dimensional case is straightforward when one of the main anisotropy axes is the vertical, which is often the case (see Chilès and Delfiner 1999, pp. 94–5, for details).

respectively). The two classical steps proposed below transform the ellipse into a circle such that anisotropy is reduced to isotropy:

- Rotate axes by the angle ϑ, so that they become parallel to the main axes of the ellipse. The new coordinates following the rotation, (x', y'), can be expressed in matrix notation as follows:

$$\begin{pmatrix} x' \\ y' \end{pmatrix} = \begin{pmatrix} \cos\vartheta & \sin\vartheta \\ -\sin\vartheta & \cos\vartheta \end{pmatrix} \begin{pmatrix} x \\ y \end{pmatrix}. \tag{3.50}$$

- Transform the ellipse into a circle with a radius equal to the major range of the ellipse. This is achieved by multiplying the coordinate y' by $\lambda = \frac{a_2}{a_1} > 1$, the ratio of anisotropy of the ellipse. As a result, the new coordinates (x^*, y^*) are obtained from the coordinates, (x', y'), by means of the following operation:

$$\begin{pmatrix} x^* \\ y^* \end{pmatrix} = \begin{pmatrix} 1 & 0 \\ 0 & \lambda \end{pmatrix} \begin{pmatrix} x' \\ y' \end{pmatrix}. \tag{3.51}$$

Therefore, by performing the transformation:

$$\begin{cases} x^* = x\cos\vartheta + y\sin\vartheta \\ y^* = -x\lambda\sin\vartheta + y\lambda\cos\vartheta \end{cases} \tag{3.52}$$

we can use isotropic theoretical semivariograms with a range equal to the major range of the directional ellipse.

Before concluding these brief notes on geometric anisotropy, we must warn that the directional behavior of the semivariogram could be a consequence of the rf under study not being (even intrinsically) stationary.

Example 3.6 (Directional spherical semivariogram) *Suppose the following North-South and West-East directional spherical semivariograms (thus, $\vartheta = \pi/2$):*

$$\gamma_{\alpha_{N-S}}\left(h_{\alpha_{N-S}}\right) = 5\left(1.5\frac{h_{\alpha_{N-S}}}{10} - 0.5\frac{h^3_{\alpha_{N-S}}}{10^3}\right) \tag{3.53}$$

$$\gamma_{\alpha_{W-E}}\left(h_{\alpha_{W-E}}\right) = 5\left(1.5\frac{h_{\alpha_{W-E}}}{20} - 0.5\frac{h^3_{\alpha_{W-E}}}{20^3}\right) \tag{3.54}$$

As the rotation of axes is not needed and the anisotropy ratio is $\lambda = \frac{a_{\alpha_{W-E}}}{a_{\alpha_{N-S}}} = 2$, $\gamma_{\alpha_{N-S}}$ can be expressed in terms of $\gamma_{\alpha_{W-E}}$ multiplying $h_{\alpha_{N-S}}$ by the anisotropy ratio:

$$\gamma_{\alpha_{N-S}}(h_{\alpha_{N-S}}) = 5\left(1.5\frac{\frac{20}{10}h_{\alpha_{N-S}}}{20} - 0.5\frac{\frac{20^3}{10^3}h^3_{\alpha_{N-S}}}{20^3}\right). \tag{3.55}$$

Thus, making the change of coordinates, $\begin{cases} x^* = x \\ y^* = \lambda y \end{cases}$, *that is,* $|\mathbf{h}^*| = \sqrt{h_1^{*2} + h_2^{*2}}$, *the spherical semivariogram* $\gamma(|\mathbf{h}^*|) = 5\left(1.5\left(\frac{|\mathbf{h}^*|}{20}\right) - 0.5\left(\frac{|\mathbf{h}^*|^3}{20}\right)\right)$ *can be used to characterize the variability in the two considered directions. In other words, the simple elongation of the X axis allows for the use of an isotropic model.*

Example 3.7 (Anisotropic exponential semivariogram) *Consider the following anisotropic exponential semivariogram:*

$$\gamma(h_1, h_2, h_3) = 5\left(1 - \exp\left\{-\sqrt{\left(\frac{h_1}{10}\right)^2 + \left(\frac{h_2}{10}\right)^2 + \left(\frac{h_3}{5}\right)^2}\right\}\right),$$

where h_1 and h_2 are distances along the horizontal axes of a Cartesian system and h_3 is the distance along the vertical axis. It can be observed that the practical range in the horizontal direction is double that in the vertical one.

The use of the following system of coordinates:

$$\begin{cases} x^* = x \\ y^* = y \\ z^* = 2z \end{cases}$$

allows us to use the isotropic exponential model:

$$\gamma(|\mathbf{h}^*|) = 5\left(1 - \exp\left(\frac{|\mathbf{h}^*|}{10}\right)\right),$$

where $\mathbf{h}^* = \sqrt{h_1^{*2} + h_2^{*2} + h_3^{*2}}$.

Zonal anisotropy cannot be corrected by means of a linear transformation of coordinates. The most common method for dealing with this type of anisotropy is as follows: when working in two dimensions and if, for example, the sill along the coordinate y is much higher than in the direction of x, we fit an isotropic model to the experimental semivariogram which considers the direction x and then add a semivariogram with geometric anisotropy, the range along the x-axis being very large (as a result, it will not affect this axis).

In the case of three dimensions, if the value of the sill were significantly higher along the vertical axis (z) than on the horizontal plane (x, y), which is common in many phenomena, an isotropic model is fitted to the horizontal plane and then another geometrically anisotropic model with very large range is added so that this second semivariogram is practically null on the horizontal plane for the entire range of distances of interest.

Pure zonal anisotropy takes place when the variability of the rf only depends on one coordinate (usually the vertical coordinate). Let ϑ be the angle between $\mathbf{h}$ and a

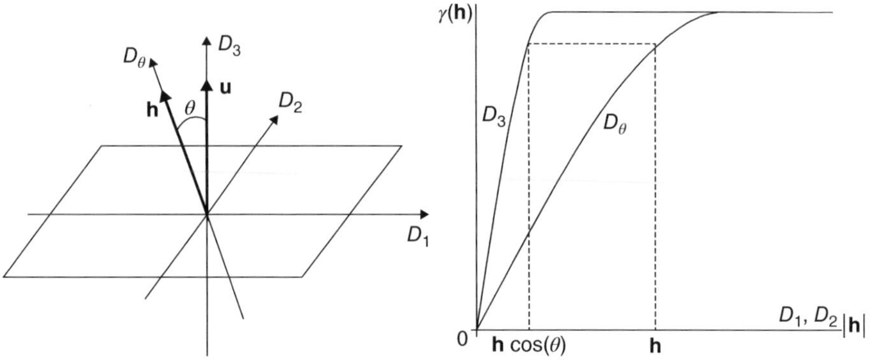

Figure 3.35 3D representation of zonal anisotropy. Left panel: Pure zonal anisotropy in vertical direction. Right panel: Directional semivariograms in horizontal directions (D_1, D_2), vertical (D_3) and in an intermediate direction (D_θ). Source: Emery (2000, p. 111). *Reproduced with permission of Xavier Emery.*

unit vector vertically oriented (see Figure 3.35) and $\gamma(\mathbf{h})$ the semivariogram associated with such a rf. Then, $\gamma(\mathbf{h}) = \gamma(|\mathbf{h}| \cos \vartheta \mathbf{u})$, and has the same sill in all directions and range $a/\cos \vartheta$ except for the horizontal plane, for which $\cos \vartheta = 0$ and, as a consequence, $\gamma(\mathbf{h}) = 0$.

The issue of anisotropy should be tackled in each specific case with great care, as the differences between the various directional semivariograms can be due to a lack of data in certain directions, to observations being heterogenous or to empirical fluctuations.

3.8 Fitting a semivariogram model

Any function that depends on distance and direction is not necessarily a valid semivariogram. In this sense, the empirical semivariogram (whether it is the classical model or one of its alternatives) cannot be used directly to obtain kriging spatial predictions, because it is defined only for a finite set of lag distances (those used in constructing it) and after joining its values at such lag distances using whichever procedure (usually a smoothing procedure), the resulting function may not necessarily fulfill the conditions that every semivariogram must meet. In particular, it does not have to fulfill the prerequisite of being conditionally negative-definite, which is an indispensable prerequisite for a valid semivariogram as this makes it possible to calculate the variance of linear combinations of spatial rv's consistently and, more specifically, it ensures that the variance of the prediction error resulting from the kriging equations is non-negative; what is more, it is not easy to test compliance with this prerequisite in the empirical case.

The solution adopted when faced with such limitations in the empirical semivariogram is to use a theoretical semivariogram that represents the spatial dependence structure shown by the data available (observed realization) as closely as possible;

or in other words, to fit a theoretical model (that ensures the post-spatial prediction process works properly) to the semivariogram representation.

But fitting a permissible or valid semivariogram is a core but not an easy task. In fact, as stated in Webster and Oliver (2001, p. 128), choosing models and fitting them to the empirical semivariogram plot are among the most controversial topics in geostatistics. The reasons why fitting models to empirical semivariogram plots is difficult include: (1) the precision of each estimate (at different distances) varies because of sample sizes differences; (2) the variation may not be the same in all directions (anisotropy); (3) the empirical plot may show much point-to-point fluctuation; and (4) most models are non-linear in one or more parameters (typically the scale parameter and anisotropies) and iterative methods have to be used to estimate the parameters. It should be added to these points the possibility that the values of the empirical semivariograms are correlated.

Fitting a semivariogram (or covariance) model can be done by eye (fitting by eye, manual fitting, fitting at first glance), using visual and graphical methods, or using statistical procedures. In addition, some literature originated in Webster and Oliver (2001) recommends a combination of both fitting by eye and statistical fitting. That is, first visualization methods can be used to choose the models that best capture the main features of the empirical semivariogram that are important for the application. Second, statistical fitting can be used to choose from among the pre-selected models and parameter estimation. In the same vein, Chilès and Delfiner (1999, p. 108) state that automatic fitting rarely provides good results, and recommend using it as a first step in manual fitting.

In practice, fitting a semivariogram model is generally done by the geostatistician rather than entirely automatically or statistically (albeit the above reasons (1)–(3) complicate the manual fitting). The fitting by eye or manual fitting could look like a strange tradition (at first glance), but, as stated in Wackernagel (2003, p. 49), it is generally not so relevant to how well the semivariogram function fits the sequence of points in the empirical semivariogram. What is really crucial is the type of continuity assumed for the regionalized variable and the stationarity hypothesis associated with the rf. These assumptions will guide the selection of an appropriate semivariogram and this has many more implications than the way the theoretical function is fitted to the empirical counterpart (see Matheron 1989) for a thorough discussion). Armstrong and Wackernagel (1988, pp. 53–4) state that the analytic form of the model does not matter very much as long as the major features of the phenomenon are respected.

Of course, knowledge about the phenomenon under study is very welcome for this purpose, though the introduction of auxiliary hypotheses carries with it the risk that kriging loses its optimality.

3.8.1 Manual fitting

When choosing the theoretical semivariogram model that best fits the empirical semivariogram, it is important to take into account that, above all, it should capture the behavior near the origin and at large distances from the latter. That is, as said above, the specific analytical format of the semivariogram does not matter as much as the fact that it respects the main features of the phenomenon. By order of importance, these features are: the nugget effect and the slope at the origin (that is, the behavior near the origin), the range and the sill (indicative of the behavior at large distances), and anisotropies.

In practice, the value of the nugget effect can be obtained from the first values of the semivariogram by extrapolating them until they cut the axis of ordinates. The behavior near the origin (for distances less than the first lag distance) is another core decision in manual fitting because it has a great influence on both kriging results and also in the numerical stability of the kriging system (see Armstrong and Wackernagel 1988, for an illustration of this point). The slope of the semivariogram at the origin can easily be derived from the first values of the empirical semivariogram. As for the behavior at large distances, the range (or practical range in case of an asymptotic sill) and the sill are usually easily detected, especially in the second-order stationary case: the distance beyond which the semivariogram stabilizes and the value of the semivariogram when it stabilizes, respectively. The variance of the sample realization will be the guiding criterion to estimate the value of that sill in the stationary case, though it tends to underestimate the sill. Modeling anisotropies is not an easy task and requires some expertise, but generally the sum of two or three models suffices to obtain a good fit.

Logically, a manual fit entails many risks stemming from subjective decisions made by the researcher. As a result, common sense and any theoretical knowledge of the problem being studied are normally extremely helpful.

3.8.2 Automatic fitting

A second way of proceeding is to automate the fitting process. From this perspective, the following methods are extremely useful:

- Least Squares methods, which include the Ordinary Least Squares method (OLS), the Generalized Least Squares method (GLS), and the Weighted Least Squares method (WLS). This approach is the most widely used in practice and fits a parametric valid model to the semivariances of the observed data at each lag distance considered (thus, it involves subjective decisions).

- Maximum Likelihood based methods, which include the Maximum Likelihood (ML) and the Restricted Maximum Likelihood (REML) methods. These methods deal with the observed data directly assuming that the rf is Gaussian.

- Composite Likelihood (CL) method, which like the Least Square methods, uses pseudo data instead of the data observed.

When using one of the foregoing methods, a parametric valid semivariogram is fitted to the observed data or pseudo data derived from them. However, it is also possible to apply a non-parametric representation (see Schabenberger and Gotway 2005, pp. 178–88, for details).

3.8.2.1 Least Squared-based fitting methods

Let us imagine that the semivariogram $\gamma(\mathbf{h})$ has been estimated for a finite set of lag distances $\mathbf{h}_i, i = 1, \ldots, k$, and that we wish to fit a semivariogram model within the theoretical semivariogram family $\gamma(\mathbf{h}, \boldsymbol{\theta})$. The vector $\boldsymbol{\theta}$ normally contains the nugget effect, the sill and the scale parameter. Let us imagine that the MoM estimator $\hat{\gamma}(\mathbf{h})$ has been used and $\hat{\gamma}(\mathbf{h}) = \left(\hat{\gamma}(\mathbf{h}_1), \hat{\gamma}(\mathbf{h}_2), \ldots, \hat{\gamma}(\mathbf{h}_k)\right)'$ is the $(k \times 1)$ vector containing the values of the empirical semivariogram at the k lag distances considered and $\gamma(\mathbf{h}, \boldsymbol{\theta})$ is the vector of the values derived by the theoretical model from the same lags.

Then, Least Squares methods, which do not require the knowledge of the probability distribution of the experimental semivariogram values, are based on the model:

$$\hat{\gamma}(\mathbf{h}) = \gamma(\mathbf{h}, \boldsymbol{\theta}) + \mathbf{e}(\mathbf{h}), \tag{3.56}$$

where $\mathbf{e}(\mathbf{h})$ is assumed to have mean $\mathbf{0}$ and a covariance matrix typically depending on $\boldsymbol{\theta}$.

In what follows, we focus on how the vector $\boldsymbol{\theta}$ is estimated with the three above-mentioned non-linear Least Squares methods, OLS, GLS, and WLS. We also outline the main merits and disadvantages of such estimation procedures.

The OLS method proceeds by minimizing the expression:

$$Q(\boldsymbol{\theta}) = \sum_{i=1}^{k} \left(\hat{\gamma}(\mathbf{h}_i) - \gamma\left(\mathbf{h}_i, \boldsymbol{\theta}\right)\right)^2. \tag{3.57}$$

The OLS fitting is easy to implement and has an attractive geometrical interpretation. However, apart from not taking into account the behavior of the semivariogram near the origin (a problem that must be overcome with knowledge and experience), it does not consider the possibility of the values $\hat{\gamma}(\mathbf{h})$ being correlated, which can have damaging effects on the estimates of the vector of parameters $\boldsymbol{\theta}$ as the amount of data increases, and its unequal dispersion.

The GLS method proceeds to obtain the vector $\boldsymbol{\theta}$ that minimizes the expression:

$$Q(\boldsymbol{\theta}) = \left(\hat{\gamma}(\mathbf{h}) - \gamma\left(\mathbf{h}, \boldsymbol{\theta}\right)\right)' \mathbf{V}^{-1} \left(\hat{\gamma}(\mathbf{h}) - \gamma\left(\mathbf{h}, \boldsymbol{\theta}\right)\right), \tag{3.58}$$

where $\mathbf{V}$ is the covariance matrix of the vector of errors $\mathbf{e}(\mathbf{h})$ in (3.56) and, as a consequence, of the vector of semivariances $\hat{\gamma}(\mathbf{h}) = \left(\hat{\gamma}(\mathbf{h}_1), \hat{\gamma}(\mathbf{h}_2), \ldots, \hat{\gamma}(\mathbf{h}_k)\right)'$.

The GLS method, as a least squares method, not only has an attractive geometrical interpretation, but incorporates the correlation of the values $\hat{\gamma}(\mathbf{h}_i)$. However, the main difficulty with the GLS estimation is the determination of the matrix $\mathbf{V}$, which typically depends on $\boldsymbol{\theta}$ and is, generally speaking, quite large. A common practice is to assume a Gaussian distribution for the rf and then proceed iteratively from an initial solution. The iterative procedure is as follows: (1) Obtain an initial solution (the OLS solution is often used as a starting point); (2) Calculate $\mathbf{V}$ with that solution (the required expressions for the classical semivariogram can easily be obtained from (3.43); (3) Compute the new vector $\boldsymbol{\theta}$ which maximizes $Q(\boldsymbol{\theta})$; and repeat the three steps of the algorithm. The expressions used in the algorithm are for Gaussian data, but can easily be generalized for transformed Gaussian data (Cressie 1993, p. 98).

WLS fitting aims to determine the vector θ that minimizes the expression:

$$Q(\theta) = \sum_{i=1}^{k} w_i \left(\hat{\gamma}(\mathbf{h}_i) - \gamma(\mathbf{h}_i;\theta)\right)^2, \tag{3.59}$$

and, as for GLS, the Gaussian distribution is usually assumed and an iterative procedure is followed. According to Pardo-Iguzquiza (1998), in the above expression, w_i is selected to coincide with:

- $\#N(\mathbf{h}_i)$, the number of pairs at each distance in such a way that larger weights are assigned to the empirical semivariogram points calculated using a larger number of pairs.
- $\left\{\gamma(\mathbf{h}_i,\theta)\right\}^{-2}$, such that the points nearest to the origin receive a larger weighting than those furthest away.
- $\#N(\mathbf{h}_i)\left\{\gamma(\mathbf{h}_i,\theta)\right\}^{-2}$, that is, they are taken as (approximately) inversely proportional to the variance of $\hat{\gamma}(\mathbf{h}_i,\theta)$. If the correlations between the $\#N(\mathbf{h}_i)$ increments are negligible, that weight can be approximated by $V\left(\hat{\gamma}(\mathbf{h}_i)\right)^{-1}$.

Although these weights are optimum when the correlations between the $\#N(\mathbf{h}_i)$ increments involved in $\hat{\gamma}(\mathbf{h}_i)$ turn out to be null (which is not generally the case), WLS can be seen as a compromise between the efficiency of the GLS method (where the off-diagonal entries of $\mathbf{V}$ are appreciable) and the simplicity of the OLS method that yields good results in practice.

3.8.2.2 Maximum Likelihood-based fitting methods

These methods deal with the observed data directly assuming they follow a multivariate Gaussian distribution. Thus, they do not require the construction of the empirical semivariogram.

As is well known, if we assume that the realization observed is associated with a Gaussian rf, then it is relatively easy to obtain the exact form of the likelihood function and maximize it numerically (Kyriakidis and Journel 2000; Mardia and Marshall, 1984).

The most popular method that relies on the Gaussian assumption is ML. Let us consider a vector of multivariate data $\mathbf{Z}$, so that:

$$\mathbf{Z} \sim G(\mathbf{X}\boldsymbol{\beta},\boldsymbol{\Sigma}), \tag{3.60}$$

$\mathbf{Z}$ being n-dimensional, $\mathbf{X}$ a matrix $n \times q$ of covariables ($q < n$; full rank), $\boldsymbol{\beta}$ a vector of size q of unknown parameters and $\boldsymbol{\Sigma}$ the matrix of observation covariances. In practice, we can assume that:

$$\boldsymbol{\Sigma} = \alpha\mathbf{V}(\boldsymbol{\theta}), \tag{3.61}$$

α being an unknown scale parameter and $\mathbf{V}(\boldsymbol{\theta})$ a matrix of standardized covariances determined by the unknown parameter vector $\boldsymbol{\theta}$ through a valid semivariogram. With $\mathbf{Z}$ defined by (3.60), the joint density function of $Z(\mathbf{s}_1), \ldots, Z(\mathbf{s}_n)$ takes the following form:

$$f(\mathbf{Z}) = (2\pi)^{-n/2}|\boldsymbol{\Sigma}|^{-1/2} \exp\left\{-\frac{1}{2}(\mathbf{Z} - X\boldsymbol{\beta})'\boldsymbol{\Sigma}^{-1}(\mathbf{Z} - X\boldsymbol{\beta})\right\}, \tag{3.62}$$

which in case of considering $\mathbf{Z}$ as a set fixed values and $f(\mathbf{Z})$ a function of $\alpha, \boldsymbol{\beta}$ and $\boldsymbol{\theta}$ is known as the likelihood function. Therefore, the negative log-likelihood will be:

$$\begin{aligned}\ell(\mathbf{Z}, \beta, \alpha, \boldsymbol{\theta}) = {} & \frac{n}{2}\log(2\pi) + \frac{n}{2}\log(\alpha) + \frac{1}{2}\log(|\mathbf{V}(\boldsymbol{\theta})|) \\ & + \frac{1}{2\alpha}(\mathbf{Z} - X\boldsymbol{\beta})'\mathbf{V}(\boldsymbol{\theta})^{-1}(\mathbf{Z} - \mathbf{X}\boldsymbol{\beta}),\end{aligned} \tag{3.63}$$

so that the estimators $\hat{\alpha}$, $\hat{\boldsymbol{\beta}}$ and $\hat{\boldsymbol{\theta}}$ are obtained by minimizing it.

Although this method is feasible to compute, the fact that it is more difficult than WLS, for example, makes it less popular. The asymptotic properties of the ML estimators were studied by Mardia and Marshall (1984), who showed that they met the requisites of asymptotic consistency and normality under certain circumstances. However, these conditions are not easy to test in practice, particularly in the case of observations on an irregular grid. Another problem is that the likelihood surface could be multimodal. In their favor, it is worth highlighting that these estimates are more efficient than those of other methods when the sample is large. However, this circumstance is not entirely clear either. Indeed, a simulation study conducted by Zimmerman and Zimmerman (1991), which compared the ML estimator to other estimators, concluded that the former was only slightly better on certain occasions. It is also in their favor that when the rf has a drift, the ML method allows for the joint estimation of both drift and scale and covariance parameters, though biases used to be certainly severe.

The idea behind REML was originally proposed by Patterson and Thompson (1971) in connection with linear models. Imagine that $Y_1, \ldots, Y_n$ are independent and normal rv's with unknown parameters μ and σ^2. As is known, the ML estimators of μ and σ^2 are $\hat{\mu} = \bar{Y} = \frac{1}{n}\sum_i Y_i$ and $\hat{\sigma}^2 = \frac{1}{n}\sum_i \left(Y_i - \bar{Y}\right)^2$, respectively. But the latter estimator is biased, resulting instead in the utilization of the unbiased estimator of σ^2, $\frac{1}{n-1}\sum_i \left(Y_i - \bar{Y}\right)^2$. Let us now imagine that instead of working with the vector $Y_1, \ldots, Y_n$, we use the joint density of $\left(Y_1 - \bar{Y}, Y_2 - \bar{Y}, \ldots, Y_{n-1} - \bar{Y}\right)$, the distribution of which does not depend on μ. Now, the ML estimator of σ^2 is $\frac{1}{n-1}\sum_i \left(Y_i - \bar{Y}\right)^2$. This idea can be extended to the general model given in (3.60). If $\mathbf{W} = \mathbf{K}'\mathbf{Z}$ is defined as a vector of $n - q$ linearly independent contrasts, i.e. the $n - q$ columns of $\mathbf{K}$ are linearly independent and, therefore, $\mathbf{K}'\mathbf{X} = \mathbf{0}$, we have that:

$$\mathbf{W} \sim G(\mathbf{0}, \mathbf{K}'\boldsymbol{\Sigma}\mathbf{K}), \tag{3.64}$$

and the negative logarithm of the likelihood function based on $\mathbf{W}$ will take the following form:

$$\begin{aligned}\ell(\mathbf{W}, \alpha, \boldsymbol{\theta}) = &\frac{n-q}{2}\log(2\pi) + \frac{n-q}{2}\log(\alpha) + \frac{1}{2}\log(|\mathbf{K}'\mathbf{V}(\boldsymbol{\theta})\mathbf{K}|)\\ &+ \frac{1}{2\alpha}\mathbf{W}'(\mathbf{K}'\mathbf{V}(\boldsymbol{\theta})\mathbf{K})^{-1}\mathbf{W}.\end{aligned} \tag{3.65}$$

As indicated by Patterson and Thompson (1971), it is possible to choose $\mathbf{K}$ such that it satisfies $\mathbf{KK}' = \mathbf{I} - \mathbf{X}(\mathbf{X}'\mathbf{X})^{-1}\mathbf{X}'$, $\mathbf{K}'\mathbf{K} = \mathbf{I}$. In this case (3.65) is simplified to:

$$\begin{aligned}\ell(\mathbf{W}, \alpha, \boldsymbol{\theta}) = &\frac{n-q}{2}\log(2\pi) + \frac{n-q}{2}\log(\alpha) - \frac{1}{2}\log(|\mathbf{X}'\mathbf{X}|)\\ &+ \frac{1}{2}\log(|\mathbf{X}'\mathbf{V}(\boldsymbol{\theta})^{-1}\mathbf{X}|) + \frac{1}{2}\log(|\mathbf{V}(\boldsymbol{\theta})|) + \frac{1}{2\alpha}G^2(\boldsymbol{\theta}),\end{aligned} \tag{3.66}$$

where $G^2(\boldsymbol{\theta}) = (\mathbf{Z} - \mathbf{X}\hat{\boldsymbol{\beta}})'\mathbf{V}(\boldsymbol{\theta})^{-1}(\mathbf{Z} - \mathbf{X}\hat{\boldsymbol{\beta}})$, and $\hat{\boldsymbol{\beta}} = (\mathbf{X}'\mathbf{V}(\boldsymbol{\theta})^{-1}\mathbf{X})^{-1}\mathbf{X}'\mathbf{V}(\boldsymbol{\theta})^{-1}\mathbf{Z}$ is the GLS estimator of $\boldsymbol{\beta}$ based on the matrix of covariances $\mathbf{V}(\boldsymbol{\theta})$.

The REML estimate of $\boldsymbol{\theta}$ is obtained by minimizing (3.66) with respect to $\boldsymbol{\theta}$ after having substituted α with $G^2(\theta)/(n-q)$, the value of alpha which minimizes (3.66).

The REML procedure yields better estimates than ML, generally speaking. In addition, since $\ell(\mathbf{W}, \alpha, \boldsymbol{\theta})$ does not depend on $\boldsymbol{\beta}$, when q is large relative to n, the REML estimator of $\boldsymbol{\theta}$ does not suffer from the frequent severe downwards bias of the ML estimator. However, it is more sensitive than the ML estimator to the incorrect specification of the vector of means.

Example 3.8 (Fitting a valid semivariogram model: Data on carbon monoxide in Madrid, week 50, 10 pm) *In Example 3.4 we constructed the empirical (or experimental) semivariogram for the average of the logarithm of carbon monoxide levels at 10 pm in the working days of the* 50*th week of* 2008 *(we called this average logCO*). Here, we fit a valid semivariogram model to such an empirical semivariogram. In light of the empirical semivariogram (see Figure 3.28(a)) and the comments made in such an example, we fit a spherical model.*

We use automatic fitting, the fitting methods being not only LS-based methods (based on the empirical semivariogram) but also ML-based methods (which deal with the observed data directly assuming they follow a multivariate Gaussian distribution and do not require the construction of the empirical semivariogram). More specifically, the fitting methods we use are OLS, WLS (the weights being $\#N(\mathbf{h}_i)\{\gamma(\mathbf{h}_i, \boldsymbol{\theta})\}^{-2}$*), ML and REML. When fitting by OLS or WLS the classical empirical semivariogram is considered. We focus on the isotropic case because, though a slight geometric anisotropy has been found, due to the scarce number of pairs in the directional empirical semivariograms, it is advisable not to take it into account. Table 3.8 displays the parameter estimates obtained with the four estimation methods. Figure 3.36 displays the fitted semivariograms. We used* `geoR` *for computations and plotting the spherical models depicted in Figure 3.36.*

Table 3.8 Estimates of the parameters of a spherical model (data on carbon monoxide in Madrid, week 50, 10 pm).

Method	Nugget effect	Partial sill	Range
OLS	0.00	0.1395	6082.918
WLS	0.00	0.1414	6175.224
ML	0.00	0.1311	5914.102
REML	0.00	0.1403	6096.484

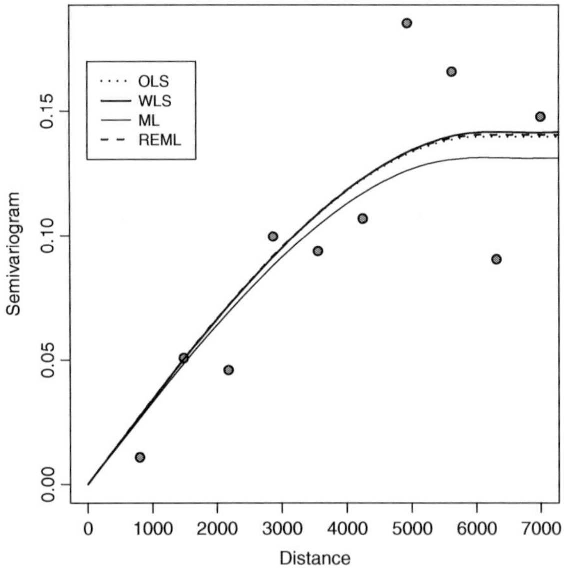

Figure 3.36 Spherical models resulting from the automatic fitting (data on carbon monoxide in Madrid, week 50, 10 pm).

As can be seen, the four methods considered suggest similar values for the parameters of the model. They coincide in agreeing that a nugget effect should not be included in the model, and the differences in the estimates of the partial sill and the practical range are not significant.

3.8.2.3 The Composite Likelihood (CL) fitting method

The CL method (Lindsay 1988) defines a general estimation method when working with large datasets.

For a measurable set of events $\{A_i \colon i \in I\}$, where I is a finite set of sub-indices, the CL function is defined by:

$$L_{CL}(\boldsymbol{\theta}) = \prod_{i \in I} f\left(Z \in A_i; \boldsymbol{\theta}\right)^{\omega_i}, \tag{3.67}$$

where $f(\cdot, \theta), \theta \in \Theta \subset \mathbb{R}^p$ is a parametric model and ω_i are convenient positive weights. Maximizing $L_{CL}(\theta)$ leads to the CL estimator.

Several types of CL have been proposed in the literature, including Vecchia (1998), Curriero and Lele (1999), Stein (2004), and Caragea and Smith (2006). We focus on the Curriero and Lele approach, an unusual CL form based on the semivariogram cloud that shares the best properties of the OLS and ML methods.

Let $Z(\mathbf{s})$ be an intrinsically stationary rf with a parametric semivariogram $\gamma(\mathbf{h}, \theta)$, where $\theta \in \boldsymbol{\Theta} \subset \mathbb{R}^p$ is an unknown p-dimensional vector of parameters.

Let us consider all the possible differences:

$$U_{ij} = Z(\mathbf{s}_i) - Z(\mathbf{s}_j), \quad i \neq j. \tag{3.68}$$

Assume that the distribution of those differences is Gaussian, $U_{ij} \sim G\left(0, \sqrt{2\gamma_{ij}(\theta)}\right)$, where $\gamma_{ij} = \gamma(\mathbf{s}_i - \mathbf{s}_j; \theta)$, as its mean is zero and its variance, which in this case would be the expectation of the squared difference, is none other than $2\gamma_{ij}(\mathbf{h})$.

Then we have that:

$$L\left(U_{ij}; \theta\right) = \frac{1}{\sqrt{2\gamma_{ij}(\theta)}\sqrt{2\pi}} e^{-\frac{1}{2}\frac{U_{ij}^2}{2\gamma_{ij}(\theta)}}$$

$$\ln L\left(U_{ij}; \theta\right) = -\ln \sqrt{2\gamma_{ij}(\theta)} - \ln \sqrt{2\pi} - \frac{1}{2}\frac{U_{ij}^2}{2\gamma_{ij}(\theta)}$$

$$= -\frac{1}{2}\ln 2\left(\gamma_{ij}(\theta)\right) - \frac{1}{2}\ln 2\pi - \frac{1}{2}\frac{U_{ij}^2}{2\gamma_{ij}(\theta)}$$

$$\ell\left(U_{ij}; \theta\right) = -\ln L\left(U_{ij}; \theta\right) = \frac{1}{2}\left(\ln 2 + \ln \gamma_{ij}(\theta) + \ln 2 + \ln \pi + \frac{U_{ij}^2}{2\gamma_{ij}(\theta)}\right),$$

hence the negative log-likelihood of one of these differences, eliminating all the constants that do not depend on θ is:

$$\ell\left(U_{ij}; \theta\right) = \ln \gamma_{ij}(\theta) + \frac{U_{ij}^2}{2\gamma_{ij}(\theta)}. \tag{3.69}$$

Assuming the differences considered above are independent, the so-called CL (in reality, it is a sum of duplicated support functions with changed signs, but its utility is the same for estimation purposes) is obtained as:

$$CL(\theta) = \sum_{i=1}^{N}\sum_{j>i}^{N} \ell(U_{ij}; \theta). \tag{3.70}$$

In order to obtain the CL estimates of $\boldsymbol{\theta}$, we proceed to minimize the function $CL(\boldsymbol{\theta})$:

$$\frac{d}{d\theta}\left(\sum_{i=1}^{N}\sum_{j>i}^{N}\ell\left(U_{ij};\boldsymbol{\theta}\right)\right)=\sum_{i=1}^{N}\sum_{j>i}^{N}\frac{d}{d\theta}\left(\ell\left(U_{ij};\boldsymbol{\theta}\right)\right)$$
$$=\sum_{i=1}^{N}\sum_{j>i}^{N}\frac{\frac{d}{d\theta}\gamma_{ij}(\boldsymbol{\theta})}{\gamma_{ij}(\boldsymbol{\theta})}-\frac{2\left(\frac{d}{d\theta}\gamma_{ij}(\boldsymbol{\theta})\right)U_{ij}^{2}}{\left(2\gamma_{ij}(\boldsymbol{\theta})\right)^{2}}$$
$$=\sum_{i=1}^{N}\sum_{j>i}^{N}\frac{\frac{d}{d\theta}\gamma_{ij}(\boldsymbol{\theta})}{\gamma_{ij}(\boldsymbol{\theta})}\left(1-\frac{U_{ij}^{2}}{2\gamma_{ij}(\boldsymbol{\theta})}\right)=0.$$

Note that a distinctive feature of the CL procedure is that the associated estimating function:

$$ee_{CL}(\boldsymbol{\theta})=\frac{d}{d\theta}\left(\sum_{i=1}^{N}\sum_{j>i}^{N}\ell\left(U_{ij};\theta\right)\right)=\sum_{i=1}^{N}\sum_{j>i}^{N}\frac{\frac{d}{d\theta}\gamma_{ij}(\boldsymbol{\theta})}{\gamma_{ij}(\boldsymbol{\theta})}\left(1-\frac{U_{ij}^{2}}{2\gamma_{ij}(\boldsymbol{\theta})}\right),\quad(3.71)$$

is unbiased, that is, $E_{\theta}\left(ee_{CL}(\boldsymbol{\theta})\right)=\mathbf{0}$, regardless of the distribution assumptions imposed on U_{ij}.

This procedure does not require the choosing of the distances to calculate the empirical semivariogram at (lag bins), as it is based on the semivariogram cloud, no matrix needs to be inverted, it is robust to poor distributional specification, and its order of computations is $O(N^2)$, between that of ML and WLS methods.

If the joint distributions of fourth order of U_{ij} were known, it would be possible to devise an optimal means of combining the individual scores $\frac{d}{d\theta}l(U_{ij};\boldsymbol{\theta})$. Hence, a further step could be taken to consider the optimal linear combinations of $V_{ij}(\boldsymbol{\theta})=\left(1-\frac{U_{ij}^{2}}{2\gamma_{ij}(\boldsymbol{\theta})}\right)$. By forming all the $V_{ij}(\boldsymbol{\theta})$ in a vector $\mathbf{V}(\boldsymbol{\theta})$, we can consider the *optimal estimating equation* (Heyde 1997):

$$ee_{OPT}(\boldsymbol{\theta})=\left(E\left(\mathbf{V}^{(1)}(\boldsymbol{\theta})\right)\right)'\left(C\left(\mathbf{V}(\boldsymbol{\theta})\right)\right)^{-1}\mathbf{V}(\boldsymbol{\theta}),\quad(3.72)$$

where $\mathbf{V}^{(1)}(\boldsymbol{\theta})$ is the matrix of partial derivatives of $\mathbf{V}(\boldsymbol{\theta})$. Using simple algebraic procedures, we can see that the generic element of $\frac{1}{2}cov(\mathbf{V}(\boldsymbol{\theta}))$ is $corr\left(U_{ij}^{2},U_{lk}^{2}\right)$, its expression in the case of zero mean stationary processes being given by:

$$Corr\left(U_{ij}^{2},U_{lk}^{2}\right)=\frac{\left(\gamma_{jk}-\gamma_{jl}+\gamma_{il}-\gamma_{ik}\right)^{2}}{4\gamma_{ij}\gamma_{lk}}\quad(3.73)$$

Indeed, as $Z\left(\mathbf{s}_1+\mathbf{h}_1\right)-Z\left(\mathbf{s}_1\right)$ and $Z\left(\mathbf{s}_2+\mathbf{h}_2\right)-Z\left(\mathbf{s}_2\right)$ form a bivariate normal distribution with zero mean, we have that (Cressie 1993, p. 93):

$$
\begin{aligned}
&Corr\left(\left(Z\left(\mathbf{s}_1+\mathbf{h}_1\right)-Z\left(\mathbf{s}_1\right)\right)^2,\left(Z\left(\mathbf{s}_2+\mathbf{h}_2\right)-Z\left(\mathbf{s}_2\right)\right)^2\right)\\
&=\left(Corr\left(\left(Z\left(\mathbf{s}_1+\mathbf{h}_1\right)-Z\left(\mathbf{s}_1\right)\right),\left(Z\left(\mathbf{s}_2+\mathbf{h}_2\right)-Z\left(\mathbf{s}_2\right)\right)\right)\right)^2\\
&=\frac{\left(C\left(\left(Z\left(\mathbf{s}_1+\mathbf{h}_1\right)-Z\left(\mathbf{s}_1\right)\right),\left(Z\left(\mathbf{s}_2+\mathbf{h}_2\right)-Z\left(\mathbf{s}_2\right)\right)\right)\right)^2}{2\gamma\left(\mathbf{h}_1\right)\cdot 2\gamma\left(\mathbf{h}_2\right)}\\
&=\frac{\left(E\left(\left(Z\left(\mathbf{s}_1+\mathbf{h}_1\right)-Z\left(\mathbf{s}_1\right)\right)\left(Z\left(\mathbf{s}_2+\mathbf{h}_2\right)-Z\left(\mathbf{s}_2\right)\right)\right)\right)^2}{2\gamma\left(\mathbf{h}_1\right)\cdot 2\gamma\left(\mathbf{h}_2\right)}\\
&=\frac{E\left(Z(\mathbf{s}_1+\mathbf{h}_1)Z(\mathbf{s}_2+\mathbf{h}_2)-Z(\mathbf{s}_1+\mathbf{h}_1)Z(\mathbf{s}_2)-Z(\mathbf{s}_1)Z(\mathbf{s}_2+\mathbf{h}_2)+Z(\mathbf{s}_1)Z(\mathbf{s}_2)\right)^2}{2\gamma(\mathbf{h}_1)\cdot 2\gamma(\mathbf{h}_2)}\\
&=\frac{\left(\gamma\left(\mathbf{s}_1-\mathbf{s}_2+\mathbf{h}_1\right)+\gamma\left(\mathbf{s}_1-\mathbf{s}_2-\mathbf{h}_2\right)-\gamma\left(\mathbf{s}_1-\mathbf{s}_2+\mathbf{h}_1-\mathbf{h}_2\right)-\gamma\left(\mathbf{s}_1-\mathbf{s}_2\right)\right)^2}{2\gamma\left(\mathbf{h}_1\right)\cdot 2\gamma\left(\mathbf{h}_2\right)},
\end{aligned}
$$

where the desired result is obtained by making the changes $\mathbf{s}_1 = \mathbf{s}_i$, $\mathbf{s}_2 = \mathbf{s}_k$, $\mathbf{s}_1 + \mathbf{h}_1 = \mathbf{s}_j$ and $\mathbf{s}_2 + \mathbf{h}_2 = \mathbf{s}_l$.

Obviously, if the U_{ij}^2 are uncorrelated, $cov(\mathbf{V}(\boldsymbol{\theta})) = \mathbf{I}$, and the associated estimating equation $ee_{CL}(\boldsymbol{\theta})$ are obtained. However, $cov(\mathbf{V}(\boldsymbol{\theta}))$ is of dimension $\left(\frac{n(n-1)}{2} \times \frac{n(n-1)}{2}\right)$ and it is computationally prohibitive to calculate its inverse for large datasets (even more so than in the case of ML).

4

Spatial prediction and kriging

4.1 Introduction

Once a theoretical semivariogram (or covariogram in the second-order stationary case) has been chosen, we are ready for spatial prediction. The method geostatistics uses for spatial prediction is termed kriging in honor of the South African mining engineer, Daniel Gerhardus Krige.

But before starting with kriging, let us give a couple of warnings. First, according to the title of the chapter, we use the term "prediction" for inference on random quantities (remember that in geostatistics data are assumed to behave according to a rf). But, for the sake of simplicity and following the criterion in most of the textbooks on geostatistics, we also will use the term "prediction" when dealing with block kriging or with the kriging of the mean (where it could be more appropriate to use the term "estimation"). Thus, we associate the term "prediction" with the result of a kriging procedure, the term "estimation" being used for inferences on unknown parameters of a model (usually a semivariogram or covariogram model). Second, we only focus on the univariate case, and kriging based not only on the rf of interest but also on other ancillary rf's correlated with such a primary rf is not studied. That is, the multivariate extension of kriging, cokriging, goes beyond the scope of this book.

Kriging aims to predict the value of a rf, $Z(\mathbf{s})$, at one or more non-observed points or blocks from a collection of data observed at n points (or blocks in the case of block prediction) of a domain D, and provides the best linear unbiased predictor (BLUP) of the regionalized variable under study at such non-observed points or blocks. Thus, the predictor support can be a point or a block. The limitation to the class of linear predictors obeys the fact that only the second-order moment of the rf is required and,

Spatial and Spatio-Temporal Geostatistical Modeling and Kriging, First Edition.
José-María Montero, Gema Fernández-Avilés, and Jorge Mateu.

Companion Website: www.wiley.com/go/montero/spatial

in general, in practice it is possible to be inferred. With more structural information, non-linear predictors can be defined. In the case of a Gaussian rf the optimal predictor and the BLUP coincide.

In this chapter, we will derive the kriging equations, which allow the prediction of the value of the rf $Z(\mathbf{s})$ at a non-observed point (or block) as a linear combination of the values of the rf at the sampled points (or blocks) or at a set of them that are close to the prediction point (we will address this issue, known as the neighborhood, in depth in Section 4.2). Obviously, we will also derive the expression of the prediction error variance, or simply the kriging variance, which indicates how accurate are the kriging predictions. We will derive the kriging equations for the case that both the observation support and the predictor are punctual. Then, we will generalize these equations for the cases of (i) point observation support and block predictor, and (ii) block observation support and block predictor. These equations are obtained by imposing on the predictor the classical conditions of unbiasedness and minimum variance, or, in other words, by imposing on the prediction error zero expectation and minimum variance (that is, minimizing the mean squared prediction error). It is nevertheless necessary to make the following clarification: the minimization of the mean-squared prediction error stems from the assumption that the semivariogram is known, which is not the case in many practical situations in which the kriging prediction is based on the empirical semivariogram (which a theoretical semivariogram is fitted to); furthermore, it is difficult to quantify the consequences of not using the true semivariogram.

More specifically, in the case of point observation support, the point kriging predictor at the non-observed point $\mathbf{s}_0$ is given by:

$$Z^*(\mathbf{s}_0) = \sum_{i=1}^{n} \lambda_i Z(\mathbf{s}_i), \tag{4.1}$$

where $Z(\mathbf{s}_i)$ are the values observed at the n points in the neighborhood of $\mathbf{s}_0$, the prediction point, and λ_i are the kriging weights obtained by imposing on the prediction error the classical conditions above referred.

In many occasions we are interested in block prediction. That is, our aim is to predict the average value of the rf being studied in a block V, given by $Z_V(\mathbf{s})$, which is assigned to the point $\mathbf{s} \in V$:

$$Z_V(\mathbf{s}) = \frac{1}{|V|} \int_V Z(\mathbf{s}')d\mathbf{s}', \tag{4.2}$$

where $|V|$ is the area or volume of V, $\mathbf{s}'$ sweeps throughout V and $\mathbf{s}$ is a point in V to which the average value of the block is assigned.

In such cases, when the observations are based on points, the predictor of the average value of the rf in V is given by:

$$Z^*(V) = \sum_{i=1}^{n} \lambda_i Z(\mathbf{s}_i), \tag{4.3}$$

In the case of a block observation support, that is, the data available are the average values in blocks v_i, the block predictor in V is:

$$Z_V^* = \sum_{i=1}^{n} \lambda_i Z_{v_i}(\mathbf{s}_i), \tag{4.4}$$

understanding $Z_{v_i}(\mathbf{s}_i)$ to be the average value in the block v_i, which is assigned to the point $\mathbf{s}_i \in v_i$:

$$Z_{v_i}(\mathbf{s}_i) = \frac{1}{|v_i|} \int_{v_i} Z(\mathbf{s}')\, d\mathbf{s}', \tag{4.5}$$

where $\mathbf{s}'$ sweeps throughout v_i.

Obviously, the quality of kriging predictions depends on the size of the sample and the quality of the data, but it also depends on (i) the location of the observations (if they are uniformly distributed in the domain under study, there will be better coverage and more information about what happens in that domain than if observations are grouped); (ii) the distance between the points (or blocks) observed and the point or block to be predicted (more trust should be placed in nearby observations than in distant observations) and (iii) the spatial continuity of the rf being studied (it is easier to predict the value of a regular rf at a point or over a block than of a rf that fluctuates markedly).

Finally, let us note that the main advantage of kriging over other spatial interpolation techniques (inverse distance method, splines, polynomial regression, among others) is that not only does it take into account geometric characteristics and the number and organization of locations, but also considers the structure of the spatial correlation that is deduced from the information available through semivariogram structures, hence yielding more reliable predictions. Therefore, the weights kriging uses are not calculated on the basis of an arbitrary rule that can be used in some cases but not others, but rather on the behavior of the function that represents the structure of spatial correlation. In this sense, it is a more flexible method than those mentioned above. Moreover: (i) Kriging makes it possible to quantify how accurate are the predictions using the prediction error variance (the kriging variance) and can yield a map of the standard deviation of the prediction errors; (ii) kriging variance does not depend on the actual realization of the rf, which acts as a probability shelter, meaning we can calculate it before learning the values of the variables at those points, providing we know the structure of the spatial dependence of that rf. This is extremely useful when designing networks of optimum observations, that is, for selecting locations to be observed that provide the least kriging variance; (iii) kriging is an exact interpolator, which means that at the points that make up part of the sample, the kriging prediction coincides with the value observed, the kriging variance therefore being zero.

In summary, the observation support can be points or blocks. In the first case the prediction support can be also points or blocks; in the second case, it can be only blocks. Whatever the case, in what follows we will focus on the different types of kriging, with special focus on the ordinary kriging (OK), if the second-order stationary hypothesis or the intrinsic one are assumed, and the universal kriging (UK), in the non-stationary (in mean) case. But first let us comment on the concept of neighborhood.

4.2 Neighborhood

Neighborhood is an important concept. As stated in the previous section the main objective of this chapter is to derive the kriging equations, which allow the prediction of the value of the rf $Z(\mathbf{s})$ at a non-observed point (or block) as a linear combination of the values of the rf at the sampled points (or blocks in the case of block prediction), as well as the kriging variance associated with such a prediction. But, when performing a kriging prediction we have two options: (i) all the observed points (or blocks) enter the prediction process, as said above; and (ii) only the points (or blocks) close to the prediction point (or block) are considered for prediction.

In the first case, the influence of the data observed far away from the prediction point (or block) will be very small. One even could think that the influence of the data beyond the range of the semivariogram will be null. However, it is not true: they influence the prediction indirectly because they are spatially correlated with data observed at points in the zone of influence; what is more, they improve the precision of the prediction. However, a "global neighborhood" requires second-order or intrinsic stationarity across the complete neighborhood, which sometimes cannot be assumed. It also leads to computational problems (time required and instability) when inverting matrices. In the second case, only the hypothesis of local second-order or intrinsic stationarity is needed. But, the size, and also the shape, of the neighborhood continue to be crucial. If the neighborhood is too small, the precision of the prediction will be affected and will depend on the values of the observed regionalization at the points included in the neighborhood. If it is too big, the hypothesis of second-order or intrinsic stationarity could not be assumed for the neighborhood. As for the shape of the neighborhood, it should reflect the anisotropy of the rf, which can be deduced from the structural analysis.

There are no rules to define the size of the neighborhood, but Webster and Oliver (2001, p. 168) suggest the following guidelines:

(i) If data are dense and the semivariogram is bounded with a small nugget effect, then the radius of the neighborhood can be set close to the range (or the practical range).

(ii) If data are sparse, points beyond the range from the prediction point could have non-negligible weights and the neighborhood should be such as to include them.

(iii) If the nugget effect is large, distant points from the prediction one will have a significant influence on the prediction and should be included in the neighborhood.

(iv) As an alternative, we can choose the nearest n points, and effectively let this number limit the neighborhood. If data are irregularly spaced, then the size of the neighborhood will vary more or less as the prediction point is moved. A maximum of $n = 20$ is usually enough.

(v) If we set a maximum radius for the neighborhood, then we may need to set a minimum number of observations, especially to cater for prediction points near the boundary of the domain under study. About seven observations is likely to be satisfactory.

(vi) Where the scatter is very uneven, it is good practice to divide the space around the prediction point into octants and take the nearest two points in each.

4.3 Ordinary kriging

4.3.1 Point observation support and point predictor

Let $Z = \{Z(\mathbf{s}),\ \mathbf{s} \in D\}$ be a second-order stationary rf with a constant but unknown mean μ and a known covariance function $C(\mathbf{h})$. Under this assumption, the ordinary kriging (OK) equations, which provide the weights needed for the kriging prediction of the rf at a non-observed point, can be expressed either in terms of the covariance function or in terms of the semivariogram. In either of these two cases, $Z = \{Z(\mathbf{s}),\ \mathbf{s} \in D\}$ is predicted in a non-observed point $\mathbf{s}_0$ using the linear predictor $Z^*(\mathbf{s}_0) = \sum_{i=1}^{n} \lambda_i Z(\mathbf{s}_i)$, imposing on the prediction error that its expectation must be zero and its variance minimum.

We first derive the OK equations in terms of the covariance functions. In order to comply with the unbiasedness of the kriging predictor, the sum of the weights must be one, as:

$$\begin{aligned} E\left(Z^*(\mathbf{s}_0) - Z(\mathbf{s}_0)\right) &= E\left(\sum_{i=1}^{n} \lambda_i Z(\mathbf{s}_i) - Z(\mathbf{s}_0)\right) = \sum_{i=1}^{n} \lambda_i E\left(Z(\mathbf{s}_i)\right) - E\left(Z(\mathbf{s}_0)\right) \\ &= \mu \sum_{i=1}^{n} \lambda_i - \mu = 0 \Longleftrightarrow \sum_{i=1}^{n} \lambda_i = 1. \end{aligned} \tag{4.6}$$

The prediction variance is then given by:

$$\begin{aligned} V\left(Z^*(\mathbf{s}_0) - Z(\mathbf{s}_0)\right) &= E\left(Z^*(\mathbf{s}_0) - Z(\mathbf{s}_0)\right)^2 \\ &= \sum_{i=1}^{n}\sum_{j=1}^{n} \lambda_i \lambda_j E\left(Z(\mathbf{s}_i)Z(\mathbf{s}_j)\right) \\ &\quad - 2\sum_{i=1}^{n} \lambda_i E\left(Z(\mathbf{s}_i)Z(\mathbf{s}_0)\right) + E\left(\left(Z(\mathbf{s}_0)\right)^2\right), \end{aligned} \tag{4.7}$$

whereby now, bearing in mind that $E(Z(\mathbf{s})) = \mu$, we have that:

$$E\left(Z(\mathbf{s}_i)Z(\mathbf{s}_j)\right) = C(\mathbf{s}_i - \mathbf{s}_j) + \mu^2, \tag{4.8}$$

$$E\left(Z(\mathbf{s}_i)Z(\mathbf{s}_0)\right) = C(\mathbf{s}_i - \mathbf{s}_0) + \mu^2, \tag{4.9}$$

$$E\left((Z(\mathbf{s}_0))^2\right) = C(\mathbf{0}) + \mu^2, \tag{4.10}$$

and therefore the expression (4.7) is transformed into:

$$\begin{aligned} V\left(Z^*(\mathbf{s}_0) - Z(\mathbf{s}_0)\right) &= \sum_{i=1}^{n}\sum_{j=1}^{n} \lambda_i\lambda_j C(\mathbf{s}_i - \mathbf{s}_j) - 2\sum_{i=1}^{n} \lambda_i C(\mathbf{s}_i - \mathbf{s}_0) \\ &+ C(\mathbf{0}) + \mu^2 \left(\sum_{i=1}^{n}\sum_{j=1}^{n} \lambda_i\lambda_j - 2\sum_{i=1}^{n} \lambda_i + 1\right), \end{aligned} \tag{4.11}$$

which under the unbiasedness condition is reduced to:

$$V\left(Z^*(\mathbf{s}_0) - Z(\mathbf{s}_0)\right) = \sum_{i=1}^{n}\sum_{j=1}^{n} \lambda_i\lambda_j C(\mathbf{s}_i - \mathbf{s}_j) - 2\sum_{i=1}^{n} \lambda_i C(\mathbf{s}_i - \mathbf{s}_0) + C(\mathbf{0}). \tag{4.12}$$

The problem of OK is to find the weights $\lambda_i, i = 1, \ldots, n$, minimizing $V\left(Z^*(\mathbf{s}_0) - Z(\mathbf{s}_0)\right)$ subject to the linear constraint $\sum_{i=1}^{n} \lambda_i = 1$. This problem is solved by the method of the Lagrange multipliers, the Lagrange function being:

$$\varphi(\lambda_i, \alpha) = V\left(Z^*(\mathbf{s}_0) - Z(\mathbf{s}_0)\right) - \alpha\left(\sum_{i=1}^{n} \lambda_i - 1\right), \tag{4.13}$$

which, by deriving it partially with respect to the weights $\lambda_i, i = 1, \ldots, n$, and the Lagrange multiplier α, and setting these partial derivatives to zero, results in the following system of $n + 1$ equations with $n + 1$ unknowns:

$$\begin{cases} \sum_{j=1}^{n} \lambda_j C(\mathbf{s}_i - \mathbf{s}_j) - \alpha = C(\mathbf{s}_i - \mathbf{s}_0), & i = 1, \ldots, n \\ \sum_{i=1}^{n} \lambda_i = 1 \end{cases}. \tag{4.14}$$

The minimized prediction variance, called the OK variance, is obtained by substituting $\sum_{j=1}^{n} \lambda_j C(\mathbf{s}_i - \mathbf{s}_j)$ with $C(\mathbf{s}_i - \mathbf{s}_0) + \alpha$, as can be deduced from (4.14), in the variance expression (4.12), which incorporates the unbiasedness condition:

$$\sigma_{OK}^2(\mathbf{s}_0) = C(\mathbf{0}) - \sum_{i=1}^{n} \lambda_i C(\mathbf{s}_i - \mathbf{s}_0) + \alpha. \tag{4.15}$$

However, when solving the system (4.14) a problem arises: Although the unknown constant mean of the random function, μ, does not expressly appear in the system, it would be necessary for it to be determined in order to estimate the covariance function. As stated in Section 3.3, the empirical covariogram uses the sample mean, $\bar{Z}$, as an estimator of μ. But in the case of spatial dependence, $\bar{Z}$, despite being an unbiased estimator of μ, is not that of minimum variance. Another consequence of using $\bar{Z}$ as an estimator of μ is that the MoM empirical covariogram is biased. An alternative to $\bar{Z}$ would involve the use of another mean that would take into account such a spatial dependence. But we would have to know the spatial covariance function to be able to calculate it, which leads to a vicious circle.

One way of avoiding this problem is to express the OK system in semivariogram terms. Since in the second-order stationary case $C(\mathbf{s}_i - \mathbf{s}_j) = C(\mathbf{0}) - \gamma(\mathbf{s}_i - \mathbf{s}_j)$, the OK system (4.14) and the OK variance (4.15) can be written as:

$$\begin{cases} \sum_{j=1}^{n} \lambda_j \gamma(\mathbf{s}_i - \mathbf{s}_j) + \alpha = \gamma(\mathbf{s}_i - \mathbf{s}_0), \qquad i = 1, \ldots, n \\ \sum_{i=1}^{n} \lambda_i = 1 \end{cases}, \tag{4.16}$$

and

$$\sigma_{OK}^2(\mathbf{s}_0) = \sum_{i=1}^{n} \lambda_i \gamma(\mathbf{s}_i - \mathbf{s}_0) + \alpha, \tag{4.17}$$

respectively.

In the case that $Z = \{Z(\mathbf{s}),\ \mathbf{s} \in D\}$ is not second-order stationary but intrinsically stationary, the OK system can only be expressed in terms of the semivariogram, as

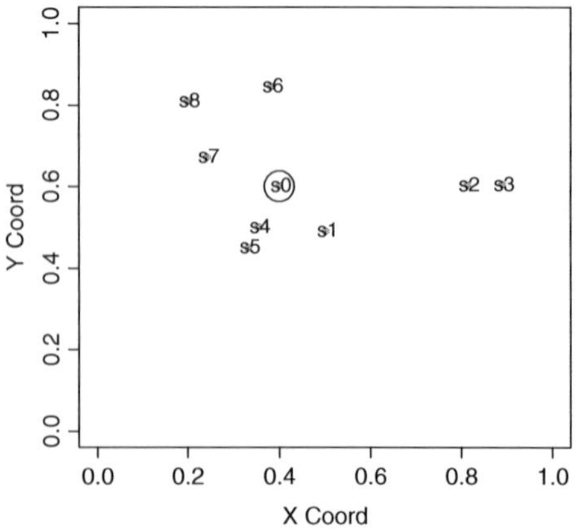

Figure 4.1 Location of eight observation points used for prediction at the non-observed point $\mathbf{s}_0$.

neither the variance nor covariance are defined. In this case, the OK system is the same as for a second-order stationary rf, but the condition $\sum_{i=1}^{n} \lambda_i = 1$ is not an unbiasedness condition but a permissibility condition.

Example 4.1 (OK at work) *In order to illustrate how OK performs, let us consider the eight points shown in Figure 4.1, where we have simulated a realization of a second-order stationary Gaussian rf with a spherical semivariogram, the range being* 0.6 *and the sill* 0.3. *Our aim is to obtain the OK prediction at the non-observed point* $\mathbf{s}_0$. *The data simulated at these eight points, together with the coordinates of such points and* $\mathbf{s}_0$ *are listed in Table 4.1.*

Table 4.2 shows inter-point distances, and Table 4.3 lists the semivariogram values for such distances.

Table 4.1 Coordinates and data values of observed points. Coordinates of prediction point.

	X-coord	Y-coord	Data value
$\mathbf{s}_1$	0.502	0.491	10.700
$\mathbf{s}_2$	0.813	0.603	11.061
$\mathbf{s}_3$	0.889	0.604	10.470
$\mathbf{s}_4$	0.354	0.500	9.845
$\mathbf{s}_5$	0.332	0.450	9.764
$\mathbf{s}_6$	0.382	0.845	10.640
$\mathbf{s}_7$	0.241	0.671	10.358
$\mathbf{s}_8$	0.200	0.810	10.928
$\mathbf{s}_0$	0.400	0.600	?

Table 4.2 Inter-point distances.

	$\mathbf{s}_0$	$\mathbf{s}_1$	$\mathbf{s}_2$	$\mathbf{s}_3$	$\mathbf{s}_4$	$\mathbf{s}_5$	$\mathbf{s}_6$	$\mathbf{s}_7$	$\mathbf{s}_8$
$\mathbf{s}_0$	0								
$\mathbf{s}_1$	0.1828	0							
$\mathbf{s}_2$	0.5056	0.4045	0						
$\mathbf{s}_3$	0.5994	0.4942	0.0937	0					
$\mathbf{s}_4$	0.1346	0.1816	0.5758	0.6680	0				
$\mathbf{s}_5$	0.2017	0.2141	0.6179	0.7083	0.0670	0			
$\mathbf{s}_6$	0.3013	0.4581	0.6054	0.6879	0.4243	0.4881	0		
$\mathbf{s}_7$	0.2130	0.3882	0.7050	0.7980	0.2510	0.2929	0.2742	0	
$\mathbf{s}_8$	0.3552	0.5380	0.7925	0.8814	0.4238	0.4696	0.2276	0.1771	0

Table 4.3 Semivariogram values.

Distance	Semivariogram values
0.0670	0.0501
0.0938	0.0698
0.1347	0.0993
0.1772	0.1290
0.1816	0.1320
0.1828	0.1329
0.2017	0.1456
0.2131	0.1531
0.2142	0.1538
0.2276	0.1625
0.2511	0.1773
0.2743	0.1914
0.2930	0.2023
0.3013	0.2070
0.3553	0.2353
0.3882	0.2505
0.4045	0.2574
0.4239	0.2650
0.4243	0.2652
0.4581	0.2768
0.4696	0.2803
0.4882	0.2853
0.4943	0.2869
0.5057	0.2895
0.5381	0.2954
0.5759	0.2993
0.5995	0.3000
0.6000 and more	0.3000

According to Table 4.3, and using the notations $\gamma_{ij} = \gamma(\mathbf{s}_i - \mathbf{s}_j)$, *we have that:*

$$\mathbf{\Gamma} = \begin{pmatrix} \gamma_{11} & \gamma_{12} & \gamma_{13} & \gamma_{14} & \gamma_{15} & \gamma_{16} & \gamma_{17} & \gamma_{18} & 1 \\ \gamma_{21} & \gamma_{22} & \gamma_{23} & \gamma_{24} & \gamma_{25} & \gamma_{26} & \gamma_{27} & \gamma_{28} & 1 \\ \gamma_{31} & \gamma_{32} & \gamma_{33} & \gamma_{34} & \gamma_{35} & \gamma_{36} & \gamma_{37} & \gamma_{38} & 1 \\ \gamma_{41} & \gamma_{42} & \gamma_{43} & \gamma_{44} & \gamma_{45} & \gamma_{46} & \gamma_{47} & \gamma_{48} & 1 \\ \gamma_{51} & \gamma_{52} & \gamma_{53} & \gamma_{54} & \gamma_{55} & \gamma_{56} & \gamma_{57} & \gamma_{58} & 1 \\ \gamma_{61} & \gamma_{62} & \gamma_{63} & \gamma_{64} & \gamma_{65} & \gamma_{66} & \gamma_{67} & \gamma_{68} & 1 \\ \gamma_{71} & \gamma_{72} & \gamma_{73} & \gamma_{74} & \gamma_{75} & \gamma_{76} & \gamma_{77} & \gamma_{78} & 1 \\ \gamma_{81} & \gamma_{82} & \gamma_{83} & \gamma_{84} & \gamma_{85} & \gamma_{86} & \gamma_{87} & \gamma_{88} & 1 \\ 1 & 1 & 1 & 1 & 1 & 1 & 1 & 1 & 0 \end{pmatrix}$$

$$= \begin{pmatrix} 0.0000 & 0.2574 & 0.2869 & 0.1320 & 0.1538 & 0.2768 & 0.2505 & 0.2954 & 1 \\ 0.2574 & 0.0000 & 0.0698 & 0.2993 & 0.3000 & 0.3000 & 0.3000 & 0.3000 & 1 \\ 0.2869 & 0.0698 & 0.0000 & 0.3000 & 0.3000 & 0.3000 & 0.3000 & 0.3000 & 1 \\ 0.1320 & 0.2993 & 0.3000 & 0.0000 & 0.0501 & 0.2652 & 0.1773 & 0.2650 & 1 \\ 0.1538 & 0.3000 & 0.3000 & 0.0501 & 0.0000 & 0.2853 & 0.2023 & 0.2803 & 1 \\ 0.2768 & 0.3000 & 0.3000 & 0.2652 & 0.2853 & 0.0000 & 0.1914 & 0.1625 & 1 \\ 0.2505 & 0.3000 & 0.3000 & 0.1773 & 0.2023 & 0.1914 & 0.0000 & 0.1290 & 1 \\ 0.2954 & 0.3000 & 0.3000 & 0.2650 & 0.2803 & 0.1625 & 0.1290 & 0.0000 & 1 \\ 1 & 1 & 1 & 1 & 1 & 1 & 1 & 1 & 0 \end{pmatrix}$$

$$\mathbf{\Gamma}_0 = \begin{pmatrix} \gamma_{10} \\ \gamma_{20} \\ \gamma_{30} \\ \gamma_{40} \\ \gamma_{50} \\ \gamma_{60} \\ \gamma_{70} \\ \gamma_{80} \\ 1 \end{pmatrix} = \begin{pmatrix} 0.1329 \\ 0.2895 \\ 0.3000 \\ 0.0993 \\ 0.1456 \\ 0.2070 \\ 0.1531 \\ 0.2353 \\ 1 \end{pmatrix} \quad and \quad \lambda = \begin{pmatrix} \lambda_1 \\ \lambda_2 \\ \lambda_3 \\ \lambda_4 \\ \lambda_5 \\ \lambda_6 \\ \lambda_7 \\ \lambda_8 \\ \alpha \end{pmatrix} = \mathbf{\Gamma}^{-1}\mathbf{\Gamma_0} = \begin{pmatrix} 0.2769 \\ 0.0570 \\ -0.0606 \\ 0.4862 \\ -0.1209 \\ 0.1729 \\ 0.2446 \\ -0.0561 \\ -0.1148 \end{pmatrix}$$

The resulting prediction at $\mathbf{s}_0$ *is obtained as:*

$$Z^*(\mathbf{s}_0) = \sum_{i=1}^{8} \lambda_i Z(\mathbf{s}_i) = 10.3254,$$

and the OK variance associated with $Z^*(\mathbf{s}_0)$ *is:*

$$\sigma^2_{OK}(\mathbf{s}_0) = \sum_{i=1}^{8} \lambda_i \gamma_{i0} + \alpha = 0.0981.$$

As can be seen, the weights of the simulated data in the kriging prediction decrease with the distance between the point where they were simulated and the prediction point. The weight for $Z(\mathbf{s}_2)$ *is the positive smallest one, since it is located at a distance from the prediction point of nearly the range of the semivariogram. The weights corresponding to* $\mathbf{s}_3, \mathbf{s}_5$ *and* $\mathbf{s}_8$ *are negative.*

This is a consequence of the screen effect. An observation is said to be screened if there is another observation between it and the prediction point. Screened observations could be assigned a negative weight in the kriging predictor, which can be certainly high when they are close to the prediction point (this is a frequent situation when dealing with irregularly spaced data).

The fact that OK can yield negative weights (and also weights greater than the unity), as in Example 4.1, cannot be considered a disadvantage because this makes it

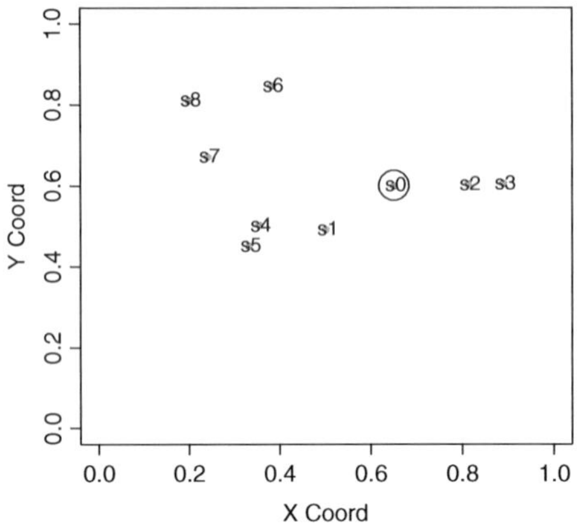

Figure 4.2 New location of the prediction point $\mathbf{s}_0$.

possible to obtain predictions larger than the largest observed value or smaller than the smallest observed one. In contrast, the procedures that impose weights in the range [0,1] can only yield predictions between the smallest and the largest observed values, which is unrealistic because it is difficult to imagine that the observed realization include both the smallest and the largest values of the rf. In addition, the positivity constraint may lead to large kriging variances.

However, it is true that negative weights could produce negative predictions, which is unrealistic in most practical applications, where the rf is non-negative. In these cases, the usual practice is to set the prediction to zero.

Example 4.2 (OK at work. Moving the prediction point) *If we move the prediction point to coordinates* (0.65, 0.60) *(Figure 4.2), the weights of* $Z(\mathbf{s}_1)$ *and* $Z(\mathbf{s}_2)$*, now the closest points to* $Z(\mathbf{s}_0)$ *increase from* 0.2769 *and* 0.0570 *to* 0.4386 *and* 0.5448*, respectively. By contrast, the weight of* $Z(\mathbf{s}_4)$ *diminishes dramatically (from* 0.4862 *to* 0.0403*), as its distance to the prediction point has increased significantly. The same can be said for* $Z(\mathbf{s}_7)$ *(*λ_7 *decreases from* 0.2446 *to* 0.0104*;* $\mathbf{s}_5$ *is now the non-screened point most distant from the prediction point) and, in a much lesser extent, for* $Z(\mathbf{s}_6)$*. The screened data* $Z(\mathbf{s}_3)$, $Z(\mathbf{s}_5)$ *and* $Z(\mathbf{s}_8)$ *keep their negative weight. Finally, the OK variance increases by nearly* 40% *because now the prediction point is surrounded by a smaller number of close observation points.*

4.3.2 Effects of a change in the model parameters

The choice of the covariance or semivariogram model has great impact in prediction and kriging variance. Thus, this choice is a core decision. But once a specific model has been inserted, the choice of the model parameters also has a noticeable influence on them. In the sequel, we illustrate the effect of the sill, the behavior near the origin, the nugget effect, range and anisotropy on both prediction and kriging variance.

4.3.2.1 Effect of a change in the sill

Two semivariograms differing only in their sill will provide the same OK weights but different OK variances. In effect, if the sill of $\gamma^{(2)}(\mathbf{h})$ is k times the sill of $\gamma^{(1)}(\mathbf{h})$, then $\gamma^{(2)}(\mathbf{h}) = k\gamma^{(1)}(\mathbf{h})$ and the kriging weights $\lambda_i^{(2)}$ obtained with $\gamma^{(2)}(\mathbf{h})$ are the same as those obtained with $\gamma^{(1)}(\mathbf{h})$: $\lambda_i^{(1)}$.

It can easily be seen that OK equations using $\gamma^{(2)}(\mathbf{h})$,

$$\begin{aligned}
\gamma_{(11)}^{(2)}\lambda_1^{(2)} + \gamma_{(12)}^{(2)}\lambda_2^{(2)} + \ldots + \gamma_{(1n)}^{(2)}\lambda_n^{(2)} + \alpha^{(2)} &= \gamma_{(10)}^{(2)} \\
\gamma_{(21)}^{(2)}\lambda_1^{(2)} + \gamma_{(22)}^{(2)}\lambda_2^{(2)} + \ldots + \gamma_{(2n)}^{(2)}\lambda_n^{(2)} + \alpha^{(2)} &= \gamma_{(20)}^{(2)} \\
&\ldots \\
\gamma_{(n1)}^{(2)}\lambda_1^{(2)} + \gamma_{(n2)}^{(2)}\lambda_2^{(2)} + \ldots + \gamma_{(nn)}^{(2)}\lambda_n^{(2)} + \alpha^{(2)} &= \gamma_{(n0)}^{(2)} \\
\lambda_1^{(2)} + \lambda_2^{(2)} + \ldots + \lambda_n^{(2)} &= 1,
\end{aligned}$$

can be written in terms of $\gamma^{(1)}(\mathbf{h})$ as follows:

$$\begin{aligned}
k\gamma_{(11)}^{(1)}\lambda_1^{(2)} + k\gamma_{(12)}^{(1)}k\lambda_2^{(2)} + \ldots + k\gamma_{(1n)}^{(1)}\lambda_n^{(1)} + \alpha^{(2)} &= k\gamma_{(10)}^{(1)} \\
k\gamma_{(21)}^{(1)}\lambda_1^{(2)} + k\gamma_{(22)}^{(1)}k\lambda_2^{(2)} + \ldots + k\gamma_{(2n)}^{(1)}\lambda_n^{(1)} + \alpha^{(2)} &= k\gamma_{(20)}^{(1)} \\
&\ldots \\
k\gamma_{(n1)}^{(1)}\lambda_1^{(1)} + k\gamma_{(n2)}^{(1)}k\lambda_2^{(1)} + \ldots + k\gamma_{(nn)}^{(1)}\lambda_n^{(1)} + \alpha^{(2)} &= k\gamma_{(n0)}^{(1)} \\
\lambda_1^{(1)} + \lambda_2^{(1)} + \ldots + \lambda_n^{(1)} &= 1,
\end{aligned}$$

that is,

$$\begin{aligned}
\gamma_{(11)}^{(1)}\lambda_1^{(2)} + \gamma_{(12)}^{(1)}\lambda_2^{(2)} + \ldots + \gamma_{(1n)}^{(1)}\lambda_n^{(1)} + \frac{\alpha^{(2)}}{k} &= \gamma_{(10)}^{(1)} \\
\gamma_{(21)}^{(1)}\lambda_1^{(2)} + \gamma_{(22)}^{(1)}\lambda_2^{(2)} + \ldots + \gamma_{(2n)}^{(1)}\lambda_n^{(1)} + \frac{\alpha^{(2)}}{k} &= \gamma_{(20)}^{(1)} \\
&\ldots \\
\gamma_{(n1)}^{(1)}\lambda_1^{(1)} + \gamma_{(n2)}^{(1)}\lambda_2^{(1)} + \ldots + \gamma_{(nn)}^{(1)}\lambda_n^{(1)} + \frac{\alpha^{(2)}}{k} &= \gamma_{(n0)}^{(1)} \\
\lambda_1^{(1)} + \lambda_2^{(1)} + \ldots + \lambda_n^{(1)} &= 1,
\end{aligned}$$

which is the OK system in terms of $\gamma^{(1)}(\mathbf{h})$, where $\alpha^{(1)} = \frac{\alpha^{(2)}}{k}$.

Therefore, the OK weights remain the same. However, the kriging variance increases by factor k:

$$\sigma_{OK}^{2(2)}(\mathbf{s}_0) = \sum_{i=1}^{n} \lambda_i^{(2)}\gamma_{i0}^{(2)} + \alpha^{(2)} = \sum_{i=1}^{n} \lambda_i^{(1)}k\gamma_{i0}^{(1)} + k\alpha^{(1)} = k\sigma_{OK}^{2(1)}(\mathbf{s}_0).$$

4.3.2.2 Effect of a discontinuity at the origin

In order to illustrate how the nugget effect influences prediction and kriging variance, let us consider the following two cases: (i) $\gamma^{(2)}(\mathbf{h})$ is formed by adding a nugget effect to $\gamma^{(1)}(\mathbf{h})$; (ii) $\gamma^{(1)}(\mathbf{h})$ has no nugget effect and $\gamma^{(2)}(\mathbf{h})$ has a nugget effect that is half of the sill.

Case (i) is the same as a change in the sill. In case (ii) the sill of $\gamma^{(2)}(\mathbf{h})$ remains constant but half of that sill is due to a nugget effect. Intuitively, the inclusion of a nugget effect, the sill remaining constant, will yield more similar kriging weights than that obtained with $\gamma^{(1)}(\mathbf{h})$. The reason is that the spatial correlation decreases at each lag distance. The limit case is when the sill coincides with the nugget effect (the pure nugget semivariogram) and all the weights are the same: $1/n$, and as a consequence the kriging predictor is the arithmetic mean, and the prediction variance $\sigma^2/n, \sigma^2$ being the variance of the underlying rf. Therefore, in the pure nugget case, what is important is the number of points; their location is irrelevant.

Example 4.3 (Effect of a discontinuity at the origin) *Let us consider the data and the spherical semivariogram used in Example 4.1. Let us call it $\gamma^{(1)}(\mathbf{h})$. Now, let $\gamma^{(2)}(\mathbf{h})$ be a spherical semivariogram with the same range and sill than $\gamma^{(1)}(\mathbf{h})$, but where half of such a sill corresponds to a nugget effect (the other half is the partial sill). Thus,*

$$\gamma^{(1)}(\mathbf{h}) = \begin{cases} 0.3\left(1.5\frac{|\mathbf{h}|}{0.6} - 0.5\left(\frac{|\mathbf{h}|}{0.6}\right)^3\right) & \text{if } |\mathbf{h}| \leq 0.6 \\ 0.3 \quad \text{if } |\mathbf{h}| > 0.6 \end{cases},$$

$$\gamma^{(2)}(\mathbf{h}) = \begin{cases} 0.15 + 0.15 \quad \left(1.5\frac{|\mathbf{h}|}{0.6} - 0.5\left(\frac{|\mathbf{h}|}{0.6}\right)^3\right) & \text{if } |\mathbf{h}| \leq 0.6 \\ 0.3 \quad \text{if } |\mathbf{h}| > 0.6 \end{cases}.$$

Table 4.4 shows the OK weights obtained in the prediction at $\mathbf{s}_0$, the OK variance and the predicted value when using $\gamma^{(1)}(\mathbf{h})$ and $\gamma^{(2)}(\mathbf{h})$. It can be seen that the weights are less dispersed when using $\gamma^{(2)}(\mathbf{h})$ (in fact, the negative weights have disappeared), which results in a higher OK variance.

4.3.2.3 Effect of a change in the behavior near the origin

The behavior of the semivariogram near the origin has remarkable effects on both the prediction value at the non-observed point and the kriging variance associated with such a prediction point.

In the kriging predictor more weight is given to the values very close to the prediction point when using a semivariogram with a parabolic behavior near the origin than when this behavior is linear. The reason is that a parabolic behavior near the origin is indicative of very regular phenomena, whereas a linear behavior near the origin indicates much more irregular ones. As a consequence of this, semivariograms

Table 4.4 OK weights, prediction and OK variance.

	$\gamma^{(1)}(\mathbf{h})$	$\gamma^{(2)}(\mathbf{h})$
λ_1	0.2769	0.2015
λ_2	0.0570	0.0563
λ_3	−0.0606	0.0242
λ_4	0.4862	0.2137
λ_5	−0.1209	0.1361
λ_6	0.1729	0.1376
λ_7	0.2446	0.1616
λ_8	−0.0562	0.0688
$Z^*(\mathbf{s}_0)$	10.3254	10.3572
$\sigma^2_{OK}(\mathbf{s}_0)$	0.0981	0.2297

with a parabolic behavior near the origin lead to lower kriging variances than semivariograms whose behavior near the origin is linear (everything else being equal, or similar).

Example 4.4 (Effect of a change in the behavior near the origin) *Recall the data of Example 4.1. Now we consider two alternative semivariogram models to predict the value of the underlying rf at* $\mathbf{s}_0$, *both of them with the same sill: (i) A spherical model with range* 0.60, *which exhibits a linear behavior near the origin; and (ii) a Gaussian model with a scale parameter of* $0.60\sqrt{3}$, *which exhibits a parabolic behavior near the origin. More specifically, these semivariograms are given by:*

$$\gamma^{(1)}(\mathbf{h}) = \begin{cases} 0.3\left(1.5\frac{|\mathbf{h}|}{0.6} - 0.5\left(\frac{|\mathbf{h}|}{0.6}\right)^3\right) & \textit{if } |\mathbf{h}| \le 0.6 \\ 0.3 \quad \textit{if } |\mathbf{h}| > 0.6 \end{cases},$$

$$\gamma^{(2)}(\mathbf{h}) = 0.3\left(1 - \exp\left(-\frac{|\mathbf{h}|}{0.6/\sqrt{3}}\right)^2\right).$$

The OK weights, the OK variance associated with $Z^*(\mathbf{s}_0)$ *and the predicted value at* $\mathbf{s}_0$ *are listed in Table 4.5 for cases (i) and (ii).*

When using the Gaussian semivariogram, it is expected that more weight is given to the values surrounding the prediction point (of course, as $\sum_{i=1}^n \lambda_i = 1$, *this implies that the more distant values receive less weight). This is a consequence of the parabolic behavior near the origin of the Gaussian model, indicative of a very regular phenomenon, which results in making more use of the closest observed values. In effect, in our case,* $Z(\mathbf{s}_4)$ *is assigned a weight of* 1.8861 *as a consequence of* $\mathbf{s}_4$ *being the closest point to the prediction one (the weight of* $Z(\mathbf{s}_4)$ *when using the spherical model is* 0.4862*). The strong increase in the weight of* $Z(\mathbf{s}_4)$ *occurred at the expense of the weights of* $Z(\mathbf{s}_1)$ *and* $Z(\mathbf{s}_7)$, *the second and third non-screened*

Table 4.5 OK weights, prediction and OK variance.

	$\gamma^{(1)}(\mathbf{h})$	$\gamma^{(2)}(\mathbf{h})$
λ_1	0.2769	0.0873
λ_2	0.0570	0.0978
λ_3	−0.0606	−0.0756
λ_4	0.4862	1.8861
λ_5	−0.1209	−1.2115
λ_6	0.1729	0.2119
λ_7	0.2446	0.1289
λ_8	−0.0562	−0.1250
$Z^*(\mathbf{s}_0)$	10.3254	10.1887
$\sigma^2_{OK}(\mathbf{s}_0)$	0.0981	0.0042

observations closer to the prediction point, which markedly reduce their influence in the prediction.

The degree to which screened observations lose their influence on the predicted value depends on the behavior of the semivariogram near the origin, that is, on the pattern of spatial continuity assumed. The more continuous the pattern of spatial continuity, the more pronounced the screen effect. Thus it is no surprise that using the Gaussian model and screen effect are closely related. In this example the negative weight of the screened observations $Z(\mathbf{s}_3), Z(\mathbf{s}_5)$ *and* $Z(\mathbf{s}_8)$ *has increased, especially that of* $Z(\mathbf{s}_5)$*, which is screened by the closest observation to the prediction point (its weight is* −0.1209 *in case (i) and* −1.2115 *in case (ii)). The close relation between Gaussian models and pronounced screen effect could produce negative predictions, which is unrealistic in most practical applications, where the rf is non-negative. Although in these cases the usual practice is to set the prediction to zero, this is an additional reason why semivariogram models with a parabolic behavior near the origin are not used in practice.*

As for the kriging variance, a parabolic behavior near the origin makes it lower, because this behavior is indicative of a very regular rf, which in turn yields more accurate predictions than when the rf presents a higher degree of irregularity. This can be checked in the example, as the OK variance reduces from 0.0981 *to* 0.0042 *as a consequence of using the Gaussian semivariogram instead of the spherical one.*

4.3.2.4 Effect of a change in the range

The range of the covariance (or semivariogram) model strongly affects both the kriging weights and the kriging variance. More specifically, the greater the range, the more heterogeneous the weights, because in a model with large range (the other model parameters remaining the same), the short-distance correlations are higher. But also the larger the range, the larger the spatial correlation between the values at the observed points and the prediction location and, as a consequence, predictions are more accurate.

Example 4.5 (Effect of a change in the range) *Let us consider again the data of Example 4.1. Now, the two alternative semivariogram models to predict the value of the underlying rf at* $\mathbf{s}_0$, *both of them reaching the same sill, are spherical (i) with range* 0.6, *and (ii) with range* 0.5:

$$\gamma^{(1)}(\mathbf{h}) = \begin{cases} 0.3\left(1.5\frac{|\mathbf{h}|}{0.6} - 0.5\left(\frac{|\mathbf{h}|}{0.6}\right)^3\right) & \textit{if } |\mathbf{h}| \leq 0.6 \\ 0.3 \;\; \textit{if } |\mathbf{h}| > 0.6 \end{cases},$$

$$\gamma^{(2)}(\mathbf{h}) = \begin{cases} 0.3\left(1.5\frac{|\mathbf{h}|}{0.5} - 0.5\left(\frac{|\mathbf{h}|}{0.5}\right)^3\right) & \textit{if } |\mathbf{h}| \leq 0.5 \\ 0.3 \;\; \textit{if } |\mathbf{h}| > 0.5 \end{cases}.$$

The OK weights, the OK variance associated with $Z^*(\mathbf{s}_0)$ *and the predicted value at* $\mathbf{s}_0$ *are listed in Table 4.6 for the cases (i) and (ii).*

In Table 4.6 it can be seen that in the case (ii) more weight is given to the values surrounding the prediction point than in the case (i) and, as a consequence, the kriging weights are more heterogeneous than in the case (i). However, the kriging variance is larger than in the case (i), because the correlation between the observation points and the prediction point is reduced.

Note that points more distant than the range (for example, $\mathbf{s}_2$*) are not assigned a zero weight. The reason is that points separated from the prediction location more than the range are spatially correlated with others whose distance to such prediction location is less than the range. This is called the relay effect (see details in Chilès and Delfiner 1999, p. 205).*

Finally, if we reduce the range of the spherical model to $a = 0.15$, *the weights range between* 0.10 *and* 0.135 *for all observations, with the exception of* $Z(\mathbf{s}_4)$ (0.2177), *and* $Z(\mathbf{s}_5)$ (0.0317)*. This is because the distance between the observed points and the prediction point (with the exception of* $\mathbf{s}_4$*) exceeds the range and thus they have*

Table 4.6 OK weights, prediction and OK variance.

	$\gamma^{(1)}(\mathbf{h})$	$\gamma^{(2)}(\mathbf{h})$
λ_1	0.2769	0.2827
λ_2	0.0570	0.0170
λ_3	−0.0606	−0.0314
λ_4	0.4862	0.4970
λ_5	−0.1209	−0.1264
λ_6	0.1729	0.1724
λ_7	0.2446	0.2511
λ_8	−0.0562	−0.0624
$Z^*(\mathbf{s}_0)$	10.3254	10.2966
$\sigma^2_{OK}(\mathbf{s}_0)$	0.0981	0.1198

a similar weight close to $1/8$, *and* $\mathbf{s}_5$ *is screened by* $\mathbf{s}_4$. *Of course, the OK variance increases dramatically (now it is* 0.3343*). If we reduce the range to* 0.1, *all the distances between the observation points and the prediction point exceed the range. As a consequence, the weights of the eight data values are around* $1/8$ *and the OK variance increases to* 0.3403. *These results are, obviously, similar to those obtained with the pure nugget effect semivariogram.*

4.3.2.5 Effect of anisotropy

Anisotropy has a great effect on kriging weights. In the case of anisotropy, the magnitude of the points located in the main direction of the ellipse will increase their weight. Obviously, the greater the anisotropy ratio, the greater the increase in the weights of such points.

Example 4.6 (Effect of anisotropy) *Using again the data of Example 4.1, we now compare the OK weights of the values obtained at the eight observed points when predicting the value of the rf at the non-observed point* $\mathbf{s}_0$. *For this purpose, we consider (i) the isotropic spherical model with* $m = 0.3$ *and* $a = 0.6$, $\gamma^{(1)}(\mathbf{h})$, *which we are using as a basis; and (ii) an anisotropic spherical model with the same sill and the same range as in (i) and with an anisotropy angle of* $\varphi = 135$ *degrees (*2.35 *radians) and an anisotropy ratio of* $\lambda = 3$, $\gamma^{(2)}(\mathbf{h})$. *Table 4.7 lists the weights obtained in the two cases, together with the corresponding predicted value at* $\mathbf{s}_0$ *and the OK variance associated with such a prediction.*

As expected, the weight of $\mathbf{s}_4$, *the highest weight in the isotropic case (*0.4862*), decreases to* 0.0777 *in the case of anisotropy, as though being close to* $\mathbf{s}_0$, *the vector linking both points is practically perpendicular to the principal direction of anisotropy. For the same reason, the weights of* $\mathbf{s}_6$ *and* $\mathbf{s}_7$ *also decrease though to a much lesser extent (from* 0.1729 *to* 0.0683 *and from* 0.2446 *to* 0.1977, *respectively), when we introduce anisotropy in the semivariogram model. By contrast, also as expected, the influence of* $\mathbf{s}_1$ *increases from* 0.2769 *in the isotropic case to* 0.5541

Table 4.7 OK weights, prediction and OK variance.

	$\gamma^{(1)}(\mathbf{h})$	$\gamma^{(2)}(\mathbf{h})$
λ_1	0.2769	0.5541
λ_2	0.0570	0.0106
λ_3	−0.0606	−0.0193
λ_4	0.4862	0.0777
λ_5	−0.1209	−0.0605
λ_6	0.1729	0.0683
λ_7	0.2446	0.1977
λ_8	−0.0562	0.1715
$Z^*(\mathbf{s}_0)$	10.3254	10.6665
$\sigma^2_{OK}(\mathbf{s}_0)$	0.0981	0.0452

in the anisotropic one, as it is in the main direction of anisotropy and close to the prediction point. Of note is that the point $\mathbf{s}_8$ *was a screened point in the isotropic case and, as a consequence, its weight was negative, whereas in the anisotropic case its weight is* 0.1715*, as it is in the direction of the major semi-axis of the ellipse (though distant from* $\mathbf{s}_0$*; otherwise its weight would increase substantially). Points* $\mathbf{s}_3$ *and* $\mathbf{s}_5$ *continue to be screened by* $\mathbf{s}_2$ *and* $\mathbf{s}_4$*, respectively, in the anisotropic case, though anisotropy has mitigated substantially the screen effect.*

Example 4.7 (OK with real data: Mapping carbon monoxide in Madrid, Spain (second week of January, 2008, at 10 am, 3 pm and 9 pm)) *In order to put OK at work with real data, we return to the database on carbon monoxide in Madrid, Spain, described in Section 1.3. In this example we construct OK maps of logCO* for the city of Madrid (along with the corresponding standard deviation prediction maps) at* 10 *am,* 3 *pm and* 9 *pm of the "typical" weekday (the mean of the weekdays) of the second week of January,* 2008. *In other words, using the notation employed in Example 3.4, we construct the OK maps of logCO* for the afore-mentioned three times. We have chosen the second week of January, when Christmas holidays have finished in Spain, children return to school, and commercial activity is higher than usual due to the after-Christmas sales, a week when cars are used heavily. The reason for such specific hours is that* 10 *am and* 9 *pm are the hours when traffic jams start disappearing (vehicle exhaust emissions are the main source of CO in the city) and, as a consequence of the emissions in the two previous hours, the smog is very widespread. These are also the times at which most of the population is exposed to CO.* 3 *pm, however, is time for lunch in Madrid and the traffic is not so busy.*

In light of the empirical semivariograms at the three times under study, the second-order stationary hypothesis can be assumed and OK can be used to construct the maps. In addition, empirical calculations do not reveal different behaviors of the empirical semivariogram depending on direction, and the isotropic hypothesis is also assumed. The valid semivariograms that best fit the empirical counterparts and represent the spatial dependencies of logCO at the three times under study are listed in Table 4.8. For the estimation of the semivariograms, we have divided half the largest distance among all points (slightly over* 14 *km) into eight bins, so that the number of pairs in each bin is large enough for the semivariances to be statistically acceptable.*

Table 4.8 Semivariogram models representing the spatial dependencies of logCO*.

	Model	Nugget	Partial sill	Range*
10 am	exponential	0	0.23	3669.35
3 pm	spherical	0	0.16	3241.44
9 pm	spherical	0	0.17	4540.54

*Practical range for the exponential model.

As expected, it can be seen in Table 4.8 that the zone of influence of the rf "logCO" is similar at* 10 *am and* 9 *pm (around* 4 *km) and a little bit lower at* 3 *pm (*3.2 *km). Since the first bin is centered at* 0.8 *km and the distance between the two closest monitoring stations is* 0.82 *km, it is no surprise that the estimated nugget effect is null. The sill is similar for the three hours studied, though it is a little bit larger at* 10 *am. The Pearson variation coefficient for logCO* at* 10 *am is* 1.5 *times that of at* 9 *pm and* 5 *times that of* 3 *pm.*

Using the above semivariograms in the OK equations results in the mappings shown in Figures 4.3, 4.4 and 4.5. We used `geoR` *library for geostatistical mapping. From such figures it can be deduced that the highest levels of CO are registered in the city center (commercial area), the financial area (the area around the main avenue in the city, which connects the city center with the North part of the city), and the industrial area of Madrid (in the South-East of the city, the area around the first kilometers of the highway A42 connecting Madrid and Toledo). On the contrary, the residential areas to the East and West of both the city center and the financial area are predicted to be the least polluted zones in Madrid. In fact, this is not true. The least polluted areas of Madrid are the peripheral areas, especially the North-West and North part of the city, which are close to the mountains. However, since they are far from the monitoring stations (the distance to them is much greater than the range*

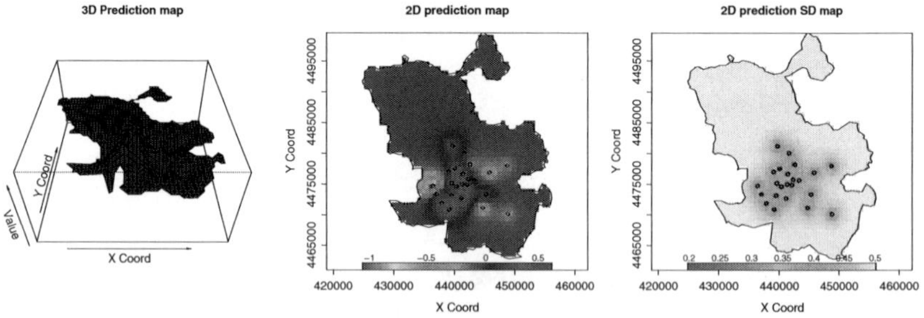

Figure 4.3 Prediction and prediction standard deviation (SD) maps of logCO: January 2008, 2nd week,* 10 *am. (See color figure in color plate section.)*

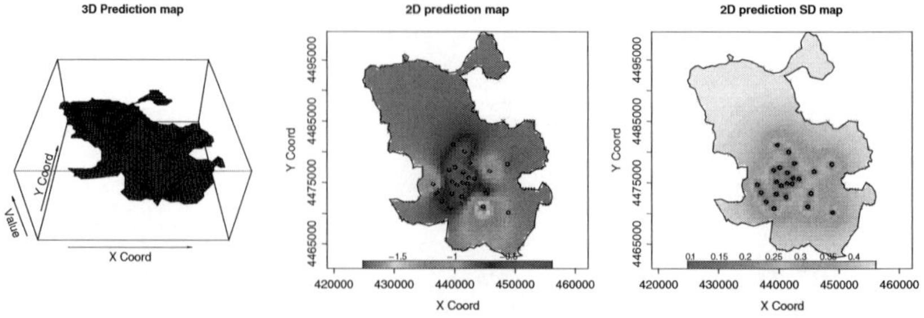

Figure 4.4 Prediction and prediction standard deviation (SD) maps of logCO: January 2008, 2nd week,* 3 *pm. (See color figure in color plate section.)*

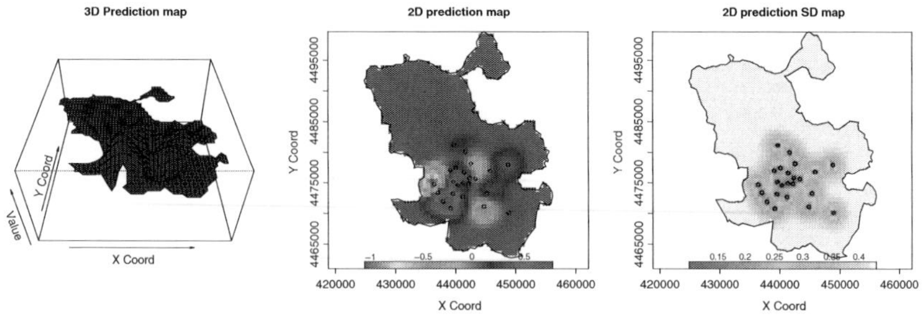

Figure 4.5 Prediction and prediction standard deviation (SD) maps of logCO: January 2008, 2nd week, 9 pm. (See color figure in color plate section.)*

of the semivariograms), the predicted value for such areas is the arithmetic mean of the values of logCO data (the estimate when spatial dependence is not present).*

As known, since kriging is an exact interpolator, the prediction standard deviation at the sites where the monitoring stations are located is zero. For sites other than those, the greater the distance to the observed sites, the larger the prediction standard deviation, which reaches its highest value (the square root of the sill plus the Lagrange multiplier) in the areas far away (a distance greater than the range) from them. For example, at 10 pm, *the prediction standard deviation at* (440000, 4475000)*, which is surrounded by most of the monitoring stations, is* 0.29. *However, when predicting at* (430000, 4490000)*, a site distant from the monitoring stations, the prediction is virtually identical to the mean of the logCO* values and the prediction standard deviation is* 0.5*, a value practically identical to the square root of the sum of the values of the sill and the Lagrange multiplier resulting from the OK equations.*

4.3.3 Point observation support and block predictor

Now, we focus on the prediction of the mean value of the rf over a block V from point observations $Z(\mathbf{s}_1), \ldots, Z(\mathbf{s}_n)$.

An initial idea to obtain the prediction of the mean value of the rf over a block when such a rf has been observed in n points of the domain of interest would be to divide the block where prediction is going to be carried out into a grid with many nodes (that is, to discretize the prediction block) and perform a point kriging at each node. Then, the mean value of the rf over the block would be calculated as the average of the point predictions made at each node of the grid. This idea is based on the additivity property of kriging: Due to the linearity of the kriging equations, the kriging prediction of a magnitude that is a linear function of the regionalized variable can be obtained from punctual kriging predictions. Thus, the average of all the kriging predictions at every point in the block coincides with the kriging prediction of the average of the block, with the constraint of using the same data for all predictions involved in the process.

However, it would be a very time-consuming task (it would be necessary to solve as many OK systems of equations as the number of nodes in the grid) and proceeding this way would not appreciate the accuracy of the block prediction. In contrast, block kriging will allow the prediction of the mean value of the rf over the block by solving only one kriging system. In addition, it will provide the kriging variance.

Under the assumption of second-order stationarity, in order to obtain the ordinary block kriging (OBK) equations, all we have to do is replace in the corresponding point OK system the value of the covariogram (or the semivariogram) for the distances between the points $\mathbf{s}_i$ and $\mathbf{s}_0$ with the average value of the covariogram $\bar{C}(\mathbf{s}_i, V)$ (or semivariogram $\bar{\gamma}(\mathbf{s}_i, V)$). We will focus later on how to approximate these average values.

In effect, as the observation support is punctual, when deriving the OBK system, the left-hand side of the first equation in (4.14) and of the equation (4.15) remains the same. However, in the right-hand side of those equations the point-to-point covariances $C(\mathbf{s}_i - \mathbf{s}_0)$ or semivariogram values $\gamma(\mathbf{s}_i - \mathbf{s}_0)$ are replaced with the point-to-block counterparts, $\bar{C}(\mathbf{s}_i, V)$ or $\bar{\gamma}(\mathbf{s}_i, V)$, respectively.

If we opt for the covariogram, then the following system of equations is obtained:

$$\begin{cases} \sum_{j=1}^{n} \lambda_j C(\mathbf{s}_i - \mathbf{s}_j) - \alpha = \bar{C}(\mathbf{s}_i, V), \quad i = 1, \ldots, n \\ \sum_{i=1}^{n} \lambda_i = 1 \end{cases}, \tag{4.18}$$

with

$$\bar{C}(\mathbf{s}_i, V) = C\left(Z(\mathbf{s}_i), Z(V)\right) = \frac{1}{|V|}\int_V C\left(Z(\mathbf{s}_i), Z(\mathbf{s}')\right)\, d\mathbf{s}' = \frac{1}{|V|}\int_V C(\mathbf{s}_i - \mathbf{s}')\, d\mathbf{s}',$$

where $\mathbf{s}'$ spans V.

The minimized prediction variance, that is, the OBK variance, is given by:

$$\sigma^2_{OBK}(V) = \bar{C}(V, V) - \sum_{i=1}^{n} \lambda_i \bar{C}(\mathbf{s}_i, V) + \alpha, \tag{4.19}$$

where $\bar{C}(V, V) = \frac{1}{|V|^2}\int_V\int_V C(\mathbf{s}' - \mathbf{s}'')d\mathbf{s}'d\mathbf{s}''$ is the average value of the covariogram in the domain V, calculated as the mean of $C(\mathbf{s}' - \mathbf{s}'')$ when $\mathbf{s}'$ and $\mathbf{s}''$ sweep independently throughout V.

If we decide to use the semivariogram, the foregoing system of equations would adopt the following form:

$$\begin{cases} \sum_{j=1}^{n} \lambda_j \gamma(\mathbf{s}_i - \mathbf{s}_j) + \alpha = \bar{\gamma}(\mathbf{s}_i, V), \quad i = 1, \ldots, n \\ \sum_{i=1}^{n} \lambda_i = 1 \end{cases}, \tag{4.20}$$

with $\bar{\gamma}(\mathbf{s}_i, V) = \frac{1}{|V|}\int_V \gamma(\mathbf{s}_i - \mathbf{s}')\, d\mathbf{s}'$.

In this case, the expression of the OBK variance is:

$$\sigma^2_{OBK}(V) = \sum_{i=1}^{n} \lambda_i \bar{\gamma}(\mathbf{s}_i, V) + \alpha - \bar{\gamma}(V, V), \tag{4.21}$$

with $\bar{\gamma}(V, V) = \frac{1}{|V|^2} \int_V \int_V \gamma(\mathbf{s}' - \mathbf{s}'') d\mathbf{s}' d\mathbf{s}''$ being the average value of the semivariogram in the domain V, calculated as the average of $\gamma(\mathbf{s}' - \mathbf{s}'')$ when $\mathbf{s}'$ and $\mathbf{s}''$ sweep independently throughout V.

Of course, if second-order stationarity is not assumed, but rather intrinsic stationarity, the only possible option would be to use the OBK system in semivariogram terms.

Finally, a pending question: How to calculate $\bar{\gamma}(\mathbf{s}_i, V)$ and $\bar{\gamma}(V, V)$, or $\bar{C}(\mathbf{s}_i, V)$ and $\bar{C}(V, V)$? These calculations are not trivial, and analytical and numerical methods are usually employed. We will briefly focus on numerical methods, since most kriging software uses them (some analytical methods can be seen in Journel and Huigbrets 1978). In what follows, it must be taken into account that both the OK system and the kriging prediction error only depend on the semivariogram model and the location of the observed points, but not on the observed values of the rf.

The general integration formula can be written as (we express it in semivariogram terms, but the same can be said for the covariogram terms):

$$\bar{\gamma}(V_1, V_2) = \sum_{i}^{m_1} \sum_{j}^{m_2} \omega_i^{(1)} \omega_j^{(2)} \gamma \left(\mathbf{s}_i'^{(1)} - \mathbf{s}_j'^{(2)} \right), \tag{4.22}$$

where $\mathbf{s}_i'^{(1)}, i = 1, \ldots, m_1$, is a set of points in V_1, $\mathbf{s}_j'^{(2)}, j = 1, \ldots, m_2$, is a set of points in V_2 and $\omega_i^{(1)}$ and $\omega_j^{(2)}$ are the weights associated with the sets of points $\{\mathbf{s}_i'^{(1)}\}$ and $\{\mathbf{s}_j'^{(2)}\}$, respectively, verifying $\sum_i^{m_1} \omega_i^{(1)} \sum_j^{m_2} \omega_j^{(2)} = 1$.

In the case of point observation support, the above expression is reduced to:

$$\bar{\gamma}(\mathbf{s}_i, V) = \sum_{j=1}^{m} \omega_j \gamma \left(\mathbf{s}_i - \mathbf{s}_j' \right), \tag{4.23}$$

where we have considered a set of m points $\{\mathbf{s}_j'\}$ in V.

Both the disposition of the points of integration and the value given to the weights depend on the numerical method used. The most popular method consists of discretizing the block into regularly spaced grid points, and give the same weight to every point.

That is, $\bar{C}(\mathbf{s}_i, V)$ and $\bar{\gamma}(\mathbf{s}_i, V)$ are approximated by the arithmetic average of the point-to-point covariance or semivariogram values, respectively, between the point $\mathbf{s}_i$ and the m points discretizing V, $Z(\mathbf{s}_1'), \ldots, Z(\mathbf{s}_m')$:

$$\bar{C}(\mathbf{s}_i, V) \simeq \frac{1}{m} \sum_{j=1}^{m} C(\mathbf{s}_i - \mathbf{s}_j'), \tag{4.24}$$

$$\bar{\gamma}(\mathbf{s}_i, V) \simeq \frac{1}{m} \sum_{j=1}^{m} \gamma(\mathbf{s}_i - \mathbf{s}_j'). \tag{4.25}$$

In the case of $\bar{\gamma}(V, V)$ or $\bar{C}(V, V)$ we proceed in the same way, that is:

$$\bar{C}(V, V) \simeq \frac{1}{m^2} \sum_{i=1}^{m} \sum_{j=1}^{m} C(\mathbf{s}'_i - \mathbf{s}'_j), \tag{4.26}$$

$$\bar{\gamma}(V, V) \simeq \frac{1}{m^2} \sum_{i=1}^{m} \sum_{j=1}^{m} \gamma(\mathbf{s}'_i - \mathbf{s}'_j). \tag{4.27}$$

Example 4.8 (Approximating $\bar{\gamma}(\mathbf{s}_i, V)$ and $\bar{\gamma}(V, V)$) *Let us consider the block $V(s)$, which has been discretized in 6 points, $\mathbf{s}'_1, \ldots, \mathbf{s}'_6$, and an observed point $\mathbf{s}_i$ (Figure 4.6). The distances between the observed point $\mathbf{s}_i$ and the six points discretizing the block, together with the corresponding point-to-point semivariogram values $\gamma(\mathbf{s}_i - \mathbf{s}'_j)$, are listed in Table 4.9. We use a spherical semivariogram model with sill $m = 1$ and range $a = 1.5$.*

According to the point-to-point semivariogram values listed in Table 4.9, the approximated point-to-block semivariogram value is:

$$\bar{\gamma}(\mathbf{s}_i, V) \simeq \frac{1}{6} \sum_{j=1}^{6} \gamma(\mathbf{s}_i - \mathbf{s}'_j) = 0.7000622.$$

The approximated value of $\bar{\gamma}(V, V)$ needed to obtain the OBK variance associated with the prediction of the mean value of the rf over the block V requires the calculation of the matrix of distances between the points discretizing the block V in order to obtain

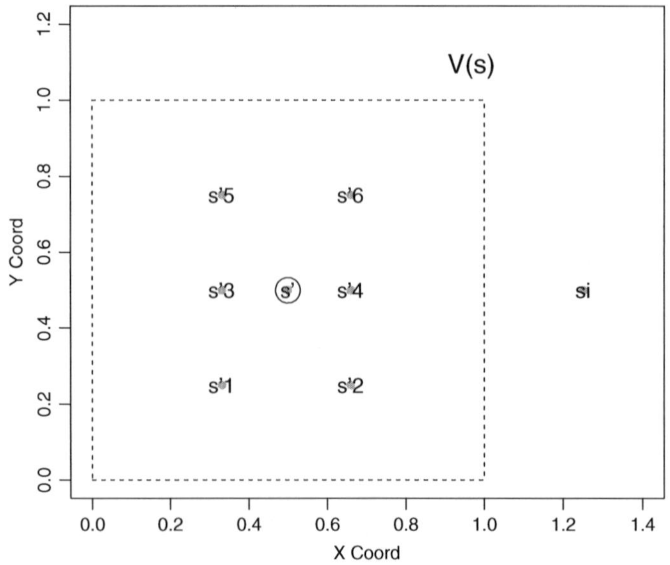

Figure 4.6 *Six points, $\mathbf{s}'_1, \ldots, \mathbf{s}'_6$, discretizing the block V, and an observed point, $\mathbf{s}_i$.*

Table 4.9 Point-to-block distances and semivariogram values.

$(\mathbf{s}_i, \mathbf{s}'_j)$	$\mathbf{h}$	$\gamma(\mathbf{h})$
$(\mathbf{s}_i, \mathbf{s}'_1)$	0.950146	0.8230689
$(\mathbf{s}_i, \mathbf{s}'_2)$	0.634647	0.5967772
$(\mathbf{s}_i, \mathbf{s}'_3)$	0.916666	0.8025545
$(\mathbf{s}_i, \mathbf{s}'_4)$	0.583333	0.5539263
$(\mathbf{s}_i, \mathbf{s}'_5)$	0.950146	0.8230689
$(\mathbf{s}_i, \mathbf{s}'_6)$	0.634647	0.5967772

Table 4.10 Distances between the points discretizing the block V.

	$\mathbf{s}'_1$	$\mathbf{s}'_1$	$\mathbf{s}'_3$	$\mathbf{s}'_4$	$\mathbf{s}'_5$	$\mathbf{s}'_6$
$\mathbf{s}'_1$	0					
$\mathbf{s}'_2$	0.3333	0				
$\mathbf{s}'_3$	0.2500	0.4167	0			
$\mathbf{s}'_4$	0.4167	0.2500	0.3333	0		
$\mathbf{s}'_5$	0.5000	0.3611	0.2500	0.4167	0	
$\mathbf{s}'_6$	0.3611	0.5000	0.4167	0.2500	0.3333	0

Table 4.11 Semivariogram values for the distances between the points discretizing the block V.

$\mathbf{h}$	$\gamma(\mathbf{h})$	Number of pairs
0	0	6
0.2500	0.2476852	8
0.3333	0.3278147	6
0.4167	0.4059807	8
0.5000	0.4814815	4
0.6009	0.5687558	4

the semivariogram values corresponding to such distances. This distances matrix and the value of the spherical semivariogram for such distances are shown in Tables 4.10 and 4.11, respectively.

According to the above semivariogram values, the approximated value of $\bar{\gamma}(V, V)$ is given by:

$$\bar{\gamma}(V, V) \simeq \frac{1}{36} \sum_{i=1}^{6} \sum_{j=1}^{6} \gamma(\mathbf{s}'_i - \mathbf{s}'_j) = 0.3165879. \tag{4.28}$$

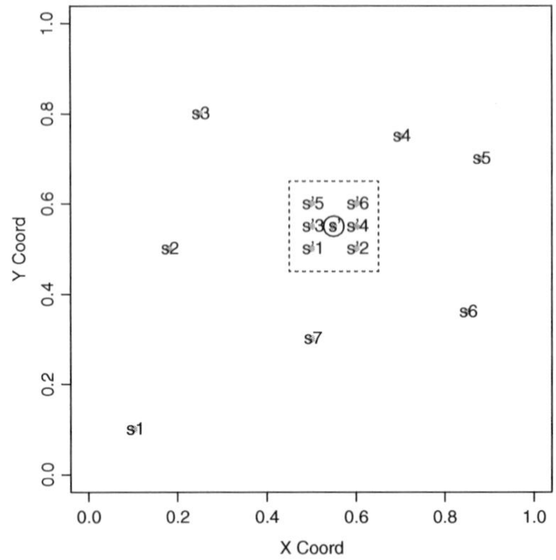

Figure 4.7 Location of seven observation points used for prediction over the block $v(\mathbf{s})$.

Other well-known methods, such as Gauss quadratures, are more accurate and do not require either discretizing the block into regularly spaced grid points or using different weights for each point.

Example 4.9 (OBK at work) *Consider the seven observation points and the prediction block v centered at $\mathbf{s}'$ depicted in Figure 4.7. The observed values of the rf at these 7 points are: $Z(\mathbf{s}_1) = 3.28, Z(\mathbf{s}_2) = 5.30, Z(\mathbf{s}_3) = 5.32, Z(\mathbf{s}_4) = 7, 18, Z(\mathbf{s}_5) = 8.08, Z(\mathbf{s}_6) = 7.05$ and $Z(\mathbf{s}_7) = 5.64$. The coordinates of such observation points, along with the coordinates of the points discretizing the prediction block, are listed in Table 4.12. Our aim is the prediction of the average value of the rf under study over the prediction block. For this purpose we use an isotropic spherical semivariogram with sill $m = 1$ and range $a = 1.5$.*

To obtain the OBK weights $\lambda_i, i = 1, \ldots, 7$ we rewrite the OBK system for point observations and block prediction in matrix terms, that is:

$$\boldsymbol{\lambda} = \boldsymbol{\Gamma}^{-1}\bar{\boldsymbol{\Gamma}}_{\mathbf{0}} \tag{4.29}$$

where:

$$\boldsymbol{\lambda} = \begin{pmatrix} \lambda_1 \\ \lambda_2 \\ \lambda_3 \\ \lambda_4 \\ \lambda_5 \\ \lambda_6 \\ \lambda_7 \\ \alpha \end{pmatrix}$$

$$\boldsymbol{\Gamma} = \begin{pmatrix} \gamma(\mathbf{s}_1-\mathbf{s}_1) & \gamma(\mathbf{s}_1-\mathbf{s}_2) & \gamma(\mathbf{s}_1-\mathbf{s}_3) & \gamma(\mathbf{s}_1-\mathbf{s}_4) & \gamma(\mathbf{s}_1-\mathbf{s}_5) & \gamma(\mathbf{s}_1-\mathbf{s}_6) & \gamma(\mathbf{s}_1-\mathbf{s}_7) & 1 \\ \gamma(\mathbf{s}_2-\mathbf{s}_1) & \gamma(\mathbf{s}_2-\mathbf{s}_2) & \gamma(\mathbf{s}_2-\mathbf{s}_3) & \gamma(\mathbf{s}_2-\mathbf{s}_4) & \gamma(\mathbf{s}_2-\mathbf{s}_5) & \gamma(\mathbf{s}_2-\mathbf{s}_6) & \gamma(\mathbf{s}_2-\mathbf{s}_7) & 1 \\ \gamma(\mathbf{s}_3-\mathbf{s}_1) & \gamma(\mathbf{s}_3-\mathbf{s}_2) & \gamma(\mathbf{s}_3-\mathbf{s}_3) & \gamma(\mathbf{s}_3-\mathbf{s}_4) & \gamma(\mathbf{s}_3-\mathbf{s}_5) & \gamma(\mathbf{s}_3-\mathbf{s}_6) & \gamma(\mathbf{s}_3-\mathbf{s}_7) & 1 \\ \gamma(\mathbf{s}_4-\mathbf{s}_1) & \gamma(\mathbf{s}_4-\mathbf{s}_2) & \gamma(\mathbf{s}_4-\mathbf{s}_3) & \gamma(\mathbf{s}_4-\mathbf{s}_4) & \gamma(\mathbf{s}_4-\mathbf{s}_5) & \gamma(\mathbf{s}_4-\mathbf{s}_6) & \gamma(\mathbf{s}_4-\mathbf{s}_1) & 1 \\ \gamma(\mathbf{s}_5-\mathbf{s}_1) & \gamma(\mathbf{s}_5-\mathbf{s}_2) & \gamma(\mathbf{s}_5-\mathbf{s}_3) & \gamma(\mathbf{s}_5-\mathbf{s}_4) & \gamma(\mathbf{s}_5-\mathbf{s}_5) & \gamma(\mathbf{s}_5-\mathbf{s}_6) & \gamma(\mathbf{s}_5-\mathbf{s}_7) & 1 \\ \gamma(\mathbf{s}_6-\mathbf{s}_1) & \gamma(\mathbf{s}_6-\mathbf{s}_2) & \gamma(\mathbf{s}_6-\mathbf{s}_3) & \gamma(\mathbf{s}_6-\mathbf{s}_4) & \gamma(\mathbf{s}_6-\mathbf{s}_5) & \gamma(\mathbf{s}_6-\mathbf{s}_6) & \gamma(\mathbf{s}_6-\mathbf{s}_7) & 1 \\ \gamma(\mathbf{s}_7-\mathbf{s}_1) & \gamma(\mathbf{s}_7-\mathbf{s}_2) & \gamma(\mathbf{s}_7-\mathbf{s}_3) & \gamma(\mathbf{s}_7-\mathbf{s}_4) & \gamma(\mathbf{s}_7-\mathbf{s}_5) & \gamma(\mathbf{s}_7-\mathbf{s}_6) & \gamma(\mathbf{s}_7-\mathbf{s}_7) & 1 \\ 1 & 1 & 1 & 1 & 1 & 1 & 1 & 0 \end{pmatrix}$$

$$\bar{\boldsymbol{\Gamma}}_{\mathbf{0}} = \begin{pmatrix} \bar{\gamma}(\mathbf{s}_1 - v(\mathbf{s}')) \\ \bar{\gamma}(\mathbf{s}_2 - v(\mathbf{s}')) \\ \bar{\gamma}(\mathbf{s}_3 - v(\mathbf{s}')) \\ \bar{\gamma}(\mathbf{s}_4 - v(\mathbf{s}')) \\ \bar{\gamma}(\mathbf{s}_5 - v(\mathbf{s}')) \\ \bar{\gamma}(\mathbf{s}_6 - v(\mathbf{s}')) \\ \bar{\gamma}(\mathbf{s}_7 - v(\mathbf{s}')) \\ 1 \end{pmatrix}$$

In order to obtain $\boldsymbol{\Gamma}$*, we need to compute the inter-point distances between the observation points (Table 4.13).*

Table 4.12 Coordinates of the observed points ($\mathbf{s}_i, i = 1, \ldots, 7$) and the points discretizing $v(\mathbf{s}')$, ($\mathbf{s}'_i, i = 1, \ldots, 6$).

	X-coord	Y-coord
$\mathbf{s}_1$	0.10	0.10
$\mathbf{s}_2$	0.18	0.50
$\mathbf{s}_3$	0.25	0.80
$\mathbf{s}_4$	0.70	0.75
$\mathbf{s}_5$	0.88	0.70
$\mathbf{s}_6$	0.85	0.36
$\mathbf{s}_7$	0.50	0.30
$\mathbf{s}'_1$	0.50	0.50
$\mathbf{s}'_2$	0.60	0.50
$\mathbf{s}'_3$	0.50	0.55
$\mathbf{s}'_4$	0.60	0.55
$\mathbf{s}'_5$	0.50	0.60
$\mathbf{s}'_6$	0.60	0.60

Table 4.13 Inter-point distances. Observation points.

	$\mathbf{s}_1$	$\mathbf{s}_2$	$\mathbf{s}_3$	$\mathbf{s}_4$	$\mathbf{s}_5$	$\mathbf{s}_6$	$\mathbf{s}_7$
$\mathbf{s}_1$	0						
$\mathbf{s}_2$	0.4079	0					
$\mathbf{s}_3$	0.7158	0.3080	0				
$\mathbf{s}_4$	0.8845	0.5769	0.4527	0			
$\mathbf{s}_5$	0.9840	0.7280	0.6378	0.1868	0		
$\mathbf{s}_6$	0.7937	0.6844	0.7440	0.4178	0.3413	0	
$\mathbf{s}_7$	0.4472	0.3773	0.5590	0.4924	0.5517	0.3551	0

Once such distances have been computed, we can build $\mathbf{\Gamma}$ *by calculating the values of an isotropic spherical semivariogram with* $m = 1$ *and* $a = 1.5$*:*

$$\Gamma(\mathbf{s}_i, \mathbf{s}_j) = \begin{pmatrix} 0 & 0.3978 & 0.6615 & 0.7820 & 0.8428 & 0.7196 & 0.4339 & 1 \\ 0.3978 & 0 & 0.3037 & 0.5485 & 0.6708 & 0.6369 & 0.3693 & 1 \\ 0.6615 & 0.3037 & 0 & 0.4390 & 0.5994 & 0.6830 & 0.5331 & 1 \\ 0.7820 & 0.5485 & 0.4390 & 0 & 0.1858 & 0.4070 & 0.4747 & 1 \\ 0.8428 & 0.6708 & 0.5994 & 0.1858 & 0 & 0.3354 & 0.5268 & 1 \\ 0.7196 & 0.6369 & 0.6830 & 0.4070 & 0.3354 & 0 & 0.3484 & 1 \\ 0.4339 & 0.3693 & 0.5331 & 0.4747 & 0.5268 & 0.3484 & 0 & 1 \\ 1 & 1 & 1 & 1 & 1 & 1 & 1 & 0 \end{pmatrix}$$

In order to obtain $\bar{\mathbf{\Gamma}}_{\mathbf{0}}$*, we have to: (i) compute the distances between every observation point and the points discretizing the prediction block (see Table 4.14); (ii) calculate the value of an isotropic spherical semivariogram with* $m = 1$ *and range* $a = 1.5$ *for such distances (Table 4.14), and (iii) compute* $\bar{\gamma}(\mathbf{s}_i - v(\mathbf{s}'))$*,* $i = 1,\ldots 7$*, as*

Table 4.14 Distances between the observation points and the points discretizing $v(\mathbf{s}')$.

	$\mathbf{s}'_1$	$\mathbf{s}'_2$	$\mathbf{s}'_3$	$\mathbf{s}'_4$	$\mathbf{s}'_5$	$\mathbf{s}'_6$
$\mathbf{s}_1$	0.5656	0.6020	0.6403	0.6403	0.6726	0.7071
$\mathbf{s}_2$	0.3200	0.3238	0.3352	0.4200	0.4229	0.4317
$\mathbf{s}_3$	0.3905	0.3535	0.3201	0.4609	0.4301	0.4031
$\mathbf{s}_4$	0.3201	0.2828	0.2500	0.2692	0.2236	0.1802
$\mathbf{s}_5$	0.4294	0.4085	0.3929	0.3440	0.3176	0.2973
$\mathbf{s}_6$	0.3769	0.3982	0.4243	0.2865	0.3140	0.3465
$\mathbf{s}_7$	0.2000	0.2500	0.3000	0.2236	0.2692	0.3162

the the average of the semivariogram values corresponding to the distances between the observation point $\mathbf{s}_i$ and the six points discretizing the prediction block.

$$\bar{\boldsymbol{\Gamma}}_{\mathbf{0}}(\mathbf{s}_i, v) = \begin{pmatrix} 0.5989 \\ 0.3673 \\ 0.3836 \\ 0.2516 \\ 0.3574 \\ 0.3506 \\ 0.2570 \\ 1 \end{pmatrix}$$

Then, the vector $\boldsymbol{\lambda}$ containing the OBK weights and the Lagrange multiplier is obtained as:

$$\boldsymbol{\lambda} = \boldsymbol{\Gamma}^{-1}\bar{\boldsymbol{\Gamma}}_{\mathbf{0}}(\mathbf{s}_i, v) = \begin{pmatrix} -0.0574 \\ 0.1394 \\ 0.1087 \\ 0.3527 \\ 0.0046 \\ 0.1094 \\ 0.3423 \\ -0.0355 \end{pmatrix}$$

so that the prediction of the average of the rf over $v(\mathbf{s}')$ is:

$$\bar{Z}_v(\mathbf{s}') = \sum_{i=1}^{7} \lambda_i Z(\mathbf{s}_i) = 6.4002.$$

Note that the OBK weights obtained above are the average of the weights obtained if we had performed the prediction at each of the points discretizing v.

The OBK variance associated with the above prediction is given by:

$$\sigma^2_{OBK}(v) = \sum_{i=1}^{n} \lambda_i \bar{\gamma}(\mathbf{s}_i, v) + \alpha - \bar{\gamma}(v, v).$$

In order to approximate the value of $\bar{\gamma}(v, v)$, we have computed the semivariogram values (Table 4.17) for the inter-point distances (Table 4.16) between the points discretizing the prediction block, so that the approximate value of $\bar{\gamma}(v, v)$ is obtained as the average of such semivariogram values.

As:

- $\bar{\gamma}(v, v) \simeq \frac{1}{36} \sum_{i,j}^{6} \gamma(\mathbf{s}'_i - \mathbf{s}'_j) = 0.0793,$

Table 4.15 Semivariogram values for distances between the observation points and the points discretizing $v(\mathbf{s}')$.

	$\mathbf{s}'_1$	$\mathbf{s}'_2$	$\mathbf{s}'_3$	$\mathbf{s}'_4$	$\mathbf{s}'_5$	$\mathbf{s}'_6$
$\mathbf{s}_1$	0.5388	0.5697	0.6014	0.6014	0.6275	0.6547
$\mathbf{s}_2$	0.3151	0.3188	0.3296	0.4090	0.4117	0.4198
$\mathbf{s}_3$	0.3816	0.3470	0.3152	0.4464	0.4183	0.3934
$\mathbf{s}_4$	0.3152	0.2794	0.2476	0.2663	0.2219	0.1794
$\mathbf{s}_5$	0.4176	0.3984	0.3839	0.3380	0.3128	0.2934
$\mathbf{s}_6$	0.3690	0.3888	0.4130	0.2830	0.3094	0.3403
$\mathbf{s}_7$	0.1988	0.2476	0.2960	0.2219	0.2663	0.3115

Table 4.16 Distances between $\mathbf{s}'_i$ and $\mathbf{s}'_j$.

	$\mathbf{s}'_1$	$\mathbf{s}'_2$	$\mathbf{s}'_3$	$\mathbf{s}'_4$	$\mathbf{s}'_5$	$\mathbf{s}'_6$
$\mathbf{s}'_1$	0					
$\mathbf{s}'_2$	0.0500	0				
$\mathbf{s}'_3$	0.1000	0.0500	0			
$\mathbf{s}'_4$	0.1000	0.1118	0.1414	0		
$\mathbf{s}'_5$	0.1118	0.1000	0.1118	0.0500	0	
$\mathbf{s}'_6$	0.1414	0.1118	0.1000	0.1000	0.0500	0

Table 4.17 Semivariogram values for distances between $\mathbf{s}'_i$ and $\mathbf{s}'_j$.

	$\mathbf{s}'_1$	$\mathbf{s}'_2$	$\mathbf{s}'_3$	$\mathbf{s}'_4$	$\mathbf{s}'_5$	$\mathbf{s}'_6$
$\mathbf{s}'_1$	0	0.0499	0.0998	0.0998	0.1115	0.1410
$\mathbf{s}'_2$	0.0499	0	0.0499	0.1115	0.0998	0.1115
$\mathbf{s}'_3$	0.0998	0.0499	0	0.1410	0.1115	0.0998
$\mathbf{s}'_4$	0.0998	0.1115	0.1410	0	0.0499	0.0998
$\mathbf{s}'_5$	0.1115	0.0998	0.1115	0.0499	0	0.0499
$\mathbf{s}'_6$	0.1410	0.1115	0.0998	0.0998	0.0499	0

- $\sum_{i=1}^{7} \lambda_i \bar{\gamma}(\mathbf{s}_i, v) = 0.2754$, *where the values of* $\bar{\gamma}(\mathbf{s}_i, v), i = 1, \ldots, 7$ *(the average of the values in the i-th row of Table 4.15) are listed in the seven first places of* $\bar{\mathbf{\Gamma}}_{\mathbf{0}}(\mathbf{s}_i, v)$, *and*
- $\alpha = -0.0355$,

we obtain

$$\sigma^2_{OBK}(v) = 0.1603.$$

4.3.3.1 Kriging of the mean

A special case of ordinary point-to-block kriging is the prediction of the unknown mean in the domain under study. This case is known as kriging of the mean. We aim to predict this unknown mean, μ, using the predictor $\mu^* = \sum_{i=1}^{n} \lambda_i Z(\mathbf{s}_i)$, $\mathbf{s}_i$, $i = 1, \ldots, n$, being the observed points. In the sequel, we assume second-order stationarity and use the covariance function.

The weights λ_i will be obtained so that the kriging prediction of the mean is unbiased and the prediction error variance is minimum. In order to comply with the unbiasedness of the kriging predictor, the sum of the weights must be one, as:

$$E[\mu^* - \mu] = E\left(\sum_{i=1}^{n} Z(\mathbf{s}_i) - \mu\right) = \sum_{i=1}^{n} \lambda_i E\left(Z(\mathbf{s}_i)\right) - \mu = \mu \sum_{i=1}^{n} \lambda_i - \mu, \tag{4.30}$$

which is zero for $\sum_{i=1}^{n} \lambda_i = 1$.

The prediction variance is then given by:

$$\begin{aligned} V\left(\mu^* - \mu\right) &= E\left(\mu^* - \mu\right)^2 = E\left(\sum_{i=1}^{n} \lambda_i Z(\mathbf{s}_i) - \mu\right)^2 \\ &= \sum_{i=1}^{n}\sum_{j=1}^{n} \lambda_i\lambda_j E\left(Z(\mathbf{s}_i)Z(\mathbf{s}_j)\right) - 2\mu \sum_{i=1}^{n} \lambda_i E\left(Z(\mathbf{s}_i)\right) + \mu^2 \\ &= \sum_{i=1}^{n}\sum_{j=1}^{n} \lambda_i\lambda_j C\left(Z(\mathbf{s}_i), Z(\mathbf{s}_j)\right) + \sum_{i=1}^{n}\sum_{j=1}^{n} \lambda_i\lambda_j \mu^2 - 2\mu^2 \sum_{i=1}^{n} \lambda_i + \mu^2, \end{aligned} \tag{4.31}$$

which under the unbiasedness condition is reduced to:

$$V\left(\mu^* - \mu\right) = \sum_{i=1}^{n}\sum_{j=1}^{n} \lambda_i\lambda_j C\left(Z(\mathbf{s}_i), Z(\mathbf{s}_j)\right) = \sum_{i=1}^{n}\sum_{j=1}^{n} \lambda_i\lambda_j C\left(\mathbf{s}_i - \mathbf{s}_j\right). \tag{4.32}$$

In order to find the weights $\lambda_i, i = 1, \ldots, n$, minimizing $V(\mu^* - \mu)$ subject to the linear constraint $\sum_{i=1}^{n} \lambda_i = 1$, we use the method of the Lagrange multipliers, the Lagrange function being:

$$\varphi(\lambda_i, \alpha) = V\left(\mu^* - \mu\right) - \alpha\left(\sum_{i=1}^{n} \lambda_i - 1\right), \tag{4.33}$$

which, by deriving it partially with respect to the weights $\lambda_i, i = 1, \ldots, n$, and the Lagrange multiplier α, and setting these partial derivatives to zero, results in the system of equations for the kriging of the mean:

$$\begin{cases} \sum_{i=1}^{n} \lambda_j C(\mathbf{s}_i - \mathbf{s}_j) - \alpha = 0, \quad \forall i = 1, \ldots, n \\ \sum_{i=1}^{n} \lambda_i = 1 \end{cases}. \tag{4.34}$$

The minimized prediction variance, or kriging of the mean variance, is obtained by setting $\sum_{i=1}^{n} \lambda_j C(\mathbf{s}_i - \mathbf{s}_j) - \alpha$ to zero in the expression of the prediction variance:

$$\sigma^2_{KM} = \sum_{i=1}^{n} \lambda_i \alpha = \alpha. \tag{4.35}$$

Now let us call again the ordinary kriging system for point observations and block prediction:

$$\begin{cases} \sum_{j=1}^{n} \lambda_j C(\mathbf{s}_i - \mathbf{s}_j) - \alpha = \bar{C}(\mathbf{s}_i, V), \quad \forall i = 1, \dots, n \\ \sum_{i=1}^{n} \lambda_i = 1 \end{cases}, \tag{4.36}$$

with $\sigma^2_{OBK}(V) = \bar{C}(V, V) - \sum_{i=1}^{n} \lambda_i \bar{C}(\mathbf{s}_i, V) + \alpha$.

If the prediction block V is large enough (that is, we can think of the region under study as a single large block for which we wish to estimate the mean of the rf) and the covariance function tends to zero with distance, then $\bar{C}(\mathbf{s}_i, V)$ and $\bar{C}(V, V)$ tend to zero and block kriging equations and block kriging variance are equivalent to the system for the kriging of the mean and the kriging of the mean variance, respectively.

However, Webster and Oliver (2001, p. 177) give some reasons against the use of the kriging of the mean approach for large regions:

(i) It is unrealistic to assume second-order stationarity throughout a large region.

(ii) The empirical covariogram (or semivariogram) is usually not well estimated for large distances.

(iii) A large number of observed points could produce kriging matrices that are too large to invert or that become unstable.

The alternative they propose is to divide the region into rectangular blocks or strata, make a kriging of the mean in each, and then compute the weighted mean of the predictions, the weights being according to the block or strata areas. However, the minimized prediction error equals the sum of minimized prediction errors in the blocks in which the region has been divided. This means considering independently the predictions in neighboring blocks, which is not the case. Details of a possible solution to this problem can be seen in Webster and Oliver (2001, pp. 177–8).

4.3.4 Block observation support and block predictor

Let us assume that the sample observations are block averages, $\{v_i,\ i = 1, \dots, n\}$, and that we wish to predict the average value of the rf at a much larger block V. In order to obtain the kriging weights, we impose the usual conditions: the kriging predictor must be unbiased and the prediction variance at the minimum.

As the sampled locations are now based on blocks, the OK equations for the case of point observation support and point predictions (which we will take as a reference) vary not only on the right-hand side, but also on the left-hand side. They do nevertheless keep the same structure.

When the underlying rf being studied is assumed to be second-order stationary, the kriging equations can be expressed either in terms of covariogram values or in terms of semivariogram values. In the first case, we arrive at the following system of equations:

$$\begin{cases} \sum_{j=1}^{n} \lambda_j \bar{C}(v_i, v_j) - \alpha = \bar{C}(v_i, V), \quad i = 1, \ldots, n \\ \sum_{i=1}^{n} \lambda_i = 1 \end{cases}, \tag{4.37}$$

where: (i) the point-to-point covariances $C(\mathbf{s}_i - \mathbf{s}_j)$ in the left-hand side of the point kriging equations have been replaced with the block-to-block covariances $\bar{C}(v_i, v_j) = \frac{1}{|v_i||v_j|} \int_{v_j} \int_{v_i} C(\mathbf{s} - \mathbf{s}')d\mathbf{s}d\mathbf{s}'$, where $\mathbf{s}$ and $\mathbf{s}'$ vary throughout v_i and v_j, respectively; and (ii) the point-to-point covariances of the right-hand term of the point kriging equations, have been substituted by the block-to-block covariances $\bar{C}(v_i, V) = \frac{1}{|v_i||V|} \int_V \int_{v_i} C(\mathbf{s} - \mathbf{s}'')d\mathbf{s}d\mathbf{s}''$, where $\mathbf{s}$ and $\mathbf{s}''$ vary throughout v_i and V, respectively, that is, by the average values of the covariogram between the points discretizing the sample blocks and those discretizing the prediction block V.

The resulting minimized prediction variance (the OBK variance) is given by:

$$\sigma^2_{OBK}(V) = \bar{C}(V, V) - \sum_{i=1}^{n} \lambda_i \bar{C}(v_i, V) + \alpha. \tag{4.38}$$

The ordinary block kriging system (4.37) and the OBK variance (4.38) associated with the prediction can be expressed in terms of the semivariogram as follows:

$$\begin{cases} \sum_{j=1}^{n} \lambda_j \bar{\gamma}(v_i, v_j) + \alpha = \bar{\gamma}(v_i, V), \quad \forall i = 1, \ldots, n \\ \sum_{i=1}^{n} \lambda_i = 1 \end{cases}, \tag{4.39}$$

where $\bar{\gamma}(v_i, v_j) = \frac{1}{|v_i||v_j|} \int_{v_j} \int_{v_i} \gamma(\mathbf{s} - \mathbf{s}')d\mathbf{s}d\mathbf{s}'$ and $\bar{\gamma}(v_i, V) = \frac{1}{|v_i||V|} \int_V \int_{v_i} \gamma(\mathbf{s} - \mathbf{s}'')d\mathbf{s}d\mathbf{s}''$, the OBK variance being:

$$\sigma^2_{OBK}(V) = \sum_{i=1}^{n} \lambda_i \bar{\gamma}(v_i, V) + \alpha - \bar{\gamma}(V, V). \tag{4.40}$$

Again, if the rf is not second-order stationary but intrinsically stationary, the only kriging system that can be used is that expressed in semivariogram terms.

Example 4.10 (Approximating $\bar{\gamma}(v_i, v_j)$) *Let us consider two blocks, $v_i(\mathbf{s})$ and $v_j(\mathbf{s}')$, which have been discretized into six points, $\mathbf{s}_1, \ldots, \mathbf{s}_6$, and $\mathbf{s}'_1, \ldots, \mathbf{s}'_6$, respectively (Figure 4.8). The distances between the six points discretizing $v_i(\mathbf{s})$ and the six points discretizing $v_j(\mathbf{s}')$ are listed in Table 4.18; the corresponding point-to-point semivariogram values, $\gamma(\mathbf{s}_i - \mathbf{s}'_j)$, are listed in Table 4.19. We use a spherical semivariogram model with sill $m = 1$ and range $a = 1.5$.*

According to the point-to-point semivariogram values listed in Table 4.19, the approximated block-to-block semivariogram value is:

$$\bar{\gamma}(v_i, v_j) \simeq \frac{1}{36} \sum_{i=1}^{6} \sum_{j=1}^{6} \gamma(\mathbf{s}_i - \mathbf{s}'_j) = 0.4223. \tag{4.41}$$

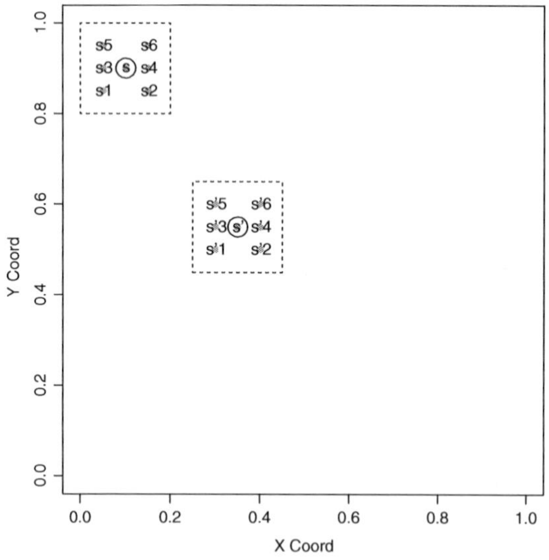

Figure 4.8 *Six points, $\mathbf{s}_1, \ldots, \mathbf{s}_6$ discretizing $v_i(\mathbf{s})$, and six points, $\mathbf{s}'_1, \ldots, \mathbf{s}'_6$ discretizing $v_j(\mathbf{s}')$.*

Table 4.18 Distances between the points discretizing $v_i(\mathbf{s})$ and $v_j(\mathbf{s}')$.

	$\mathbf{s}'_1$	$\mathbf{s}'_1$	$\mathbf{s}'_3$	$\mathbf{s}'_4$	$\mathbf{s}'_5$	$\mathbf{s}'_6$
s_1	0.4301	0.3905	0.3535	0.4949	0.4609	0.4301
s_2	0.3807	0.3354	0.2915	0.4301	0.3905	0.3535
s_3	0.4716	0.4301	0.3905	0.5315	0.4949	0.4609
s_4	0.4272	0.3807	0.3354	0.4716	0.4301	0.3905
s_5	0.5147	0.4716	0.4301	0.5700	0.5315	0.4949
s_6	0.4743	0.4272	0.3807	0.5147	0.4716	0.4301

Table 4.19 Semivariogram values for distances between the points discretizing $v_i(\mathbf{s})$ and $v_j(\mathbf{s}')$ (columns 1–6). The values for $\bar{\gamma}(\mathbf{s}_i, v_j)$ are shown is the last column. and those for $\bar{\gamma}(\mathbf{s}'_j, v_i)$ are listed in the last row.

	$\mathbf{s}'_1$	$\mathbf{s}'_1$	$\mathbf{s}'_3$	$\mathbf{s}'_4$	$\mathbf{s}'_5$	$\mathbf{s}'_6$	$\bar{\gamma}(\mathbf{s}_i, v_j)$
s_1	0.4183	0.3816	0.3470	0.4770	0.4464	0.4183	0.4148
s_2	0.3726	0.3298	0.2878	0.4183	0.3816	0.3470	0.3562
s_3	0.4561	0.4183	0.3816	0.5092	0.4770	0.4464	0.4481
s_4	0.4156	0.3726	0.3298	0.4561	0.4183	0.3816	0.3957
s_5	0.4945	0.4561	0.4183	0.5426	0.5092	0.4770	0.4829
s_6	0.4585	0.4156	0.3726	0.4945	0.4561	0.4183	0.4359
$\bar{\gamma}(\mathbf{s}'_j, v_i)$	0.5102	0.3957	0.3562	0.4829	0.4148	0.5158	

This result could have been obtained by averaging the six point-to-block semivariograms $\bar{\gamma}(\mathbf{s}_i, v_j)$ or $\bar{\gamma}(\mathbf{s}'_j, v_i)$.

4.4 Simple kriging: the special case of known mean

Sometimes the mean of the rf under study is known from previous experience or can be assumed from the nature of the phenomenon being analyzed. In such cases, it seems reasonable to take advantage of that knowledge to improve predictions.

If μ, the mean of the underlying second-order stationary rf is known, in order to simplify calculations we can define another rf $Y = \{Y(\mathbf{s}) = Z(\mathbf{s}) - \mu,\ \mathbf{s} \in D\}$ so that $E(Y(\mathbf{s})) = 0$. It is now for this new rf that we intend to make a prediction at the point $\mathbf{s}_0$ by means of the linear predictor:

$$Y^*(\mathbf{s}_0) = \sum_{i=1}^{n} \lambda_i Y(\mathbf{s}_i), \tag{4.42}$$

and once we have obtained the kriging weights, λ_i, the predicted value of $Z(\mathbf{s})$ at $\mathbf{s}_0$ is obtained as:

$$\begin{aligned} Z^*(\mathbf{s}_0) &= \mu + \sum_{i=1}^{n} \lambda_i \left(Z(\mathbf{s}_i) - \mu\right) \\ &= \sum_{i=1}^{n} \lambda_i Z\left(\mathbf{s}_i\right) + \left(1 - \sum_{i=1}^{n} \lambda_i\right)\mu, \end{aligned} \tag{4.43}$$

where the factor $k = \left(1 - \sum_{i=1}^{n} \lambda_i\right)$ is known as the weighting of the mean in simple kriging (SK).

The weights λ_i are obtained in the same way as in OK, bearing in mind that now there is no need to impose the non-bias condition, thereby arriving at the following system of n equations with n unknowns:

$$\sum_{j=1}^{n} \lambda_j C(\mathbf{s}_i - \mathbf{s}_j) = C(\mathbf{s}_i - \mathbf{s}_0), \quad i = 1, \ldots, n. \tag{4.44}$$

This system is called the SK system, the minimized prediction variance, or the SK variance, being:

$$\sigma^2_{SK}(\mathbf{s}_0) = V\left(Y^*(\mathbf{s}_0) - Y(\mathbf{s}_0)\right) = C(\mathbf{0}) - \sum_{i=1}^{n} \lambda_i C(\mathbf{s}_i - \mathbf{s}_0). \tag{4.45}$$

Note that the SK variance is less than $C(\mathbf{0}) = V(Z)$, also called the dispersion variance. Note also that in the SK case the mean of the rf does not need to be estimated and, as a consequence, the covariances involved in (4.44) can be estimated with the experimental covariogram.

Equations (4.44) and (4.45) can also be expressed in terms of the semivariogram, when the rf under study is second-order stationary, by using the relation between the covariogram and the semivariogram. Of course, prediction by SK is not an option for rf's that are not second-order stationary but intrinsically stationary.

The simple block kriging (SBK) equations can be obtained in a similar fashion to those in the previous case. If the observation support is punctual, the SBK equations are:

$$\sum_{j=1}^{n} \lambda_j C(\mathbf{s}_i - \mathbf{s}_j) = \bar{C}(\mathbf{s}_i, V), \ \forall i = 1, \ldots, n, \tag{4.46}$$

the resulting SBK variance being:

$$\sigma^2_{SBK}(V) = \bar{C}(V, V) - \sum_{i=1}^{n} \lambda_i \bar{C}(\mathbf{s}_i, V). \tag{4.47}$$

Similarly, in case of averaged observations and block predictor, the SBK system and the SBK variance are:

$$\sum_{j=1}^{n} \lambda_j \bar{C}(v_i, v_j) = \bar{C}(v_i, V), \ \forall i = 1, \ldots, n, \tag{4.48}$$

and

$$\sigma^2_{SBK}(V) = \bar{C}(V, V) - \sum_{i=1}^{n} \lambda_i \bar{C}(v_i, V), \tag{4.49}$$

respectively.

4.5 Simple kriging with an estimated mean

There is an important relation between SK, OK and kriging of the mean: The OK predictor coincides with the SK predictor where the mean of the underlying rf is replaced by the kriging of the mean predictor. As a consequence, the OK variance can be split into two terms: the SK variance and a factor representing the increase in the prediction variance as a consequence of the kriging of the mean.

Let us consider the case of point support for both observations and also predictions. Replacing μ by the kriging of the mean predictor in the SK predictor (4.42) we obtain:

$$Z^*(\mathbf{s}_0) = \sum_{i=1}^{n} Z(\mathbf{s}_i)(\lambda_i^{sk} + k\lambda_i^{km}), \tag{4.50}$$

where $k = 1 - \sum_{i=1}^{n} \lambda_i^{sk}$, λ_i^{sk} represent the SK weights and λ_i^{km} those obtained in the kriging of the mean procedure.

Let us check now that (4.50) is exactly the expression of the OK predictor. In effect, the sum of weights is equal to 1:

$$\sum_{i=1}^{n}(\lambda_i^{sk} + k\lambda_i^{km}) = \sum_{i=1}^{n} \lambda_i^{sk} + k = \sum_{i=1}^{n} \lambda_i^{sk} + \left(1 - \sum_{i=1}^{n} \lambda_i^{sk}\right) = 1. \tag{4.51}$$

In addition, it is verified that $\sum_{j=1}^{n} \lambda_j^{ok} C(\mathbf{s}_i - \mathbf{s}_j) - \alpha^{ok} = C(\mathbf{s}_i - \mathbf{s}_0)$. In effect, substituting λ_j^{ok} with $\lambda_j^{sk} + k\lambda_j^{km}$ we have:

$$\begin{aligned} \sum_{j=1}^{n} \lambda_j^{ok} C(\mathbf{s}_i - \mathbf{s}_j) &= \sum_{i=1}^{n}(\lambda_j^{sk} + k\lambda_j^{km}) C(\mathbf{s}_i - \mathbf{s}_j) \\ &= \sum_{j=1}^{n} \lambda_j^{sk} C(\mathbf{s}_i - \mathbf{s}_j) + \sum_{j=1}^{n} k\lambda_j^{km} C(\mathbf{s}_i - \mathbf{s}_j), \end{aligned} \tag{4.52}$$

where the first term of the right-hand side is $C(\mathbf{s}_i - \mathbf{s}_0)$, see (4.44), and the second is $k\alpha^{km}$, see (4.34). Thus,

$$\sum_{j=1}^{n} \lambda_j^{ok} C(\mathbf{s}_i - \mathbf{s}_j) = C(\mathbf{s}_i - \mathbf{s}_0) + k\alpha^{km}, \tag{4.53}$$

and setting $k\alpha^{km}$ to α^{ok} these weights satisfy the OK system:

$$\sum_{j=1}^{n} \lambda_j^{ok} C(\mathbf{s}_i - \mathbf{s}_j) - \alpha^{ok} = C(\mathbf{s}_i - \mathbf{s}_0). \tag{4.54}$$

As a consequence, (4.50) is nothing but the expression of the OK predictor.

Replacing λ_i^{ok} by $\lambda_i^{sk} + k\lambda_i^{km}$ in the expression of the OK variance (4.15), we obtain:

$$\begin{aligned}\sigma_{OK}^2(\mathbf{s}_0) &= \sum_{i=1}^{n} \lambda_i^{ok} C(\mathbf{s}_i - \mathbf{s}_0) + \alpha^{ok} \\ &= \sum_{i=1}^{n} (\lambda_i^{sk} + k\lambda_i^{km}) C(\mathbf{s}_i - \mathbf{s}_0) + \alpha^{ok} \\ &= \sigma_{SK}^2(\mathbf{s}_0) + k^2\sigma_{KM}^2,\end{aligned} \tag{4.55}$$

where the second term of the right-hand side of the above expression represents the increase in the prediction variance that is a consequence of having no knowledge about the value of the mean.

The relation between SK, OK and the kriging of the mean, as well as the splitting of the OK variance into the SK variance and a factor representing the increase in the prediction variance as a consequence of the kriging of the mean, can easily be generalized for the block kriging case. For example, in the case of point observation support and prediction in a block V, we have:

$$\begin{aligned}\sigma_{OBK}^2(V) &= \sum_{i=1}^{n} \lambda_i^{obk} \bar{C}(\mathbf{s}_i, V) + \alpha^{obk} \\ &= \sum_{j=1}^{n} (\lambda_i^{sbk} + k\lambda_i^{km}) \bar{C}(\mathbf{s}_i, V) + \alpha^{obk} \\ &= \sigma_{SBK}^2(V) + k^2\sigma_{KM}^2.\end{aligned} \tag{4.56}$$

4.6 Universal kriging

4.6.1 Point observation support and point predictor

Let $Z = \{Z(\mathbf{s}),\ \mathbf{s} \in D\}$ be a non-stationary rf in mean, that is, $E(Z(\mathbf{s})) = \mu(\mathbf{s})$ is a spatial function (it depends on location) that is known in geostatistical literature as *drift*. This situation already entails a difference in regard to the situation of stationarity: the classical estimator of the semivariogram used in the cases of second-order or intrinsic stationarity is no longer unbiased. Indeed,

$$\gamma(\mathbf{h}) = \frac{1}{2}E\left((Z(\mathbf{s}+\mathbf{h}) - Z(\mathbf{s}))^2\right) - \frac{1}{2}(\mu(\mathbf{s}+\mathbf{h}) - \mu(\mathbf{s}))^2. \tag{4.57}$$

This circumstance warns of the danger in operating with a classical semivariogram calculated directly using the data from the observed regionalization in a non-stationary context, which is a serious limitation when making inferences.

In this case, in order to be able to continue using the classical estimator, which is unbiased in the case of stationarity, a natural way to proceed would be, paraphrasing

Emery (2000, p. 233), to extract "something stationary" from non-stationary rf's. This could be achieved by breaking down the rf into the sum of two components: one deterministic, $\mu(\mathbf{s})$, and the other stochastic, $e = \{e(\mathbf{s}),\ \mathbf{s} \in D\}$, which can be treated as an intrinsically stationary rf (its mean is constant, in fact we know it is zero, but not its variance):

$$Z(\mathbf{s}) = \mu(\mathbf{s}) + e(\mathbf{s}). \tag{4.58}$$

Let us assume that the drift, albeit unknown, can be expressed locally by means of an expression such as:

$$\mu(\mathbf{s}) = \sum_{h=1}^{p} a_h f_h(\mathbf{s}), \tag{4.59}$$

where $\{f_h(\mathbf{s}),\ h = 1, \ldots, p\}$ are p known functions, a_h are constant coefficients, but obtained with a moving neighborhood that can differ from one neighborhood to another, and p is the number of terms used in the approximation. This is the reason why in universal kriging (UK) not all available observations are used in the prediction at the unobserved location $\mathbf{s}_0$, but only a number of them $(n(\mathbf{s}_0) < n)$ surrounding $\mathbf{s}_0$.

The expression (4.59) is only considered valid locally, because if we could assume that the drift accepted an expression of that nature globally, the most suitable way to proceed would be to krige the residuals (which, as we will see later, consists of estimating the global drift and performing OK on the resulting residuals).

In this case, as in the previous ones (OK, SK), the equations from which we obtain the weights λ_i are obtained by imposing on the prediction error the conditions of zero expectation and minimum variance. By imposing the condition of an expected value of zero on the prediction error, we achieve that:

$$\sum_{i=1}^{n(\mathbf{s}_0)} \lambda_i \mu(\mathbf{s}_i) = \mu(\mathbf{s}_0), \tag{4.60}$$

which, taking into account the expression of the drift proposed initially, can be re-written as:

$$\sum_{i=1}^{n(\mathbf{s}_0)} \lambda_i \sum_{h=1}^{p} a_h f_h(\mathbf{s}_i) = \sum_{h=1}^{p} a_h f_h(\mathbf{s}_0), \tag{4.61}$$

or as:

$$\sum_{h=1}^{p} a_h \sum_{i=1}^{n(\mathbf{s}_0)} \lambda_i f_h(\mathbf{s}_i) = \sum_{h=1}^{p} a_h f_h(\mathbf{s}_0). \tag{4.62}$$

This equation is verified providing:

$$\sum_{i=1}^{n(\mathbf{s}_0)} \lambda_i f_h(\mathbf{s}_i) = f_h(\mathbf{s}_0), \forall h = 1, \ldots, p, \tag{4.63}$$

which are the p unbiasedness conditions, equivalent to the only condition of unbiasedness, $\sum_{i=1}^{n(\mathbf{s}_0)} \lambda_i = 1$, in the OK procedure. Note that each of these equations or conditions has the effect of eliminating an unknown coefficient for the drift. It is for this reason that they are also known as "elimination conditions" in the geostatistical literature.

The prediction variance, bearing in mind the expression (4.58), can be expressed as:

$$\begin{aligned} V\left(Z^*(\mathbf{s}_0) - Z(\mathbf{s}_0)\right) &= E\left(\left(\sum_{i=1}^{n} \lambda_i e(\mathbf{s}_i) - e(\mathbf{s}_0)\right)^2\right) \\ &= \sum_{i=1}^{n}\sum_{j=1}^{n} \lambda_i \lambda_j E\left(\left(e(\mathbf{s}_i) - e(\mathbf{s}_0)\right)\left(e(\mathbf{s}_j) - e(\mathbf{s}_0)\right)\right) . \end{aligned} \tag{4.64}$$

This expression resembles that used to obtain the OK equations, but with the residuals $e(\mathbf{s}_i)$ instead of the variables $Z(\mathbf{s}_i)$. Therefore, proceeding in similar fashion, we obtain the expression of the prediction variance in terms of the semivariogram of the residuals:

$$V\left(Z^*(\mathbf{s}_0) - Z(\mathbf{s}_0)\right) = 2\sum_{i=1}^{n} \lambda_i \gamma_e(\mathbf{s}_i - \mathbf{s}_0) - \sum_{i=1}^{n}\sum_{j=1}^{n} \lambda_i \lambda_j \gamma_e(\mathbf{s}_i - \mathbf{s}_j). \tag{4.65}$$

Hence, this is the expression that must be minimized under the conditions of unbiasedness mentioned previously. In order to do so, the following Lagrange function is constructed:

$$\varphi(\lambda_i, \alpha_h) = \frac{1}{2} V\left(Z^*(\mathbf{s}_0) - Z(\mathbf{s}_0)\right) - \sum_{h=1}^{p} \alpha_h \left(\sum_{i}^{n(\mathbf{s}_0)} \lambda_i f_h(\mathbf{s}_i) - f_h(\mathbf{s}_0)\right), \tag{4.66}$$

which, by deriving it partially with respect to λ_i, $i = 1, \ldots, n(\mathbf{s}_0)$, and α_h, $h = 1, \ldots, p$, and setting these partial derivatives to zero, provides the following system of UK equations:

$$\begin{cases} \sum_{j=1}^{n(\mathbf{s}_0)} \lambda_j \gamma_e(\mathbf{s}_i - \mathbf{s}_j) + \sum_{h=1}^{p} \alpha_h f_h(\mathbf{s}_i) = \gamma_e(\mathbf{s}_i - \mathbf{s}_0), \quad i = 1, \ldots, n(\mathbf{s}_0) \\ \sum_{i=1}^{n(\mathbf{s}_0)} \lambda_i f_h(\mathbf{s}_i) = f_h(\mathbf{s}_0), \quad h = 1, \ldots, p. \end{cases} \tag{4.67}$$

Note that in this case, $i = 1, \ldots, n(\mathbf{s}_0)$, instead of $i = 1, \ldots, n$. This is due, as noted previously, to the drift expression only being valid locally. As a result, $n(\mathbf{s}_0)$ represents the number of observations in an area of $\mathbf{s}_0$ ($n(\mathbf{s}_0) \leq n$).

The existence and uniqueness of the solution of the UK equations require the functions f_h to be linearly independent in the observed regionalization, that is,

$$\sum_{h=1}^{p} v_h f_h(\mathbf{s}_i) = 0,\ i = 1, \cdots, n(\mathbf{s}_0) \iff v_h = 0,\ h = 1, 2, \cdots, p, \tag{4.68}$$

a condition that is frequently met as $n(\mathbf{s}_0)$ is normally considerably larger than p.[1]

The minimized prediction variance, or UK variance, is as follows:

$$\sigma^2_{UK}(\mathbf{s}_0) = \sum_{i=1}^{n(\mathbf{s}_0)} \lambda_i \gamma_e(\mathbf{s}_i - \mathbf{s}_0) + \sum_{h=1}^{p} \alpha_h f_h(\mathbf{s}_0), \tag{4.69}$$

which is obtained by merely substituting the expression $\sum_{j=1}^{n(\mathbf{s}_0)} \lambda_j \gamma_e(\mathbf{s}_i - \mathbf{s}_j)$ by $\gamma_e(\mathbf{s}_i - \mathbf{s}_0) - \sum_{h=1}^{p} \alpha_h f_h(\mathbf{s}_i)$ in the prediction variance equation, (4.65), as can be deduced from Equation (4.67).

We can see that UK variance does not depend on the specific values of the observed regionalization, as the coefficients a_h (which can be different for each area, resulting in the prediction with local kriging not being continuous) are not even expressly calculated. In UK, what we need to know is the structure of the drift that is, the values $f_h(\mathbf{s}_i)$, not the coefficients a_h. For this reason, we only need the structure of the drift and the semivariogram $\gamma_e(\mathbf{h})$ to calculate the UK variance. We can also see how UK provides a prediction of $Z(\mathbf{s}_0)$ without having to estimate $\mu(\mathbf{s})$. The drift is estimated implicitly.

However, now that we have discussed the main feature of this method, we must highlight that UK shows, among others, the following practical problem: the semivariogram that appears in UK equations is that of $e = \{e(\mathbf{s}) = Z(\mathbf{s}) - \mu(\mathbf{s}),\ \mathbf{s} \in D\}$, which is unknown and cannot be estimated directly using the regionalization available due to not having the value of the drift at each point, meaning we do not have the value of $e(\mathbf{s}_i)$ either, hence making it impossible to construct the semivariogram $\gamma_e(\mathbf{h})$. There are several possible solutions to this problem, of which we highlight the following:

(i) To assume that $\gamma_Z \approx \gamma_e$ in a small prediction environment, as in this case, in view of the relationship between the respective semivariograms,

$$\begin{aligned}\gamma_Z(\mathbf{h}) &= \frac{1}{2}E\left((Z(\mathbf{s}+\mathbf{h}) - Z(\mathbf{s}))^2\right) \\ &= \frac{1}{2}(\mu(\mathbf{s}+\mathbf{h}) - \mu(\mathbf{s}))^2 + \frac{1}{2}E\left((e(\mathbf{s}+\mathbf{h}) - e(\mathbf{s}))^2\right) \\ &= \frac{1}{2}(\mu(\mathbf{s}+\mathbf{h}) - \mu(\mathbf{s}))^2 + \gamma_e(\mathbf{h}),\end{aligned} \tag{4.70}$$

[1] Regardless of whether $n(\mathbf{s}_0) \geq p$, there are degenerate configurations (collinear points in a two-dimensional space or coplanar points in a three-dimensional space) in which the system of UK equations becomes singular.

the first term of the right-hand side of (i) will be small, as $\mu(\mathbf{s})$ and $\mu(\mathbf{s}+\mathbf{h})$ will be similar. Under this assumption, therefore, we will replace γ_e by γ_Z in the UK equations, obtaining the weights λ_i necessary to predict $Z(\mathbf{s}_0)$ by means of $Z^*(\mathbf{s}_0) = \sum_{i=1}^{n(\mathbf{s}_0)} \lambda_i Z(\mathbf{s}_i)$.

(ii) To proceed with the so-called *iterated universal kriging* procedure, which consists, as in the previous case, of applying the UK equations to the semivariogram of the data (it is assumed that $\gamma_Z \approx \gamma_e$). We thus obtain initial weights to estimate the drift. By eliminating the estimated drift from the observations we obtain the residuals:

$$R(\mathbf{s}_i) = Z(\mathbf{s}_i) - \mu^*(\mathbf{s}_i). \tag{4.71}$$

Using these residuals as a basis, we calculate their empirical semivariogram and fit a valid semivariogram. Using this valid semivariogram in the UK equations we obtain new weights to estimate the drift. At this point, we would repeat the process until the solutions do not vary substantially.

(iii) Another possibility that is certainly used in practice is to assume *a priori* the form of semivariogram γ_e, to solve the UK equations. These equations include the following: they are set equal to the variance obtained from the semivariogram γ_e (that is, to its sill) thereby obtaining an estimate of the drift at each location.[2] The residuals are calculated by subtracting the estimates of the drift from the corresponding observations,

$$R(\mathbf{s}_i) = Z(\mathbf{s}_i) - \mu^*(\mathbf{s}_i), \tag{4.72}$$

and the empirical semivariogram of the residuals is constructed and later compared to that which was assumed initially: if their structures coincide, it is because the semivariogram model assumed *a priori* was correct; if they differ significantly, we will have to go back to the beginning and try another γ_e. We proceed in this way until the empirical semivariogram of the residuals does not differ substantially from the theoretical semivariogram γ_e assumed initially. Once we have reached this point, it can be accepted that γ_e suitably represents the stationary part of the rf under study, resulting in the solution of the UK equations that incorporate γ_e, considered suitable to model the stationary part of the rf, providing the weights λ_i necessary to determine $Z^*(\mathbf{s}_0)$.

We can see how the first of the alternatives may be subject to certain theoretical objections, while the other two are not easy to implement in practice. For this reason, in Sections 4.7 and 4.8 we present the algorithms for other kriging procedures that

[2] This procedure estimates the drift at that point because by placing the sill in the term on the right of the system, instead of the value of the semivariogram at the distance between that point and the observed locations, we are taking for granted that the point is beyond the range or scope.

overcome some of the limitations of UK: both direct and iterative residual kriging and median-polish kriging. All three perform an "explicit" estimation of the drift (which is considered to be a large-scale variation).[3] Another way of proceeding consists of proposing that the k-order increments of the rf under study are stationary (the intrinsic model of order k), but this method has not been used a great deal due to its theoretical and practical limitations (details of this procedure can be found in Samper and Carrera 1996, Chapter 10; Emery 2000, pp. 256–7; Cressie 1993, pp. 299–309 and Chilès and Delfiner 1999, Chapter 4).

4.6.2 Point observation support and block predictor

As in the case of OBK, if the observations are based on points and the prediction is made over a block, the point-to-point semivariograms used in the right-hand side of the first set of equations of system (4.67) are replaced with point-to-block semivariograms, so that the universal block kriging (UBK) system can be written as follows:

$$\begin{cases} \sum_{j=1}^{n(V)} \lambda_j \gamma_e(\mathbf{s}_i - \mathbf{s}_j) + \sum_{h=1}^{p} \alpha_h f_h(\mathbf{s}_i) = \bar{\gamma}_e(\mathbf{s}_i, V), \quad i = 1, \ldots, n(V) \\ \sum_{i=1}^{n(V)} \lambda_i f_h(\mathbf{s}_i) = \bar{f}_h(V), \quad h = 1, \ldots, p \,, \end{cases} \tag{4.73}$$

where $n(V)$ refers to the local neighborhood and $\bar{f}_h(V) = \frac{1}{|V|}\int_V f_h(\mathbf{s})d\mathbf{s}, \ h = 1, \ldots, p$, is the average value of the functions f_h over the block V.

The resulting minimized prediction variance (UBK variance with point observation support) is given by:

$$\sigma^2_{UBK}(V) = \sum_{i=1}^{n(V)} \lambda_i \bar{\gamma}_e(\mathbf{s}_i, V) + \sum_{h=1}^{p} \alpha_h \bar{f}_h(V). \tag{4.74}$$

4.6.3 Block observation support and block predictor

When both the observations and the prediction are based on blocks, point-to-point semivariograms must be replaced by block-to-block semivariograms. This way, the UK system based on point observations and point predictor is transformed to give:

$$\begin{cases} \sum_{j=1}^{n(V)} \lambda_j \bar{\gamma}_e(v_i, v_j) + \sum_{h=1}^{p} \alpha_h \bar{f}_h(v_i) = \bar{\gamma}_e(v_i, V), \quad i = 1, \ldots, n(V) \\ \sum_{i=1}^{n(V)} \lambda_i \bar{f}_h(v_i) = \bar{f}_h(V), \quad h = 1, \ldots, p \,, \end{cases} \tag{4.75}$$

$\bar{f}_h(v_i) = \frac{1}{|v_i|}\int_{v_i} f_h(\mathbf{s})d\mathbf{s}$ being the average value of the functions f_h on the block v_i, $i = 1, \ldots, n(V)$.

[3] Strictly speaking, UK does not require the drift to only be valid locally.

The minimized prediction variance for block observation support adopts the expression:

$$\sigma^2_{UBK}(V) = \sum_{i=1}^{n(V)} \lambda_i \bar{\gamma}_e(v_i, V) + \sum_{h=1}^{p} \alpha_h \bar{f}_h(V). \tag{4.76}$$

4.6.4 Kriging and exact interpolation

If the prediction point coincides with a point where the rf has been observed, the OK system is, in matrix terms:[4]

$$\begin{pmatrix} \gamma(\mathbf{s}_0 - \mathbf{s}_0) & \gamma(\mathbf{s}_0 - \mathbf{s}_1) & \cdots & \gamma(\mathbf{s}_0 - \mathbf{s}_n) & 1 \\ \gamma(\mathbf{s}_1 - \mathbf{s}_0) & \gamma(\mathbf{s}_1 - \mathbf{s}_1) & \cdots & \gamma(\mathbf{s}_1 - \mathbf{s}_n) & 1 \\ \vdots & \vdots & \vdots & \vdots & \vdots \\ \gamma(\mathbf{s}_n - \mathbf{s}_0) & \gamma(\mathbf{s}_n - \mathbf{s}_1) & \cdots & \gamma(\mathbf{s}_n - \mathbf{s}_n) & 1 \\ 1 & 1 & 1 & 1 & 0 \end{pmatrix} \begin{pmatrix} \lambda_0 \\ \lambda_1 \\ \vdots \\ \lambda_n \\ 1 \end{pmatrix} = \begin{pmatrix} \gamma(\mathbf{s}_0 - \mathbf{s}_0) \\ \gamma(\mathbf{s}_0 - \mathbf{s}_1) \\ \vdots \\ \gamma(\mathbf{s}_0 - \mathbf{s}_n) \\ 1 \end{pmatrix}$$

and the right-hand side of the OK system coincides with one column of the matrix in the left-hand side of the system (in this case the first column). As a consequence, the solution of the OK system is $\lambda_0 = 1$ and all the other weights and the Lagrange multiplier equal to zero, so that $Z^*(\mathbf{s}_0) = Z(\mathbf{s}_0)$.

This is the only solution, because of the non-singularity of the matrix in the left-hand side of the system. It can easily be seen that, in this case, $\sigma^2_{OK} = \gamma(\mathbf{s}_0 - \mathbf{s}_0)$.

The same can be said for UK equations: the UK predictor always coincides with the observations of the rf at the sampled points, the UK variance associated with these points being 0.

In the presence of a nugget effect, kriging continues to be an exact interpolator, but in this case a discontinuity appears in the prediction at each observed point.

4.7 Residual kriging

According to Samper and Carrera (1996, p. 167), kriging the residuals can be considered an alternative to UK and its ambiguities when choosing the drift (that, in addition, is not defined explicitly) and estimating the semivariogram. The basic hypothesis of residual kriging is to assume the drift to be known and perform an OK on the residuals. In practice the drift is estimated by least squares methods and OK is performed on $R(\mathbf{s}) = Z(\mathbf{s}) - \mu^*(\mathbf{s})$. Obviously, this procedure is not exempt from problems: for example, the estimation of the drift is translated into a bias when estimating the semivariogram.

More specifically, in this section we present several versions of the alternative kriging of the residuals that overcome some of the limitations of UK. Following Samper and Carrera (1996, Ch. 9), we begin presenting the direct residual kriging, where the

[4] We use the OK system in semivariogram terms, but we could have used it in covariogram terms.

structure of the errors does not affect the estimation of the drift. Then, we describe the iterative residual kriging, which takes into account the interaction between the estimation of both the drift and the semivariogram of the residuals. Finally, we outline a modified version of the iterative residual kriging. For the sake of simplicity, we focus on the case of point observation support and point prediction.

4.7.1 Direct residual kriging

The so-called direct residual kriging procedure, proposed by Volpi and Gambolati (1978) and Gambolati and Volpi (1979), is the OK of the residuals resulting from an OLS estimation of the drift. More than a solution to estimate γ_e, it is another way of proceeding from the beginning that considers the order of the drift to be known. The steps to follow in order to perform direct residual kriging are as follows:

(i) The drift (whose shape is assumed to be known on the basis of existing knowledge of the phenomenon being studied) is estimated by OLS.

(ii) The residuals $R(\mathbf{s}_i) = Z(\mathbf{s}_i) - \mu^*(\mathbf{s}_i)$ are calculated.

(iii) The empirical semivariogram of the residuals is constructed and a valid model is fitted to it.

(iv) OK is performed on the residuals obtained, achieving a prediction of the residual at the desired point $\mathbf{s}_0$ of the domain under study, $R^*(\mathbf{s}_0)$.

(v) The estimate of the drift at that point is incorporated into (ii), obtaining:

$$Z^*(\mathbf{s}_0) = R^*(\mathbf{s}_0) + \mu^*(\mathbf{s}_0). \tag{4.77}$$

The prediction variance is that of the OK estimator given by the expression (4.17).

However, there is an important internal contradiction in this procedure: While OLS assumes the residuals are independent, we are also assuming they are spatially correlated through a given semivariogram. The drift could have been estimated using *splines* or a moving average, but as was the case with OLS, neither is an optimum estimator of the drift. This is the price that must be paid for not knowing the semivariogram of the rf under study.

In addition to this, it has to be taken into account that drift estimation errors result in bias when estimating the semivariogram: even if the optimum estimator of the drift is used, the theoretical semivariogram of residuals is always lower than the semivariogram of real errors. Therefore, the empirical semivariogram obtained tends to underestimate real variability. Furthermore, it is not generally stationary (it not only depends on the distance h between locations, but also on the locations themselves) and displays an excessively short range that can lead to the unjustified belief that the regionalization observed is due to the aggregation of the drift and a nugget effect.

Wackernagel (2003, pp. 85–6) proposes a similar procedure for the second-order stationary case combining a kriging of the mean and an OK of the residual. This procedure is named *kriging of the residual* and assumes the mean is uncorrelated with the residual and that kriging weights sum up to zero (instead to one), which has the effect of removing the mean. The resulting OK weights are composed of the weights of the kriging of the mean and the kriging of the residual. The same can be said for the Lagrange multiplier and kriging variances.

4.7.2 Iterative residual kriging

This method, elaborated by Binsariti (1980), Fennessy and Neuman (1982) and Neumann and Jaccbscn (1984), consists of adjusting the drift by OLS, calculating the semivariogram of residuals, repeating these two steps iteratively and finally performing OK on the residuals. The difference in regard to direct residual kriging is that the drift is adjusted in two stages: in the first stage, the degree of the polynomial is identified such that the semivariogram of the residuals is isotropic and stationary. The second stage proceeds with GLS estimation with the covariance matrix of the residuals. This overcomes the inconsistency inherent in direct residual kriging, but not the considerations referring to the semivariogram analysis of the residuals.

The algorithm is as follows (we follow Díaz-Viera 2002, p. 50):

(i) The drift is assumed to be of the order $p = 1$.

(ii) OLS is used to estimate the drift.

(iii) The residuals $R(\mathbf{s}_i) = Z(\mathbf{s}_i) - \mu^*(\mathbf{s}_i)$ are calculated.

(iv) The empirical semivariogram of the residuals is constructed.

(v) We check if the theoretical semivariogram that approximates the empirical semivariogram obtained is stationary. If it is stationary, we proceed with step (vi); if not, the previous steps are repeated, increasing the order of the drift model by one.

(vi) The covariance matrix of the residuals,

$$\left(V_R\right)_{ij} = \sigma_R^2 - \gamma_R(\mathbf{s}_i - \mathbf{s}_j),$$

is calculated.

(vii) The drift is estimated using GLS, obtaining $\mu^*(\mathbf{s}_i)$.

(viii) The residuals are calculated: $R(\mathbf{s}_i) = Z(\mathbf{s}_i) - \mu^*(\mathbf{s}_i)$.

(ix) The empirical semivariogram of the residuals is constructed.

(x) We check that both the empirical semivariogram and also the drift $\mu^*(\mathbf{s}_i)$ obtained in step (vii) record "stable values." If this is the case, we proceed with step (xi); otherwise, we return to step (vi).

(xi) Using the empirical semivariogram of the residuals as a basis, a valid semivariogram model is selected. This valid semivariogram is used to perform OK on the residuals, obtaining a prediction of the residual at the desired point $\mathbf{s}_0$ in the domain under study, $R^*(\mathbf{s}_0)$.

(xii) The predicted value of the drift at that point is incorporated into this prediction, obtaining:

$$Z^*(\mathbf{s}_0) = R^*(\mathbf{s}_0) + \mu^*(\mathbf{s}_0).$$

4.7.3 Modified iterative residual kriging

As seen above, the stop criterion used in iterative residual kriging is based on the comparison of the values of the drift and the empirical semivariogram in two successive iterations. However, the classical empirical semivariogram has a poor behavior when the number of observations is not large enough, which leads to an undesirable variability for large lag distances; in addition, it is very sensitive to outliers. This is the reason why Samper (1986) states that a stop criterion based on the empirical semivariogram of the residuals cannot be the best stop criterion.

Samper (1986) developed a modification of the iterative residual kriging which only varies in the second phase. More specifically, instead of obtaining the covariance matrix of the residuals, the proposal of Samper (1986) is to estimate $\left(V_R\right)_{ij}$ directly from the residuals maximizing the likelihood of cross-validation errors (details can be seen in Carrera and Samper 1989, pp. 132–43). The stop criterion is then based on the convergence of the parameters of the drift and the semivariogram, rather than on the convergence of the empirical semivariogram of the residuals.

4.8 Median-Polish kriging

The above procedures fit parametric models to the drift to detrend the data before attempting the analysis of the spatial correlation structure existing in the residuals. A non-parametric approach to the detrending of the data can also be performed. This non-parametric approach is specially useful when the drift is difficult to model in a functional way. The most popular of these non-parametric detrending procedures is median-polish. Once median-polish has been used to estimate the drift,[5] a valid semivariogram model is fitted to the empirical semivariogram of the residuals and OK is performed on them. Finally, prediction at a non-observed point is obtained by adding both the estimation of the drift at such a point and the corresponding OK residual: $\tilde{Z}(\mathbf{s}_0) = \tilde{\mu}(\mathbf{s}_0) + \tilde{R}(\mathbf{s}_0)$. This procedure is called median-polish kriging and the symbol "$\sim$" indicates that the estimation of $\mu(\mathbf{s}_0)$ and the prediction of $R(\mathbf{s}_0)$ and $Z(\mathbf{s}_0)$ have been obtained by using this procedure.

Typically, median-polish is used for lattice data, but it can be used with non-lattice data by overlaying a grid on the observations.

With two or more dimensions, it is natural to assume that the drift, $\mu(\mathbf{s})$, can be broken down into the sum of its directional components and a constant value.

[5] Cressie (1999) does not use the term "drift" but "trend." However, we use the term "drift" because, as stated in Chilès and Delfiner (1999, p. 57), a drift can be understood as a space-varying mean and represents a trend in the data.

Normally, as $\mathbf{s} \in D \subset \mathbb{R}^2$, we can decompose $\mu(\mathbf{s})$ as follows:

$$\mu(\mathbf{s}) = a + c(x) + r(y), \quad \mathbf{s} = (x, y)' \in \mathbb{R}^2 . \tag{4.78}$$

Furthermore, if the locations $\{\mathbf{s}_i, i = 1,..., n\}$ are on the nodes of a $p \times q$ grid mesh (if the data are unevenly distributed, they are assigned to the node that is nearest to an overlapping imaginary grid mesh), $\{ (x_l, y_k)', k = 1, \ldots, p; l = 1, \ldots, q \}$, the notation can be simplified by writing $\mathbf{s}_i = (x_l, y_k)'$, which implies that:

$$\mu(\mathbf{s}_i) = a + r_k + c_l, \tag{4.79}$$

r_k indicating the row effect, c_l the column effect and a the overall effect.

In $\mathbb{R}^2$, the median-polish algorithm, proposed by Tukey (1977), permits the non-parametric estimation of the additive effects of (4.79). This algorithm provides the estimation of the overall effect, $\tilde{a}$, the row effects $\{\tilde{r}_k, k = 1,..., p\}$ and the column effects $\{\tilde{c}_l, l = 1,..., q\}$ from the original data $\{z_{k,l}, k = 1, \ldots, p; l = 1, \ldots, q\}$ expressed in a matrix of dimension $p \times q$.

The skeleton of the algorithm is as follows:

Using a matrix $p \times q$ with original observations as a basis, another matrix is constructed with dimension $(p + 1) \times (q + 1)$ which will contain the row effects in the p first places of the last column, the column effects in the q first places of the last row, the overall effect in the place $(p + 1) \times (q + 1)$ and the values of the residuals of the smoothing process in the $p \times q$ first places.

$$\begin{pmatrix} z_{11}^{(\infty)} & z_{12}^{(\infty)} & \cdots & z_{1q}^{(\infty)} & \tilde{r}_1 \\ z_{21}^{(\infty)} & z_{22}^{(\infty)} & \cdots & z_{2q}^{(\infty)} & \tilde{r}_2 \\ \cdots & \cdots & \cdots & \cdots & \cdots \\ z_{p1}^{(\infty)} & z_{p2}^{(\infty)} & \cdots & z_{pq}^{(\infty)} & \tilde{r}_p \\ \tilde{c}_1 & \tilde{c}_2 & \cdots & \tilde{c}_q & \tilde{a} \end{pmatrix}. \tag{4.80}$$

(i) We start with a matrix of dimension $(p + 1) \times (q + 1)$ which contains the $p \times q$ original data and the last row and column that are null. That is, in order to begin the algorithm, the following are assumed as initial values:

$$z_{k,l}^{(0)} = \begin{cases} z_{k,l}, & k = 1, \ldots, p; \ l = 1, \ldots, q \\ 0, & \text{otherwise.} \end{cases} \tag{4.81}$$

At this point, the iterative stage commences.

(ii) For the odd iterations $i = 1, 3, 5,...$

$$z_{kl}^{(i)} = z_{kl}^{(i-1)} - Med\left\{ z_{kl}^{(i-1)}; l = 1, \ldots, q \right\} \quad \text{if } k = 1, \ldots, p + 1; l = 1, \ldots, q$$

$$z_{k,q+1}^{(i)} = z_{k,q+1}^{(i-1)} + Med\left\{ z_{kl}^{(i-1)}; l = 1, \ldots, q \right\} \quad \text{if } k = 1, \ldots, p + 1 ,$$

and for the even iterations $i = 2, 4, 6, \ldots$

$$z_{kl}^{(i)} = z_{kl}^{(i-1)} - Med\left\{z_{kl}^{(i-1)}; k = 1, \ldots, p\right\} \quad \text{if } k = 1, \ldots, p; l = 1, \ldots, q+1$$

$$z_{p+1,l}^{(i)} = z_{p+1,l}^{(i-1)} + Med\left\{z_{kl}^{(i-1)}; k = 1, \ldots, p\right\} \quad \text{if } l = 1, \ldots, q+1\,,$$

where $Med\{\cdot\}$ is the median of the observed values.
The first iteration, which is performed on the initial matrix, consists of subtracting the median of the row from each item and adding the amount subtracted (median of the row) to the p first places in the last column (all of which was initially null). In the second iteration, we proceed in similar fashion, but by column, that is, we subtract the median of the column from each previous iteration, including in this case the items in the column $q + 1$, adding this amount (the median of the column) to the row $p + 1$. Iterations are carried out accordingly until convergence is achieved. We can start with row or column subtracting indifferently.

(iii) The iterative process explained in step (ii) stops when an identical matrix to the previous matrix is obtained, within a pre-established level of tolerance. Once the iterative process has concluded, that is, once the iterative algorithm has converged, the estimated effects are as follows:

$$\tilde{a} \equiv z_{p+1,q+1}^{(\infty)},$$

$$\tilde{r}_k \equiv z_{k,q+1}^{(\infty)} \quad \text{for } k = 1, \ldots, p,$$

$$\tilde{c}_l \equiv z_{p+1,l}^{(\infty)} \quad \text{for } l = 1, \ldots, q\,,$$

such that the original values can be decomposed into:

$$z_{k,l} = \tilde{a} + \tilde{r}_k + \tilde{c}_l + z_{k,l}^{(\infty)} \quad \text{for } k = 1, \ldots, p \text{ and } l = 1, \ldots, q. \tag{4.82}$$

Therefore, once the foregoing iterative process has stopped, the original matrix is replaced by another of dimension $(p + 1)$ x $(q+1)$ which contains the $p \times q$ residuals $z_{k,l}^{(\infty)}$, the p row effects, the q column effects and the overall effect $\tilde{a}$.

Once the median-polish algorithm has been performed, the following steps must be followed to carry out the median-polish kriging procedure:

(iv) The empirical semivariogram of the residuals is constructed.

(v) Using the empirical semivariogram of the residuals as a basis, a valid semivariogram is selected.

(vi) This theoretical semivariogram is used to perform OK on the residuals, obtaining a prediction of the residual at the desired point $\mathbf{s}_0$ in the domain under study, $\tilde{R}(\mathbf{s}_0)$.

(vii) The non-parametric estimation of the drift at that point is incorporated into this prediction (global effect + row effect + column effect), obtaining:

$$\tilde{Z}(\mathbf{s}_0) = \tilde{\mu}(\mathbf{s}_0) + \tilde{R}(\mathbf{s}_0).$$

The additivity in $\tilde{\mu}(\mathbf{s}_i)$ allows interpolating planes to be defined uniquely. Thus, the estimation of μ at the point $\mathbf{s}_0$ can be obtained by:

$$\tilde{\mu}(\mathbf{s}_0) = \tilde{a} + \tilde{r}_k + \left(\frac{y_0 - y_k}{y_{k+1} - y_k}\right)(\tilde{r}_{k+1} - \tilde{r}_k) + \tilde{c}_l + \left(\frac{x_0 - x_l}{x_{l+1} - x_l}\right)(\tilde{c}_{l+1} - \tilde{c}_l), \quad (4.83)$$

where $\mathbf{s}_0 = (x_0, y_0)', x_l \leq x_0 \leq x_{l+1}, y_k \leq y_0 \leq y_{k+1}$.

Of course, there are other types of polishing. Instead of medians, one might choose to polish with means (if there are no gross outliers), trimmed means, or any type of weighted average of the data values, the decomposition procedure remaining the same. However, median-polish is the most popular choice due to its robustness to outliers and its ability to avoid linear dependencies in the residuals.

Finally, the kriging variance associated with median-polish kriging predictions is defined (with little justification) as the OK variance based on the median-polish residuals.

Example 4.11 (Median-polish kriging at work. Coal-ash data) *We are going to illustrate how median-polish kriging performs using the classic data on the percentage of coal-ash for the Robena Mine Property in Greene County, Pennsylvania, which were collected to analyze coal-ash content in the Pittsburgh coal seam (Gomez and Hazen 1970, Tables 19 and 20). As in Cressie (1993), we used as a total the* 208 *coal-ash core samples at locations with west coordinates greater that* 64,000 *ft, which define an approximately square grid with* 2,500 *ft spacing. The coal-ash data can be found as the built-in data set coalash in the* `gstat` *package in R (Pebesma 2004). We used this package to perform this example. Figure 4.9 shows the observed values (coal-ash percentages) and the location where they were observed. The 3D scatterplot of the data is displayed in Figure 4.10. The location of the grid has been reoriented so that top can be identified with north, bottom with south, left with west, and right with east. We used* `gstat` *and* `geoR` *R-libraries for figures and calculations.*

A contour plot of the coal-ash percentages is displayed in Figure 4.11. This plot can be very useful to check if there is a drift over the spatial area. The plotting of a trend or drift surface using spatial interpolation via triangulation (Figure 4.12) can also be used for this purpose. Another way to look for drifts across rows (y) and columns (x) is to generate side-to-side boxplots across rows and columns, because

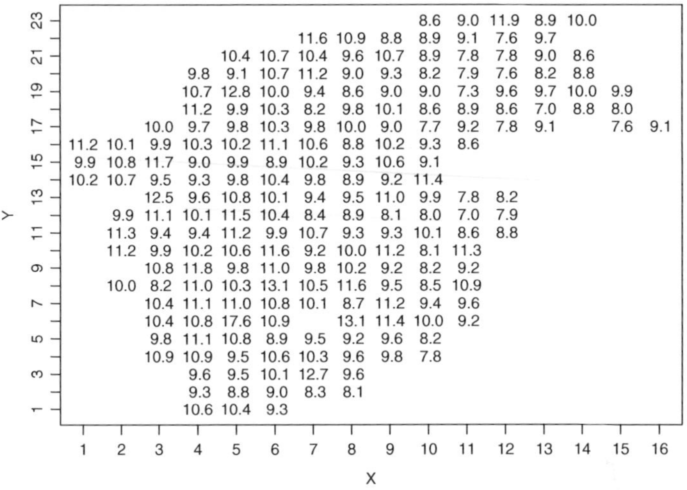

Figure 4.9 Coal-ash data. Location (reoriented) and observed values.

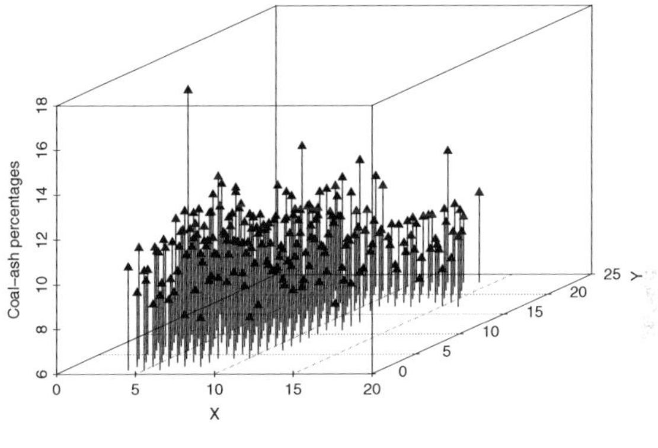

Figure 4.10 Coal-ash data. 3D scatterplot.

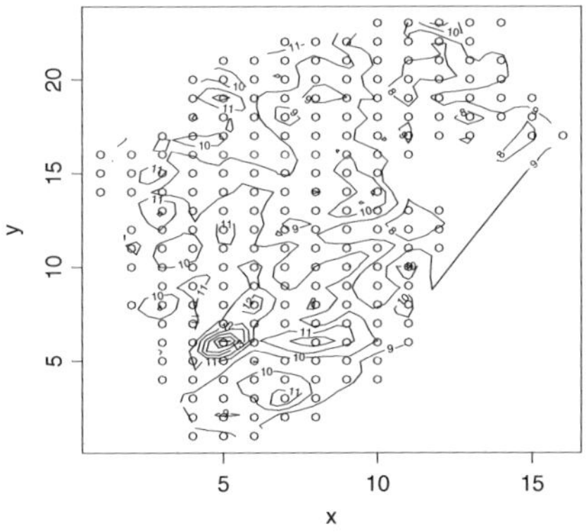

Figure 4.11 Contour plot of coal-ash percentages.

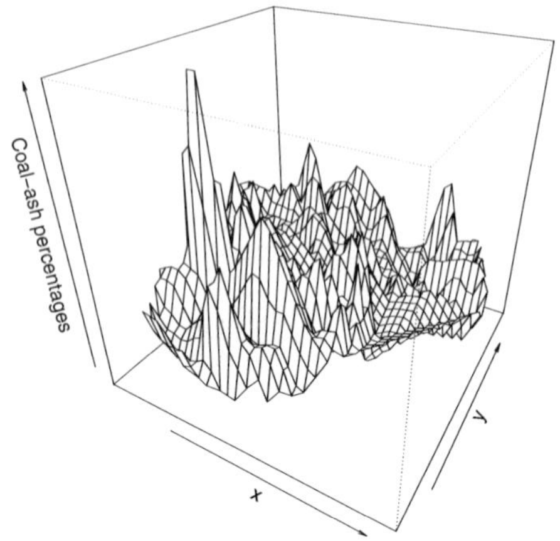

Figure 4.12 Coal-ash percentages surface interpolation (via triangulation).

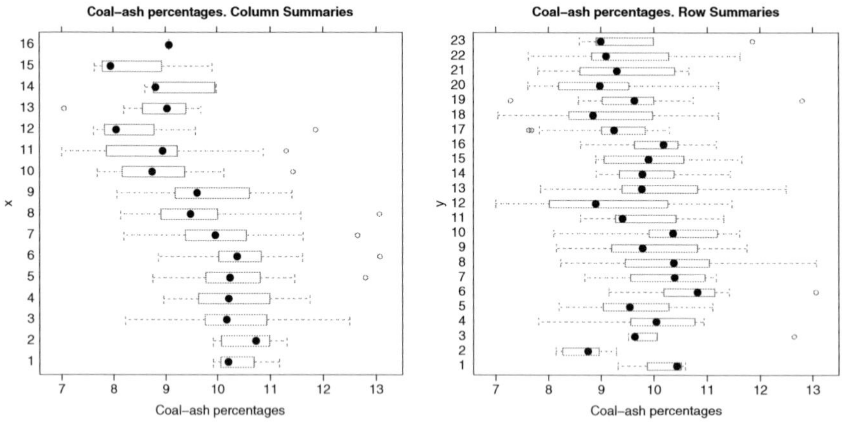

Figure 4.13 Coal-ash percentages: Column and row summaries.

this allows us to see the median percentages as well as the degree of variation and shape across rows and columns (Figure 4.13).[6]

From the above figures, there appears to be a linear drift in the east-west direction, but no drift (or at most a very slight drift) in the north-south direction. Taking this into account, now we proceed to the median-polish kriging prediction of the coal-ash percentage at the non-observed point located in column 7 and row 6, which has been represented with a cross in Figure 4.14.

[6] Although we are directly focusing on drift, previous analysis of indicator maps, stem-and-leaf plots, and boxplots are a great help in identifying outliers.

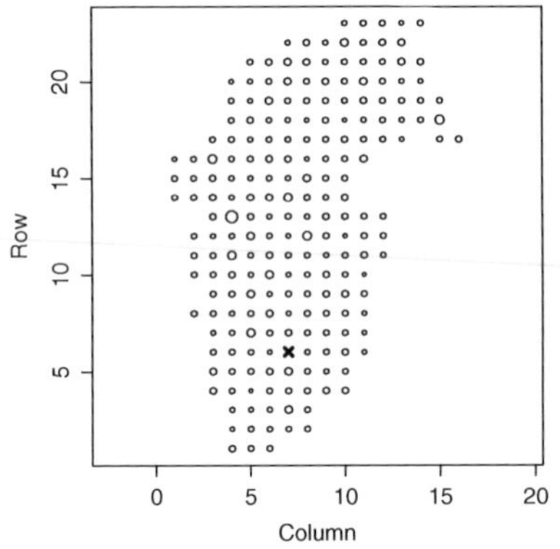

Figure 4.14 Prediction point.

Table 4.20 Row effects.

Row:												
	1	2	3	4	5	6	7	8	9	10	11	12
	13	14	15	16	17	18	19	20	21	22	23	
	0.22	−1.46	−0.36	0.11	−0.26	0.89	0.69	0.21	0.33	0.33	−0.25	−0.62
	−0.12	−0.40	−0.09	0.00	−0.28	−0.08	0.39	−0.54	0.15	0.23	0.13	

Table 4.21 Column effects.

Column:								
	1	2	3	4	5	6	7	8
	9	10	11	12	13	14	15	16
	0.93	1.17	0.33	0.67	0.55	0.76	0.46	−0.14
	0.14	−1.00	−0.79	−1.25	−0.75	−0.36	−1.64	−0.34

For this purpose, we first perform the median-polish algorithm, which yields an overall effect of 9.66125 *and the row and column effects listed in Tables 4.20 and 4.21, respectively.*

In light of the above effects, the median-polish estimation of the drift at the prediction point is calculated using (4.83), the estimate being $\tilde{\mu}(\mathbf{s}_0) = 11.04$.

After removing the overall, row and column effects from the original data, we obtain the residuals shown in Figure 4.15. Figure 4.16 displays the grayscale plots of the original data, the median-polish drift, and the median-polish residuals.

The second step in median-polish kriging is to perform an OK of these residuals. As can be seen in Figure 4.17, the classical empirical semivariogram shows that after removing the median-polish effects from the original data, the residuals have no spatial dependence.

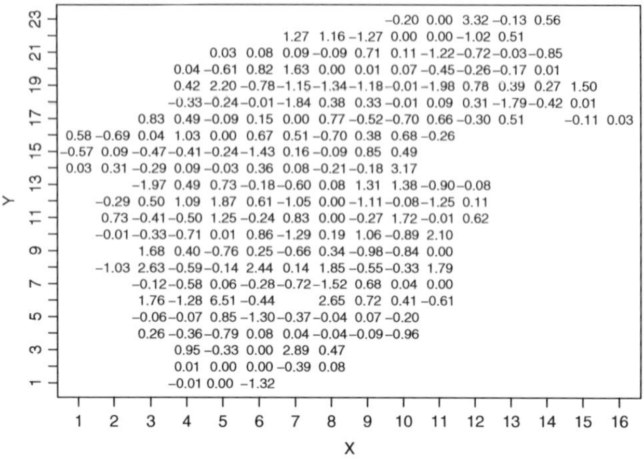

Figure 4.15 Coal-ash percentages: Median-polish residuals.

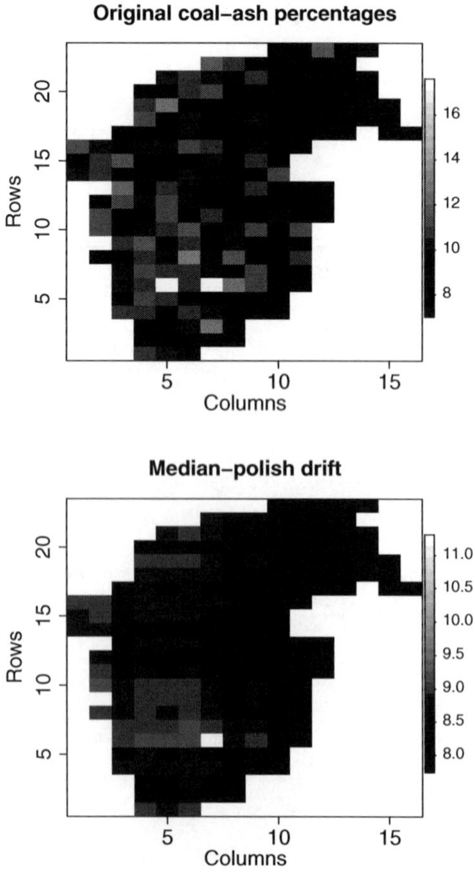

Figure 4.16 Coal-ash data: Original data, median-polish drift and median-polish residuals (the lighter the color, the higher the value).

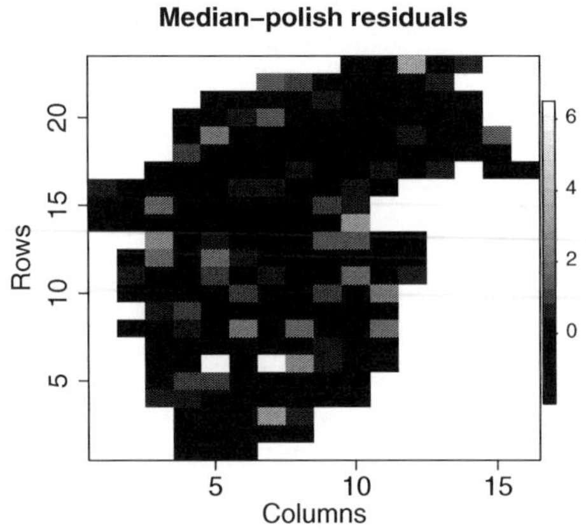

Figure 4.16 (continued).

In fact, using weighted least squares, the valid semivariogram that best fits the empirical one is the pure nugget model,

$$\gamma(\mathbf{h}) = \begin{cases} 1.15 & \text{if } \mathbf{h} > 0 \\ 0 & \text{if } \mathbf{h} = 0 \end{cases}. \tag{4.84}$$

It is important to note that the pure nugget semivariogram of the residuals is not a direct consequence of the median-polish procedure, but only a particularity of this dataset. As stated in Cressie (1993, p. 191), at the scale of 1 *grid spacing of* 2500 *ft. no spatial dependence is detected. The real behavior of the micro-scale variation may be far from white noise, but there is no way to tell without taking more observations close together.*

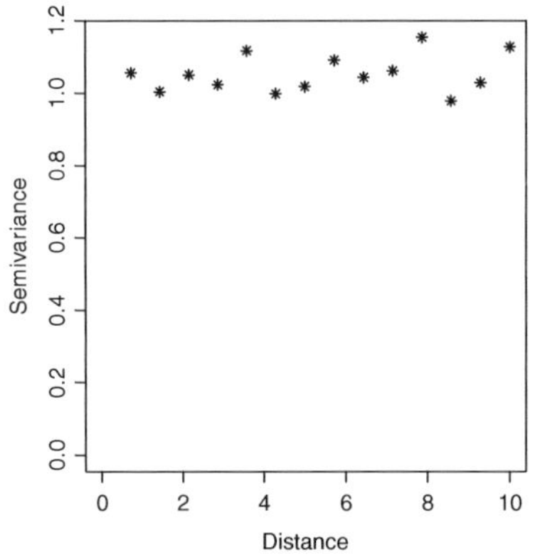

Figure 4.17 Coal-ash residuals: Classical semivariogram.

The OK prediction of the residual at the prediction point $\mathbf{s}_0$ *(column 7, row 6) is* $\tilde{R}(\mathbf{s}_0) = \sum_{i=1}^{208} \lambda_i R(\mathbf{s}_i) = 0.10735$. *Finally, the non-parametric estimation of the drift at that point is incorporated into this prediction (global effect + row effect + column effect), obtaining:*

$$\tilde{Z}(\mathbf{s}_0) = \tilde{\mu}(\mathbf{s}_0) + \tilde{R}(\mathbf{s}_0) = 11.04 + 0.11 = 11.15,$$

the kriging variance associated with the prediction (that is, the OK variance based on the median-polish residuals), being 1.15552.

4.9 Cross-validation

In the kriging process, a valid semivariogram model has been chosen (using any of the procedures detailed in Section 3.8) and used, and some other assumptions have been made. However, we should validate these assumptions, especially the one about the type of semivariogram model and its parameters. Otherwise, the results obtained from the kriging process could lead us to erroneous conclusions.

The usual statistical approach to such a validation is testing. However, the construction of statistical tests in geostatistics is a difficult challenge, due precisely to the existing spatial correlation between the observed values (and also to other factors including the large variability of the empirical semivariogram at large distances).

The use of statistical tests being discarded, another possibility is the use of the chosen semivariogram model for kriging and to check how well it performs (this is why it was decided to include this section after having studied the kriging equations). In order to carry out such a validation procedure, the sample should be divided into two sets: the "fitting" set and the validation set. The "fitting" set is used to construct the empirical semivariogram and to choose the theoretical one that best fits the empirical counterpart. The validation set is to evaluate the performance of the chosen semivariogram model in kriging by comparing the observed data in this set with the kriging predictions resulting from the use of such a semivariogram. If the fitting process is deficient, the results obtained from the subsequent kriging procedure will not be as good as we would like. But, unfortunately, in most situations the empirical information available is not enough to allow a validation set, and a cross-validation (CV), or leave-one-out process, is the most widely used procedure to validate the semivariogram model selected.

The CV process consists of:

(i) Obtaining the kriging prediction $\hat{Z}(\mathbf{s}_i)$ at each sample point $\mathbf{s}_i$, $i = 1, \ldots, n$ (as if the sample value at such points were unknown) from the observations at the $n - 1$ remaining points (or from a set of neighboring observations). The prediction variance at each sample point, $\hat{\sigma}^2(\mathbf{s}_i)$, is also calculated.

(ii) Calculating the following diagnostic statistics from the results obtained in (i):

- the mean prediction error:

$$ME = \frac{1}{n}\sum_{i}^{n}\left(Z(\mathbf{s}_i) - \hat{Z}(\mathbf{s}_i)\right). \tag{4.85}$$

- the mean squared prediction error:

$$MSE = \frac{1}{n}\sum_{i}^{n}\left(Z(\mathbf{s}_i) - \hat{Z}(\mathbf{s}_i)\right)^2. \tag{4.86}$$

- the mean squared standardized prediction error:

$$MSDE = \frac{1}{n}\sum_{i}^{n}\left(\frac{Z(\mathbf{s}_i) - \hat{Z}(\mathbf{s}_i)}{\hat{\sigma}(\mathbf{s}_i)}\right)^2. \tag{4.87}$$

Note that if $\gamma(\mathbf{h})$ is a valid semivariogram, then the kriging prediction error is a rv with zero mean (it follows from the unbiasedness property of kriging), variance $\hat{\sigma}^2(\mathbf{s}_i)$, and covariance (between the values at non-sampled points $\mathbf{s}_\alpha$ and $\mathbf{s}_\beta$):

$$\begin{aligned} C\left(\hat{Z}(\mathbf{s}_\alpha) - Z(\mathbf{s}_\alpha), \hat{Z}(\mathbf{s}_\beta) - Z(\mathbf{s}_\beta)\right) &= \gamma(\mathbf{s}_\alpha - \mathbf{s}_\beta) - \sum_{i=1}^{n}\sum_{i=1}^{n} -\lambda_i^\alpha \lambda_j^\beta \gamma(\mathbf{s}_i - \mathbf{s}_j) \\ &+ \sum_{i}^{n} \lambda_i^\alpha \gamma(\mathbf{s}_i - \mathbf{s}_\alpha) + \sum_{i}^{n} \lambda_j^\beta \gamma(\mathbf{s}_j - \mathbf{s}_\alpha), \end{aligned} \tag{4.88}$$

where λ_i^α and λ_j^β are the kriging weights used in the prediction of $Z(\mathbf{s}_\alpha)$ and $Z(\mathbf{s}_\beta)$, respectively.

Thus, if the semivariogram choice is successful:

- ME should be approximately 0, which is indicative of non-systematic prediction errors, no matter what the fitted semivariogram is, kriging predictions are unbiased and the ME value is expected to tend to 0.
- MSE should be small.
- MSDE should be approximately 1, that is, the prediction errors are compatible with the kriging prediction variance. It is also desirable that "bad," or in of Emery's terms (2000, p. 118), "non-robust" predictions, understanding "good" or "robust" to be those which yield a standardized error that belongs to an arbitrarily fixed interval $[-\epsilon, +\epsilon]$, do not to exceed, let's say 5%.
  Obviously, no matter what the type of semivariogram model considered to represent the structure of the spatial dependence of the phenomenon being studied, at isolated points, the squared prediction error will be greater than in the case of others surrounded by neighboring points. This "isolation effect" is corrected

by dividing the squared prediction error by the estimated kriging variance, obtaining the standardized error which, due to the way it is built, has unit variance regardless of the geometrical configuration of the observed realization, and is a good indicator of the accuracy the prediction is expected to provide.

- In addition, the correlation coefficient between the predicted and observed values should be close to 1, and the standardized errors should not depend on the magnitude of the predicted values (no conditional bias).

As stated in Cressie (1993, p. 104), if the semivariogram model has been successfully cross-validated, one can be confident that prediction based on such a semivariogram is approximately unbiased and that the mean-square prediction error is about right. CV cannot prove that the fitted model is correct, merely that it is not grossly incorrect (there is no reason for rejecting it). Thus, it must be made clear that the success of CV does not guarantee that the semivariogram chosen (fitted) is correct; what it does guarantee is that it is not incorrect.

Finally, a word of caution: Although the CV method is extremely useful when deciding between two candidate semivariogram models, it is not a good idea to generalize the method as if it were an automatic procedure for fitting the semivariogram (see details in Chilès and Delfiner 1999, p. 112).

Example 4.12 (Validating the semivariogram: Data on carbon monoxide in Madrid, week 50, 10 pm) *In Example 3.4 we decided that the spherical model was our first candidate to represent the spatial dependencies observed in the data on logCO* at* 10 *pm in the* 50*th week of* 2008, *and this is why in Example 3.8 we fitted (automatic fitting) a spherical semivariogram to the empirical counterpart. But we also commented that the exponential model could also be a good choice to represent such dependencies.*

This is why here, in order to illustrate the cross-validating process, and using the WLS method, we have fitted an exponential model to the empirical semivariogram and, in addition, two more candidate models with a parabolic behavior near the origin. This was made using `geoR`. *Figure 4.18 shows the spherical, exponential, cubic and Gaussian models fitted to the empirical semivariogram, and Table 4.22 lists the values of the diagnostic statistics ME, MSE and MSDE for each model when performing OK on logCO* data. As expected, ME is close to zero for the four models considered. The MSDE of both the cubic and Gaussian models (which have a parabolic behavior near the origin) is around the unity. However, in the case of the two models exhibiting a linear behavior near the origin (exponential and spherical) there is evidence that the OK variances underestimate the true prediction variances (especially in the case of the spherical semivariogram). As for MSE, the lowest values are obtained with the cubic and Gaussian models, and the largest one with the exponential model.*

In addition, (i) the correlation coefficient between the predicted and observed values when using the spherical model is 0.75, *similar to that obtained with the Gaussian and cubic models,* 0.78 *and* 0.79 *respectively, and greater than the obtained when using the exponential semivariogram* (0.69)*; and (ii) the standardized errors obtained*

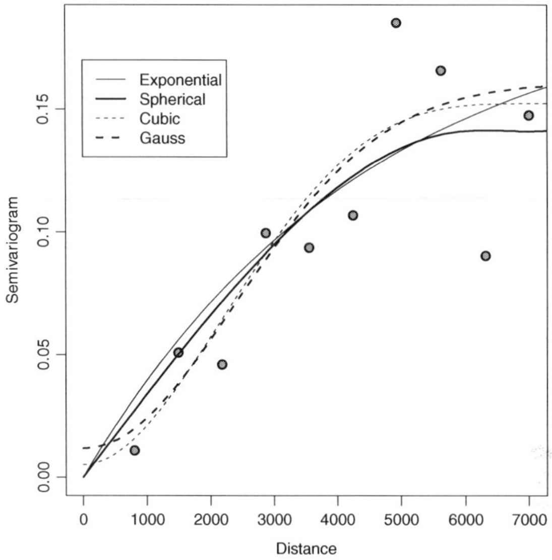

Figure 4.18 Exponential, spherical, cubic, and Gaussian models resulting from the WLS fitting. (Data on carbon monoxide in Madrid, week 50, 10 *pm).*

Table 4.22 ME, MSE and MSDE for five semivariogram models. Ordinary kriging of logCO*.

Model	ME	MSE	MSDE
Exponential	−0.02098	0.07267	0.82163
Spherical	−0.02520	0.06556	0.76523
Cubic	−0.03091	0.05132	1.06757
Gaussian	−0.02875	0.05245	0.97867

do not depend on the magnitude of the predicted values, whatever the model is. Although initially the cross-validation procedure was carried out to confirm that the spherical model was the right choice and to discard the exponential one, as a consequence of the above comments, both the cubic and the Gaussian models could be better than the spherical model when it comes to representing the spatial dependencies existing in logCO data at* 10 *pm of the* 50*th week of* 2008. *Maybe the initial assumption on the linear behavior of the semivariogram near the origin was wrong and a cubic or a Gaussian semivariogram are better choices. Or maybe not, and these slight differences between the models are due to the fact that we have only* 23 *monitoring stations and the two closest ones are separated by nearly* 1 *km. In any case, it is clear that when opting for a semivariogram with a linear behavior near the origin, the best choice is the spherical model.*

The above comments illustrate why choosing models and fitting them are among the most controversial topics in geostatistics and manual fitting has many supporters.

4.10 Non-linear kriging

We have presented so far the most popular kriging procedures, all of them aiming to predict the expected value (or the average value) of the rf under study at a non-observed point (or block), or more generally, to obtain prediction maps of such a rf over the domain of interest, D. These procedures use a linear predictor, but cannot give an answer to a series of questions including the estimation of the rf distribution rather than simply an expected value at a point or over a block.

In fact, as is well known, the BLUP of a rv Z that can be obtained from n rv's $Z_1, \ldots, Z_n$ is $E(Z|Z_1, \ldots, Z_n)$, which is non-linear in $Z_1, \ldots, Z_n$ and requires the joint probability distribution of the $n+1$ rv's to be known. This joint distribution rarely can be estimated from the available data, the Gaussian case being an exception because in this case the conditional expectation is a linear function of $Z_1, \ldots, Z_n$. In general,

$$Z^* = E(Z|Z_1, \ldots, Z_n) = f(Z_1, \ldots, Z_n), \tag{4.89}$$

where the SK predictor is a particular case:

$$Z^*_{SK} = \sum_{i=1}^{n} \lambda_i Z_i, \tag{4.90}$$

which is optimal when the $n+1$ rv's follow a multivariate Gaussian distribution. If the covariance function is known but the mean is unknown, the role of the optimal predictor is given to the OK predictor.

In light of the above considerations, in this section non-linear predictors of the form:

$$Z^* = \sum_{i=1}^{n} f_i(Z_i), \tag{4.91}$$

where each function $f_i(\cdot)$ depends only on the rv Z_i, are proposed.

As stated in Cressie (1993, p. 278), non-linear geostatistics focuses on techniques that use non-linear functions of the observed data to obtain or approximate $E(Z|Z_1, \ldots, Z_n)$ and $P(Z \geq z|Z_1, \ldots, Z_n)$. These non-linear techniques often require assumptions for which no methods of verification are available and they can yield computationally complex solutions.

Among the spectrum of non-linear techniques provided by non-linear geostatistics, in what follows we briefly outline two of them: Disjunctive kriging and kriging indicator, which are, perhaps, the most interesting and well-known non-linear kriging methods.

4.10.1 Disjunctive kriging

Disjunctive kriging (DK), proposed by Matheron (1976), is a non-linear kriging procedure which in general has some advantages over linear kriging methods and does not require knowledge of the $n+1$ joint probability distributions necessary for the conditional expectation but only of the bivariate distributions. In addition, through DK it is possible to obtain an estimate of the conditional probability that a measured indicator variable exceeds some tolerance level or cut-off, which makes DK an interesting decision-making tool based on quantitative information.

Consider a second-order stationary rf, $Z(\mathbf{s})$, which has been observed on a point support at locations $\mathbf{s}_1, \ldots, \mathbf{s}_n$, and a transformed rf, $Y(\mathbf{s})$, in general its Gaussian transform. The DK predictor uses the family of functions of $Y(\mathbf{s})$ that minimizes the prediction variance. Under a particular bivariate assumption, an isofactorial family of functions can be found. More specifically, under the bivariate Gaussian assumption, this family is the Hermite polynomials. However, other transformations can be done and different orthogonal polynomials must be used. For a uniform transformation, Legendre polynomials are used. If the transformation is exponential, Laguerre polynomials are appropriate.

Following Yates and Warrick (1986), applying the Gaussian transform to each $Z(\mathbf{s}_i)$ produces a standard Gaussian rv, $Y(\mathbf{s}_i)$. Then, the DK predictor can be written as:

$$Z^*_{DK}(\mathbf{s}_0) = \sum_{i=1}^{n} f_i\left(Y(\mathbf{s}_i)\right) = \sum_{i=1}^{n}\sum_{k=0}^{\infty} f_{ik} H_{ik}\left(Y(\mathbf{s}_i)\right), \tag{4.92}$$

where each function involved in the predictor depends on only one standardized observed value $Y(\mathbf{s}_i)$,$f_i(Y(\mathbf{s}_i))$ is a function to be determined and expressed as a series of Hermite polynomials, $H_k(Y(\mathbf{s}))$ (see Rivoirard 1994, for a full explanation and definition of Hermite polynomials), and f_{ik} is a constant which depends on both i and k.

As ever, in order to obtain the BLUP, we impose on the prediction error the conditions of zero expectation and minimum variance, that is: $E\left(Z(\mathbf{s}_0) - Z^*_{DK}(\mathbf{s}_0)\right)^2$ minimum. This minimum occurs when $Z(\mathbf{s}_0) - Z^*_{DK}(\mathbf{s}_0)$ is perpendicular to any function $f(Z(\mathbf{s}_i))$ in the hyper-plane defined by the measurable functions $f_i(Z(\mathbf{s}_i))$ involved in the definition of the predictor.

Thus, using the perpendicular projection, that is, taking into account that $Z(\mathbf{s}_0) - Z^*_{DK}(\mathbf{s}_0)$ and $f(Z(\mathbf{s}_i))$ are orthogonal, we have:

$$E\left(\left(Z(\mathbf{s}_0) - Z^*_{DK}(\mathbf{s}_0)\right) f_i(Z(\mathbf{s}_i))\right) = 0, \tag{4.93}$$

that is,

$$E\left(Z(\mathbf{s}_0) f_i(Z(\mathbf{s}_i))\right) = E\left(Z^*_{DK}(\mathbf{s}_0) f_i(Z(\mathbf{s}_i))\right), \tag{4.94}$$

which can be written in terms of the conditional expectation as follows:

$$E\left(Z(\mathbf{s}_0)|Z(\mathbf{s}_j)\right) = E\left(Z^*_{DK}(\mathbf{s}_0)|Z(\mathbf{s}_j)\right) \quad j = 1, \ldots, n. \tag{4.95}$$

The transformation of $Z(\mathbf{s})$ to a standardized Gaussian rf can be done using Hermite polynomials. Therefore, the transformation function $\phi(Y(\mathbf{s}))$ can be written as:

$$\phi(Y(\mathbf{s})) = Z(\mathbf{s}) = \sum_{k=0}^{\infty} C_k H_k(Y(\mathbf{s})), \tag{4.96}$$

where $Y(\mathbf{s})$ is assumed to be bivariate Gaussian and the coefficients C_k are determined as follows:

$$C_k = \frac{\int_{-\infty}^{\infty} \phi(y) H_k(y) \exp\left(-y^2/2\right) dy}{k!(2\pi)^{1/2}}, \tag{4.97}$$

$\phi(y)$ being the function represented by the infinite series of Hermite polynomials.

However, for DK, $\phi(y)$ is not generally known, and the above integral cannot be evaluated analytically. An alternative is Hermite integration (Abramowitz and Stegun 1965), which uses only a few special points:

$$C_k = \frac{1}{k!(2\pi)^{1/2}} \sum_{i=1}^{J} w_i \phi(v_i) H_k(v_i) \exp\left(-v_i^2/2\right), \tag{4.98}$$

where the abscissas, v_i, and the weight factors, w_i, can be found in Abramowitz and Stegun (1965).

Now, Equation (4.95) can be written as:

$$E\left(\phi(Y(\mathbf{s}_0)) | Y(\mathbf{s}_j)\right) = \sum_{i=1}^{n} E\left(f_i(Y(\mathbf{s}_i)) | Y(\mathbf{s}_j)\right), \quad j = 1, \ldots, n. \tag{4.99}$$

Since the bivariate joint density of two standardized Gaussian variables, X and Y, is given by:

$$f(x, y) = \sum_{k=0}^{\infty} (\rho_{XY})^k H_k(x) H_k(y) g(x) g(y)/k!, \tag{4.100}$$

ρ_{XY} representing the correlation coefficient between X and Y, and g being the standard Gaussian density function, the conditional expectation of a function $\phi(X)$, given by $\phi(y) = \sum_{k=0}^{\infty} C_k H_k(y)$, can be computed as follows:

$$E\left(\phi(X) | Y\right) = \sum_{k=0}^{\infty} (\rho_{XY})^k C_k H_k(Y). \tag{4.101}$$

Applying this last result to Equation (4.99), we obtain:

$$\sum_{k=0}^{\infty} C_k H_k(Y(\mathbf{s}_j)) \left((\rho_{0j})^k - \sum_{i=1}^{n} f_{ik} (\rho_{ij})^k / C_k \right) = 0, \quad j = 1, \ldots, n, \tag{4.102}$$

where the infinite series are truncated to K terms.

Finally, defining $b_{ik} = f_{ik}/C_k$ and taking into account that Equation (4.102) must be satisfied for all k, we have the following results:

1. $Z_{DK}^*(\mathbf{s}_0) = \sum_{k=0}^{\infty} C_k H_k^*(Y(\mathbf{s}_0))$.
2. $H_k^*(Y(\mathbf{s}_0))$ is obtained as a weighted sum of Hermite polynomials of the observed values:

$$H_k^*(Y(\mathbf{s}_0)) = \sum_{i=1}^{n} b_{ik} H_k(Y(\mathbf{s}_j)). \tag{4.103}$$

3. The weights b_{ik} are determined by solving the following simple kriging system:

$$\sum_{i=1}^{n} b_{ik}(\rho_{ij})^k = (\rho_{0j})^k, \quad j = 1, \dots, n. \tag{4.104}$$

In matrix notation:

$$\begin{pmatrix} (\rho_{11})^k & \cdots & (\rho_{n1})^k \\ \vdots & \ddots & \vdots \\ (\rho_{1n})^k & \cdots & (\rho_{nn})^k \end{pmatrix} \begin{pmatrix} b_{1k} \\ \vdots \\ b_{nk} \end{pmatrix} = \begin{pmatrix} (\rho_{01})^k \\ \vdots \\ (\rho_{0n})^k \end{pmatrix}$$

Note that (i) the term for the mean is not present, since the mean of the Hermite polynomial is 0 for all $k > 0$; (ii) the simple kriging prediction of the polynomial of degree 0 is 1 by definition; and (iii) only the correlation function (or the semivariogram) of the Gaussian transformed values is needed in order to perform all the krigings required.

4. The DK prediction variance is:

$$\sigma^2_{DK} = \sum_{k=1}^{K} k! C_k^2 \left(1 - \sum_{i=1}^{n} b_{ik}(\rho_{0i})^k \right). \tag{4.105}$$

Yates and Warrick (1986, p. 620) summarize the above process to obtain a prediction of a rf at a non-observed location $\mathbf{s}_0$ as follows:

1. Transform the original data, $Z(\mathbf{s})$, into a new variable, $Y(\mathbf{s})$, which is assumed to be uni- and bivariate Gaussian.
2. Once the values of $Y(\mathbf{s})$ are known, the coefficients C_k are calculated from Equations (4.97) or (4.98).
3. Calculate the values of $Y(\mathbf{s}_i)$ for each observed point $(\mathbf{s}_1, \dots, \mathbf{s}_n)$ by inverting Equation (4.96), $k = 1, \dots, K$. This is necessary because a truncated series is used for the transform. This way, it is assured that the DK system interpolates exactly at the observed points.
4. Compute the sample mean, C_0, and variance, $\sum_{k=1}^{\infty} k! C_k^2$, to verify (i) that appropriate values C_k in step 2 were calculated, and (ii) to aid in determining the number of $k's$ necessary. It is also advisable to plot $\phi[Y(\mathbf{s}_i)]$ against the original data, in order to help to determine the number of values C_k necessary.
5. Calculate the sample correlation function using the transformed data $Y(\mathbf{s}_i)$. If the semivariogram for $Z(\mathbf{s}_i)$ is available, then the following relation can be used:

$$\rho(\mathbf{h}) = 1 - \gamma(\mathbf{h})/C(\mathbf{0}). \tag{4.106}$$

The remaining steps are required to calculate a prediction at each observed point and the entire sequence is repeated for each prediction desired:

(i) Set $k = 0$.

(ii) Calculate $H_k^*[Y(\mathbf{s}_0)]$ in Equation (4.103) by solving the simple kriging system (4.104), using the values of $H_k[Y(\mathbf{s}_i)]$.

(iii) Calculate the kth term of $Z_{DK}^*(\mathbf{s}_0) = \sum_{k=0}^{\infty} C_k H_k^*[Y(\mathbf{s}_0)]$.

(iv) Increment k by one.

(v) Repeat the sequence for all K. Note that as k increases, $(\rho_{0i})^k$ and b_{ik} go to zero, so that K do not need to be very large.

4.10.2 Indicator kriging

The main objective of the indicator kriging is the estimation of the distribution of the rf under study at a non-observed point (or block) in the domain considered. The starting point of the method is an indicator function $I(\mathbf{s}, z_T)$ defined, at a location $\mathbf{s}$ and for the cut-off z_T, as a binary function that assumes the value 0 or 1 under the following conditions:

$$I(\mathbf{s}, z_T) = \begin{cases} 1 & \text{if } Z(\mathbf{s}) \leq z_T \\ 0 & \text{if } Z(\mathbf{s}) > z_T \end{cases}, \tag{4.107}$$

where it follows that:

$$P(I(\mathbf{s}, z_T) = 1) = P(Z(\mathbf{s}) \leq z_T) = F_{Z(\mathbf{s})}(z_T) \tag{4.108}$$

$$P(I(\mathbf{s}, z_T) = 0) = P(Z(\mathbf{s}) > z_T) = 1 - F_{Z(\mathbf{s})}(z_T). \tag{4.109}$$

There is an indicator function $I(\mathbf{s}, z_T)$ for each value of the cut-off z_T.

Assuming that the rf under study has second-order stationary indicators, the following expressions can immediately be deduced from Equations (4.108) and (4.109):

$$E\left(I(\mathbf{s}, z_T)\right) = F_{Z(\mathbf{s})}(z_T), \tag{4.110}$$

$$V\left(I(\mathbf{s}, z_T)\right) = F_{Z(\mathbf{s})}(z_T)\left(1 - F_{Z(\mathbf{s})}(z_T)\right), \tag{4.111}$$

$$\begin{aligned} C_{z_T}(\mathbf{h}) &= E\left(I(\mathbf{s}, z_T) I(\mathbf{s}+\mathbf{h}, z_T)\right) - E\left(I(\mathbf{s}, z_T)\right) E\left(I(\mathbf{s}+\mathbf{h}, z_T)\right) \\ &= P\left(I(\mathbf{s}, z_T) I(\mathbf{s}+\mathbf{h}, z_T) = 1\right) - P\left(I(\mathbf{s}, z_T) = 1\right) P\left(I(\mathbf{s}+\mathbf{h}, z_T) = 1\right) \\ &= F_{Z(\mathbf{s}),\, Z(\mathbf{s}+\mathbf{h})}\left(z_T, z_T\right) - \left(F_{Z(\mathbf{s})}(z_T)\right)^2, \end{aligned} \tag{4.112}$$

$$\begin{aligned}\gamma_{z_T}(\mathbf{h}) &= C_{z_T}(\mathbf{0}) - C_{z_T}(\mathbf{h}) \\ &= V\left(I(\mathbf{s}_0, z_T)\right) - C_{z_T}\left(\mathbf{h}, z_T\right) \\ &= F_{Z(\mathbf{s})}(z_T) - F_{Z(\mathbf{s}),\, Z(\mathbf{s}+\mathbf{h})}\left(z_T, z_T\right).\end{aligned} \tag{4.113}$$

The ordinary indicator kriging (OIK) predictor at the non-observed point $\mathbf{s}_0$ is given by:

$$I^*\left(\mathbf{s}_0, z_T\right) = \sum_{i=1}^{n} \lambda_i(z_T) I\left(\mathbf{s}_i, z_T\right), \tag{4.114}$$

which is an estimate of $P\left(Z(\mathbf{s}_0) \leq z_T | I\left(\mathbf{s}_1, z_T\right), \ldots, I\left(\mathbf{s}_1, z_T\right)\right)$ and then has to verify:

- $0 \leq I^*\left(\mathbf{s}_0, z_T\right) \leq 1$
- $I^*\left(\mathbf{s}_0, z_T\right) \leq I^*\left(\mathbf{s}_0, z_T^{'}\right)$ if $z_T \leq z_T^{'}$.

The OIK weights λ_i are obtained as usual, that is, by imposing on the prediction error to have zero expectation and minimum variance. The first condition implies $\sum_{i=1}^{n} \lambda_i(z_T) = 1$. The second leads to the following OIK system:

$$\begin{cases} \sum_{i=1}^{n} \lambda_i(z_T)\gamma_{z_T}\left(\mathbf{s}_i - \mathbf{s}_j\right) + \omega(z_T) = \gamma_{z_T}\left(\mathbf{s}_0 - \mathbf{s}_j\right), \quad j = 1, \ldots, n, \quad z_T \in \mathbb{R} \\ \sum_{i=1}^{n} \lambda_i(z_T) = 1 \end{cases}, \tag{4.115}$$

where $\omega(z_T)$ is the Lagrange multiplier. The OIK variance associated with $I^*\left(\mathbf{s}_0, z_T\right)$ is given by:

$$\sigma^2_{OIK}(\mathbf{s}_0, z_T) = \sum_{i=1}^{n} \lambda_i(z_T)\gamma_{z_T}(\mathbf{s}_0 - \mathbf{s}_j) + \omega(z_T). \tag{4.116}$$

Both the OIK system and the OIK variance can easily be extended to the case of block prediction replacing the point-to-point semivariograms with point-to-block or block-to-block semivariograms, depending on the observation support.

The above equations must be solved for different values of the cut-off z_T, which implies the estimation and modeling of the same number of semivariograms. This procedure is not exempt from difficulties. For example, the estimates could not be monotonic in z_T. Some solutions to this problem (basically ad hoc modifications to the indicator kriging) are given in Cressie (1993, p. 283), but there is still no guarantee that the estimated conditional probabilities are compatible with an underlying joint distribution.

Finally, note that two advantages of the indicator kriging are:

(i) it makes no assumptions about the distribution of the rf under study; and

(ii) the [0 – 1] transformations of the data make $I^*\left(\mathbf{s}_0, z_T\right)$ robust to outliers.

However, as said in Cressie (1993, p. 281),

- it requires the estimation and modeling of many indicator semivariograms;
- the indicator kriging system of equations is very large; and
- it gives a worse approximation to the conditional expectation than disjunctive kriging.

5

Geostatistics and spatio-temporal random functions

5.1 Spatio-temporal geostatistics

Geostatistical research has typically analyzed rf's in nature. However, nowadays studies are also conducted in other disciplines, such as social sciences, which share a common feature: they are performed in space, but researchers normally also monitor time to a certain extent. For example, if the topic under study is air pollution due to particulate matter in a given city, we will have a set of monitoring stations distributed spatially throughout the city that will measure the concentration of the pollutant and the measurements taken by each of them will be monitored over time. As we will see later, when modeling and predicting a given phenomenon, we obtain significant benefits from considering how it evolves in both space and time rather than only considering its spatial distribution at a given time of reference (a merely spatial process, such as those studied in the previous chapters), or its evolution over time at a given location (a merely temporal process).

Proof of the usefulness of spatio-temporal statistical modeling is the spectacular growth observed over the last few years in the number of practical applications in a wide variety of branches of science, such as monitoring environmental pollution, climatology, social sciences, geology, biology, medicine, archaeology or any other branch of science concerned with studying phenomena that occur in both space and time. One example of this exponential growth is that Cressie (1993), one of the referents in spatial geostatistics, devoted only three pages to the analysis of

Spatial and Spatio-Temporal Geostatistical Modeling and Kriging, First Edition.
José-María Montero, Gema Fernández-Avilés, and Jorge Mateu.

Companion Website: www.wiley.com/go/montero/spatial

spatio-temporal data. However, we can now find in the literature summaries of the main spatio-temporal modeling techniques, together with numerous practical applications.

5.2 Spatio-temporal continuity

As spatial continuity is a property that characterizes the relationship between different locations in space, spatio-temporal continuity is a property that characterizes the relationship between observations at different locations in a spatio-temporal domain. Let us consider two observations $Z(\mathbf{s}_i, t_i)$ and $Z(\mathbf{s}_j, t_j)$. Their relationship will normally depend on the distance between them $(\mathbf{s}_i - \mathbf{s}_j, t_i - t_j)$.

Figure 5.1 displays a simple example in which three measurements $Z(\mathbf{s}_1, t_1)$, $Z(\mathbf{s}_2, t_2)$ and $Z(\mathbf{s}_3, t_3)$ are mutually separated by $(\mathbf{h}, u)_{12}$, $(\mathbf{h}, u)_{13}$ and $(\mathbf{h}, u)_{23}$. It is normal for the relationship between $Z(\mathbf{s}_1, t_1)$ and $Z(\mathbf{s}, t)_2$ to be stronger than between $Z(\mathbf{s}_1, t_1)$ and $Z(\mathbf{s}_3, t_3)$, as $\mathbf{h}_{12} < \mathbf{h}_{13}$ and $\mathbf{h}_{12} < \mathbf{h}_{13}$.

Generalizing the spatial case, the spatio-temporal characterization includes (we follow Luo 1998 in the rest of this first subsection):

1. The description of the spatio-temporal continuity along with the empirical spatio-temporal covariance or semivariogram (see Chapter 8), which will provide the empirical basis for the hypothesis of spatio-temporal continuity and for fitting a valid spatio-temporal covariogram or semivariogram model.

2. The investigation of the hypotheses made in modeling the spatio-temporal continuity (including those about the stationarity of the rf, the behavior of the covariance function/semivariogram at the origin, the existence of anisotropies,

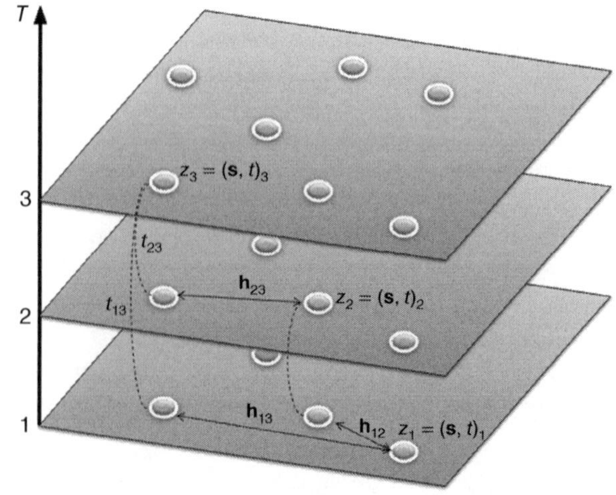

Figure 5.1 A spatio-temporal dataset on $\mathbb{R}^2 \times T$: 7 spatial locations observed at 3 *moments in time (adapted from Luo 1998).*

and the separability of the covariance function/semivariogram, among others) that could reflect and summarize adequately the characteristics of the empirical covariance function or semivariogram, which, as in the spatial case, provide the foundation of the covariance or semivariogram model fitting.

3. The fitting of a valid covariogram or semivariogram model to characterize the spatio-temporal continuity of the process.

Thus, as in the merely spatial case, according to the characteristics of the empirical covariance function (or semivariogram), some hypotheses for the space-time continuity are proposed and, according to Luo (1998), the study of such hypotheses can be imaginatively regarded as the "bridge" between the empirical covariance functions (or semivariograms) and their valid (or theoretical) counterparts.

As in the spatial case, the spatio-temporal structural analysis is the starting point in the spatio-temporal kriging process; then, a valid covariance or semivariogram model is fitted to the empirical counterpart resulting from the structural analysis stage; and finally the semivariogram values are included in the spatio-temporal kriging equations in order to obtain the weights used in the spatio-temporal predictor.

However, since in the spatio-temporal context valid covariance functions are given great importance and are considered a key part of this book, and taking into account that the reader is already familiar with spatial kriging, we proceed in a slightly different manner than in the spatial part of the book. After defining some relevant spatio-temporal concepts and presenting the main properties of the spatio-temporal covariance function and semivariograms (in this chapter), Chapter 6 deals with both the empirical spatio-temporal covariance function and the semivariogram. Chapter 7 studies a long list of valid spatio-temporal covariance models that can be potentially used in the kriging equations depending on the hypotheses made in the modeling of the spatio-temporal continuity. Chapter 7 is very long and this is why we decided to include in Chapter 6 the fitting of a valid model to the empirical covariogram or semivariogram. Finally, Chapter 8 is devoted to extending the spatial kriging equations to the spatio-temporal framework.

5.3 Relevant spatio-temporal concepts

Every spatio-temporal location can be seen as a point on $\mathbb{R}^d \times \mathbb{R}$, with $\mathbb{R}^d$ being the d-dimensional Euclidean space and $\mathbb{R}$ the time dimension. While from a mathematical viewpoint $\mathbb{R}^d \times \mathbb{R} = \mathbb{R}^{d+1}$, from a physical and social viewpoint, it would make no sense to consider spatial and temporal aspects in the same way, due to the significant differences between the two axes of coordinates. Therefore, while the time axis is ordered intrinsically (as it exists in the past, present and future), the same does not occur with the spatial coordinates. For example, in the case of the time axis, observations are generally taken in one direction (past), the aim being to extrapolate their value for the future. Meanwhile, in the case of spatial axes, the value of an unobserved location is normally interpolated. Neither the scales nor the units of distance in time and space are generally comparable. Notwithstanding, as we will see later, the main difficulty in analyzing spatio-temporal rf's is to choose the model of covariance that best fits the observations from those which are valid.

The aim of this section is to introduce the most relevant concepts in the analysis of spatio-temporal rf's. Let $(\mathbf{s}_i, t_i), i = 1, 2, \dots, n$ be a set of spatio-temporal locations. The set of spatio-temporal observations is represented by $Z(\mathbf{s}, t)$.

Extending geostatistics to the spatio-temporal case implies that the observations

$$\{Z(\mathbf{s}, t): \quad \mathbf{s} \in D, \quad t \in T\}, \tag{5.1}$$

with $D \subset \mathbb{R}^2$ and $T \subset \mathbb{R}$, have been captured in the spatio-temporal domain. In keeping with the standards of classical geostatistics, the spatio-temporal distribution of the observations is modeled as a Gaussian distribution. This distribution will be perfectly characterized by defining its first and second-order moments, that is, its expectation and covariance function.

Definition 5.3.1 *A spatio-temporal rf $Z(\mathbf{s}, t)$ is said to be Gaussian if the random vector $\mathbf{Z} = (Z(\mathbf{s}_1,t_1), \dots, Z(\mathbf{s}_n,t_n))'$, for any set of spatio-temporal locations $\{(\mathbf{s}_1,t_1), \dots, (\mathbf{s}_n,t_n)\}$, follows a multivariate Gaussian distribution.*

Under regularity conditions[1] the existence of the two first moments of the rf is guaranteed.

The mean of the spatio-temporal rf is defined as the function:

$$E\left(Z(\mathbf{s}, t)\right) = \mu\left(\mathbf{s}, t\right), \tag{5.2}$$

and the covariance function as:

$$C((\mathbf{s}_i, t_i), (\mathbf{s}_j, t_j)) = C(Z(\mathbf{s}_i, t_i), Z(\mathbf{s}_j, t_j)), \tag{5.3}$$

for any two pairs of spatio-temporal locations $(\mathbf{s}_i, t_i)$ and $(\mathbf{s}_j, t_j)$ on $\mathbb{R}^d \times \mathbb{R}$.

Definition 5.3.2 *A spatio-temporal rf $Z(\mathbf{s}, t)$ is strictly stationary if its probability distribution is translation invariant. In other words, if, in the case of any two given vectors, $\mathbf{h}$ and u:*

$$Z(\mathbf{s}_1,t_1), Z(\mathbf{s}_2,t_2), \dots, Z(\mathbf{s}_n,t_n) \tag{5.4}$$

and

$$Z(\mathbf{s}_1 + \mathbf{h}, t_1 + u), Z(\mathbf{s}_2 + \mathbf{h}, t_2 + u), \dots, Z(\mathbf{s}_n + \mathbf{h}, t + u) \tag{5.5}$$

have the same multivariate distribution function.

Second-order stationarity is a less demanding condition than strict stationarity.

---

[1] Throughout this chapter, we will assume a spatio-temporal rf $Z(\mathbf{s}, t)$ on $\mathbb{R}^d \times \mathbb{R}$ which fulfills the *property of regularity*, that is, that $V(Z(\mathbf{s}, t)) < \infty, \forall (\mathbf{s}, t) \in \mathbb{R}^d \times \mathbb{R}$, which therefore assures the existence of the two first moments.

Definition 5.3.3 *A spatio-temporal rf $Z(\mathbf{s}, t)$ is second-order stationary if:*

$$E(Z(\mathbf{s}, t)) = \mu(\mathbf{s}, t) = \mu, \quad \forall(\mathbf{s}, t) \in \mathbb{R}^2 \times \mathbb{R} \tag{5.6}$$

$$C\left(Z(\mathbf{s}_i, t_i), Z(\mathbf{s}_j, t_j)\right) = C(\mathbf{h}, u) \quad \forall(\mathbf{s}, t) \in \mathbb{R}^2 \times \mathbb{R} \tag{5.7}$$

that is, if it has a constant mean and the covariance function depends on **h** *and* u.

Definition 5.3.4 *A spatio-temporal rf $Z(\mathbf{s}, t)$ is intrinsically stationary if the increment process $Z(\mathbf{s} + \mathbf{h}, t + u) - Z(\mathbf{s}, t)$ is second-order stationary in space-time for any fixed* $(\mathbf{h}, t)$.

Definition 5.3.5 *The spatio-temporal rf $Z(\mathbf{s}, t)$ is said to have a spatially stationary covariance function if, for any two pairs $(\mathbf{s}_i, t_i)$ and $(\mathbf{s}_j, t_j)$ on $\mathbb{R}^d \times \mathbb{R}$, the covariance $C((\mathbf{s}_i, t_i), (\mathbf{s}_j, t_j))$ only depends on the distance between the locations $(\mathbf{s}_i - \mathbf{s}_j)$ and the times t_i and t_j.*

Definition 5.3.6 *The spatio-temporal rf $Z(\mathbf{s}, t)$ is said to have a temporally stationary covariance function if, for any two pairs $(\mathbf{s}_i, t_i)$ and $(\mathbf{s}_j, t_j)$ on $\mathbb{R}^d \times \mathbb{R}$, the covariance $C((\mathbf{s}_i, t_i), (\mathbf{s}_j, t_j))$ only depends on the distance between the times $(t_i - t_j)$ and the spatial locations $\mathbf{s}_i$ and $\mathbf{s}_j$.*

Definition 5.3.7 *If the spatio-temporal rf $Z(\mathbf{s}, t)$ has a stationary covariance function in both spatial and temporal terms, then it is said to have a stationary covariance function. In this case, the covariance function can be expressed as*

$$C((\mathbf{s}_i, t_i), (\mathbf{s}_j, t_j)) = C(\mathbf{h}, u), \tag{5.8}$$

$\mathbf{h} = s_i - \mathbf{s}_j$ *and* $u = t_i - t_j$ *being the distances in space and time, respectively.*

Note that if a rf has a stationary covariance function, then its variance does not depend on the spatio-temporal location, as

$$V(Z(\mathbf{s}, t)) = C(\mathbf{0}, 0) = \sigma^2, \quad \forall(\mathbf{s}, t) \in \mathbb{R}^d \times \mathbb{R},$$

where $C(\mathbf{0}, 0) \geq 0$ receives the name of *rf a priori variance*.

In Figure 5.2(a), the pairs of spatio-temporal locations linked with arrows are at the same spatial distance $\mathbf{s}_i - \mathbf{s}_j$ but at different time distances $t_i - t_j$. Thus, under the hypothesis of a spatially stationary spatio-temporal covariance, the covariance between the pairs of variables at the spatio-temporal locations linked with arrows will depend on t_i and t_j but, from the spatial perspective, will only depend on $\mathbf{s}_i - \mathbf{s}_j$, which is the same in the three cases (the specific spatial location of the variables being irrelevant). Having said that, since the pairs (t_i, t_j) are different for the three pairs of spatio-temporal locations the three covariances will be different.

In (b), in the pairs of spatio-temporal locations linked with arrows, $t_i - t_j$ is the same but $\mathbf{s}_i - \mathbf{s}_j$ is different in the three cases. Thus, if the spatio-temporal covariance function governing the process is a temporally stationary covariance function, the

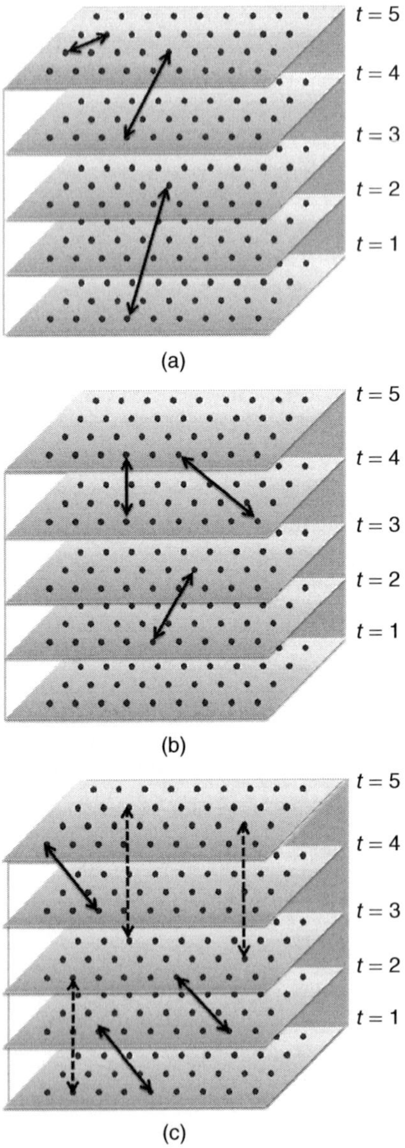

Figure 5.2 (a) Three pairs of spatio-temporal locations with the same $\mathbf{s}_i - \mathbf{s}_j$*. (b) Three pairs of spatio-temporal locations with the same* $t_i - t_j$*. (c) Two sets of three pairs of variables, with both the same* $\mathbf{s}_i - \mathbf{s}_j$ *and also* $t_i - t_j$ *for the three pairs of each set.*

covariance between such pairs of variables will depend on the specific locations $\mathbf{s}_i$ and $\mathbf{s}_j$ but, from the temporal point of view will only depend on $t_i - t_j$ and not on the specific instants of time. Although $t_i - t_j$ is always the same, since the pairs of spatial locations are different in the three cases, the three corresponding covariances will be different.

Finally, in (c), both $\mathbf{s}_i - \mathbf{s}_j$ and $t_i - t_j$ are the same for the three pairs of spatio-temporal locations linked with continuous arrows and for the other three linked with dashed ones. Thus, under the hypothesis of spatial and temporal stationarity, the covariances between the three pairs of variables associated with locations linked with continuous arrows will be the same; also it will be the same covariance between the pairs of variables corresponding to locations linked with dashed arrows.

Definition 5.3.8 *A spatio-temporal rf* $Z(\mathbf{s}, t)$ *has a separable covariance function if there is a purely spatial covariance function* $C_s(\mathbf{s}_i, \mathbf{s}_j)$ *and a purely temporal covariance function* $C_t(t_i, t_j)$ *such that*

$$C((\mathbf{s}_i, t_i), (\mathbf{s}_j, t_j)) = C_s(\mathbf{s}_i, \mathbf{s}_j)C_t(t_i, t_j) \tag{5.9}$$

for any pair of spatio-temporal locations $(\mathbf{s}_i, t_i)$ *and* $(\mathbf{s}_j, t_j) \in \mathbb{R}^d \times \mathbb{R}$. *If this breakdown is not possible, the covariance function will be called non-separable.*

Breaking down the covariance into the product of purely spatial and purely temporal covariances permits more efficient inferences in terms of computing. It is for this reason that the separable models have been the most widely used in geostatistical applications, even in situations when this hypothesis was not justified by the very nature of the process under analysis.

Definition 5.3.9 *A spatio-temporal rf* $Z(\mathbf{s}, t)$ *has fully symmetric covariance function if*

$$C((\mathbf{s}_i, t_i), (\mathbf{s}_j, t_j)) = C\left((\mathbf{s}_i, t_j), (\mathbf{s}_j, t_i)\right) \tag{5.10}$$

for any pair of spatio-temporal locations $(\mathbf{s}_i, t_i)$ *and* $(\mathbf{s}_j, t_j) \in \mathbb{R}^d \times \mathbb{R}$.

Some atmospheric, environmental and geophysical processes are often influenced by air or sea currents, giving rise to a lack of full symmetry. These types of "transport effects" are well known in the literature and have been addressed by Gneiting (2002), Stein (2005a) and Huang and Hsu (2004), among others. Note that separability is a particular case of full symmetry (see Figure 5.6) and, as such, any test to verify full symmetry can be used to reject separability.

In the case of stationary spatio-temporal covariance functions, the condition of full symmetry reduces to:

$$C(\mathbf{h}, u) = C(\mathbf{h}, -u) = C(-\mathbf{h}, u) = C(-\mathbf{h}, -u), \quad \forall (\mathbf{h}, u) \in \mathbb{R}^d \times \mathbb{R}. \tag{5.11}$$

Note that, due to the symmetry of the covariance function, $C(\mathbf{h}, u) = C(-\mathbf{h}, -u)$ and $C(\mathbf{h}, -u) = C(-\mathbf{h}, u)$ will always be fulfilled in the case of stationary rf's and,

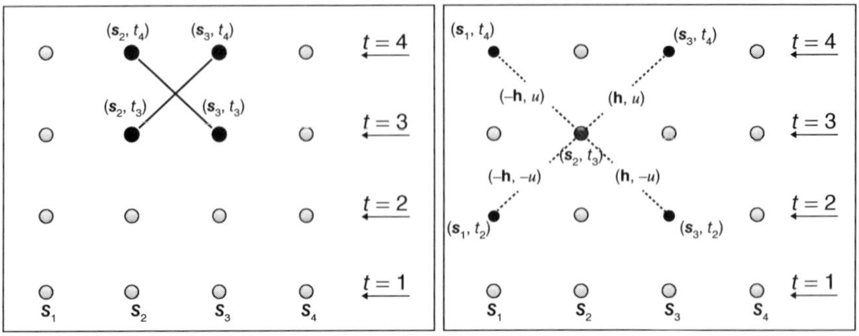

(a) (b)

Figure 5.3 (a) Two pairs of spatio-temporal locations with the same covariance under the assumption of full symmetry. (b) Five spatio-temporal locations such that, under the hypotheses of stationarity and full symmetry, the covariance between the peripheral locations and the central one is the same.

therefore, verifying full symmetry comes down to confirming that $C(\mathbf{h}, u) = C(\mathbf{h}, -u)$ (or $C(-\mathbf{h}, u) = C(-\mathbf{h}, -u)$).

Consider the two pairs of spatio-temporal locations linked with a line in Figure 5.3(a) (the representation is on $\mathbb{R} \times T$ for the sake of simplicity). Under the full symmetry assumption, the covariance between $Z(\mathbf{s}_2, t_4)$ and $Z(\mathbf{s}_3, t_3)$ is the same as the covariance between $Z(s_2, t_3)$ and $Z(s_3, t_4)$. In (b) consider the spatio-temporal locations $(\mathbf{s}_2, t_3)$, $\left(\overbrace{\mathbf{s}_2 - \mathbf{h}}^{s_1}, \overbrace{t_3 + u}^{t_4}\right)$, $\left(\overbrace{\mathbf{s}_2 + \mathbf{h}}^{s_3}, \overbrace{t_3 + u}^{t_4}\right)$, $\left(\overbrace{\mathbf{s}_2 - \mathbf{h}}^{s_1}, \overbrace{t_3 - u}^{t_2}\right)$ and $\left(\overbrace{\mathbf{s}_2 + \mathbf{h}}^{s_3}, \overbrace{t_3 - u}^{t_2}\right)$. Then, if the rf is stationary, under the assumption of full symmetry the covariance between the pairs of variables $Z(\mathbf{s}_2, t_3)$ and $Z(\mathbf{s}_1, t_4), Z(\mathbf{s}_2, t_3)$ and $Z(\mathbf{s}_3, t_4), Z(\mathbf{s}_2, t_3)$ and $Z(\mathbf{s}_1, t_2)$, and $Z(\mathbf{s}_2, t_3)$ and $Z(\mathbf{s}_3, t_2)$ is the same.

Example 5.1 (Full symmetry) *Consider the spatio-temporal dataset displayed in Figure 5.4, which is composed of* 48 *observations measured at the* 16 *nodes of a* 500 *meter regularly spaced grid on three successive instants of time.*

In the case of a stationary spatio-temporal covariance function, the property of full symmetry, for $\mathbf{h} = 1500$ *and* $u = 1$ *implies that the covariance of the spatio-temporal pairs joined with a discontinuous black line in Figure 5.5 (left panel), that is,* $C(\mathbf{s} + 1500, t + 1) = C(\mathbf{s} - 1500, t - 1)$ *coincides with the covariance of the spatio-temporal pairs joined with a discontinuous black line in Figure 5.5 (right panel), that is to say* $C(\mathbf{s} + 1500, t - 1) = C(\mathbf{s} - 1500, t + 1)$.

In other words, the covariance of the two joint distributions in Table 5.1 must be the same.

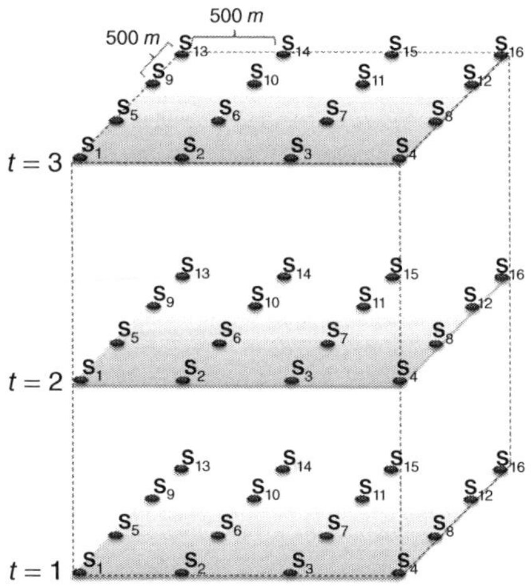

Figure 5.4 Regularly spaced grid.

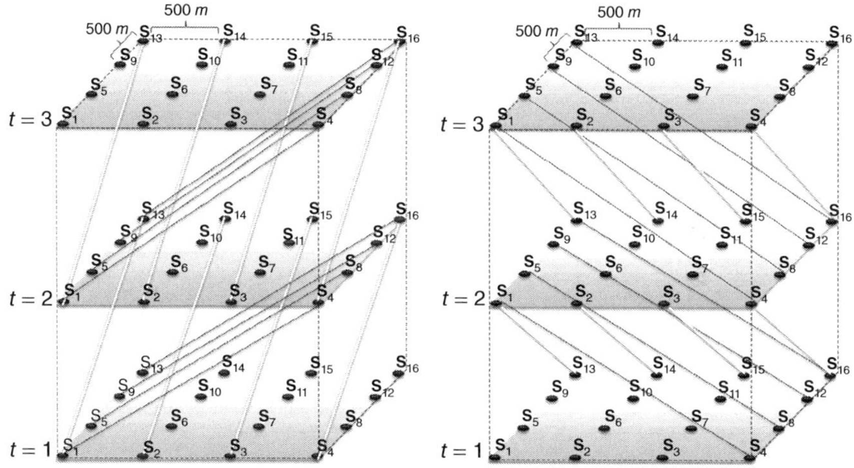

Figure 5.5 Full symmetry.

Of course, for a covariance function to be fully symmetric $C(\mathbf{h}, u)$ *must coincide with* $C(\mathbf{h} - u)$ *for all values of both* $\mathbf{h}$ *and* u.

Definition 5.3.10 *A spatio-temporal rf has a compactly supported covariance function if, for any pair of spatio-temporal locations* $(\mathbf{s}_1, t_1)$ *and* $(\mathbf{s}_2, t_2) \in \mathbb{R}^d \times \mathbb{R}$, *the covariance function* $C((\mathbf{s}_1, t_1), (\mathbf{s}_2, t_2))$ *tends towards zero when the spatial or temporal distance is sufficiently large.*

Those rf's with a compactly supported covariance function are appealing from a computing perspective. For this reason, some authors impose this condition when

Table 5.1 Full symmetry.

$Z(\mathbf{s}, t)$	$Z(\mathbf{s} + 1500, t + 1)$	$Z(\mathbf{s}, t)$	$Z(\mathbf{s} + 1500, t - 1)$
$Z(\mathbf{s}_1, t_2)$	$Z(\mathbf{s}_4, t_3)$	$Z(\mathbf{s}_1, t_3)$	$Z(\mathbf{s}_4, t_2)$
$Z(\mathbf{s}_5, t_2)$	$Z(\mathbf{s}_8, t_3)$	$Z(\mathbf{s}_5, t_3)$	$Z(\mathbf{s}_8, t_2)$
$Z(\mathbf{s}_9, t_2)$	$Z(\mathbf{s}_{12}, t_3)$	$Z(\mathbf{s}_9, t_3)$	$Z(\mathbf{s}_{12}, t_2)$
$Z(\mathbf{s}_{13}, t_2)$	$Z(\mathbf{s}_{16}, t_3)$	$Z(\mathbf{s}_{13}, t_3)$	$Z(\mathbf{s}_{16}, t_2)$
$Z(\mathbf{s}_1, t_2)$	$Z(\mathbf{s}_{13}, t_3)$	$Z(\mathbf{s}_1, t_3)$	$Z(\mathbf{s}_{13}, t_2)$
$Z(\mathbf{s}_2, t_2)$	$Z(\mathbf{s}_{14}, t_3)$	$Z(\mathbf{s}_2, t_3)$	$Z(\mathbf{s}_{14}, t_2)$
$Z(\mathbf{s}_3, t_2)$	$Z(\mathbf{s}_{15}, t_3)$	$Z(\mathbf{s}_3, t_3)$	$Z(\mathbf{s}_{15}, t_2)$
$Z(\mathbf{s}_4, t_2)$	$Z(\mathbf{s}_{16}, t_3)$	$Z(\mathbf{s}_4, t_3)$	$Z(\mathbf{s}_{16}, t_2)$
$Z(\mathbf{s}_1, t_1)$	$Z(\mathbf{s}_4, t_2)$	$Z(\mathbf{s}_1, t_2)$	$Z(\mathbf{s}_4, t_1)$
$Z(\mathbf{s}_5, t_1)$	$Z(\mathbf{s}_8, t_2)$	$Z(\mathbf{s}_5, t_2)$	$Z(\mathbf{s}_8, t_1)$
$Z(\mathbf{s}_9, t_1)$	$Z(\mathbf{s}_{12}, t_2)$	$Z(\mathbf{s}_9, t_2)$	$Z(\mathbf{s}_{12}, t_1)$
$Z(\mathbf{s}_{13}, t_1)$	$Z(\mathbf{s}_{16}, t_2)$	$Z(\mathbf{s}_{13}, t_2)$	$Z(\mathbf{s}_{16}, t_1)$
$Z(\mathbf{s}_1, t_1)$	$Z(\mathbf{s}_{13}, t_2)$	$Z(\mathbf{s}_1, t_2)$	$Z(\mathbf{s}_{13}, t_1)$
$Z(\mathbf{s}_2, t_1)$	$Z(\mathbf{s}_{14}, t_2)$	$Z(\mathbf{s}_2, t_2)$	$Z(\mathbf{s}_{14}, t_1)$
$Z(\mathbf{s}_3, t_1)$	$Z(\mathbf{s}_{15}, t_2)$	$Z(\mathbf{s}_3, t_2)$	$Z(\mathbf{s}_{15}, t_1)$
$Z(\mathbf{s}_4, t_1)$	$Z(\mathbf{s}_{16}, t_2)$	$Z(\mathbf{s}_4, t_2)$	$Z(\mathbf{s}_{16}, t_1)$

kriging is applied to datasets whose high spatial resolution makes them difficult to analyze.

Figure 5.6 reproduces the schematic illustration of the relationships between separable, fully symmetric, stationary, and compactly supported covariances within the general class of (stationary or non-stationary) spatio-temporal covariance functions.

Definition 5.3.11 *A stationary spatio-temporal covariance function* $C(\mathbf{h}, u)$ *defined on* $\mathbb{R}^d \times \mathbb{R}$ *is said to verify Taylor's (1938) hypothesis if there is a velocity vector* $V \in \mathbb{R}^d$ *such that*

$$C(\mathbf{0}, u) = C(Vu, 0), \quad \forall u \in \mathbb{R}. \tag{5.12}$$

As can be observed, Taylor's hypothesis focuses exclusively on the existence of a certain type of relationship between the marginal covariance functions, $C(\mathbf{0}, u)$ and $C(\mathbf{h}, 0)$, of the rf.

Definition 5.3.12 *A stationary spatio-temporal rf* $Z(\mathbf{s}, t)$ *has a spatially isotropic covariance function if*

$$C(\mathbf{h}, u) = C(\|\mathbf{h}\|, u), \quad \forall (\mathbf{s}, t) \in \mathbb{R}^d \times \mathbb{R}. \tag{5.13}$$

Definition 5.3.13 *A stationary spatio-temporal rf* $Z(\mathbf{s}, t)$ *has a temporally isotropic (or symmetric) covariance function if*

$$C(\mathbf{h}, u) = C(\mathbf{h}, |u|), \quad \forall (\mathbf{s}, t) \in \mathbb{R}^d \times \mathbb{R}. \tag{5.14}$$

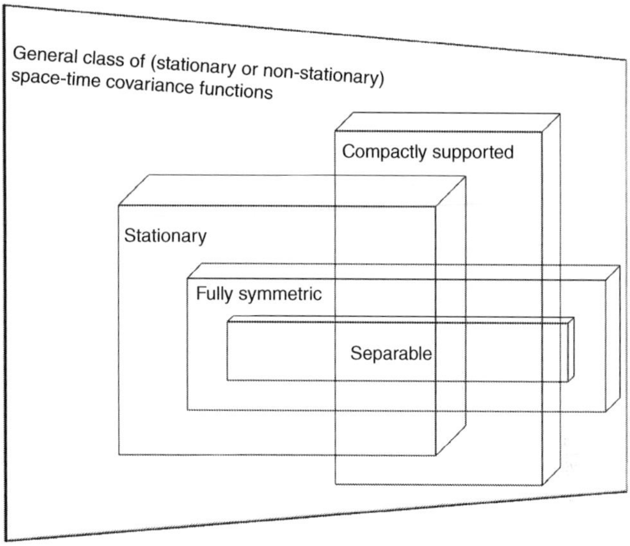

Figure 5.6 Relationships between the different types of spatio-temporal covariance functions (adapted from Gneiting et al. 2007).

Note that if the covariance function of a stationary rf is isotropic in space and time, then it is fully symmetric.

As stated in Gneiting *et al.* (2007, p. 155), semivariograms are occasionally used instead of covariance functions to model the second-order structure of a spatio-temporal rf.

Definition 5.3.14 *The spatio-temporal variogram is defined as the function*

$$2\gamma((\mathbf{s}_i,t_i),(\mathbf{s}_j,t_j)) = V(Z(\mathbf{s}_i,t_i) - Z(\mathbf{s}_j,t_j)), \tag{5.15}$$

while half this quantity is called a semivariogram.

In the case of a rf with a constant mean,

$$\gamma((\mathbf{s}_i,t_i),(\mathbf{s}_j,t_j)) = \frac{1}{2}E\big(Z(\mathbf{s}_i,t_i) - Z(\mathbf{s}_j,t_j)\big)^2. \tag{5.16}$$

Whenever it is possible to define the covariance function and the semivariogram, they will be related by means of the following expression:

$$\gamma((\mathbf{s}_i,t_i),(\mathbf{s}_j,t_j)) = \frac{1}{2}V\left(Z(\mathbf{s}_i,t_i)\right) + \frac{1}{2}V\left(Z(\mathbf{s}_j,t_j)\right) - C\left((\mathbf{s}_i,t_i),(\mathbf{s}_j,t_j)\right). \tag{5.17}$$

Definition 5.3.15 *A spatio-temporal rf $Z(\mathbf{s},t)$ is said to have an intrinsically stationary semivariogram in space if, for any pair of spatio-temporal locations $(\mathbf{s}_i,t_i)$ and*

$(\mathbf{s}_j,t_j)$ *on* $\mathbb{R}^d \times \mathbb{R}$, *the semivariogram* $\gamma((\mathbf{s}_i,t_i),(\mathbf{s}_j,t_j))$ *only depends on* $\mathbf{h} = \mathbf{s}_i - \mathbf{s}_j$ *and the times* t_i *and* t_j.

Definition 5.3.16 *A spatio-temporal rf* $Z(\mathbf{s},t)$ *is said to have an intrinsically stationary semivariogram in time if, for any pair of spatio-temporal locations* $(\mathbf{s}_i,t_i)$ *and* $(\mathbf{s}_j,t_j)$ *on* $\mathbb{R}^d \times \mathbb{R}$, *the semivariogram* $\gamma((\mathbf{s}_i,t_i),(\mathbf{s}_j,t_j))$ *only depends on* $u = t_i - t_j$ *and the spatial locations* $\mathbf{s}_i$ *and* $\mathbf{s}_j$.

Definition 5.3.17 *If the spatio-temporal rf* $Z(\mathbf{s},t)$ *has an intrinsically stationary semivariogram in both space and time, then it is said to have an intrinsically stationary semivariogram. In this case, the semivariogram can be expressed as*

$$\gamma((\mathbf{s}_i,t_i),(\mathbf{s}_j,t_j)) = \gamma(\mathbf{h},u), \tag{5.18}$$

$\mathbf{h} = s_i - \mathbf{s}_j$ *and* $u = t_i - t_j$ *representing the distance in space and time, respectively. The constraints* $\gamma(\cdot,u)$ *and* $\gamma(\mathbf{h},\cdot)$ *are called purely spatial and purely temporal semivariograms, respectively.*

Definition 5.3.18 *A spatio-temporal rf* $Z(\mathbf{s},t)$ *is said to be intrinsically stationary if it has a constant mean and an intrinsically stationary semivariogram.*

Intrinsic stationarity is less restrictive than second-order stationarity, as, given a second-order stationary rf $Z(\mathbf{s},t)$ with a covariance function $C(\mathbf{h},u)$, then, due to (5.17), it satisfies the following property:

$$\begin{aligned} V\left(Z(\mathbf{s}_i,t_i) - Z(\mathbf{s}_j,t_j)\right) &= V(Z(\mathbf{s}_i,t_i)) + V(Z(\mathbf{s}_j,t_j)) - 2C((\mathbf{s}_i,t_i),(\mathbf{s}_j,t_j)) \\ &= 2C(\mathbf{0},0) - 2C(\mathbf{s}_i - \mathbf{s}_j, t_i - t_j) \end{aligned}$$

and for this reason the rf $Z(\mathbf{s},t)$ will be intrinsically stationary with a semivariogram

$$\gamma(\mathbf{h},u) = C(\mathbf{0},0) - C(\mathbf{h},u). \tag{5.19}$$

This expression shows the relationship of the semivariogram with the covariance function under stationarity. The reciprocal will generally not be true, as the relationship of the covariance function with the semivariogram is only verified if the semivariogram $\gamma(\mathbf{h},u)$ is bounded, which is not necessarily the case for the intrinsically stationary semivariograms. While the property of separability makes sense for spatio-temporal covariance functions, it does not in the case of the spatio-temporal semivariogram, as the product of semivariograms does not assure a valid semivariogram. In contrast, it is possible to generalize other properties such as full symmetry or spatial or temporal isotropy.

Another widely used function when modeling implicit spatio-temporal dependence in stationary rf's is the correlation function.

Definition 5.3.19 *Let $Z(\mathbf{s}, t)$ be a spatio-temporal second-order stationary rf with a priori variance $\sigma^2 = C(\mathbf{0},0) > 0$. The correlation function of this rf is defined as*

$$\rho(\mathbf{h}, u) = \frac{C(\mathbf{h}, u)}{C(\mathbf{0}, 0)}. \tag{5.20}$$

Evidently, if $\rho(\mathbf{h}, u)$ is a correlation function on $\mathbb{R}^d \times \mathbb{R}$, then its marginal functions $\rho(\mathbf{0}, u)$ and $\rho(\mathbf{h}, 0)$ will respectively be the spatial correlation function on $\mathbb{R}^d$ and the temporal correlation function on $\mathbb{R}$.

Definition 5.3.20 *A spatio-temporal rf is ergodic if the dependence between observations tends to* 0 *when the separation between them increases (both in terms of space and time).*

Ergodicity is a necessary property to be able to estimate the characteristics of a rf using one sole realization, the reason being that if the rf is not ergodic, when the size of the realization increases, no additional information is acquired, due to all the observations being highly dependent.

5.4 Properties of the spatio-temporal covariance and semivariogram

In the previous section we defined the covariance function and the semivariogram for a spatio-temporal rf. In the first part of this text, it was stated that the semivariogram was the tool usually used to model the spatial variability of a rf. However, when dealing with spatio-temporal rf's, the covariance function is used more often to model the spatio-temporal variability of the rf. Notwithstanding, it must be taken into account that whenever it is possible to define both functions, due to (5.17), we can shift easily from one to another.

A *necessary and sufficient condition* for a real-valued function $C((\mathbf{s}_i, t_i), (\mathbf{s}_j, t_j))$, defined on $\mathbb{R}^d \times \mathbb{R}$ to be a covariance function is for it to be symmetric, $C((\mathbf{s}_i, t_i), (\mathbf{s}_j, t_j)) = C((\mathbf{s}_j, t_j), (\mathbf{s}_i, t_i))$, and positive-definite, that is,

$$\sum_{i=1}^{n} \sum_{j=1}^{n} a_i a_j C((\mathbf{s}_i, t_i), (\mathbf{s}_j, t_j)) \geq 0 \tag{5.21}$$

for any $n \in \mathbb{N}$, and for any $(\mathbf{s}_i, t_i) \in \mathbb{R}^d \times \mathbb{R}$ and $a_i \in \mathbb{R}, i = 1, \ldots, n$. The condition (5.21) is *sufficient* if the covariance function can take complex values. In practical applications, it is much more difficult to verify this condition in the spatio-temporal case than in the merely spatial context.

Similarly, a *necessary and sufficient* condition for a non-negative function of real values $\gamma((\mathbf{s}_i, t_i), (\mathbf{s}_j, t_j))$ defined on $\mathbb{R}^d \times \mathbb{R}$ to be a semivariogram is that it is a symmetric function, $\gamma((\mathbf{s}_i, t_i), (\mathbf{s}_j, t_j)) = \gamma((\mathbf{s}_j, t_j), (\mathbf{s}_i, t_i))$, and conditionally negative-definite, that is,

$$\sum_{i=1}^{n} \sum_{j=1}^{n} a_i a_j \gamma((\mathbf{s}_i, t_i), (\mathbf{s}_j, t_j)) \leq 0 \tag{5.22}$$

for any $n \in \mathbb{N}$, and for any $(\mathbf{s}_i, t_i) \in \mathbb{R}^d \times \mathbb{R}$ and $a_i \in \mathbb{R}, i = 1, \ldots, n$ with $\sum_{i=1}^{n} a_i = 0$.

The results from Schoenberg (1938) led to the following theorem characterizing the spatio-temporal semivariogram:

Theorem 5.4.1 (Berg *et al.* 1984, p. 74) *Let $\gamma((\mathbf{s}_i, t_i), (\mathbf{s}_j, t_j))$ be a function defined on $\mathbb{R}^d \times \mathbb{R}$, with $\gamma((\mathbf{s}, t), (\mathbf{s}, t)) = 0, \forall (\mathbf{s}, t) \in \mathbb{R}^d \times \mathbb{R}$. Then the following statements are equivalent:*

- $\gamma((\mathbf{s}_i,t_i), (\mathbf{s}_j,t_j))$ *is a semivariogram on* $\mathbb{R}^d \times \mathbb{R}$.
- $\exp\left(-\theta\gamma((\mathbf{s}_i,t_i), (\mathbf{s}_j, t_j))\right)$ *is a covariance function on* $\mathbb{R}^d \times \mathbb{R}, \forall \theta > 0$.
- $\omega((\mathbf{s}_i, t_i), (\mathbf{s}_j, t_j)) = \gamma((\mathbf{s}_i, t_i), (\mathbf{0},0)) + \gamma((\mathbf{s}_j, t_j), (\mathbf{0},0)) - \gamma((\mathbf{s}_i, t_i), (\mathbf{s}_j, t_j))$ *is a covariance function on* $\mathbb{R}^d \times \mathbb{R}$.

In the case of stationarity, we can deduce from the foregoing results that a function of real values $C(\mathbf{h}, u)$ defined on $\mathbb{R}^d \times \mathbb{R}$ is a stationary covariance function if, and only if, it is an even function ($C(\mathbf{h}, u) = C(-\mathbf{h}, -u)$) and positive-definite, that is,

$$\sum_{i=1}^{n} \sum_{j=1}^{n} a_i a_j C(\mathbf{h}_i, u_i) \geq 0, \tag{5.23}$$

for any $n \in \mathbb{N}$, and for any $(\mathbf{h}_i, u_i) \in \mathbb{R}^d \times \mathbb{R}$ and $a_i \in \mathbb{R}, i = 1, \ldots, n$.

In the same way, a non-negative function of real values $\gamma(\mathbf{h}, u)$ defined on $\mathbb{R}^d \times \mathbb{R}$ is an intrinsically stationary semivariogram if, and only if, it is an even function ($\gamma(\mathbf{h}, u) = \gamma(-\mathbf{h}, -u)$) and conditionally negative-definite, that is,

$$\sum_{i=1}^{n} \sum_{j=1}^{n} a_i a_j \gamma(\mathbf{h}_i, u_i) \leq 0, \tag{5.24}$$

for any $n \in \mathbb{N}$, and for any $(\mathbf{h}_i, u_i) \in \mathbb{R}^d \times \mathbb{R}$ and $a_i \in \mathbb{R}, i = 1, \ldots, n$, with $\sum_{i=1}^{n} a_i = 0$.

Theorem 5.4.1 will also have a stationary equivalent, according to which, given a function $\gamma(\mathbf{h}, u)$ defined on $\mathbb{R}^d \times \mathbb{R}$, with $\gamma(\mathbf{0}, 0) = 0$, the following statements are equivalent:

- $\gamma(\mathbf{h}, u)$ is an intrinsically stationary semivariogram on $\mathbb{R}^d \times \mathbb{R}$.
- $\exp(-\theta\gamma(\mathbf{h}, u))$ is a stationary covariance function on $\mathbb{R}^d \times \mathbb{R}, \forall \theta > 0$.

- $\omega((\mathbf{s}_i, t_i), (\mathbf{s}_j, t_j)) = \gamma(\mathbf{s}_i, t_i) + \gamma(\mathbf{s}_j, t_j) - \gamma(\mathbf{s}_i - \mathbf{s}_j, t_i - t_j)$ is a covariance function on $\mathbb{R}^d \times \mathbb{R}$.

Having now defined the spatio-temporal covariance function and semivariogram, the rest of this section generalizes the properties of the spatial covariance function and spatial semivariogram to the spatio-temporal case, considering them as functions on $\mathbb{R}^{d+1} \equiv \mathbb{R}^d \times \mathbb{R}$.

1. If $C_1(\mathbf{h}, u)$ is a covariance function defined on $\mathbb{R}^d \times \mathbb{R}$ and $b > 0$, then
$$C(\mathbf{h}, u) = bC_1(\mathbf{h}, u) \tag{5.25}$$
is also a covariance function defined on $\mathbb{R}^d \times \mathbb{R}$.

2. Let $C_1(\mathbf{h}, u)$ and $C_2(\mathbf{h}, u)$ be two covariance functions defined on $\mathbb{R}^d \times \mathbb{R}$, then
$$C(\mathbf{h}, u) = C_1(\mathbf{h}, u) + C_2(\mathbf{h}, u) \tag{5.26}$$
is a covariance function defined on $\mathbb{R}^d \times \mathbb{R}$.

3. First property of stability: given two covariance functions $C_1(\mathbf{h}, u)$ and $C_2(\mathbf{h}, u)$ on $\mathbb{R}^d \times \mathbb{R}$, then
$$C(\mathbf{h}, u) = C_1(\mathbf{h}, u)C_2(\mathbf{h}, u) \tag{5.27}$$
is also a covariance function defined on $\mathbb{R}^d \times \mathbb{R}$.

4. Second property of stability: let $\mu(a)$ be a bounded non-negative measure defined on $U \subset \mathbb{R}$ and let $C(\mathbf{h}, u \mid a)$ be a covariance function defined on $\mathbb{R}^d \times \mathbb{R}$, for each $a \in V \subset U$, which can be integrated into the set $V \subset U$, then
$$C(\mathbf{h}, u) = \int_V C(\mathbf{h}, u \mid a) d\mu(a) \tag{5.28}$$
is a covariance function on $\mathbb{R}^d \times \mathbb{R}$.

5. Another interesting property that makes it possible to construct new covariance functions is that they remain valid or permissible on the edge, that is, if $\{C_n(\mathbf{h}, u), n = 1, \dots,\}$ is a sequence of covariance functions on $\mathbb{R}^d \times \mathbb{R}$ converging to $C(\mathbf{h}, u), \forall (\mathbf{h}, u) \in \mathbb{R}^d \times \mathbb{R}$, then
$$C(\mathbf{h}, u) = \lim_{n \to \infty} C_n(\mathbf{h}, u) \tag{5.29}$$
is a covariance function on $\mathbb{R}^d \times \mathbb{R}$.

6. In the case of a stationary spatial covariance function $C_s(\mathbf{h})$ defined on $\mathbb{R}^d$, and a stationary temporal covariance function $C_t(u)$ defined on $\mathbb{R}$, the functions
$$C(\mathbf{h}, u) = C_s(\mathbf{h} + \boldsymbol{\theta} u)$$
$$C(\mathbf{h}, u) = C_t(u + \boldsymbol{\theta}'\mathbf{h}),$$
with $\boldsymbol{\theta} \in \mathbb{R}^d$, are stationary covariance functions on $\mathbb{R}^d \times \mathbb{R}$.

7. Let $\boldsymbol{\theta}$ be a random vector on $\mathbb{R}^d$. Then

$$C(\mathbf{h}, u) = E_{\boldsymbol{\theta}}\left(C_s(h + \boldsymbol{\theta} u)\right) \tag{5.30}$$

$$C(\mathbf{h}, u) = E_{\boldsymbol{\theta}}\left(C_t(u + \boldsymbol{\theta}'\mathbf{h})\right) \tag{5.31}$$

are stationary covariance functions on $\mathbb{R}^d \times \mathbb{R}$ providing the foregoing expectations exist.

8. Bochner's (1933) theorem states that a function $C(\mathbf{h}, u)$ remains defined on $\mathbb{R}^d \times \mathbb{R}$ and is a stationary covariance function if, and only if, it has the following form:

$$C(\mathbf{h}, u) = \int\int e^{i(\boldsymbol{\omega}'\mathbf{h}+\tau u)} dF(\boldsymbol{\omega}, \tau), \quad (\mathbf{h}, u) \in \mathbb{R}^d \times \mathbb{R}, \tag{5.32}$$

where the function F is the distribution function of a non-negative finite measure defined on $\mathbb{R}^d \times \mathbb{R}$, which is known as a spectral distribution function. Therefore, the class of stationary spatio-temporal covariance functions on $\mathbb{R}^d \times \mathbb{R}$ is identical to the class of Fourier transforms of distribution functions of non-negative finite measures on that domain.
If, in addition, the function C is integrable, then the spectral distribution function F is absolutely continuous and the representation (5.32) simplifies to

$$C(\mathbf{h}, u) = \int\int e^{i(\boldsymbol{\omega}'\mathbf{h}+\tau u)} f(\boldsymbol{\omega}, \tau) d\boldsymbol{\omega} d\tau, \quad (\mathbf{h}, u) \in \mathbb{R}^d \times \mathbb{R}, \tag{5.33}$$

where f is a non-negative, continuous and integrable function that is known as a spectral density function. The covariance function C and the spectral density function f then form a pair of Fourier transforms, and

$$f(\boldsymbol{\omega}, \tau) = (2\pi)^{-d-1} \int\int e^{-i(\boldsymbol{\omega}'\mathbf{h}+\tau u)} C(\mathbf{h}, u) d\mathbf{h} du. \tag{5.34}$$

Definition (5.3.8) showed that a stationary spatio-temporal covariance function C was separable if there were two stationary covariance functions $C_s(\mathbf{h})$ and $C_t(u)$ that are purely spatial and purely temporal, respectively, such that $C(\mathbf{h}, u) = C_s(\mathbf{h})C_t(u), \forall(\mathbf{h}, u) \in \mathbb{R}^d \times \mathbb{R}$.
Then, applying the spectral decomposition above, Definition (5.3.8) will be equivalent to the spectral distribution function being decomposed as the product of a spectral distribution function on the spatial domain and a spectral distribution function on the temporal domain.

9. The decomposition (5.32) can be specialized for fully symmetric covariance functions. Let $C(\cdot, \cdot)$ be a continuous function defined on $\mathbb{R}^d \times \mathbb{R}$, then $C(\cdot, \cdot)$ is a fully symmetric stationary covariance function if, and only if, the following decomposition is possible:

$$C(\mathbf{h}, u) = \int\int \cos(\boldsymbol{\omega}'\mathbf{h}) \cos(\tau u) dF(\boldsymbol{\omega}, \tau), \quad (\mathbf{h}, u) \in \mathbb{R}^d \times \mathbb{R}, \tag{5.35}$$

where F is the non-negative and symmetric spectral distribution function defined on $\mathbb{R}^d \times \mathbb{R}$.

10. If the spectral density function f also exists, then f is fully symmetric, that is:

$$f(\boldsymbol{\omega}, \tau) = f(\boldsymbol{\omega}, -\tau) = f(-\boldsymbol{\omega}, \tau) = f(-\boldsymbol{\omega}, -\tau), \quad \forall(\boldsymbol{\omega}, \tau) \in \mathbb{R}^d \times \mathbb{R}. \quad (5.36)$$

All the foregoing results apply to continuous functions. In practice, fitted stationary spatio-temporal covariance functions often include a nugget effect, that is, discontinuity at the origin. In the spatio-temporal case, the nugget effect can be exclusively spatial, exclusively temporal, or spatio-temporal and will therefore be given by

$$C(\mathbf{h}, u) = a\delta_{(\mathbf{h},u)=(\mathbf{0},0)} + b\delta_{\mathbf{h}=\mathbf{0}} + c\delta_{u=0}, \quad (5.37)$$

where a, b, c are non-negative constants and δ is an indicator function.

11. The product and sum of continuous spatio-temporal covariance functions with a nugget effect of the type (5.37), yield valid covariance models. Cressie and Huang (1999) provide a theorem for characterizing the class of stationary spatio-temporal covariance functions under the additional hypothesis of integrability.

12. Let $C(\cdot, \cdot)$ be a continuous, bounded, symmetric and integrable function defined on $\mathbb{R}^d \times \mathbb{R}$, then $C(\cdot, \cdot)$ is a stationary covariance function if, and only if, for a given $u \in \mathbb{R}$,

$$C_{\boldsymbol{\omega}}(u) = \int e^{-i\boldsymbol{\omega}'\mathbf{h}} C(\mathbf{h}, u) d\mathbf{h}, \quad (5.38)$$

is a covariance function for every $\boldsymbol{\omega} \in \mathbb{R}^d$ except, for at most a set of null Lebesgue measures. Gneiting (2002) generalizes this result for C defined on $\mathbb{R}^d \times \mathbb{R}^l$, from which the previous statement is a particular case for $l = 1$. Both the covariance function and the spectral density function are important tools for characterizing random stationary spatio-temporal fields. Mathematically speaking, both functions are closely related as a pair of Fourier transforms. Furthermore, the spectral density function is particularly useful in situations where there is no explicit expression of the covariance function.

6

Spatio-temporal structural analysis (I): empirical semivariogram and covariogram estimation and model fitting

6.1 Introduction

Generalizing the concepts presented in the structural analysis of the merely spatial case (Chapter 3), the key question of this chapter is: How do we express in a function (a covariance function or a semivariogram) the structure of the spatio-temporal dependence or correlation present in the realization observed? This is a key issue in the subsequent process of spatio-temporal kriging prediction, as the success of the spatio-temporal kriging methods depends on the functions yielding information about the spatio-temporal dependence detected.

We can deduce such a structure of spatio-temporal dependence from the observed realization we have in practice, thus constructing the experimental or empirical covariance function or semivariogram. But they may not fulfill (they usually do not do so) the necessary requisites for a function to be a valid model. Thus, a collection of valid models is needed (the larger, the better) to fit one of them to the empirical counterpart. Methods for fitting a spatio-temporal permissible model to

Spatial and Spatio-Temporal Geostatistical Modeling and Kriging, First Edition.
José-María Montero, Gema Fernández-Avilés, and Jorge Mateu.

Companion Website: www.wiley.com/go/montero/spatial

the empirical counterpart are also needed, as well as procedures to validate a model or compare candidate models.

The spatio-temporal covariograms and semivariograms were defined in the Chapter 5. And since the collection of valid models we present in this book is very comprehensive, we study them in Chapter 7. In this chapter we focus on the construction of empirical spatio-temporal covariograms and semivariograms and the procedures to fit one of the spatio-temporal valid models included in the next chapter to such an experimental covariogram or semivariogram. Basically we generalize the concepts presented in the merely spatial case. As in the spatial structural analysis, we also deal with the validation of a spatio-temporal covariogram or semivariogram model and comparison of several candidate models.

6.2 The empirical spatio-temporal semivariogram and covariogram

The empirical estimation of the covariance function or the semivariogram of a spatio-temporal process can be generalized naturally using the procedures for merely spatial processes illustrated in Sections 3.3 and 3.6, respectively.

Let $Z(\cdot,\cdot)$ be an intrinsically stationary process observed on a set of n spatio-temporal locations $\{(\mathbf{s}_1, t_1), \ldots, (\mathbf{s}_n, t_n)\}$. Two very popular alternatives (there are obviously more, see the literature references in Section 3.6) to estimate the semivariogram $\gamma(\cdot,\cdot)$ (and its covariance function $C(\cdot,\cdot)$ if the process is also second-order stationary) using the observed values are the classical estimator proposed by Matheron (1989) and the robust estimator proposed by Cressie and Hawkins (1980).

The classical estimator is obtained by implementing the MoM, which for the semivariogram of the process is given in its most general form by:

$$\hat{\gamma}(\mathbf{h}(l), u(k)) = \frac{1}{2\#N(\mathbf{h}(l), u(k))} \sum_{(\mathbf{s}_i,t_i),(\mathbf{s}_j,t_j)\in N(\mathbf{h}(l),u(k))} (Z(\mathbf{s}_i, t_i) - Z(\mathbf{s}_j, t_j))^2, \qquad (6.1)$$

where $N(\mathbf{h}(l), u(k)) = \{(\mathbf{s}_i, t_i)(\mathbf{s}_j, t_j) : \mathbf{s}_i - \mathbf{s}_j \in T(\mathbf{h}(l)), t_i - t_j \in T(u(k))\}$, $T(\mathbf{h}(l))$ being a tolerance region on $\mathbb{R}^d$ around $\mathbf{h}(l)$ and $T(u(k))$ a tolerance region on $\mathbb{R}$ around $u(k)$, and $\#N(\mathbf{h}(l), u(k))$ the number of different elements in $N(\mathbf{h}(l),u(k))$, with $l = 1, \ldots, L$ and $k = 1, \ldots, K$.

Generally speaking, the areas $T(\mathbf{h}(l))$ and $T(u(k))$ are chosen so as to yield disjoint sets with a sufficient number of elements to generate stable estimates. If it is reasonable to assume the hypothesis of isotropy for the spatial process under analysis, we can choose the area of spatial tolerance around each of the values $\mathbf{h}(l)$ as $[h(l) - d_l/2, h(l) + d_l/2]$, d_l being the spatial tolerance used. It is also common for the temporal component to take values in $\mathbb{Z}$, in which case the empirical semivariogram is calculated for $u(k) = 0, 1, \ldots$, obtained as the subsequent differences in time at which the process is observed.

In the case of the covariance function, the MoM estimator takes the following form:

$$\hat{C}(\mathbf{h}(l), u(k)) = \frac{1}{\#N(\mathbf{h}(l), u(k))} \sum_{\substack{(\mathbf{s}_i,t_i),(\mathbf{s}_j,t_j) \\ \in N(\mathbf{h}(l),u(k))}} (Z(\mathbf{s}_i, t_i) - \bar{Z})(Z(\mathbf{s}_j, t_j) - \bar{Z}), \tag{6.2}$$

where $\bar{Z} = \frac{1}{n}\sum_i Z(\mathbf{s}_i, t_i)$ is an estimator of the mean μ of the rf and $N(\mathbf{h}(l), u(k))$ is defined as previously.

Although the classical estimation method has the advantage of being easy to calculate, it also has some practical drawbacks, such as not being robust in the case of extreme values. In order to avoid this problem, Cressie and Hawkins (1980) define the following semivariogram estimator for the spatio-temporal case:

$$\hat{\gamma}(\mathbf{h}(l), u(k)) = \left(\frac{1}{2\#N(\mathbf{h}(l), u(k))} \sum_{\substack{(\mathbf{s}_i,t_i),(\mathbf{s}_j,t_j) \\ \in N(\mathbf{h}(l),u(k))}} |Z(\mathbf{s}_i, t_i) - Z(\mathbf{s}_j, t_j)|^{\frac{1}{2}} \right)^4 \times \left(0.457 + \frac{0.494}{\#N(\mathbf{h}(l), u(k))} \right)^{-1}.$$

As in the spatial case, these estimators of the covariance function or semivariogram of the rf do not generally speaking fulfill the condition of being positive-definite or conditionally negative-definite, respectively. For this reason, in practice we select a parametric model of covariance or semivariogram that we already know is valid, by estimating the parameters of the model that best fits the values of the empirical estimator.

As said previously, for this purpose Chapter 7 provides a detailed description of the main spatio-temporal valid models of covariance and semivariogram used in the literature, in order to have a sufficiently large and flexible array of models available that enables us to analyze any given natural or social phenomenon.

Example 6.1 (Calculating spatio-temporal distances and $\#N(\mathbf{h}, u)$ for such distances) *Let us suppose that we have a set of spatio-temporal data measured in three moments of time, t_1, t_2 and t_3, on a regular 4×4, 500 meters spaced grid (see Figure 6.1).*

Assuming an isotropic, stationary rf, and opting for not using tolerance regions, the empirical classical spatio-temporal semivariogram will be given by:

$$\hat{\gamma}(\mathbf{h}, u) = \frac{1}{2\#N(\mathbf{h}, u)} \sum_{N(\mathbf{h},u)} (Z(\mathbf{s}_i, t_i) - Z(\mathbf{s}_j, t_j))^2,$$

where $N(\mathbf{h}, u) = \{(\mathbf{s}_i, t_i)(\mathbf{s}_j, t_j)\colon \mathbf{s}_i - \mathbf{s}_j = \mathbf{h}$ and $t_i - t_j = u\}$.

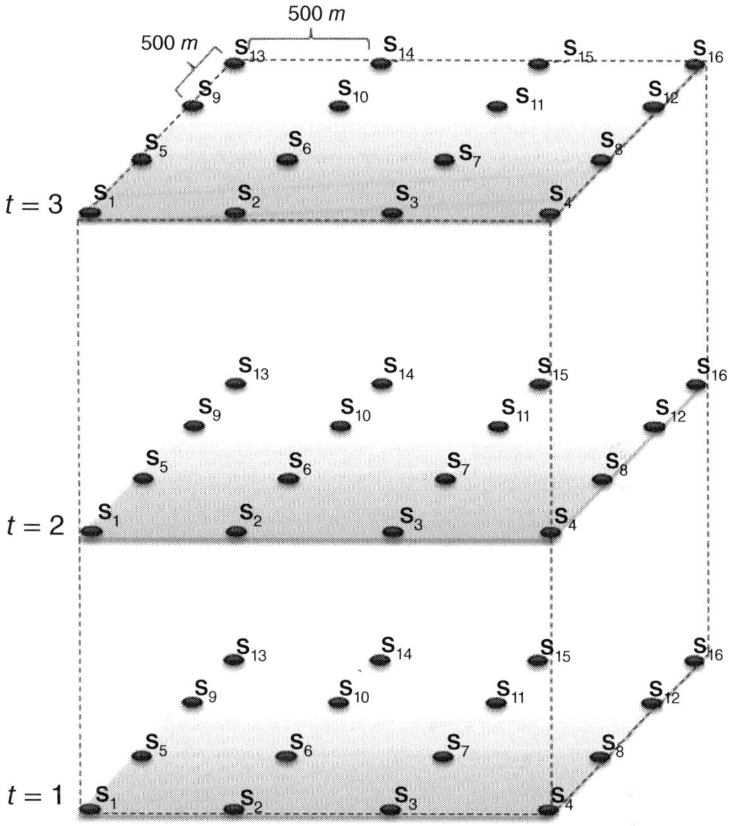

Figure 6.1 *Regular* 500 *meters spaced grid.*

The set of possible spatial distances between the pairs of the locations considered in the grid is { 0, 500, 707.11, 1000, 1119.03, 1414.21, 1500, 1581.14, 1802.78, 2000, 2121.32 }, *whereas the lag times are:* 0, 1 *and* 2. *Thus, combining the distance and the time lags,* $11 \times 3 = 33$ *spatio-temporal distances are initially obtained.*

Computing the semivariances for each spatio-temporal distance $(\mathbf{h}, u)$ *we can obtain the empirical classical spatio-temporal semivariogram. In a similar way, we can obtain the empirical classical spatio-temporal covariogram.*

$$\hat{C}(\mathbf{h}, u) = \frac{1}{\#N(\mathbf{h}, u)} \sum_{N(\mathbf{h},u)} (Z(\mathbf{s}_i, t_i) - \bar{Z})(Z(\mathbf{s}_j, t_j) - \bar{Z}),$$

where $\bar{Z} = \frac{1}{n}\sum_i Z(\mathbf{s}_i, t_i)$ *is an estimator of the mean* μ *of the process and* $N(\mathbf{h}, u)$ *is defined as previously.*

For example, there are $16 \times 3 = 48$ *spatio-temporal points at distance* (0, 0)*, that is,* $\#N(0,0) = 48$*, the* 48 *points included in the spatio-temporal dataset.*

There are $16 \times 2 = 32$ *spatio-temporal points at distance* (0, 1)*:*

$$\{(\mathbf{s}_1,t_3),(\mathbf{s}_1,t_2)\},\{(\mathbf{s}_2,t_3),(\mathbf{s}_2,t_2)\},\ldots,\{(\mathbf{s}_{16},t_3),(\mathbf{s}_{16},t_2)\}$$

$$\{(\mathbf{s}_1,t_2),(\mathbf{s}_1,t_1)\},\{(\mathbf{s}_2,t_2),(\mathbf{s}_2,t_1)\},\ldots,\{(\mathbf{s}_{16},t_2),(\mathbf{s}_{16},t_1)\}$$

There are $16 \times 1 = 16$ *spatio-temporal points at distance (0,2):*

$$\{(\mathbf{s}_1,t_3),(\mathbf{s}_1,t_1)\},\{(\mathbf{s}_2,t_3),(\mathbf{s}_2,t_1)\},\ldots,\{(\mathbf{s}_{16},t_3),(\mathbf{s}_{16},t_1)\}$$

Computing the semivariogram values corresponding to the above spatio-temporal distances the purely temporal, empirical semivariogram is obtained. It is also easy to check that there are $24 \times 3 = 72$ *spatio-temporal points at distance (500,0),* $16 \times 3 = 48$ *at distance (1500,0),... Computing the semivariogram values for the spatio-temporal distances* $(\mathbf{h}, 0)$ *it is obtained the purely spatial, empirical semivariogram.*

*When both spatial and time lags are non-null we must be more cautious when identifying the spatio-temporal points that are separated by a specific distance. For example, for a spatio-temporal distance of (*1500,2*) we can count* 16 *pairs of points separated by such a spatio-temporal distance (see Figure 6.2):*

$$\{(\mathbf{s}_1,t_3),(\mathbf{s}_4,t_1)\},\{(\mathbf{s}_4,t_3),(\mathbf{s}_1,t_1)\},\{(\mathbf{s}_5,t_3),(\mathbf{s}_8,t_1)\},\{(\mathbf{s}_8,t_3),(\mathbf{s}_5,t_1)\},$$
$$\{(\mathbf{s}_9,t_3),(\mathbf{s}_{12},t_1)\},\{(\mathbf{s}_{12},t_3),(\mathbf{s}_9,t_1)\},\{(\mathbf{s}_{13},t_3),(\mathbf{s}_{16},t_1)\},\{(\mathbf{s}_{16},t_3),(\mathbf{s}_{13},t_1)\},$$
$$\{(\mathbf{s}_1,t_3),(\mathbf{s}_{13},t_1)\},\{(\mathbf{s}_{13},t_3),(\mathbf{s}_1,t_1)\},\{(\mathbf{s}_2,t_3),(\mathbf{s}_{14},t_1)\},\{(\mathbf{s}_{14},t_3),(\mathbf{s}_2,t_1)\},$$
$$\{(\mathbf{s}_3,t_3),(\mathbf{s}_{15},t_1)\},\{(\mathbf{s}_{15},t_3),(\mathbf{s}_3,t_1)\},\{(\mathbf{s}_4,t_3),(\mathbf{s}_{16},t_1)\},\{(\mathbf{s}_{16},t_3),(\mathbf{s}_4,t_1)\}.$$

Example 6.2 (Constructing and visualizing the empirical spatio-temporal semivariogram with R) *In Example 6.1 we have illustrated how difficult it is both to identify the number of different spatio-temporal distances and to calculate the number of pairs separated by such distances. In that example, we considered a small number of spatial locations (only* 16*, regularly spaced) and moments in time (only* 3*). Even in such a case, the manual construction of the empirical semivariogram (or covariogram) is such a very hard and time-consuming task that in practice the only way to carry it out is by using some kind of spatio-temporal geostatistical software. This is why in this example we proceed to the construction (and visualization) of an empirical spatio-temporal semivariogram (as well as the merely spatial and temporal ones), using the R packages* `CompRandFld` *and* `RandomFields`*.*

Let us consider a spatio-temporal database consisting of 1200 *data simulated at* 200 *points irregularly spaced on a* $[0,10] \times [0,10]$ *grid at* 6 *instants of time. Figure 6.3 depicts such a spatio-temporal simulated database (the larger and darker*

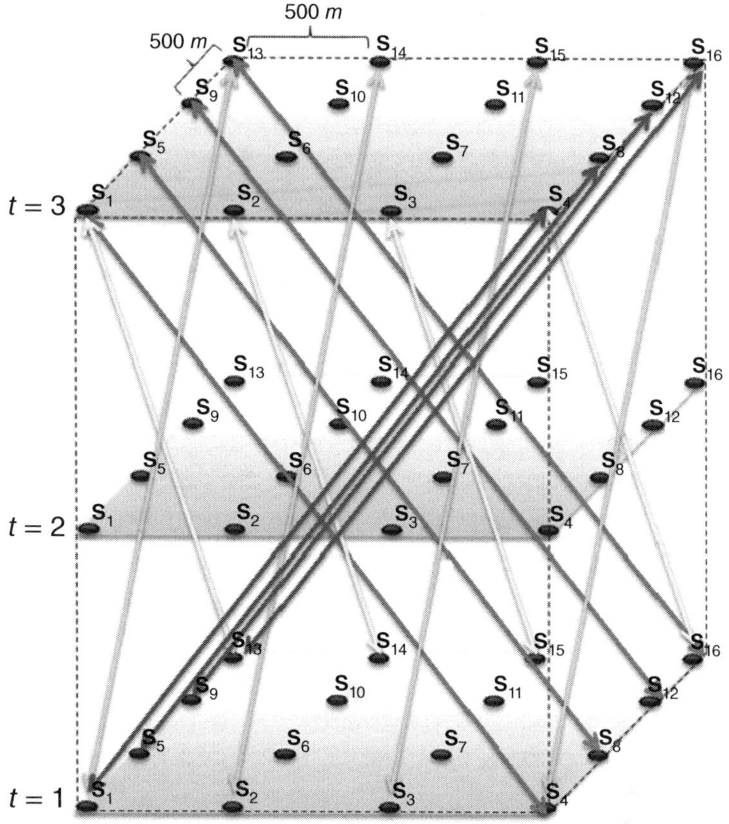

Figure 6.2 Pairs of points separated by a spatio-temporal distance of (1500,2). (See color figure in color plate section.)

the circle, the greater the simulated value). The simulation was performed using as a generating data process a spatio-temporal rf with Gneiting covariance function without a nugget effect, zero mean, unitary sill, and the rest of the parameters being: scaling parameters of space and time: 0–9 *and* 1, *respectively; spatial and temporal smoothing parameters:* 1; *space-time interaction parameter:* 0.5. *The expression of such covariance function is as follows (see Section 7.11):*

$$C(\mathbf{h}, u) = \frac{1}{(0.9|u|^2+1)^{1/2}} \exp\left(-\frac{\|\mathbf{h}\|^2}{\left(|u|^2+1\right)^{1/2}}\right), \quad (\mathbf{h}, u) \in \mathbb{R}^d \times \mathbb{R}. \qquad (6.3)$$

For the case of 200 *points irregularly spaced on a* $[0, 10] \times [0, 10]$ *grid at* 6 *instants of time, the R package* `CompRandFld` *sets* 12 *spatial and* 4 *temporal bins (apart of*

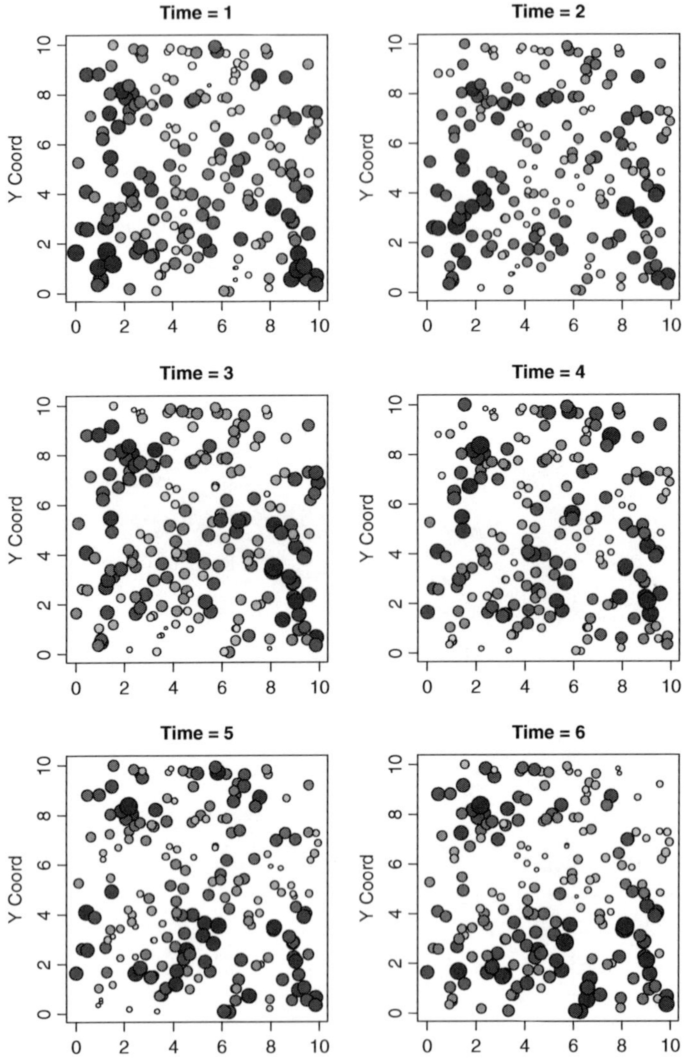

Figure 6.3 1200 *data simulated with the Gneiting non-separable covariance function (6.3) at* 200 *points irregularly spaced on a* $[0, 10] \times [0, 10]$ *grid at* 6 *instants of time.*

the origin). The number of bins in space is as in most of packages; as for time bins, it can be appreciated the semivariances or semivariogram values involving both $t = 6$ *and* $t = 1$ *(that is, a bin of five) are not taken into account.*

The semivariogram values for the set of combinations of spatial and temporal bins are listed in Table 6.1, the number of pairs of spatio-temporal points used for computing such semivariogram values being presented in Table 6.2. In both tables

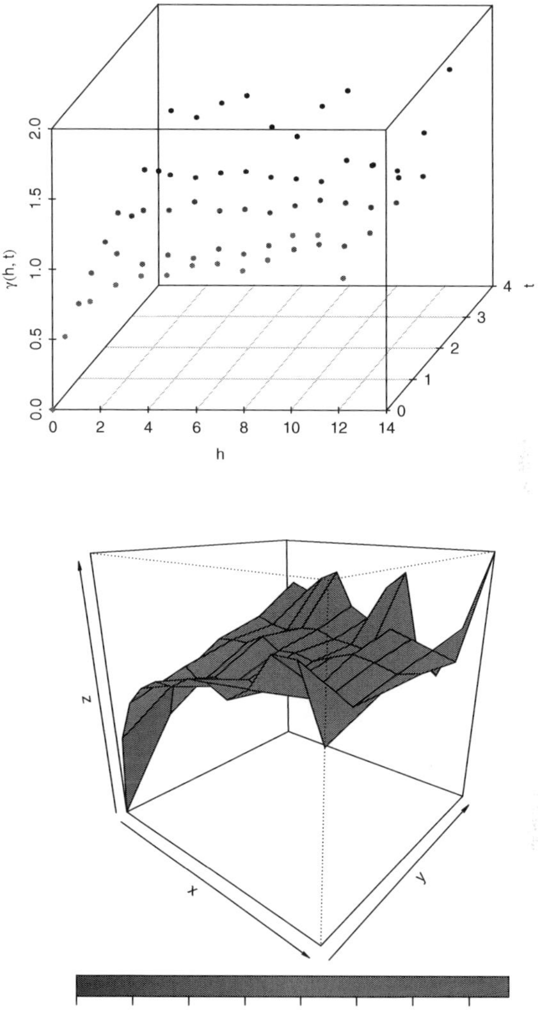

Figure 6.4 The empirical spatio-temporal semivariogram. Upper panel: The point plot version. Bottom panel: The smoothed version.

the semivariogram values (and number of pairs) for the merely spatial and temporal semivariograms are shadowed. Note that the number of pairs does not decrease with the spatial or spatio-temporal distance, due to the fact that the simulation was performed on a set of points irregularly spaced on a $[0, 10] \times [0, 10]$ *grid.*

Figure 6.4 depicts the empirical spatio-temporal semivariogram and a smoothed version of it. Figure 6.5 includes the plots of the purely spatial (left panel) and purely temporal (right panel) semivariograms.

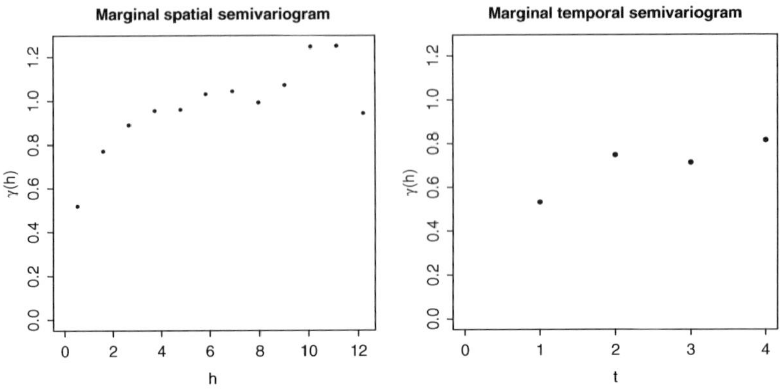

Figure 6.5 Empirical purely spatial (left panel) and temporal (right panel) semivariograms.

Table 6.1 Semivariogram values for the combinations of the 13 spatial and 5 temporal bins.

space \ time	1	2	3	4	5
1	0.00	0.53	0.75	0.71	0.81
2	0.52	0.75	0.96	1.04	1.24
3	0.77	0.89	0.97	1.01	1.19
4	0.89	0.81	0.98	0.99	1.30
5	0.95	0.88	1.04	1.02	1.35
6	0.96	0.86	0.97	1.03	1.13
7	1.03	0.92	0.98	0.99	1.06
8	1.04	0.89	0.96	0.98	1.28
9	0.99	0.95	1.01	0.96	1.39
10	1.07	0.92	1.05	1.11	0.86
11	1.24	0.96	1.03	1.08	0.77
12	1.25	0.95	1.00	1.04	1.09
13	0.94	1.04	1.03	1.00	1.54

6.3 Fitting spatio-temporal semivariogram and covariogram models

As stated in Section 5.3, any function that depends on spatio-temporal distance and direction is not necessarily a valid covariogram or semivariogram. This is the case of the empirical covariogram or semivariogram, which do not necessarily fulfill the requisites to be a valid (and useful) covariogram or semivariogram (in fact, they never

Table 6.2 Number of pairs used for computing the semivariogram values listed in Table 6.1.

space \ time	1	2	3	4	5
1	0	1000	800	600	400
2	3642	6070	28350	24820	5230
3	10620	4856	22680	19856	4184
4	14838	3642	17010	14892	3138
5	17010	2428	11340	9928	2092
6	17982	17700	29970	19690	1380
7	17172	14160	23976	15752	1104
8	14892	10620	17982	11814	828
9	11814	7080	11988	7876	552
10	7284	24730	28620	12140	290
11	3138	19784	22896	9712	232
12	828	14838	17172	7284	174
13	174	9892	11448	4856	116

do it) and cannot be used for spatio-temporal kriging prediction. This is the reason why they must first be replaced by a theoretical (valid) model.

As in the spatial case (Section 3.8), fitting a semivariogram (or covariogram) model can be done by eye (manual fitting), using statistical procedures (automatic fitting), or combining both approaches. Whichever procedure used, what is really important is that the theoretical model chosen in the fitting process captures the type of continuity assumed for the regionalized variable and the hypotheses on stationarity and separability associated with it, because it has more implications than the approach (manual, automatic or combined) used for fitting. Thus, expertise and knowledge about the phenomenon we are studying are welcome in the fitting process.

In manual fitting, we proceed as in the spatial case: trying to capture adequately the main features of the phenomenon (the nugget effect and the behavior near the origin, the range and the sill – both indicative of the behavior at large distances – and anisotropies). As for automatic fitting, any of the least squares (OLS, GLS, and WLS) or maximum likelihood (ML, REML and CL) based methods presented in Section 3.8 can easily be generalized to the spatio-temporal case. All we have to do is to incorporate the temporal argument into the analysis. This is why in this section we focus only on a modification of the Curriero and Lele (1999) particular form of CL to estimate a covariance function based on the semivariogram cloud (see Section 3.8.2), which shares some of the best properties of the LS and ML based methods. In particular, it does not depend on the choice of lag bins considered in the semivariogram, no matrix needs to be inverted, it is robust to poor distributional specification and its order of computation is $O(N^2)$, between that of ML and WLS.

The modification referred above was proposed by Bevilacqua *et al.* (2012), who partially extended it to the spatio-temporal case, and aim to enhance the efficiency of estimates and reduce the computation burden. This issue is of vital importance in the spatio-temporal framework, as normally the number of observations is large enough to make ML method prohibitive in terms of computation. More specifically, the modification by Bevilacqua *et al.* (2012) basically consists of the following modified (weighted) version of (3.70):

$$WCL(\boldsymbol{\theta}, \mathbf{d}) = \frac{1}{W_{N,\mathbf{d}}} \sum_{i=1}^{N} \sum_{j>i}^{N} \ell(U_{ij}; \boldsymbol{\theta}) w_{ij}(\mathbf{d}), \tag{6.4}$$

where N is the number of observed sites $\mathbf{s}_i$, $\mathbf{d} = (d_{\mathbf{s}}, d_t)'$ is a vector including two distances, one spatial and the other temporal, $w_{ij}(\mathbf{d})$ are non-negative weights that depend on the distances $\mathbf{s}_i - \mathbf{s}_j$ and $t_i - t_j$ and $W_{N,\mathbf{d}} = \sum_{i=1}^{N} \sum_{j>i}^{N} w_{ij}(\mathbf{d})$.

Note that the weights $w_{ij}(\mathbf{d})$ do not depend on $\boldsymbol{\theta}$ to preserve the unbiasedness of the associated estimating equation:

$$ee_{WCL}(\boldsymbol{\theta}) = \frac{1}{W_{N,\mathbf{d}}} \sum_{i=1}^{N} \sum_{j>i}^{N} \frac{\gamma_{ij}^{(1)}(\boldsymbol{\theta})}{\gamma_{ij}(\boldsymbol{\theta})} \left(1 - \frac{U_{ij}^2}{2\gamma_{ij}(\boldsymbol{\theta})}\right) w_{ij}(\mathbf{d}) = 0, \tag{6.5}$$

where, for the sake of easier reading, $\gamma_{ij}^{(1)}(\boldsymbol{\theta}) = \gamma^{(1)}(\mathbf{s}_i - \mathbf{s}_j; t_i - t_j; \boldsymbol{\theta}) = \frac{d}{d\boldsymbol{\theta}}\gamma_{ij}(\boldsymbol{\theta}))$.

As can be noted, the proposal by Bevilacqua *et al.* (2012) comes under the framework of the Weighted Composite Likelihood (WCL), a general estimation method defined by Lindsay (1988) when working with large sets of observations (see Section 3.8.2) that has been applied in a large number of fields of science in recent years.

Before continuing with the proposal by Bevilacqua *et al.* (2012), we will stop to remind readers that the semivariogram is traditionally estimated using the empirical semivariogram. In line with Genton (1998), a practical rule is to consider only half the maximum distance in the data and, at least, 30 points. This golden rule is based on tradition and is at least questionable. However, this rule suggests that not considering large distances in the estimation could improve the accuracy of the estimates. Continuing in the same vein, the search for estimators based on differences, considering an optimal distance (in some sense) in the estimation process is interesting in terms of computation and efficiency. For this reason, it is necessary to define objective criteria when determining this optimal distance. It is probably better to build estimators based on the semivariogram cloud, as in general the empirical semivariogram is not unbiased.

Initially, in Equation (6.5), there are many options when choosing the weights w_{ij}. For example, one possibility would be to choose a weighting function that decreases smoothly as distance increases. However, in order to obtain the asymptotic distribution of the WCL estimator $\hat{\theta}_{N,\mathbf{d}}$ it is required, for technical reasons, weighting functions with compact support. Furthermore, a cut-off weighting function has evident computational advantages in comparison, for example, to one with a smooth profile.

It is for these reasons that the proposal by Bevilacqua *et al.* (2012) is to consider cut-off weights, that is, $w_{ij} = 1$ if $\|\mathbf{s}_i - \mathbf{s}_j\| \le d_{\mathbf{s}}, |t_i - t_j| \le d_t$ and 0 otherwise, d_s and d_t being spatial and temporal distances, respectively, selected accordingly for the chosen purpose. Hence, the WCL estimator (that can be proved to be consistent and asymptotically Gaussian under increasing domain asymptotics) depends on $\mathbf{d} = (d_{\mathbf{s}}, d_t)'$ through the weights w_{ij}.

Obviously, the question now is how to select $\mathbf{d} = (d_{\mathbf{s}}, d_t)'$. The Godambe information matrix (Godambe 1991) is defined as:

$$\mathbf{G}_N(\theta, \mathbf{d}) = \mathbf{H}_N(\theta, \mathbf{d})\mathbf{J}_N(\theta, \mathbf{d})^{-1}\mathbf{H}'_N(\theta, \mathbf{d}), \tag{6.6}$$

with:

$$\begin{aligned}
\mathbf{H}_N(\theta, \mathbf{d}) &= -E(ee^{(1)}_{WCL}(\theta, \mathbf{d}))\\
&= -E\left(\frac{1}{W_{N,\mathbf{d}}}\sum_{i=1}^{N}\sum_{j>i}^{N}\frac{\gamma^{(1)}_{ij}(\theta)}{\gamma_{ij}(\theta)}\left(1-\frac{U^2_{ij}}{2\gamma_{ij}(\theta)}\right)w_{ij}(\mathbf{d})\right)^{(1)}\\
&= \frac{1}{W_{N,\mathbf{d}}}\sum_{i=1}^{N}\sum_{j>i}^{N}\left(\frac{\gamma^{(2)}_{ij}(\theta)\gamma_{ij}(\theta) - \gamma^{(1)}_{ij}(\theta)\gamma^{(1)}_{ij}(\theta)}{\gamma^2_{ij}(\theta)}\left(1-\frac{U^2_{ij}}{2\gamma_{ij}(\theta)}\right)\right.\\
&\qquad\left. + \frac{\gamma^{(1)}_{ij}(\theta)}{\gamma_{ij}(\theta)}\frac{2\gamma^{(1)}_{ij}(\theta)'U^2_{ij}}{4\gamma_{ij}(\theta)}\right)w_{ij}(\mathbf{d})\\
&= \frac{1}{W_{N,\mathbf{d}}}\sum_{i=1}^{N}\sum_{j>i}^{N}\frac{\gamma^{(1)}_{ij}(\theta)\gamma^{(1)}_{ij}(\theta)'}{\gamma^2_{ij}(\theta)}w_{ij}(\mathbf{d}),
\end{aligned}$$

because if the rf is Gaussian, $E(U^2_{ij}) = 2\gamma_{ij}(\theta)$, and

$$\begin{aligned}
\mathbf{J}_N(\theta, \mathbf{d}) &= E(e_{WCL}(\theta, \boldsymbol{d})e_{WCL}(\theta, \boldsymbol{d})')\\
&= \frac{2}{W^2_{N,\mathbf{d}}}\sum_{i=1}^{N}\sum_{j>i}^{N}\sum_{l}^{N}\sum_{k>l}^{N}\frac{\gamma^{(1)}_{ij}(\theta)\gamma^{(1)}_{lk}(\theta)'}{\gamma_{ij}(\theta)\gamma_{lk}(\theta)}Corr(U^2_{ij}, U^2_{lk})w_{ij}(\mathbf{d})w_{lk}(\mathbf{d}).
\end{aligned} \tag{6.7}$$

Note that $\mathbf{G}_N(\theta, \mathbf{d})$ depends on $\boldsymbol{d}$ through w_{ij}. Note also that if the rf is Gaussian, then:

$$C(U^2_{ij}U^2_{lk})w_{ij} = 2(\gamma_{il}(\theta) - \gamma_{jl}(\theta) + \gamma_{jk}(\theta) - \gamma_{ik}(\theta)),$$

hence it is easy to evaluate $\mathbf{J}_N(\theta, \mathbf{d})$.

The inverse of $\mathbf{G}_N(\theta, \mathbf{d})$ is an approximation of the asymptotic variance of the WCL estimator. Hence, in order to enhance statistical efficiency, it seems appropriate to choose $\mathbf{d}$ such that $\mathbf{G}_N^{-1}(\boldsymbol{\theta}, \boldsymbol{d})$ minimizes in the partial order of the non-negative definite matrices. Bevilacqua *et al.* (2012) seek the optimum $\mathbf{d}^*$ such that:

$$\mathbf{d}^* = \arg\min_{\mathbf{d} \in D} tr(\mathbf{G}^{-1}(\boldsymbol{\theta}, \mathbf{d})) \tag{6.8}$$

where D is a finite set of pairs of distances in space and time. If the data are observed on a regular grid in space and time, it is easy to define D. Otherwise, it can be very useful to inspect the semivariogram cloud when choosing the lags in D.

Since in (6.8) $\mathbf{d}^*$ depends on $\boldsymbol{\theta}$ and requires $\mathbf{G}_N(\boldsymbol{\theta}, \mathbf{d})$ to be calculated (which is costly in computational terms for large data sets), Bevilacqua *et al.* (2012) recommend estimating $\boldsymbol{\theta}$ by WLS, as a preliminary and consistent estimation before estimating $\mathbf{G}_N(\boldsymbol{\theta}, \mathbf{d})$ using subsampling techniques.

Example 6.3 (Fitting a spatio-temporal semivariogram) *Let us consider a spatio-temporal database consisting of* 300 *simulated data in* 50 *locations irregularly spaced on a* $[0, 1] \times [0, 1]$ *grid at six consecutive instants of time. The simulation was performed using as a generating data process a spatio-temporal mean-zero Gaussian rf with a doubly exponential covariance function (see Equation (7.26) and Figure 7.27 in the next chapter) without the nugget effect, unitary variance, and the space and time scaling parameters* $a_s = 0.5$ *and* $a_t = 2$*, respectively; that is to say:*

$$C(\mathbf{h}, u) = \exp\left(-\|\mathbf{h}\|/0.5 - |u|/2\right), \qquad (\mathbf{h}, u) \in \mathbb{R}^d \times \mathbb{R}. \tag{6.9}$$

Figure 6.6 depicts such a spatio-temporal simulated database (the larger and darker the circle, the greater the simulated value).

The computation was performed with `CompRandFld` *package in R software, which only uses ML-based fitting methods. Since our simulated database is not a massive database we used (full) ML for fitting purposes, the estimates of the parameters being* 1.3320 *for the sill, 0.6555 for the spatial scaling parameter and* 2.2887 *the time scaling one (the estimates using REML are practically the same).*

Figure 6.7 shows the empirical spatio-temporal semivariogram (upper panel) and the theoretical spatio-temporal semivariogram corresponding to the doubly exponential covariance function with the above estimates of the sill and spatial and time scaling parameters (bottom panel).

6.4 Validation and comparison of spatio-temporal semivariogram and covariogram models

As in the spatial case, in the spatio-temporal kriging process, a valid model indicating the spatio-temporal structure of the dependence has been chosen (using any of the procedures mentioned in the previous section) and used, and some other assumptions have been made. However, in order not to obtain results from the kriging process

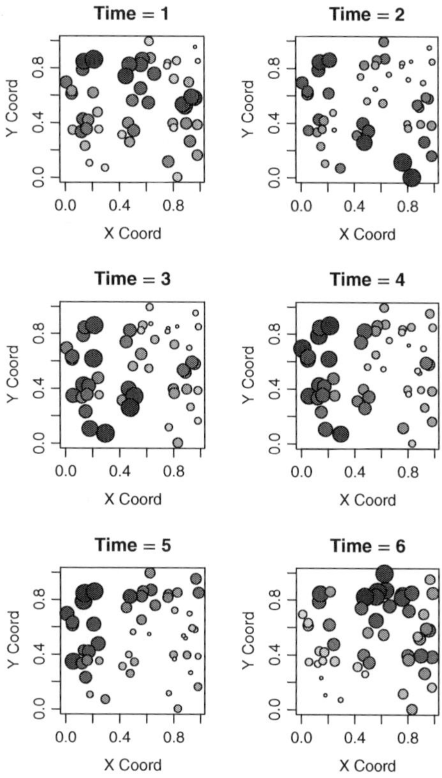

Figure 6.6 300 *spatio-temporal data simulated with a spatio-temporal Gaussian rf with zero mean and doubly exponential covariance function (6.9) in* 50 *irregularly spaced sites on a* $[0, 1] \times [0, 1]$ *grid at* 6 *instants of time.*

that could lead us to erroneous conclusions, we should validate these assumptions, especially the one about the type of model and its parameters.

Let us assume that the values of the process of interest $Z(\mathbf{s}, t)$ have been observed on a given set of n spatio-temporal locations. As in Section 4.9 for the spatial case, if the size of the set permits, it can be divided into two subsets, one for modeling (the fitting set) comprising $n - k$ locations and another for validating, made up of the remaining k locations. Using the fitting set as a basis, a spatio-temporal model will be fitted to the $n - k$ observed data. This model is used to obtain, for each of the k locations $(\mathbf{s}_i, t_i)$ of the validating set $\{Z(\mathbf{s}_1, t_1), \ldots, Z(\mathbf{s}_k, t_k)\}$, the prediction $\hat{Z}(\mathbf{s}_i, t_i)$ and the prediction error $\sigma^2(\mathbf{s}_i, t_i)$ resulting from the spatio-temporal kriging procedure. After obtaining the k predictions and their associated prediction errors, we can calculate the spatio-temporal version of the synthetic statistics listed in Section 4.9, ME (4.85), MSE (4.86), and MSDE (4.87), which will allow the prediction power of the model to be evaluated.

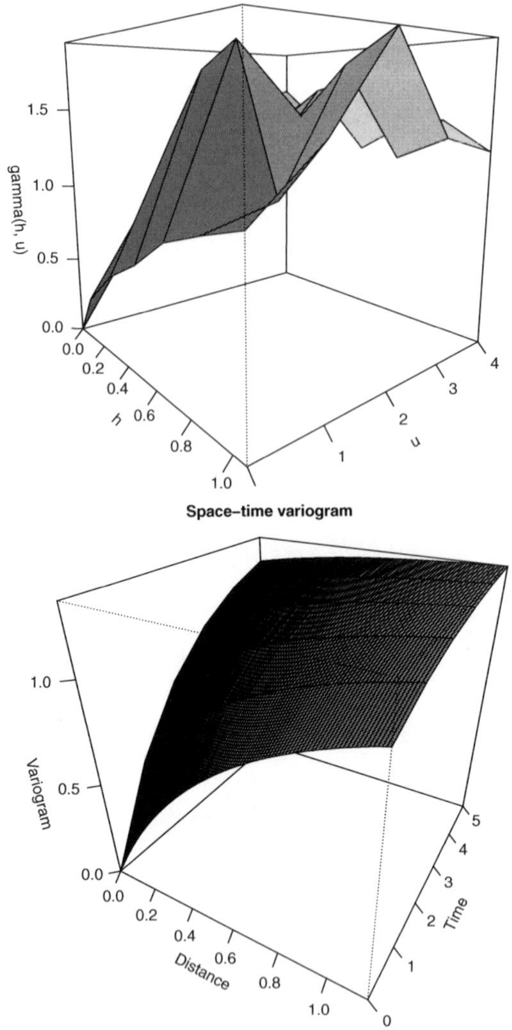

Figure 6.7 Empirical spatio-temporal semivariogram corresponding to the data in Figure 6.6 (upper panel) and the fitted semivariogram corresponding to the doubly exponential covariance function (bottom panel).

However, generally, there will not be sufficient observations to define a large enough fitting set and also a large enough validating set. In this case, cross-validation tools can be used, as we saw in Section 4.9 for the spatial case. Cross-validation or leave-one-out procedure evaluates the prediction power of the selected model by calculating for each of the observations $Z(\mathbf{s}_i, t_i),\ i = 1, 2, \ldots, n$ the predicted values $\hat{Z}_{-i}(\mathbf{s}_i, t_i)$, where the subscript $-i$ indicates that the observation $Z(\mathbf{s}_i, t_i)$ has been removed from the dataset in order to be predicted, and the square errors associated to those predictions $\sigma^2_{-i}(\mathbf{s}_i, t_i)$ obtained on the basis of the set of all the remaining observations,

$$\{Z(\mathbf{s}_1, t_1), \ldots, Z(\mathbf{s}_{i-1}, t_{i-1}), Z(\mathbf{s}_{i+1}, t_{i+1}), \ldots, Z(\mathbf{s}_n, t_n)\}.$$

Although it is desirable for all the above-mentioned synthetic statistics to take the expected values, in practice, an improvement in one can lead to a worsening of another. It is therefore recommended to conduct an integral analysis of all of them. As stated in Section 4.9 for the spatial case, ME should be approximately 0, MSE should be small, MSDE should be approximately 1, the correlation coefficient between the predicted and observed values should be close to 1, and the standardized errors should not depend on the magnitude of the predicted values (no conditional bias).

Finally, a word of caution: As stated in Section 4.9, CV cannot prove that the fitted model is correct; merely that there is no reason for rejecting it. In addition, though the cross-validation method is extremely helpful in deciding between two candidate semivariogram models, it should not be used as if it were an automatic procedure for fitting a permissible covariogram or semivariogram model to its empirical counterpart.

7

Spatio-temporal structural analysis (II): theoretical covariance models

7.1 Introduction

As will be appreciated in Chapter 8, as in the merely spatial case, the key stage of the spatio-temporal kriging prediction procedure is also choosing the function (covariogram or semivariogram) that models the structure of the spatio-temporal dependence of the data. However, while the semivariogram is normally chosen for this purpose in the spatial case, in a spatio-temporal framework the covariance function is the most commonly chosen tool. For this reason, in this chapter we present a survey of the valid or permissible covariance models that are most widely used in spatio-temporal situations, as well as presenting new strategies that are more in keeping with reality than those above and which tackle new challenges that have not, as yet, been studied. In other words, the objective is to make a large and flexible set of models available to researchers in order to allow them to model the majority of physical and also, as far as possible, social processes both efficiently and operationally.

By referring to a valid spatio-temporal covariogram model, we are implicitly stating that the covariance function must be positive-definite. The purely spatial and temporal covariance models have been widely studied and there is a long list of those guaranteeing positive-definiteness, which can be used to model spatial or spatio-temporal dependence.

Spatial and Spatio-Temporal Geostatistical Modeling and Kriging, First Edition.
José-María Montero, Gema Fernández-Avilés, and Jorge Mateu.

Companion Website: www.wiley.com/go/montero/spatial

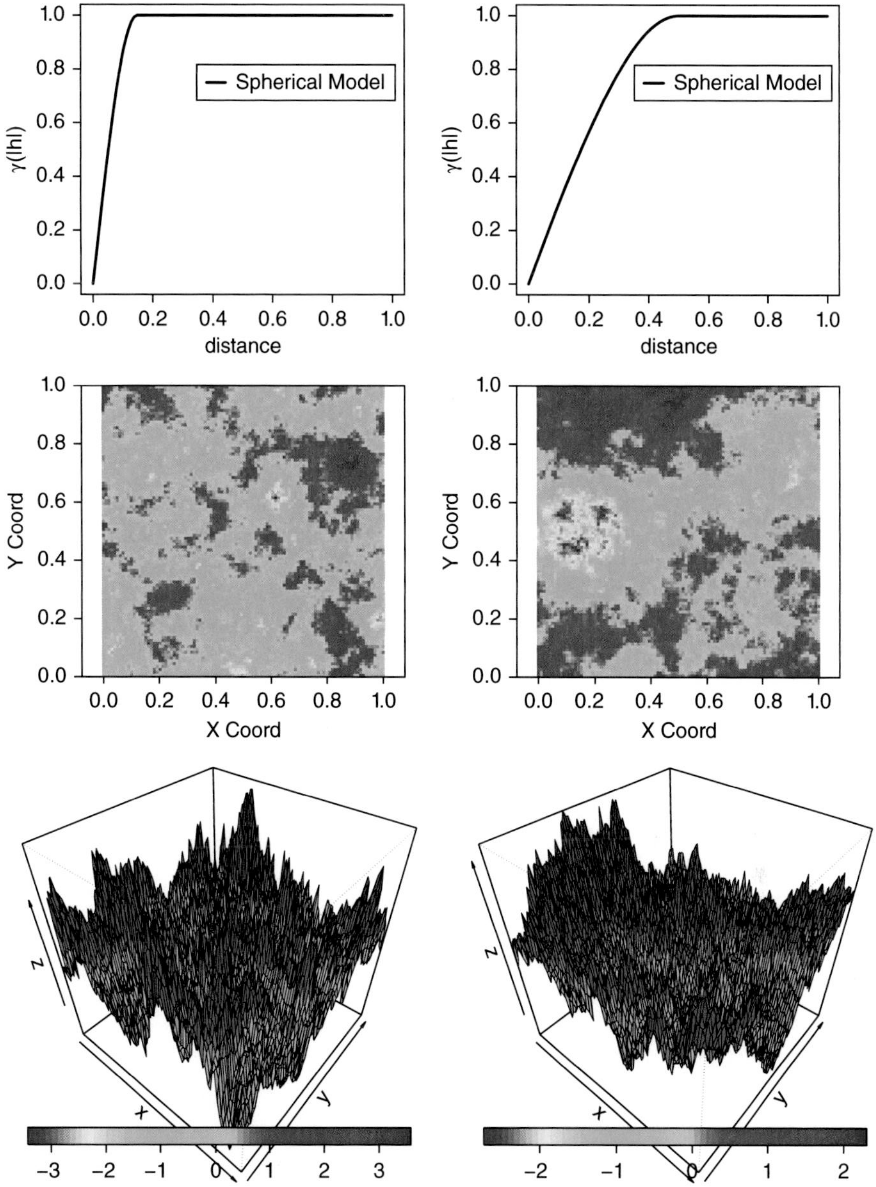

Figure 3.8 Upper panel: Spherical model. Left: $a = 0.15$, $m = 1$. Right: $a = 0.50$, $m = 1$. Middle panel: Simulation of a rf having a spherical semivariogram (2D representation). Left: $a = 0.15$, $m = 1$. Right: $a = 0.50$, $m = 1$. Bottom panel: Simulation of a rf having a spherical semivariogram (3D representation). Left: $a = 0.15$, $m = 1$. Right: $a = 0.50$, $m = 1$.

Spatial and Spatio-Temporal Geostatistical Modeling and Kriging, First Edition.
José-María Montero, Gema Fernández-Avilés, and Jorge Mateu.

Companion Website: www.wiley.com/go/montero/spatial

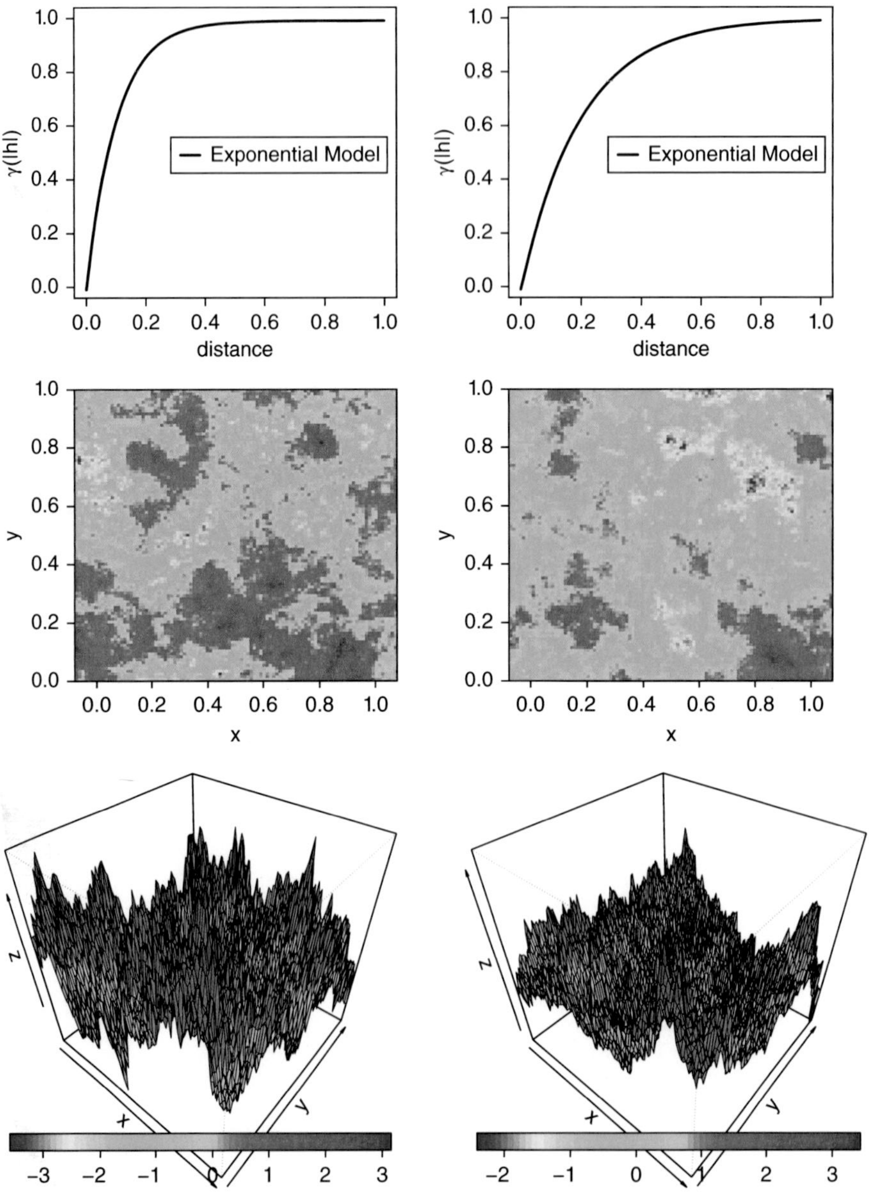

Figure 3.10 Upper panel: Exponential model. Left: $a = 0.10$, $m = 1$. Right: $a = 0.25$, $m = 1$. Middle panel: Simulation of a rf having an exponential semivariogram (2D representation). Left: $a = 0.10$, $m = 1$. Right: $a = 0.25$, $m = 1$. Bottom panel: Simulation of a rf having an exponential semivariogram (3D representation). Left: $a = 0.10$, $m = 1$. Right: $a = 0.25$, $m = 1$.

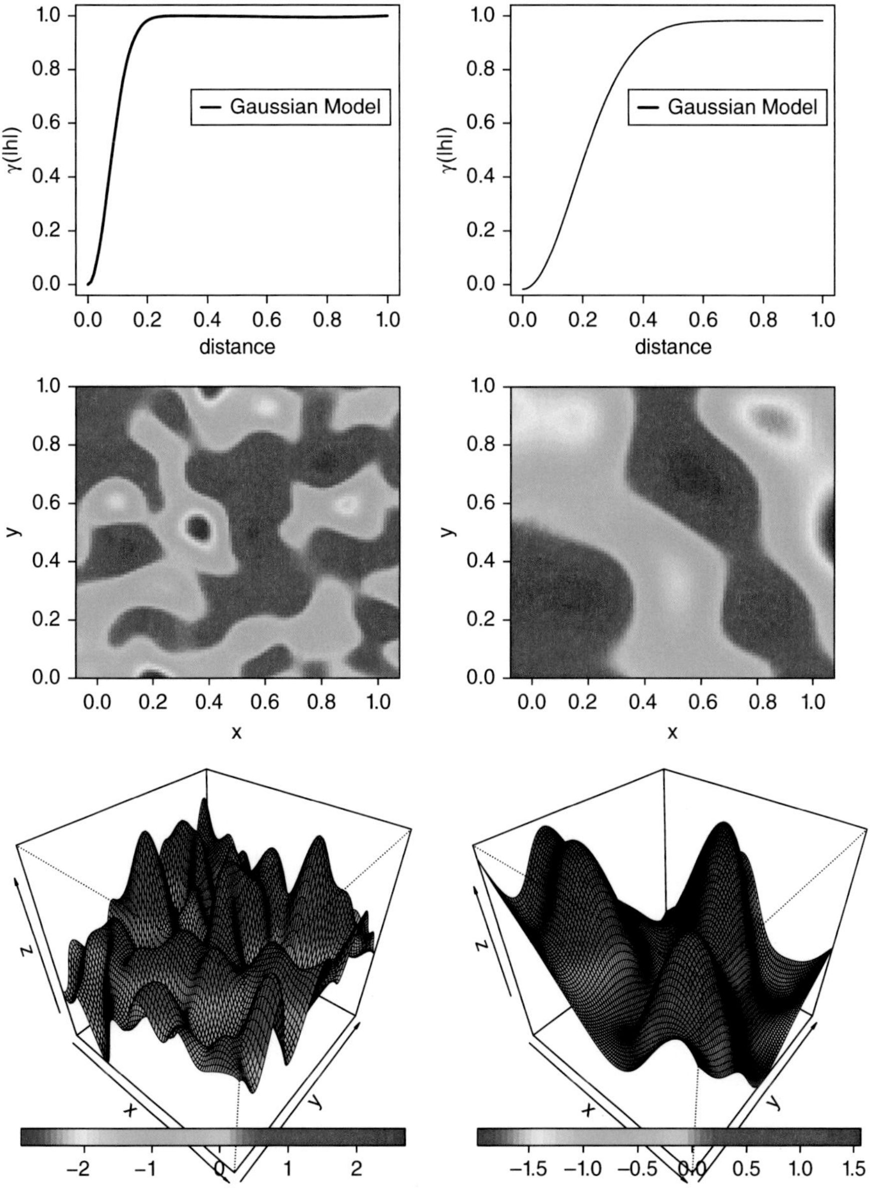

Figure 3.11 Upper panel: Gaussian model. Left: a = 0.15, m = 1. Right: a = 0.25, m = 1. Middle panel: Simulation of a rf having a Gaussian semivariogram (2D representation). Left: a = 0.10, m = 1. Right: a = 0.25, m = 1. Bottom panel: Simulation of a rf having a Gaussian semivariogram (3D representation). Left: a = 0.10, m = 1. Right: a = 0.25, m = 1.

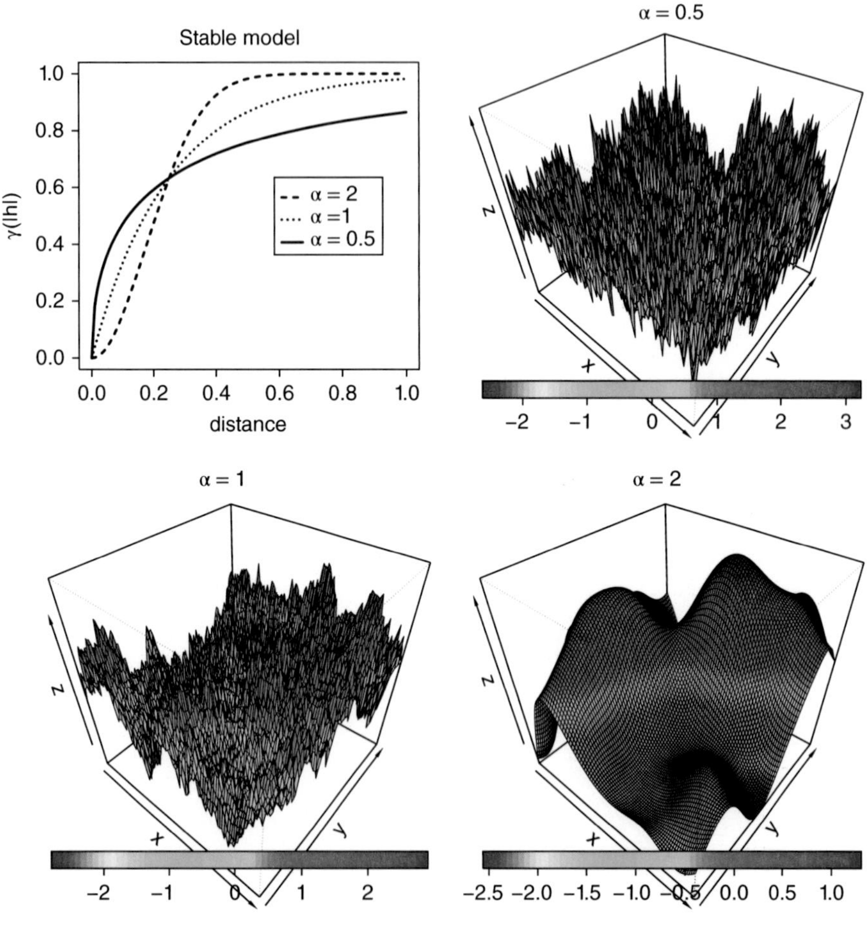

Figure 3.15 *Upper panel, left: Stable model with the same sill ($m = 1$) and scale parameter ($a = 0.25$) but different shape parameter α. Upper panel, right: 3D representation of a simulation of a rf having a stable semivariogram ($a = 0.25$, $m = 1$, $\alpha = 0.5$). Bottom panel, left: 3D representation of a simulation of a rf having a stable semivariogram ($a = 0.25$, $m = 1$, $\alpha = 1$). Bottom panel, right: 3D representation of a simulation of a rf having a stable semivariogram ($a = 0.25$, $m = 1$, $\alpha = 2$).*

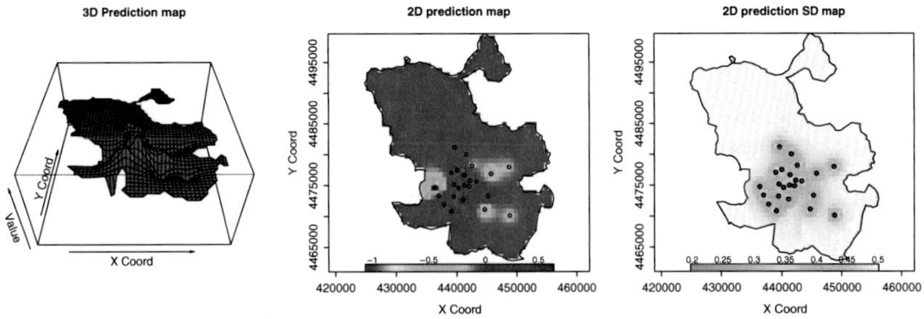

Figure 4.3 Prediction and prediction standard deviation (SD) maps of logCO*: January 2008, 2nd week, 10 am.

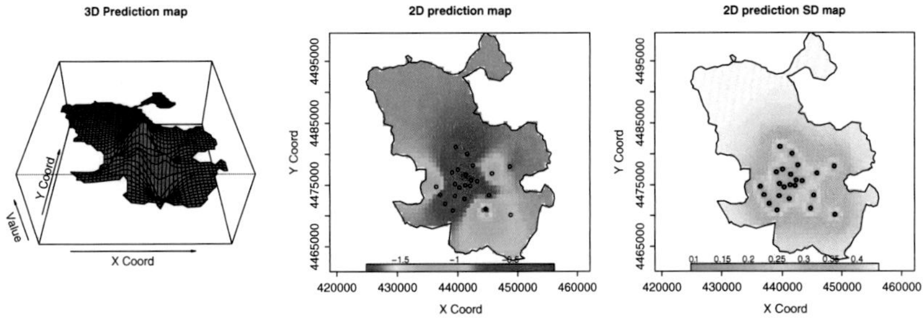

Figure 4.4 Prediction and prediction standard deviation (SD) maps of logCO*: January 2008, 2nd week, 3 pm.

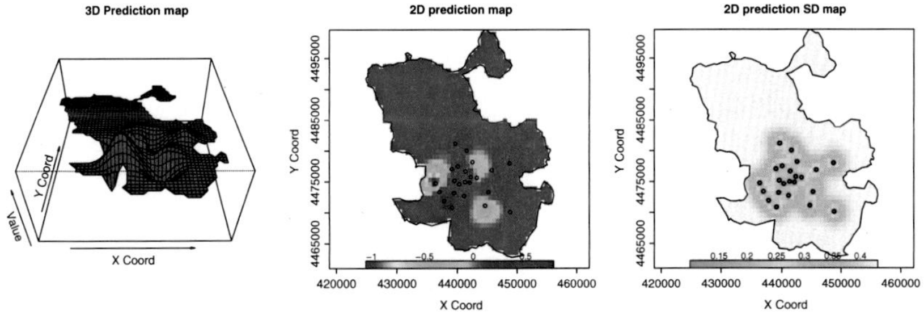

Figure 4.5 Prediction and prediction standard deviation (SD) maps of logCO*: January 2008, 2nd week, 9 pm.

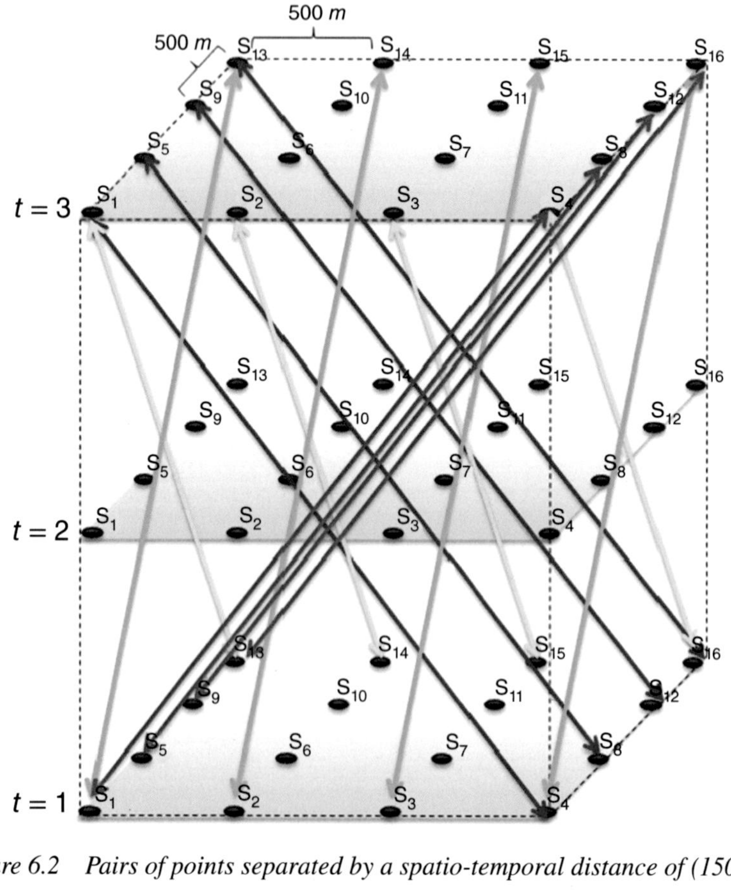

Figure 6.2 Pairs of points separated by a spatio-temporal distance of (1500,2).

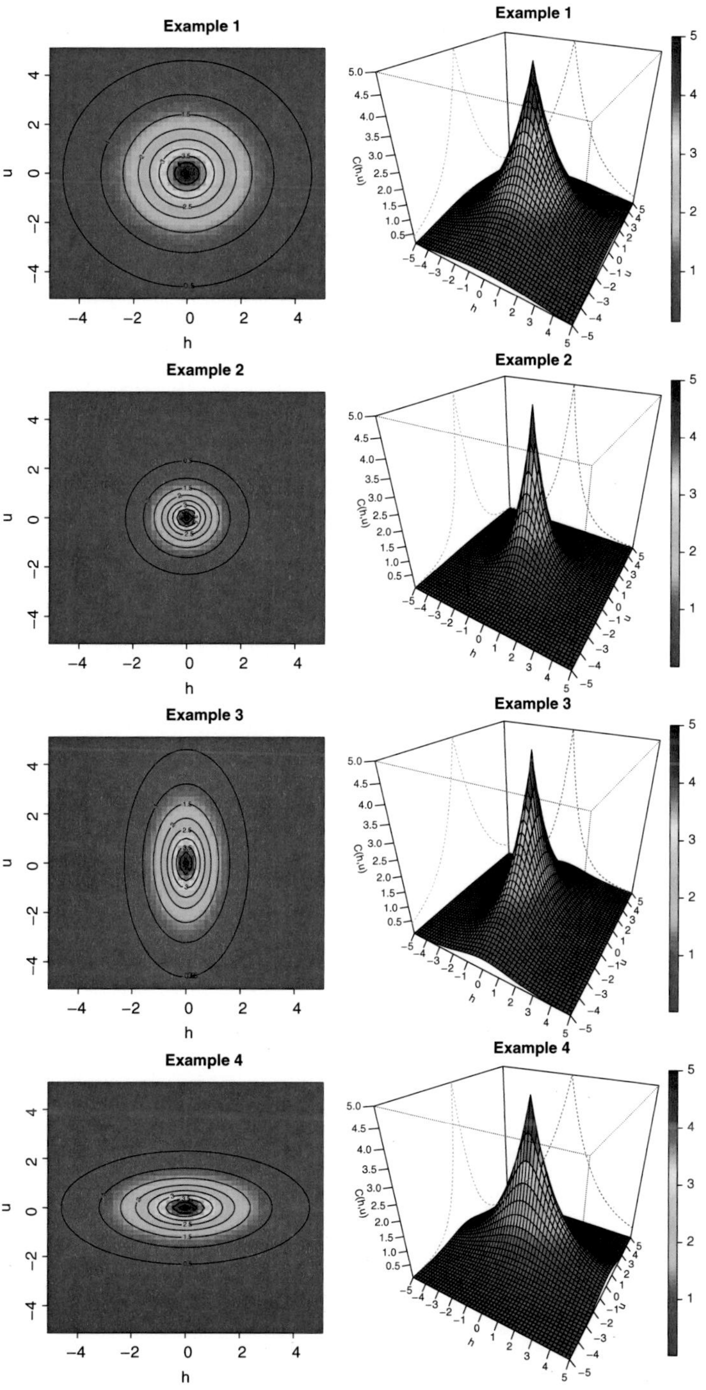

Figure 7.2 2D and 3D different representations of the exponential metric model (7.6).

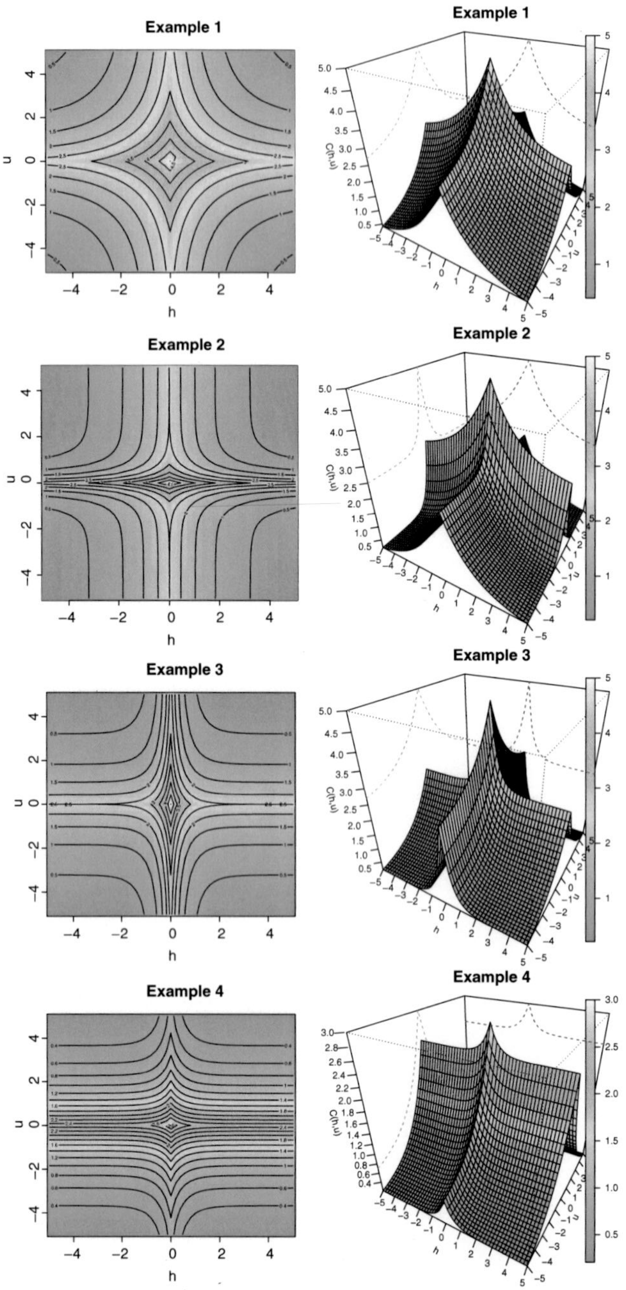

Figure 7.3 *2D and 3D different representations of the exponential sum model (7.13).*

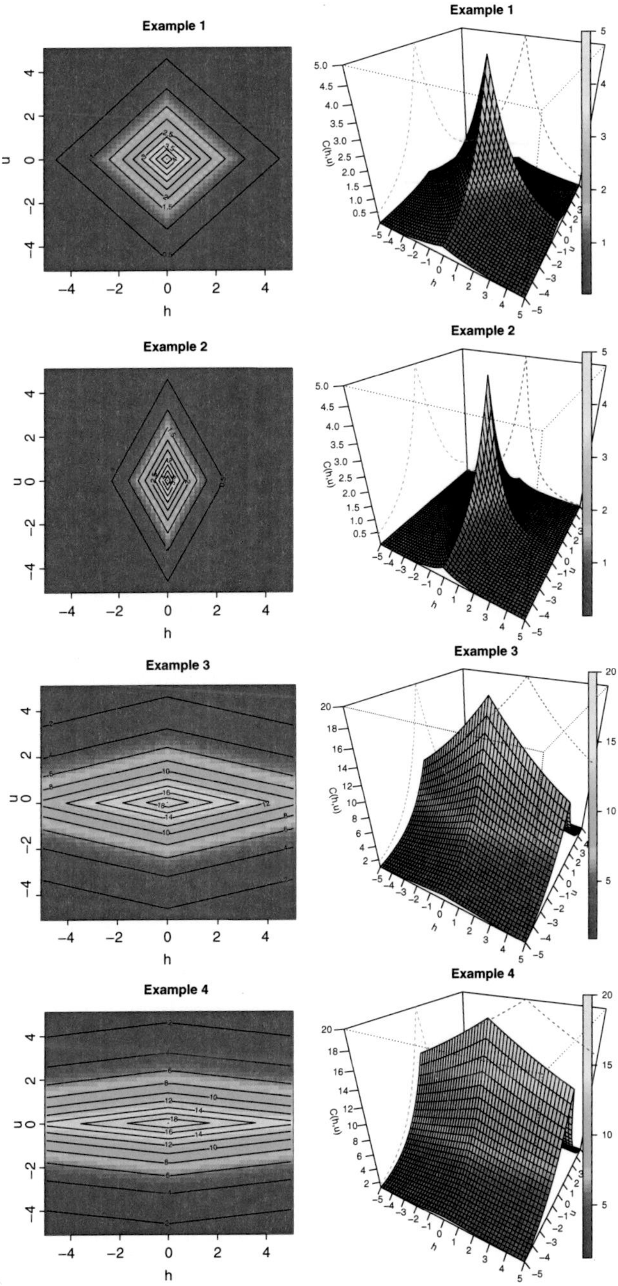

Figure 7.6 2D and 3D different representations of the product model (7.35).

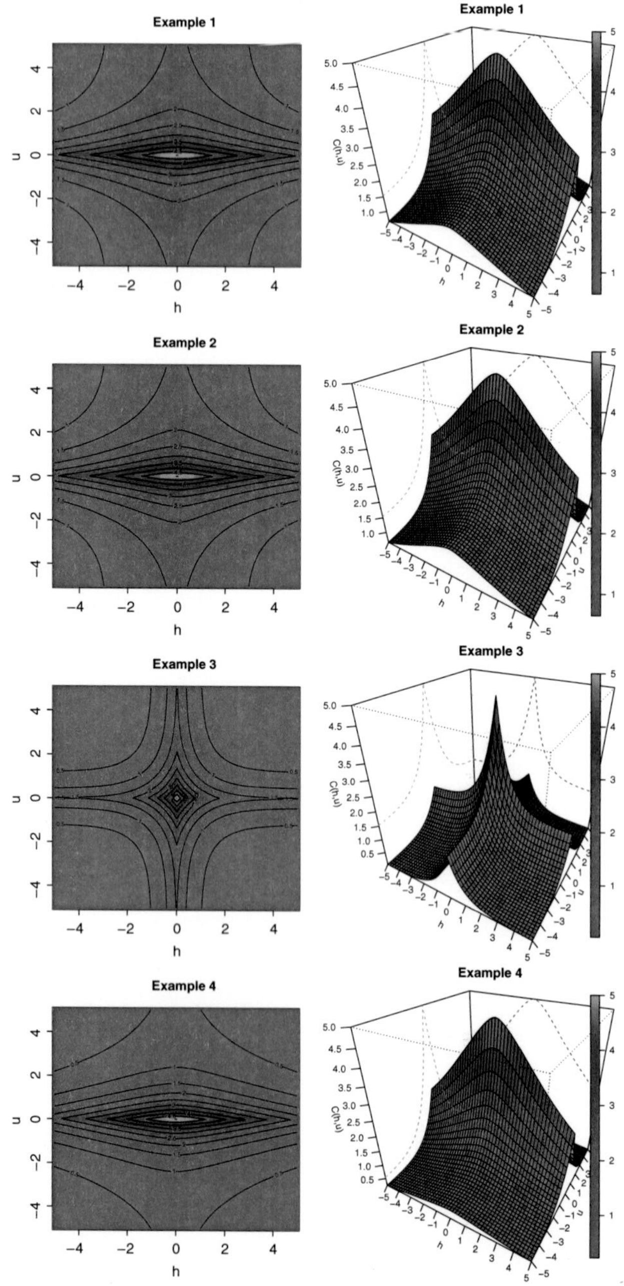

Figure 7.7 *2D and 3D different representations of the product-sum model (7.35).*

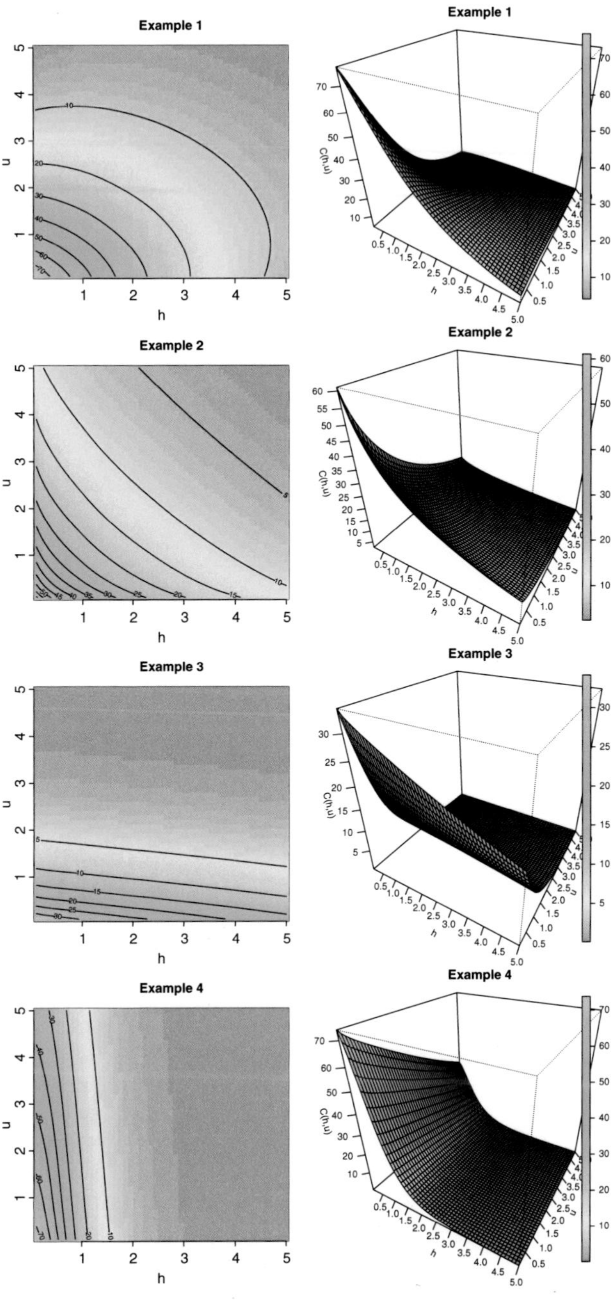

Figure 7.8 *2D and 3D different representations of the model based on mixtures (7.37).*

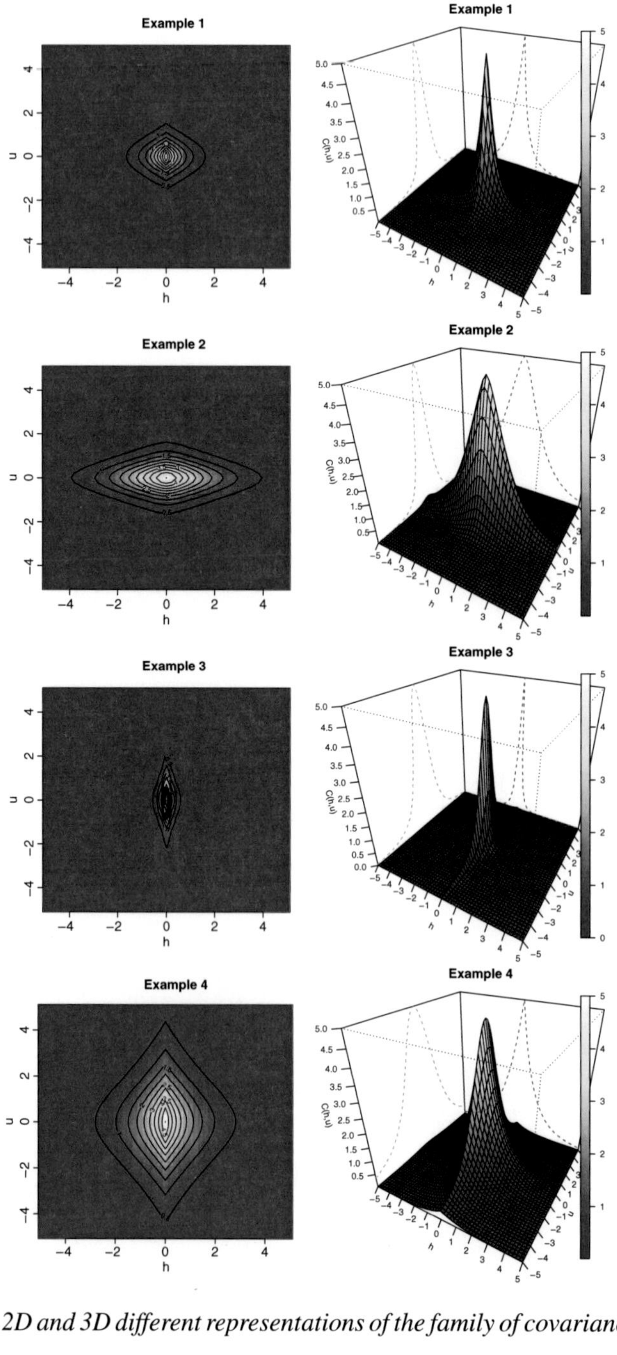

Figure 7.9 *2D and 3D different representations of the family of covariance functions (7.76).*

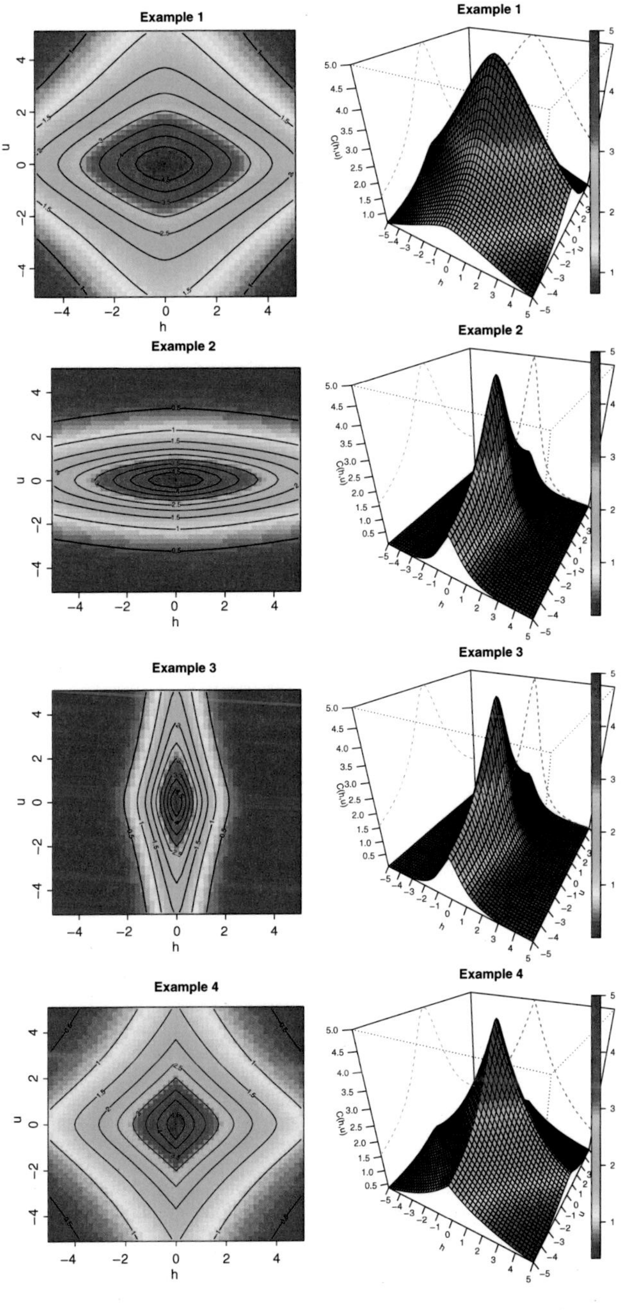

Figure 7.10   2D and 3D different representations of the family of covariance functions (7.81).

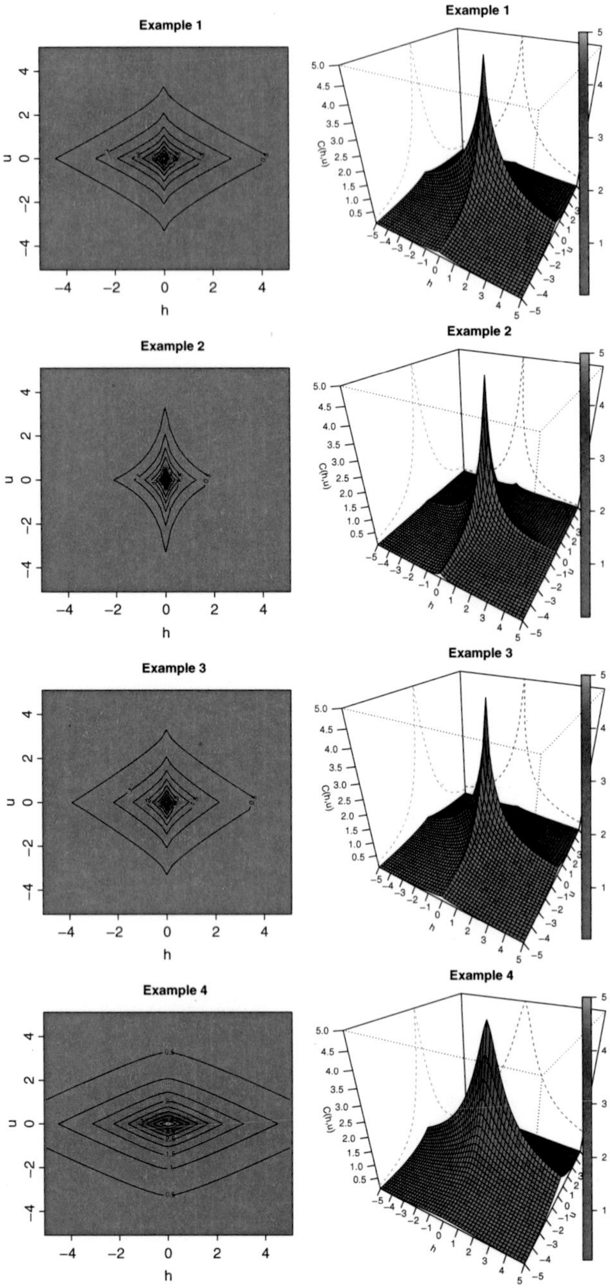

Figure 7.12 *2D and 3D different representations of the generalized product-sum model (7.138).*

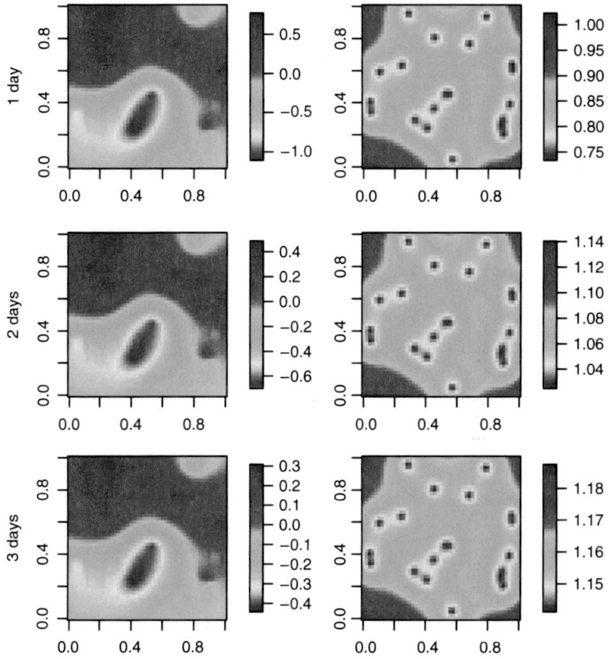

Figure 8.2 Spatio-temporal SK prediction maps (left panel) and prediction variance maps (right panel): 1-day, 2-day and 3-day time horizons.

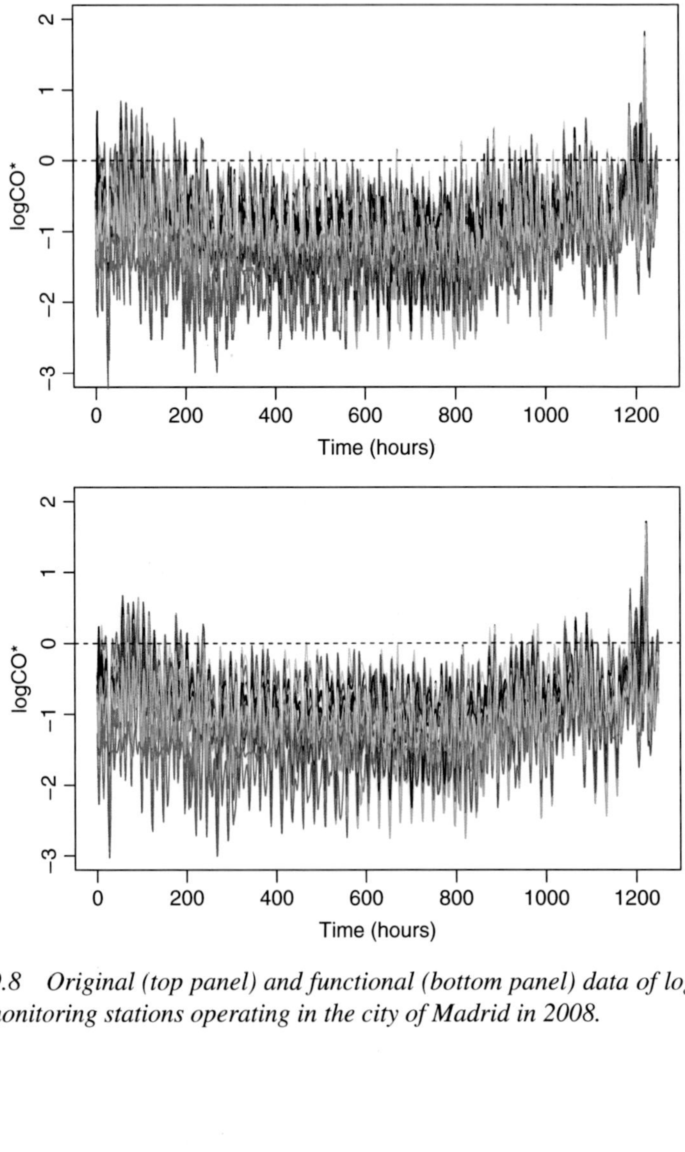

Figure 9.8 Original (top panel) and functional (bottom panel) data of logCO for the 23 monitoring stations operating in the city of Madrid in 2008.*

However, this is not the case in the spatio-temporal scenario, in which constructing valid spatio-temporal covariance models is one of the main research activities. In addition, while it is difficult to demonstrate that a spatial or temporal function is positive-definite, it is even more so when seeking to determine valid spatio-temporal covariance models.

For this reason, many authors began to study how to combine valid spatial and temporal models to obtain (valid) spatio-temporal covariance models.

By way of introduction, and before proceeding to discuss conventional and new generation covariance models in detail, the first approximations to modeling spatio-temporal dependence using covariance functions were nothing more than generalizations of the stationary models used in the spatial scenario.

In this sense, early studies often modeled spatio-temporal covariance using metric models, that is, by defining a metric in space and time that allowed researchers to directly use isotropic models that are valid in the spatial case (see, for example, the pioneer research by Dimitrakopoulos and Luo 1994). Such metric models were characterized by being non-separable, isotropic and stationary.

A word of caution is needed when dealing with isotropic models in the spatio-temporal case. Unlike in the previous chapters, in this chapter we use the notation $\|\mathbf{h}\|$ instead of $|\mathbf{h}|$ to denote the norm or module of a vector $\mathbf{h}$. The reason is that, unlike in the literature on spatial geostatistics, in the literature on spatio-temporal geostatistics $\|\mathbf{h}\|$ is the notation commonly used to denote the norm of a vector $\mathbf{h}$.

The next step in this initial stage consisted of configuring spatio-temporal covariance functions by means of the sum or product of a spatial covariance and a temporal covariance, both of which were stationary, giving rise to separable, isotropic and stationary models.

Later, realizing the limitations of the two procedures detailed above in terms of capturing the spatio-temporal dependence that really exists in the large majority of the phenomena studied, interest shifted towards including the interaction of space and time, present in most of these phenomena, in covariance models, giving rise to the so-called non-separable models (while remaining isotropic and stationary). It is worth highlighting the non-separable models developed by Jones and Zhang (1997), Cressie and Huang (1999), Brown *et al.* (2000), De Cesare *et al.* (2001a,b, 2002), De Iaco *et al.* (2001, 2002a,b, 2003), Myers *et al.* (2000), Gneiting (2002), Ma (2002, 2003a,b,c, 2005a,b,c), Fernández-Casal *et al.* (2003), Kolovos *et al.* (2004), and Stein (2005a), among others.

Development continued with the search for non-separable spatio-temporal, spatially anisotropic and/or temporally asymmetrical models such as those described in Fernández-Casal *et al.* (2003), Stein (2005b), Porcu *et al.* (2006), Mateu (2007), Mateu *et al.* (2007), among others.

Finally, we can cite some recent approaches to the problem of modeling non-stationary covariance functions, such as those made by Ma (2002, 2003c), Fuentes *et al.* (1999), Stein (2005a), Chen *et al.* (2006), Porcu *et al.* (2007), Porcu *et al.* (2009a), and Porcu *et al.* (2010), among others.

In the following sections we present and illustrate the most important spatio-temporal covariance models, the structure of the chapter following that of Martinez (2008, Chapter 4) and the content reproducing basically that of the original works.

7.2 Combined distance or metric model

The combined distance or metric model is that in which the covariance function is given by:

$$C(\mathbf{h}, u) = C\left(\|\mathbf{h}\| + c|u|\right), \ (\mathbf{h}, u) \in \mathbb{R}^d \times \mathbb{R}, \quad c > 0, \tag{7.1}$$

where $\|\mathbf{h}\| + c|u|$ is a distance on $\mathbb{R}^d \times \mathbb{R}$ and c is a positive constant.

If we consider $\|\mathbf{h}\|^2 + c|u|^2$ (Dimitrakopoulos and Luo 1994) instead of $\|\mathbf{h}\| + c|u|$ as a distance, which in topological terms is equivalent to the above, the alternative metric model is:

$$C(\mathbf{h}, u) = C\left(\|\mathbf{h}\|^2 + c|u|^2\right), \ (\mathbf{h}, u) \in \mathbb{R}^d \times \mathbb{R}, \quad c > 0. \tag{7.2}$$

We can also adopt $\sqrt{c_1^2\|\mathbf{h}\|^2 + c_2^2|u|^2}$ as a distance, giving rise to:

$$C(\mathbf{h}, u) = C\left(\sqrt{c_1^2\|\mathbf{h}\|^2 + c_2^2|u|^2}\right), \quad (\mathbf{h}, u) \in \mathbb{R}^d \times \mathbb{R}. \tag{7.3}$$

As can be observed, regardless of the metrics under consideration, the hypothesis underlying the models of combined distance is that the spatio-temporal covariance has a uniform structure across the entire domain (exponential, spherical, Gaussian …), which implies that the marginal spatial covariance function obeys the same type of model as the marginal temporal covariance function, and that both have the same sill (if the model is bounded), even though they may have different ranges. Furthermore, in the typology in question, c (in the first two types) and c_1 and c_2 (in the third), the constants that define the spatio-temporal metrics are nothing other than the anisotropic factors (geometric or elliptic anisotropy) between the "spatial distance" and "time lag" axes that determine the different spatial and temporal ranges.

In the third case, in semivariogram terms we have:

$$\begin{aligned}\gamma(\mathbf{h}, u) &= \gamma\left(\sqrt{c_1^2\|\mathbf{h}\|^2 + c_2^2|u|^2}\right) \\ &= C(\mathbf{0}, 0) - C\left(\sqrt{c_1^2\|\mathbf{h}\|^2 + c_2^2|u|^2}\right), \ (\mathbf{h}, u) \in \mathbb{R}^d \times \mathbb{R}.\end{aligned} \tag{7.4}$$

We can see that the semivariogram above is a valid semivariogram model insofar as

$$\exp\left(-\alpha\sqrt{c_1^2\|\mathbf{h}\|^2 + c_2^2|u|^2}\right) \tag{7.5}$$

is positive-definite for any given value of α.

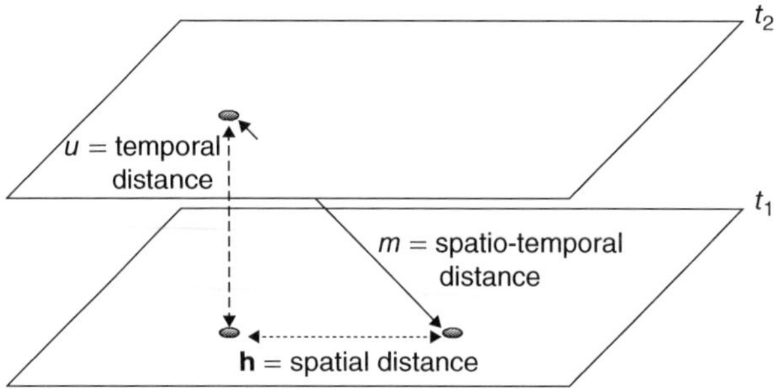

Figure 7.1 Representation of a spatio-temporal distance in the metric model.

As can be appreciated, this type of model is based on a spatio-temporal metrics (see Figure 7.1), a spatio-temporal distance, which makes it possible to use valid isotropic models directly in the merely spatial context. $\mathbb{R}^2 \times T$ is simply considered a $d+1$-dimensional space. For example, for $C(\mathbf{h}, u) = C\left(\sqrt{c_1^2\|\mathbf{h}\|^2 + c_2^2|u|^2}\right)$, $(\mathbf{h}, u) \in \mathbb{R}^d \times \mathbb{R}$, with $c_1 = c_2 = 1$, the model can be considered isotropic.

Example 7.1 (Metric model) *The exponential metric model is given by:*

$$C(\mathbf{h}, u) = C(\mathbf{0}, 0) \exp\left(-\frac{\sqrt{c_1^2\|\mathbf{h}\|^2 + c_2^2|u|^2}}{a}\right), \tag{7.6}$$

the marginal covariance functions being:

$$C(\mathbf{h}, 0) = C(\mathbf{0}, 0) \exp -\left(\frac{\sqrt{c_1^2\|\mathbf{h}\|^2}}{a}\right) = C(\mathbf{0}, 0) \exp\left(\frac{\|\mathbf{h}\|}{\frac{a}{c_1}}\right) \tag{7.7}$$

and

$$C(\mathbf{0}, u) = C(\mathbf{0}, 0) \exp\left(-\frac{\sqrt{c_2^2|u|^2}}{a}\right) = C(\mathbf{0}, 0) \exp\left(-\frac{|u|}{\frac{a}{c_2}}\right) \tag{7.8}$$

Figure 7.2 shows the covariance function associated with the Equation (7.6), for different combinations of parameters. The first example represents the covariance function obtained with $C(\mathbf{0}, 0) = 5, a = 1, c_1 = 2$ and $c_2 = 2$, whereas the second example corresponds to that obtained with $C(\mathbf{0}, 0) = 5, a = 2, c_1 = 2$ and $c_2 = 2$, with $c_1 = c_2$ being maintained in both representations, that is, maintaining the isotropic nature of the spatio-temporal rf. However, the third example is the covariance function obtained with $C(\mathbf{0}, 0) = 5, a = 2, c_1 = 1$ and $c_2 = 2$ (larger

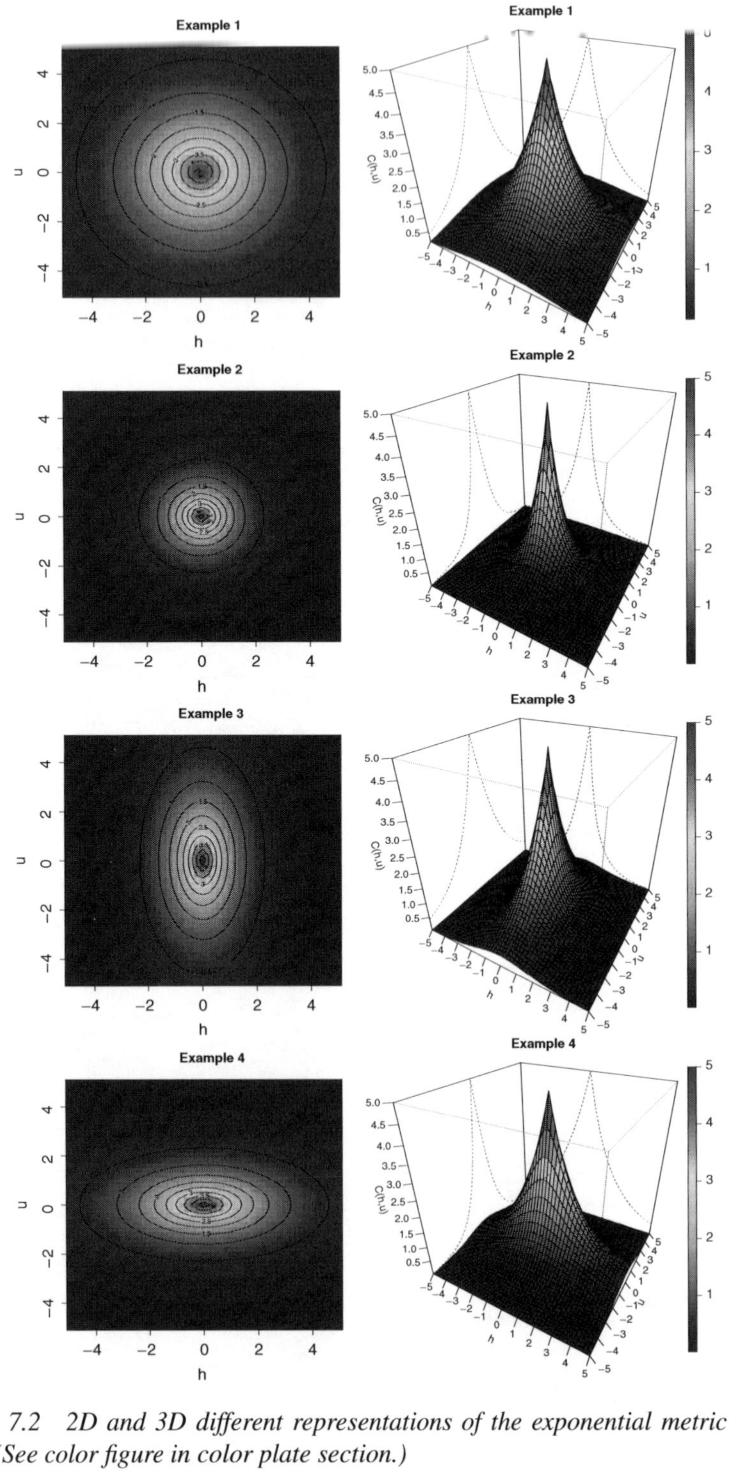

Figure 7.2 2D and 3D different representations of the exponential metric model (7.6). (See color figure in color plate section.)

temporal range) and the fourth is that configured with $C(\mathbf{0}, 0) = 5, a = 2, c_1 = 2$ and $c_2 = 1$ (larger spatial range). Using a dotted line we have also included the representations of the marginal covariance functions $C(\mathbf{0}, u)$ (in green) and $C(\mathbf{h}, 0)$ (in red) associated to each function.

Despite the conceptual simplicity of this model and how efficient it is in terms of computation, it is very limited when applied to studying real processes.

7.3 Sum model

The sum model, introduced by Rouhani and Hall (1989), obeys a type of separability that consists of decomposing the spatio-temporal covariance into the sum of a purely spatial covariance function and a purely temporal covariance function. For this reason it is known as a zonal model (referring to how zonal anisotropy is treated by summing covariance functions) and also as a linear model:

$$C((\mathbf{s}_i, t_i)(\mathbf{s}_j, t_j)) = C_s(\mathbf{s}_i, \mathbf{s}_j) + C_t(t_i, t_j), \quad \forall \mathbf{s}_i, \mathbf{s}_j \in D \subset \mathbb{R}^d, \forall t_i, t_j \in T \subset \mathbb{R}, \tag{7.9}$$

C_s being a spatial covariance function defined on $\mathbb{R}^d$ and C_t a temporal covariance function defined on $\mathbb{R}$.

If we consider two stationary covariance functions, C_s and C_t, then the sum model is given by:

$$C(\mathbf{h}, u) = C_s(\mathbf{h}) + C_t(u), \quad (\mathbf{h}, u) \in \mathbb{R}^d \times \mathbb{R}, \tag{7.10}$$

C_s being a spatial covariance function defined on $\mathbb{R}^d$ and C_t a temporal covariance function defined on $\mathbb{R}$.

The semivariogram version of this type of separability is given in the stationary case by:

$$\gamma(\mathbf{h}, u) = \gamma_s(\mathbf{h}) + \gamma_t(u), \quad (\mathbf{h}, u) \in \mathbb{R}^d \times \mathbb{R} \tag{7.11}$$

This model was used widely in the initial stages of modeling spatio-temporal covariance functions, mainly due to its simplicity. However, the drawback is that the sum of the two covariance models, one spatial and the other temporal, is not generally speaking positive-definite, but rather only positive semi-definite (Myers and Journel 1990; Rouhani and Myers 1990; Dimitrakopoulos and Luo 1994). This limitation implies that the matrix of covariances in the kriging equations may not be invertible for certain sets of locations.

With regard to this issue, Schlather (1999) discusses the conditions under which a finite linear combination of positive-definite covariance functions with non-negative scalars,

$$C = a_1 C_s + a_2 C_t,$$

yields a positive-definite function.

The question that arises now is: Can the sum model overcome the limitation of covariance functions generally not being positive-definite? In more formal terms, Is there a function $G(\cdot, \cdot)$ on $\mathbb{R}^2$ such that

$$\gamma_s(\mathbf{h}) + \gamma_t(u) + G(\gamma_s(\mathbf{h}) + \gamma_t(u)) \tag{7.12}$$

is a strictly positive-definite function? The answer is yes and it implies combining a sum model with a metric model.

Example 7.2 (Cressie and Majure 1997) *One example of how this model is used can be found in Cressie and Majure (1997), where the authors study the spatio-temporal trend of the logarithm of nitrate concentration in a river in Texas. In order to do so, they model variation on a small scale with a sum variogram of two (one-dimensional) stationary variograms, one spatial and the other temporal but both exponential. The covariance function associated with such a spatio-temporal variogram model is:*

$$C(\mathbf{h}, u) = C_s(\mathbf{0}) \exp\left(-\frac{\|\mathbf{h}\|}{c_s}\right) + C_t(0) \exp\left(-\frac{|u|}{c_t}\right), \quad (\mathbf{h}, u) \in \mathbb{R}^d \times \mathbb{R}. \tag{7.13}$$

Figure 7.3 shows four examples of the model (7.13) obtained simply by using different values in the parameters of the spatio-temporal covariance function. The first example shows the covariance function obtained for $C_s(\mathbf{0}) = 2.5, c_s = 2, C_t(0) = 2.5$ and $c_t = 2$. The second example shows that obtained for $C_s(\mathbf{0}) = 2.5, c_s = 2, C_t(0) = 5/2$ and $c_t = 0.5$. The third example is the covariance generated for $C_s(\mathbf{0}) = 2.5$, $c_s = 0.5, C_t(0) = 2.5$ and $c_t = 2$ and, finally, the fourth example is specified with $C_s(\mathbf{0}) = 0.5, c_s = 0.5, C_t(0) = 5/2$ and $c_t = 2$.

7.4 Combined metric-sum model

This type of model emerges as a solution to the problem suffered by the sum or zonal model. The reason is that if we combine a sum model with a metric model, the limitation of the sum model in relation to the permissibility condition can be overcome.

In semivariogram terms, this combined model is given by:

$$\gamma_{st}(\mathbf{h}, u) = \gamma_s(\mathbf{h}) + \gamma_t(u) + \gamma(\|\mathbf{h}\| + a\,|u|) \tag{7.14}$$

$$\gamma_{st}(\mathbf{h}, u) = \gamma_s(\mathbf{h}) + \gamma_t(u) + \gamma(\|\mathbf{h}\|^2 + a|u|^2) \tag{7.15}$$

$$\gamma_{st}(\mathbf{h}, u) = \gamma_s(\mathbf{h}) + \gamma_t(u) + \gamma\left(\sqrt{\|\mathbf{h}\|^2 + a|u|^2}\right) \tag{7.16}$$

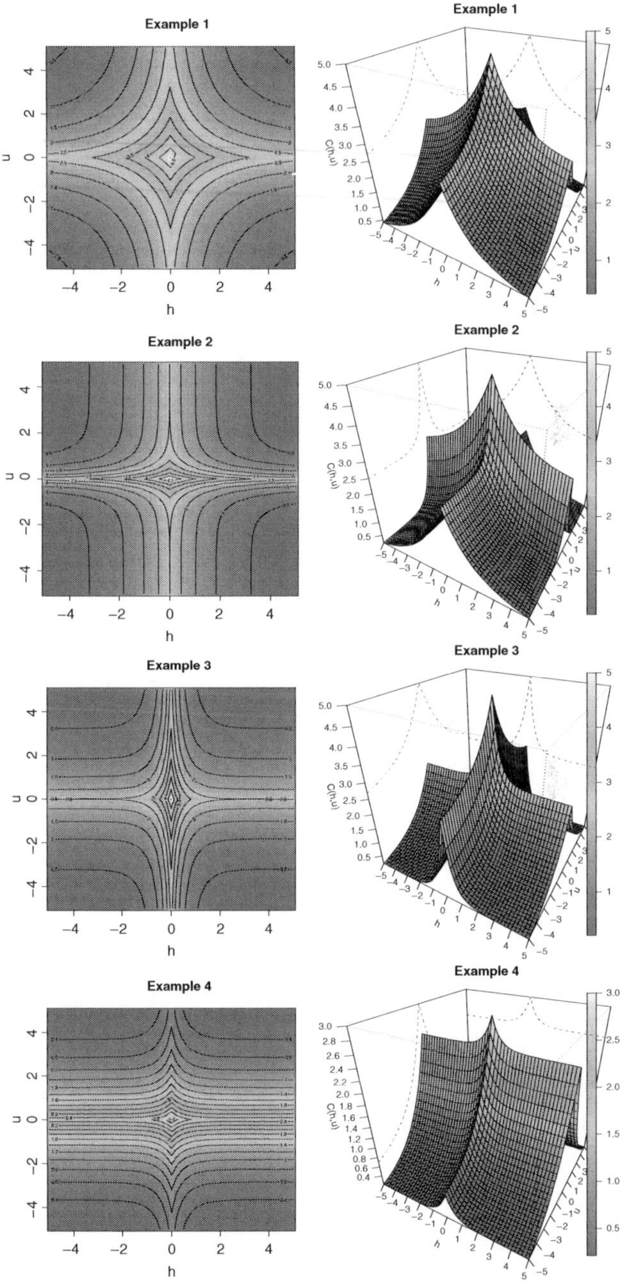

Figure 7.3 2D and 3D different representations of the exponential sum model (7.13). *(See color figure in color plate section.)*

Example 7.3 *Using the Examples 7.1 and 7.2 as a basis, it is easy to obtain the expression of the exponential metric-sum model, which is given by:*

$$C(\mathbf{h}, u) = C_s(\mathbf{0}) \exp\left(-\frac{\|\mathbf{h}\|}{c_s}\right) + C_t(0) \exp\left(-\frac{|u|}{c_t}\right) + C(\mathbf{0}, 0) \exp\left(-\frac{\sqrt{c_1^2\|\mathbf{h}\|^2 + c_2^2|u|^2}}{a}\right), \quad (\mathbf{h}, u) \in \mathbb{R}^d \times \mathbb{R}. \tag{7.17}$$

Figure 7.4 shows two examples of the model (7.17), the first in covariance terms (upper panel) and the second in semivariogram terms (bottom panel). Both examples were constructed using Example 7.1 in Figure 7.2 and Example 7.2 in Figure 7.3 as a basis. The parameters considered were: $C_s(\mathbf{0}) = 5$, $c_s = 2, C_t(0) = 5, c_t = 2, C(\mathbf{0}, 0) = 5, c_1 = 1$ and $c_2 = 1, a = 2$.

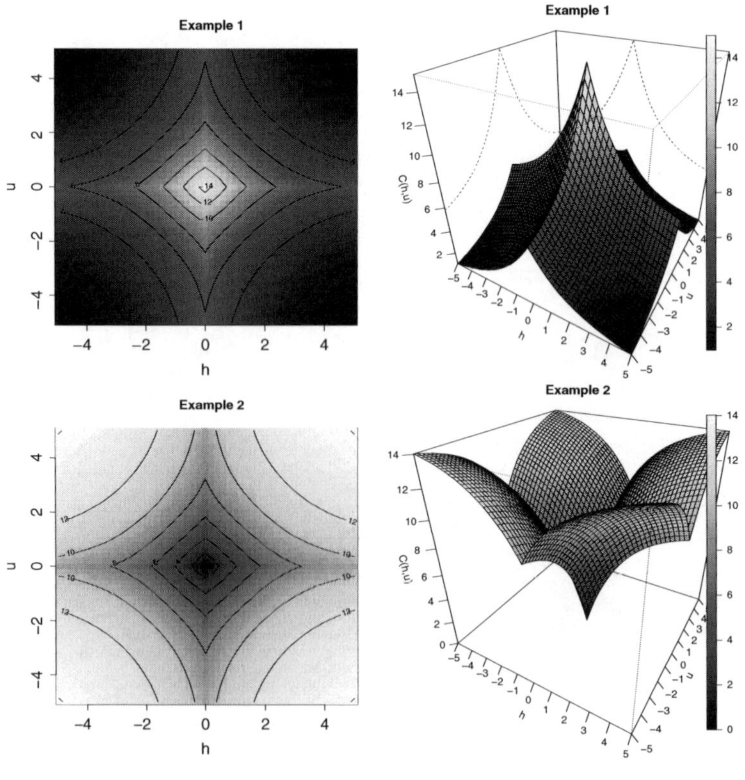

Figure 7.4 2D and 3D representations of the exponential metric-sum model (7.17). *Upper panel: Covariance function; Bottom panel: Semivariogram.*

7.5 Product model

It is one of the most used separable covariogram models in the literature (Rodríguez-Iturbe and Mejia (1974), or De Cesare *et al.* (1997), among others). Notwithstanding, there have been many occasions when it has been used more due to the simplicity of the model itself than for how well it fits the data being studied. It is generally defined, in covariogram terms, as:

$$C\left((\mathbf{s}_1,t_1),(\mathbf{s}_2,t_2)\right) = C_s(\mathbf{s}_1,\mathbf{s}_2)C_t(t_1,t_2), \quad \forall \mathbf{s}_i,\mathbf{s}_j \in D \subset \mathbb{R}^d, \forall t_i,t_j \in T \subset \mathbb{R}. \tag{7.18}$$

and in the stationary case

$$C(\mathbf{h},u) = C_s(\mathbf{h})C_t(u), \qquad (\mathbf{h},u) \in \mathbb{R}^d \times \mathbb{R}, \tag{7.19}$$

C_s and C_t being positive-definite functions.

It is important to highlight that the product model has the following peculiarities:

(i) This definition of separability implies that in the case of any given spatial and/or temporal lags:

$$\left\{\begin{array}{l} C(\mathbf{h}_1,u) = C_s(\mathbf{h_1})C_t(u) \\ C(\mathbf{h}_2,u) = C_s(\mathbf{h}_2)C_t(u) \end{array}\right\} \to C(\mathbf{h}_1,u) \propto C(\mathbf{h}_2,u), \quad \forall u \in \mathbb{R} \tag{7.20}$$

$$\left\{\begin{array}{l} C(\mathbf{h},u_1) = C_s(\mathbf{h})C_t(u_1) \\ C(\mathbf{h},u_2) = C_s(\mathbf{h})C_t(u_2) \end{array}\right\} \to C(\mathbf{h},u_1) \propto C(\mathbf{h},u_2), \quad \forall \mathbf{h} \in \mathbb{R}^d. \tag{7.21}$$

Therefore, the structure of the temporal covariation is the same for any spatial separation. Likewise, the structure of the spatial dependence maintains the same form for the different time lags.

(ii) The fact that the covariance function is separable does not mean the related semivariogram is. Nevertheless, while the sum or the product of a spatial semivariogram and a temporal semivariogram is not necessarily a valid semivariogram model, the combination of the two obtained by means of the expression (7.22) verifies such a condition.[1]

$$\gamma(\mathbf{h},u) = C_s(\mathbf{0})\gamma_t(u) + C_t(0)\gamma_s(\mathbf{h}) - \gamma_s(\mathbf{h})\gamma_t(u). \tag{7.22}$$

(iii) The model (7.22) has interesting and useful practical properties. By setting $u = 0$ and $\mathbf{h} = \mathbf{0}$ separately, and taking into account that $\gamma_t(0) = \gamma_s(\mathbf{0}) = 0$, we have:

$$\gamma(\mathbf{h},0) = C_t(0)\gamma_s(\mathbf{h}), \tag{7.23}$$

$$\gamma(\mathbf{0},u) = C_s(\mathbf{0})\gamma_t(u). \tag{7.24}$$

The sills $C_s(\mathbf{0})$ and $C_t(0)$ must have already been determined.

[1] See proof in Appendix D.

(iv) As in practice we do not observe the marginal rf's but rather the spatio-temporal rf, the condition of separability, in spatio-temporal terms, can be expressed as:

$$C(\mathbf{h}, u) = \frac{C(\mathbf{h}, 0)C(\mathbf{0}, u)}{C(\mathbf{0},0)}, \quad (\mathbf{h}, u) \in \mathbb{R}^d \times \mathbb{R}, \tag{7.25}$$

$C(\mathbf{0}, 0) = C_s(\mathbf{0})C_t(0) = \sigma^2 > 0$ being the *a priori* variance of the rf.[2]

(v) Mitchell *et al.* (2005) characterize an alternative to this model using the Kronecker product. Let $\mathbf{U}$ and $\mathbf{V}$ be the matrices of variances and covariances of dimension s and p for space and time respectively, and let $\mathbf{\Sigma}$ be the matrix of variances and covariances of dimension $s \times p$ of the spatio-temporal rf. The covariance of the overall rf is said to be separable if, and only if, $\mathbf{\Sigma} = \mathbf{U} \otimes \mathbf{V}$, where $\otimes$ represents the Kronecker product of both matrices. Note that $\mathbf{U}$ and $\mathbf{V}$ are unique only up to constant multiples since $a\mathbf{U} \otimes (1/a)\mathbf{V} = \mathbf{U} \otimes \mathbf{V}, \forall a \neq 0$.

As mentioned previously, one of the main advantages of separability is that far fewer calculations are required. For example, in the case of a problem with 50 locations observed at 100 different moments in time where the aim is to predict the value of the rf on a new unobserved spatio-temporal location, it is necessary to invert the variance-covariance matrix, $\mathbf{\Sigma}$, which implies inverting a matrix of 5000×5000 for a non-separable model. In the case of a separable model, due to the properties of the Kronecker product, we only need to invert a matrix of 50×50 and a matrix of 100×100, as $\mathbf{\Sigma}^{-1} = (\mathbf{U} \otimes \mathbf{V})^{-1} = \mathbf{U}^{-1} \otimes \mathbf{V}^{-1}$.

As a result of this advantage, there are several tests in the literature to verify the suitability of this hypothesis when analyzing a given spatio-temporal process. Some of the tests proposed in the literature are specific to certain models, such as that by Brown *et al.* (2000) for "blur-generated" models, which uses ML to verify the hypothesis of whether or not the parameters that mark the separability of this model are zero. Others are general procedures, such as that proposed by Mitchell *et al.* (2005, 2006), who adapt a likelihood ratio test based on repeated multivariate measures to the spatio-temporal context by subsampling. Fuentes (2003) and Fuentes *et al.* (2005) proposed another non-parametric procedure based on the spectral representation of the spatio-temporal rf.

(vi) One interesting feature of the product model is the screen effect it creates when the mean of the rf is known. More specifically, given the observations at a moment in time t, all the remaining observations in the same spatial locations but at different times will be irrelevant when predicting at the moment t using the simple kriging procedure. That is: "when the present is known, there is no information in the past." But, it is worth remembering that this is only the case with simple kriging.

[2] See proof in Appendix D.

(vii) Finally, it is worth mentioning that only the family of spatio-temporal covariances is closed in with respect to the sum and the product to date, whereas the family of spatio-temporal semivariograms is only closed with respect to the sum. However, a recent paper by Porcu and Schilling (2011) demonstrated that, under certain conditions, the family of semivariogram functions is also closed with respect to the product.

Figure 7.5 represents the product model (7.27) simulated with $C(\mathbf{0},0)=8, a_s=2$ and $a_t=2$. The left panel shows the marginal spatial covariance function and the spatial covariance functions for different time lags. The right one depicts the marginal temporal covariance function and the temporal covariance functions for a series of spatial distances.

Example 7.4 *One simple example of a stationary product model is the doubly exponential model, which is given by*

$$C(\mathbf{h},u)=C(\mathbf{0},0)\exp(-\|\mathbf{h}\|/a_s-|u|/a_t), \quad (\mathbf{h},u)\in\mathbb{R}^d\times\mathbb{R}, \tag{7.26}$$

a_s *and* a_t *being two positive space and time scaling parameters, respectively and* σ^2 *the a priori variance of the rf.*

Figure 7.6 shows four particular examples of this model. The first example has been simulated using $C(\mathbf{0},0)=5, a_s=2$ and $a_t=2$, and the second using $C(\mathbf{0},0)=5, a_s=1$ and $a_t=2$. Note that if we consider $1/a_t=-\log(\rho)$, the reparameterization allows us to write (7.26) as

$$C(\mathbf{h},u)=C(\mathbf{0},0)\exp\left(-\frac{\|\mathbf{h}\|}{a_s}\right)\exp\left(|u|\log(\rho)\right)=C(\mathbf{0},0)\rho^{|u|}\exp\left(-\frac{\|\mathbf{h}\|}{a_s}\right), \tag{7.27}$$

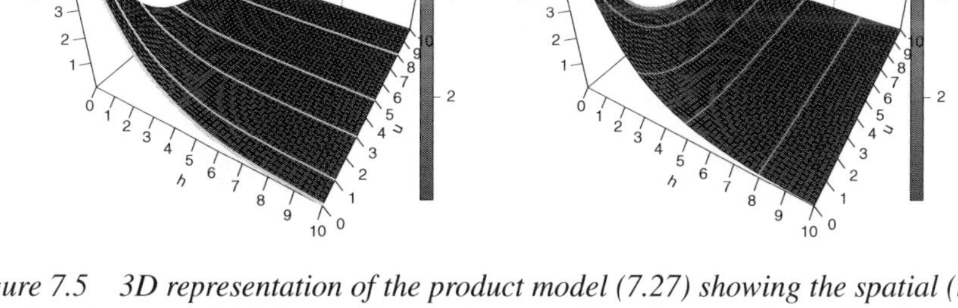

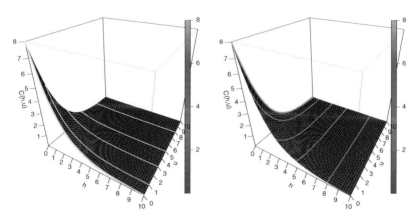

Figure 7.5 3D representation of the product model (7.27) showing the spatial (left panel) and temporal (right panel) margins as well as the covariance functions for different spatial distances and time lags.

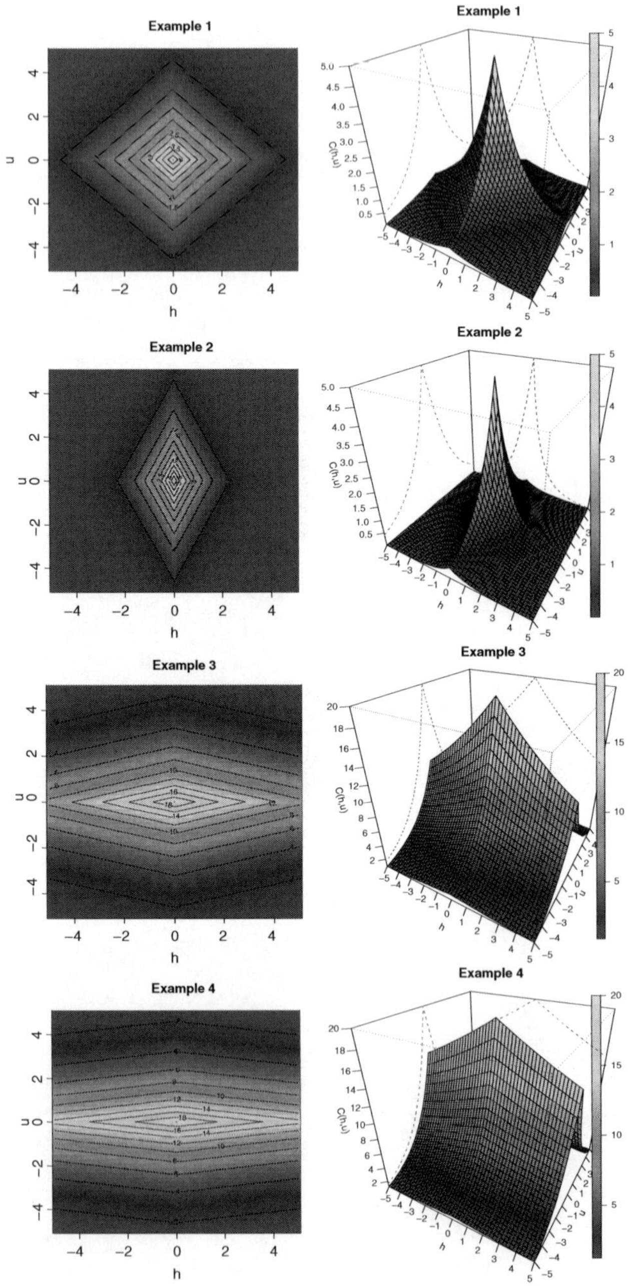

Figure 7.6 2D and 3D different representations of the product model (7.27). (See color figure in color plate section.)

which is the product of a temporal correlation model of a time series AR(1) and an exponential spatial model. Both the third example, simulated using $C(\mathbf{0},0) = 20, a_s = 8$ and $a_t = 2$, and the fourth, generated with $C(\mathbf{0},0) = 20, a_s = 16$ and $a_t = 2$, illustrate the equation (7.20) of the Property (i) of the product model discussed above. This property can also be observed in Figure 7.6.

7.6 Product-sum model

As we have seen, the sum and the product are basic combinations that can be used to generate spatio-temporal covariance structures. Each one, separately, has its disadvantages or limitations that a combination of the two does not. As stated by De Cesare *et al.* (2001a,b), De Iaco *et al.* (2001) and De Iaco *et al.* (2002a), among others, the stationary product-sum model is given by:

$$C(\mathbf{h},u) = k_1 C_s(\mathbf{h}) C_t(u) + k_2 C_s(\mathbf{h}) + k_3 C_t(u), \tag{7.28}$$

or equivalently in semivariogram terms:[3]

$$\gamma(\mathbf{h},u) = \left(k_2 + k_1 C_t(0)\right) \gamma_s(\mathbf{h}) + \left(k_3 + k_1 C_s(0)\right) \gamma_t(u) - k_1 \gamma_s(\mathbf{h}) \gamma_t(u), \tag{7.29}$$

where C_s and C_t are covariance functions, γ_s and γ_t are the corresponding semivariograms and $k_1 > 0,\ k_2 \geq 0,\ k_3 \geq 0$ are constants that guarantee validity. $C(\mathbf{0},0)$ is the sill of γ, called the global sill, $C_s(\mathbf{0})$ is the sill of γ_s and $C_t(0)$ is the sill of γ_t (partial sills). It is sufficient to assume second-order stationarity to ensure these semivariograms have a sill.

We can see how both the sum model ($k_1 = 0$) and also the product model ($k_2 = k_3 = 0$) are cases specific to the product-sum model.

The product-sum model gives rise to the following relationships:

$$\gamma(\mathbf{h},0) = \gamma_s(\mathbf{h}) \left(k_2 + k_1 C_t(0)\right) = k_s \gamma_s(\mathbf{h}) \tag{7.30}$$

$$\gamma(\mathbf{0},u) = \gamma_t(u) \left(k_3 + k_1 C_s(\mathbf{0})\right) = k_t \gamma_t(u). \tag{7.31}$$

That is, the spatial and temporal marginal semivariograms are proportional to the merely spatial and temporal semivariograms. In this sense $[k_2 + k_1 C_t(0)] = k_s$ and $[k_3 + k_1 C_s(\mathbf{0})] = k_t$ can be seen as coefficients of proportionality between the spatio-temporal semivariograms $\gamma(\mathbf{h},0)$ and $\gamma(\mathbf{0},u)$ and the spatial and temporal semivariograms $\gamma_s(\mathbf{h})$ and $\gamma_t(u)$, respectively. As in the case of the product model, estimating and modeling $\gamma(\mathbf{h},0)$ is equivalent to estimating and modeling $\gamma_s(\mathbf{h})$. Likewise, estimating and modeling $\gamma(\mathbf{0},u)$ is equivalent to estimating and modeling $\gamma_t(u)$.

---

[3] See proof in Appendix D.

These two relationships can be simplified by imposing three constraints:

$$k_2 + k_1 C_t(0) = 1 \tag{7.32}$$

$$k_3 + k_1 C_s(\mathbf{0}) = 1 \tag{7.33}$$

$$k_1 + k_2 + k_3 = 1, \tag{7.34}$$

which makes it easier to estimate and model $\gamma(\mathbf{h}, 0)$ and $\gamma(\mathbf{0}, u)$ using $\gamma_s(\mathbf{h})$ and $\gamma_t(u)$, that is, by determining k_1, k_2 and k_3.

Example 7.5 (Martínez 2008) *Consider the product-sum model built through a spatial covariance function defined by the Matérn correlation function:*

$$C(\mathbf{h}) = \left(\frac{2^{1-\nu}}{\Gamma(\nu)}\right)(b\|\mathbf{h}\|)^{\nu} K_{\nu}(b\|\mathbf{h}\|)$$

and a temporal covariance function defined by the autocorrelation function of a process AR(1) taking the form:

$$C_t(u) = \alpha^{|u|}.$$

These covariance functions yield the following model:

$$C(\mathbf{h}, u) = k_1 \frac{2^{1-\nu}}{\Gamma(\nu)}(b\|\mathbf{h}\|)^{\nu} K_{\nu}(b\|\mathbf{h}\|)\alpha^{|u|} + k_2 \frac{2^{1-\nu}}{\Gamma(\nu)}(b\|\mathbf{h}\|)^{\nu} K_{\nu}(b\|\mathbf{h}\|) + k_3 \alpha^{|u|}, \tag{7.35}$$

where $k_1 > 0, k_2 \geq 0$ and $k_3 \geq 0$ are the weightings of the linear combination that defines the covariance model, $\alpha \in (0, 1]$ is a temporal smoothing parameter, $b > 0$ is a spatial scaling parameter and $\nu > 0$ a spatial smoothing parameter.

Figure 7.7 shows different representations of the family of covariances (7.35) for a set of possible values of the parameters that define it. More specifically, if $\boldsymbol{\theta} = (k_1, k_2, k_3, \alpha, b, \nu)'$ is the vector of parameters that define the model (7.35), the covariance functions represented correspond to $\boldsymbol{\theta} = (2, 1.5, 1.5, 0.4, 0.3, 1)'$ for the first example, $\boldsymbol{\theta} = (2, 1.5, 1.5, 0.8, 0.3, 1)'$ for the second, $\boldsymbol{\theta} = (2, 1.5, 1.5, 0.4, 1.1, 0.5)'$ for the third and $\boldsymbol{\theta} = (4, 0.5, 0.5, 0.4, 0.3, 1)'$ for the fourth.

7.7 Porcu and Mateu mixture-based models

The idea underlying the construction of spatio-temporal covariance models generated by mixtures is the linear combination of both the spatial and temporal covariance functions and also the non-separable spatio-temporal covariance functions. Porcu and Mateu (2007) propose the following generalization of the product-sum model by De Iaco *et al.* (2001):

$$C(\mathbf{h}, u) = \left(\lambda_{12} C_1(\mathbf{h}, u) C_2(\mathbf{h}, u) + \lambda_3 C_3(\mathbf{h}) + \lambda_4 C_4(u)\right)^{\xi}, \tag{7.36}$$

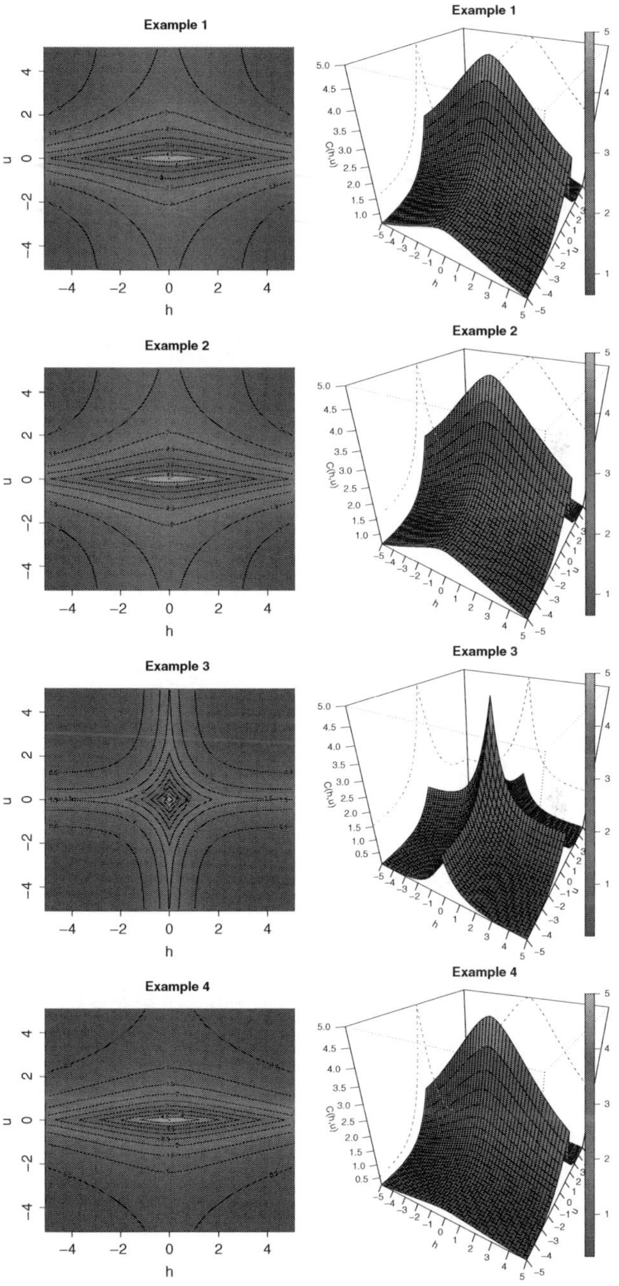

Figure 7.7 2D and 3D different representations of the product-sum model (7.35). *(See color figure in color plate section.)*

where C_1 and C_2 are non-separable spatio-temporal covariance functions and C_3 and C_4 are merely spatial and temporal covariance functions, respectively. The constants λ_{12}, λ_3 and λ_4 are non-negative weightings and ξ is a natural number that acts as an external smoothing parameter.

By combining spatio-temporal covariance functions with other merely spatial or temporal functions, we can modify the degree of smoothing beyond the origin of the resulting covariance function, which eliminates some of the undesirable features of some covariance functions discussed previously in the literature, such as the lack of differentiability beyond the origin (Stein 2004).

Example 7.6 (Model based on mixtures) *One interesting example of the generalization by Porcu and Mateu (2007) is as follows:*

$$C(\mathbf{h}, u) = \frac{k}{B_{u,\mathbf{h}}}(\alpha_1 \alpha_2 \, \|\mathbf{h}\| \, |u|)^{\upsilon} \times \kappa_{\upsilon}\left(\frac{\alpha_2 \, \|\mathbf{h}\|}{\left(\alpha_1 |u|^{2\alpha_1} + 1\right)^{\beta/2}}\right) \kappa_{\upsilon}\left(\frac{a_1 \, |u|}{\left(a_2 \|\mathbf{h}\|^{2\alpha_2} + 1\right)^{\beta/2}}\right) \quad (7.37)$$

where $(\mathbf{h}, u) \in \mathbb{R}^d \times \mathbb{R}, \upsilon = \epsilon + \beta/2$ *and* $k = \frac{\sigma^4}{2^{2(\upsilon-1)}(\Gamma(\upsilon))^2}$ *with* $\sigma_1^2 = \sigma_2^2 = \sigma^2$. *Furthermore,* $B_{u,\mathbf{h}} = (1 + a_1 |u|^{2\alpha_1} + a_2 \|\mathbf{h}\|^{2\alpha_2} + a_1 a_2 |u|^{2\alpha_1} \|\mathbf{h}\|^{2\alpha_2})$. *Here* $a_1, a_2, \alpha_1, \alpha_2, \beta, \upsilon$ *are non-negative parameters and* $\kappa_{\upsilon}()$ *is the modified Bessel function of the second kind of order* υ.

Figure 7.8 provides different representations of the family of covariances (7.37). More specifically, if $\boldsymbol{\theta} = (a_1, a_2, \alpha_1, \alpha_2, \beta, \nu)'$ is the vector of parameters that define the model in (7.37), the covariance functions represented correspond to $\boldsymbol{\theta} = (0.6, 0.5, 0.5, 0.5, 1, 1)'$ for the first example, $\boldsymbol{\theta} = (0.2, 0.2, 0.5, 0.2, 0.2, 0.2)'$ for the second, $\boldsymbol{\theta} = (0.1, 0.2, 0.5, 0.2, 0.2, 1)'$ for the third and $\boldsymbol{\theta} = (0.2, 1, 0.2, 1, 0.2, 1)'$ for the fourth.

7.8 General product-sum model

De Iaco *et al.* (2001, 2002b) believe the constraints imposed on the product-sum model, $k_2 + k_1 C_t(0) = 1$ and $k_3 + k_1 C_s(\mathbf{0}) = 1$, are unnecessary and, moreover, impose an unnecessary form of symmetry on the model. More specifically, they make the impact of the components of spatial and temporal correlation symmetric.

In this sense, they show that the three coefficients of the model can be written in terms of the sills and the parameters k_s and k_t. In this particular case, it would lead us to model $\gamma_s(\mathbf{h})$ and $\gamma_t(u)$, using the models of $\gamma\,(\mathbf{h}, 0)$ and $\gamma\,(\mathbf{0}, u)$ respectively. In addition, these two parameters can be combined to make one sole parameter.

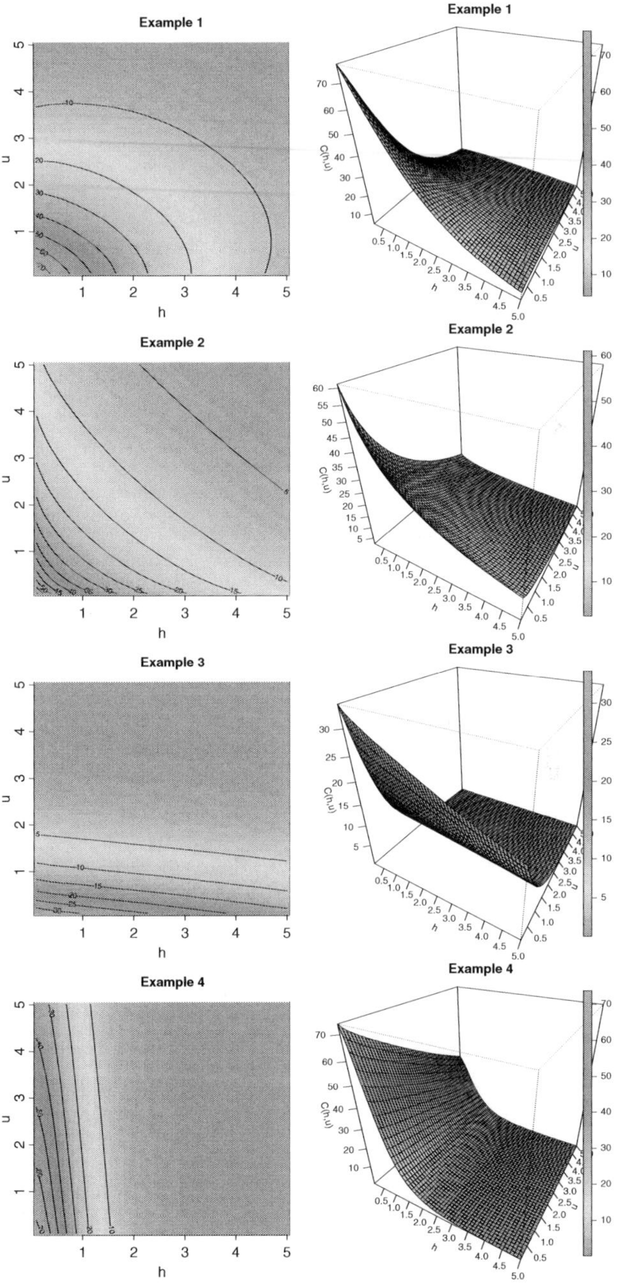

Figure 7.8 2D and 3D different representations of the model based on mixtures (7.37). (See color figure in color plate section.)

As regards the first issue, having defined k_s and k_t as:

$$k_s = k_2 + k_1 C_t(0) \tag{7.38}$$

$$k_t = k_3 + k_1 C_s(\mathbf{0}), \tag{7.39}$$

and taking into account that

$$C(\mathbf{0}, 0) = k_1 C_s(\mathbf{0}) C_t(0) + k_2 C_s(\mathbf{0}) + k_3 C_t(0), \tag{7.40}$$

then, k_1, k_2 and k_3 can be expressed in terms of the values of the sills $C(\mathbf{0}, 0), C_s(\mathbf{0})$, $C_t(0)$ and the coefficients k_s and k_t:

$$k_1 = \frac{k_s C_s(\mathbf{0}) + k_t C_t(0) - C(\mathbf{0}, 0)}{C_s(\mathbf{0}) C_t(0)} \tag{7.41}$$

$$k_2 = \frac{C(\mathbf{0}, 0) - k_t C_t(0)}{C_s(\mathbf{0})} \tag{7.42}$$

$$k_3 = \frac{C(\mathbf{0}, 0) - k_s C_s(\mathbf{0})}{C_t(0)}. \tag{7.43}$$

In order for the covariance function to be positive-definite, the following must be true: $k_1 > 0, \quad k_2 \geq 0$ and $k_3 \geq 0$. As a consequence, the validity of the covariograms and semivariograms that belong to the product-sum model is related to the sill of the spatial and temporal components.

As far as the second issue is concerned, combining the parameters k_s and k_t to form one sole parameter, k, the expression of the spatio-temporal semivariogram, $\gamma(\mathbf{h}, u)$, can be simplified using the expressions (7.30) and (7.31):

$$\begin{aligned}
\gamma(\mathbf{h}, u) &= \gamma_s(\mathbf{h}) \left(k_2 + k_1 C_t(0)\right) + \gamma_t(u) \left(k_3 + k_1 C_s(\mathbf{0})\right) - k_1 \gamma_s(\mathbf{h}) \gamma_t(u) \\
&= \gamma_s(\mathbf{h}) k_s + \gamma_t(u) k_t - k_1 \gamma_s(\mathbf{h}) \ \gamma_t(u) \\
&= \gamma(\mathbf{h}, 0) + \gamma(\mathbf{0}, u) - k_1 \frac{\gamma(\mathbf{h}, 0)}{k_s} \frac{\gamma(\mathbf{0}, u)}{k_t} \\
&= \gamma(\mathbf{h}, 0) + \gamma(\mathbf{0}, u) - k \gamma(\mathbf{h}, 0) \gamma(\mathbf{0}, u),
\end{aligned} \tag{7.44}$$

where $k = \frac{k_1}{k_s k_t}$ can be expressed as:[4]

$$k = \frac{k_s C_s(\mathbf{0}) + k_t C_t(0) - C(\mathbf{0}, 0)}{\left(k_s C_s(\mathbf{0})\right) \left(k_t C_t(0)\right)}. \tag{7.45}$$

The asymptotic behavior of $\gamma(\mathbf{h}, u), \gamma(\mathbf{h}, 0)$ and $\gamma(\mathbf{0}, u)$ suggests that these three semivariograms do not have the same sill. See Theorem 7.8.1 to this effect.

---

[4] See proof in Appendix D.

Theorem 7.8.1[5] **(De Iaco *et al.* 2001)** *Let Z be a spatio-temporal second-order stationary rf. Using the spatio-temporal covariance expression* $C(\mathbf{h}, u) = k_1 C_s(\mathbf{h}) C_t(u) + k_2 C_s(\mathbf{h}) + k_3 C_t(u)$ *as a basis, assuming that C is continuous in the space-time domain and taking into account that* $\gamma(\mathbf{h}, 0) = \left(k_2 + k_1 C_t(0)\right) \gamma_s(\mathbf{h}) = k_s \gamma_s(\mathbf{h})$ *and* $\gamma(\mathbf{0}, u) = \left(k_3 + k_1 C_s(\mathbf{0})\right) \gamma_t(u) = k_t \gamma_t(u)$*, we obtain:*

$$\lim_{\mathbf{h} \to \infty} \lim_{u \to \infty} \gamma(\mathbf{h}, u) = C(\mathbf{0}, 0) \tag{7.46}$$

$$\lim_{\mathbf{h} \to \infty} \gamma(\mathbf{h}, 0) = k_s C_s(\mathbf{0}) \tag{7.47}$$

$$\lim_{u \to \infty} \gamma(\mathbf{0}, u) = k_t C_t(0). \tag{7.48}$$

By estimating and modeling $\gamma(\mathbf{h}, 0)$ and $\gamma(\mathbf{0}, u)$ we obtain the sill values $k_s C_s(\mathbf{0})$ and $k_t C_t(0)$ defined in (7.47) and (7.48). Therefore, the only parameter left to estimate is k, which depends on the global sill. Theorem 7.8.2 shows a necessary and sufficient condition, in terms of bounding the parameter k, to ensure the validity of:

$$\gamma(\mathbf{h}, u) = \gamma_{st}(\mathbf{h}, 0) + \gamma(\mathbf{0}, u) - k\gamma(\mathbf{h}, 0)\gamma(\mathbf{0}, u). \tag{7.49}$$

Theorem 7.8.2[6] **(De Iaco *et al.* 2001)** *Let Z be a spatio-temporal second-order stationary rf. Suppose that the covariance function* $C(\mathbf{h}, u) = k_1 C_s(\mathbf{h}) C_t(u) + k_2 C_s(\mathbf{h}) + k_3 C_t(u)$ *is continuous in the space-time domain. Suppose that the parameter k takes the value* $k = \frac{k_1}{k_s k_t}$ *in the spatio-temporal semivariogram* $\gamma(\mathbf{h}, u) = \gamma(\mathbf{h}, 0) + \gamma(\mathbf{0}, u) - k\gamma(\mathbf{h}, 0)\gamma(\mathbf{0}, u)$*. Then,* $k_1 > 0$*,* $k_2 \geq 0$ *and* $k_3 \geq 0$ *if, and only if, k fulfills the following condition:*

$$0 \leq k \leq \frac{1}{\max\left\{sill(\gamma(\mathbf{h}, 0)); sill(\gamma(\mathbf{0}, u))\right\}}. \tag{7.50}$$

The two following corollaries of Theorem 7.8.2 are of interest:

Corollary 1 *If* $\gamma(\mathbf{0}, u)$ *or* $\gamma(\mathbf{h}, 0)$ *are unbounded, then there is no choice for k that fulfills the inequality*

$$0 \leq k \leq \frac{1}{\max\left\{sill(\gamma(\mathbf{h}, 0)); sill(\gamma(\mathbf{0}, u))\right\}} \tag{7.51}$$

such that

$$\gamma(\mathbf{h}, u) = \gamma(\mathbf{h}, 0) + \gamma(\mathbf{0}, u) - k\gamma(\mathbf{h}, 0)\gamma(\mathbf{0}, u) \tag{7.52}$$

is a valid spatio-temporal semivariogram.

---

[5] See proof in Appendix D.
[6] See proof in Appendix D.

Corollary 2 *If $\gamma_s(\mathbf{h})$ or $\gamma_t(u)$ are unbounded, then*

$$\gamma(\mathbf{h}, u) = \left(k_2 + k_1 C_t(0)\right) \gamma_s(\mathbf{h}) + \left(k_3 + k_1 C_s(\mathbf{0})\right) \gamma_t(u) - k_1 \gamma_s(\mathbf{h}) \gamma_t(u) \tag{7.53}$$

is not a valid spatio-temporal semivariogram for any coefficients k_1, k_2 and k_3.

Finally, it is worth indicating that both the product-sum covariance model and also its generalization are non-separable and generally non-integrable; hence, they cannot be obtained using the Cressie-Huang representation (see Cressie and Huang 1999). In addition, they do not require spatio-temporal metrics and are more flexible than the product model when it comes to estimating and modeling structures of spatio-temporal dependence.

7.9 Integrated product and product-sum models

De Iaco *et al.* (2002b) apply the second stability property of covariance functions, discussed in Matérn (1960, p. 10) and Chilès and Delfiner (1999, p. 60) (see Theorem 7.9.1), to product and product-sum models to generate new valid stationary models of the covariance function and semivariogram: the integrated product model and the integrated product-sum model.

These models will be mixtures of valid spatial and temporal covariance functions and cannot generally be obtained by the Cressie-Huang representation (Cressie and Huang 1999).

Theorem 7.9.1 (De Iaco *et al.* 2002b) *Let $\mu(a)$ be a positive measure over $U \subseteq \mathbb{R}$ and let $C_s(\mathbf{h}, a)$ and $C_t(u, a)$ be covariance functions on $D \subset \mathbb{R}^d$ and $T \subset \mathbb{R}_+$, respectively, for each $a \in V \subseteq U$.*

1. Integrated product model: *If $C_s(\mathbf{h}, a)C_t(u, a)$ is integrable with respect to the measure μ over V for each $\mathbf{h}$ and u, then for any given $k > 0$*

$$C(\mathbf{h}, a) = \int_V k C_s(\mathbf{h}, a) C_t(u, a) d\mu(a) \tag{7.54}$$

is a covariance function on $\mathbb{R}^d \times T$. This result can be rewritten in semivariogram terms:

$$\gamma(\mathbf{h}, u) = \int_V k(C_t(\mathbf{0}, a)\gamma_s(\mathbf{h}, a) + C_s(\mathbf{0}, a)\gamma_t(u, a) - \gamma_s(\mathbf{h}, a)\gamma_t(u, a) d\mu(a). \tag{7.55}$$

2. Integrated product-sum model: *Likewise, if $k_1 C_s(\mathbf{h}, a)C_t(u, a) + k_2 C_s(\mathbf{h}, a) + k_3 C_t(u, a)$ is integrable with respect to the measure μ over V for each $\mathbf{h}$ and u, given $k_1 > 0, k_2 \geq 0$ and $k_3 \geq 0$, then*

$$C(\mathbf{h}, u) = \int_V \left(k_1 C_s(\mathbf{h}, a)C_t(u, a) + k_2 C_s(\mathbf{h}, a) + k_3 C_t(u, a)\right) d\mu(a) \tag{7.56}$$

is a covariance function on $\mathbb{R}^d \times T$. *This result can be rewritten in semivariogram terms:*

$$\gamma(\mathbf{h}, u) = \int_V \left(k_2 + k_1 C_t(0, a)\right) \gamma_s(\mathbf{h}, a) + \left(k_3 + k_1 C_s(\mathbf{0}, a)\right) \gamma_t(\mathbf{h}, a) - k_1 \gamma_s(\mathbf{h}, a) \gamma_t(\mathbf{h}, a) d\mu(a). \tag{7.57}$$

According to De Iaco *et al.* (2002b) the following clarifications are of particular interest:

(i) In general, product-sum covariance models are not integrable on **h** and u, are non-separable and do not correspond to the use of a spatio-temporal metric. As a result, integrated product-sum models have the same characteristics as product-sum models.

(ii) Although the product model is separable and integrable, its integrated form can give rise to non-separable and non-integrable models, as can be seen in the example at the end of this section. Therefore, this type of model cannot generally be obtained from the Cressie-Huang representation.

(iii) Since the complex exponential can be written as:

$$e^{i\mathbf{h}'\omega} = cos(\mathbf{h}'\omega) + isin(\mathbf{h}'\omega), \tag{7.58}$$

if $\rho(\omega, u)k(\omega)$ is symmetric with respect to the origin on $\mathbb{R}^n$, then the Cressie-Huang representation can be seen as a special case of the integrated product model:

$$C(\mathbf{h}, u) = \int_V kC_s(\mathbf{h}, a) C_t(u, a) d\omega(a). \tag{7.59}$$

Therefore,

$$C(\mathbf{h}, u) = \int_{\mathbb{R}^n} e^{i\mathbf{h}'\omega} \rho(\omega, u) k(\omega) d\omega = \int_{\mathbb{R}^n_+} C_s(\mathbf{h}, \omega) C_t(u, \omega) k(\omega) d\omega, \tag{7.60}$$

where $k(\omega)$ is defined, positive and integrable on $\mathbb{R}^n_+ = \mathbb{R}_+ \times \ldots \times \mathbb{R}_+$, n times, and $C_s(\mathbf{h}, \omega)$ is only positive semi-definite for each $\omega \in \mathbb{R}_+$. Only in this special case can non-separable Cressie-Huang covariance models be rewritten in semivariogram terms as in (7.57) and take advantage of the fact that semivariograms are null at the origin. Note that all the models obtained by the Cressie-Huang representation fulfill the property of symmetry. In addition, spatio-temporal covariance models $\gamma(\mathbf{h}, u), \gamma(\mathbf{0}, u)$ and $\gamma(\mathbf{h}, 0)$ obtained from that representation have the same sill.

(iv) In non-separable Cressie-Huang models, merely spatial and temporal models can only be considered by adding them to the model considered, whereas in the case of the product-sum model and the integrated product-sum model, merely spatial and temporal structures are part of the models due to their construction itself.

Finally, we provide an example of these constructions below. In the example we assume that the measure μ is generated by an absolute continuous function, F, its distribution function, which means there will be a function f (its density function), such that $dF(a) = f(a)da$ almost everywhere.

Example 7.7 (De Iaco *et al.* 2002b. Example 3) *Let the density function associated with the exponential distribution of the parameter* β:

$$f(a,\beta) = \beta e^{-\beta a}, \quad a > 0, \beta > 0. \tag{7.61}$$

Consider the following spatial and temporal covariance functions, valid for any given $a \in V = [0,\infty)$:

$$C_s(\mathbf{h}, a, \omega) = cos\,(a(\omega\|\mathbf{h}\|)), \quad a > 0, \omega \in \mathbb{R} \tag{7.62}$$

$$C_t(u, a, c, \delta) = e^{\left(-\frac{au^\delta}{c}\right)}, \quad 1 \le \delta \le 2, \quad a > 0, \quad c > 0. \tag{7.63}$$

Then, the integrated product model is obtained as follows:

$$\begin{aligned} C(\mathbf{h}, u) &= \int_0^\infty ke^{\left(-\frac{au^\delta}{c}\right)} cos\,(a(\omega\|\mathbf{h}\|))\, \beta e^{-\beta a} da \\ &= k\beta \int_0^\infty e^{-a\left(\frac{u^\delta}{c}+\beta\right)} cos\,(a(\omega\|\mathbf{h}\|))\, da \\ &= \frac{k\beta\left(\frac{u^\delta}{c}+\beta\right)}{\left(\frac{u^\delta}{c}+\beta\right)^2 + (\omega\|\mathbf{h}\|)^2}, \end{aligned}$$

which is a spatio-temporal covariance function that depends on the vector of parameters $\boldsymbol{\theta} = (c, k, \omega, \beta, \delta)'$.

The integrated product-sum model emerges as follows:

$$\begin{aligned} C(\mathbf{h}, u) &= \int_0^\infty \left(k_1 e^{\left(-\frac{au^\delta}{c}\right)} cos\,(a(\omega\|\mathbf{h}\|)) + k_2 cos\,(a(\omega\|\mathbf{h}\|)) + k_3 e^{\left(-\frac{au^\delta}{c}\right)}\right) \\ &\quad \times \beta e^{-\beta a} da \\ &= k_1 \frac{\beta\left(\frac{u^\delta}{c}+\beta\right)}{\left(\frac{u^\delta}{c}+\beta\right)^2 + (\omega\|\mathbf{h}\|)^2} + k_2 \frac{\beta^2}{\beta^2 + (\omega\|\mathbf{h}\|)^2} + k_3 \frac{\beta}{\frac{u^\delta}{c}+\beta}, \end{aligned}$$

which is a spatio-temporal covariance function that depends on the vector of parameters $\boldsymbol{\theta} = \left(c, k_1, k_2, k_3, \omega, \beta, \delta\right)'$.

Of course, the two models above can be expressed in semivariogram terms. De Iaco *et al.* (2002b) provide other additional examples to that above.

7.10 Models proposed by Cressie and Huang

Cressie and Huang (1999) start out by criticizing separable models for not taking into account spatio-temporal interaction and then proceed to construct parametric models of non-separable spatio-temporal covariance functions. In order to do so, they use the equivalence of the validity of the covariance function and the fact that the rf has a spectral distribution function as a basis (see examples in Matérn 1960, p. 12). Matérn, whose results in the spatial field are generalized by Cressie and Huang in the spatio-temporal framework, constructs covariance functions by directly inverting spectral densities. Cressie and Huang use this same procedure (and sometimes even the same formulas) in the spatio-temporal scenario. The problem they encounter is that despite the strength of the procedure, it is restricted to a series of covariance functions that are integrable and for which there is an analytical solution of the corresponding Fourier integral on $\mathbb{R}^d$, which is quite small. The procedure is performed as follows (Cressie and Huang 1999, pp.1331–2).

Let $C(\mathbf{h}, u)$ be the spatio-temporal covariance function of a stationary rf. If C is assumed to be continuous and its spectral distribution function has a spectral density $g(\boldsymbol{\omega}, \tau) \geq 0$, then according to Bochner's theorem:

$$C(\mathbf{h}, u) = \int\int e^{i(\mathbf{h}'\boldsymbol{\omega}+u\tau)} g(\boldsymbol{\omega}, \tau) d\boldsymbol{\omega} d\tau. \tag{7.64}$$

Moreover, if $C(\cdot, \cdot)$ is integrable, then

$$g(\omega, \tau) = \frac{1}{(2\pi)^{d+1}} \int\int e^{-i(\mathbf{h}'\omega+u\tau)} C(\mathbf{h}, u) d\mathbf{h} du = \frac{1}{2\pi} \int e^{-iu\tau} \mathbf{h}(\boldsymbol{\omega}, u) du, \tag{7.65}$$

where

$$\mathbf{h}(\boldsymbol{\omega}, u) = \frac{1}{(2\pi)^d} \int e^{-i\mathbf{h}'\boldsymbol{\omega}} C(\mathbf{h}, u) d\mathbf{h} = \int e^{iu\tau} g(\boldsymbol{\omega}, \tau) d\tau. \tag{7.66}$$

From (7.64) and (7.65) the expression of the covariance function can be written as:

$$C(\mathbf{h}, u) = \int e^{i\mathbf{h}'\boldsymbol{\omega}} \mathbf{h}(\boldsymbol{\omega}, u) d\boldsymbol{\omega}, \tag{7.67}$$

so that the key point for the construction of the covariance function is the specification of appropriate models for $\mathbf{h}(\boldsymbol{\omega}, u)$.

Cressie and Huang (1999) assume that $\mathbf{h}(\boldsymbol{\omega}, u)$ can be broken down into the product:

$$\mathbf{h}(\boldsymbol{\omega}, u) = k(\boldsymbol{\omega})\rho(\boldsymbol{\omega}, u), \tag{7.68}$$

where $k(\boldsymbol{\omega})$ is the spectral density of a merely spatial process and $\rho(\boldsymbol{\omega}, \cdot)$ is a temporal autocorrelation function. The following two conditions are satisfied in (7.67):

(i) For each $\boldsymbol{\omega} \in \mathbb{R}^d$, $\rho(\boldsymbol{\omega}, \cdot)$ is a continuous autocorrelation function, $\int \rho(\boldsymbol{\omega}, u)du < \infty$ and $k(\boldsymbol{\omega}) > 0$.

(ii) $\int k(\boldsymbol{\omega})d\boldsymbol{\omega} < \infty$.

Then the expression (7.65) transforms into:

$$g(\boldsymbol{\omega}, \tau) = \frac{1}{2\pi} k(\boldsymbol{\omega}) \int e^{-iu\tau} \rho(\boldsymbol{\omega}, u)du, \tag{7.69}$$

which, for (i), is greater than 0.

Furthermore,

$$\int \int g(\boldsymbol{\omega}, \tau)d\boldsymbol{\omega}d\tau = \int k(\boldsymbol{\omega})d\boldsymbol{\omega}, \tag{7.70}$$

which, for (ii), is finite.

Hence, assuming that $\mathbf{h}(\boldsymbol{\omega}, u) = k(\boldsymbol{\omega})\rho(\boldsymbol{\omega}, u)$ and that (i) and (ii) are verified, we have that

$$C(\mathbf{h}, u) = \int e^{i\mathbf{h}'\boldsymbol{\omega}} \mathbf{h}(\boldsymbol{\omega}, u)d\boldsymbol{\omega} \tag{7.71}$$

is a valid, stationary and continuous spatio-temporal covariance model on $\mathbb{R}^d \times \mathbb{R}$.

Note that it is not at all difficult to see that any given continuous, stationary and integrable spatio-temporal covariance function can be written in the form:

$$\mathbf{h}(\boldsymbol{\omega}, u) = k(\boldsymbol{\omega})\rho(\boldsymbol{\omega}, u),$$

fulfilling conditions (i) and (ii). In order to do so, it is sufficient to define:

$$\rho(\boldsymbol{\omega}, u) = \frac{\mathbf{h}(\boldsymbol{\omega}, u)}{\int g(\boldsymbol{\omega}, \tau)d\tau}$$

and

$$k(\boldsymbol{\omega}) = \int g(\boldsymbol{\omega}, \tau)d\tau.$$

In summary, this procedure makes it possible to obtain valid spatio-temporal covariance models by selecting two functions, $\rho(\boldsymbol{\omega}, u)$ and $k(\boldsymbol{\omega})$, which fulfill conditions (i) and (ii) and allow the integral (7.71) to be analytically evaluated, the last condition being a merely practical constraint. The covariance functions constructed are generally non-separable, the product model being a particular case of the expression (7.71) when $\rho(\boldsymbol{\omega}, u)$ is considered as a function that is merely dependent on u.

In their article, Cressie and Huang present seven different examples obtained using this procedure, although Gneiting (2002) later proves that examples 5 and 6 are not valid models due to not fulfilling the necessary conditions. Examples 1, 2, 3, 4 and 7 are shown below.

Example 7.8 (Cressie and Huang 1999. Example 1) *Consider the functions*

$$\rho(\boldsymbol{\omega}, u) = \exp\left(-\frac{\|\boldsymbol{\omega}\|^2 u^2}{4}\right)\exp\left(-\delta|u|^2\right), \qquad k(\boldsymbol{\omega}) = \exp\left(-\frac{c_0\|\boldsymbol{\omega}\|^2}{4}\right),$$

with $\delta > 0$ and $c_0 > 0$, which fulfill conditions (i) and (ii). For (7.67), we have that

$$C(\mathbf{h}, u) \propto \int \exp\left(i\mathbf{h}'\boldsymbol{\omega}\right)\exp\left(-\frac{c_0\|\boldsymbol{\omega}\|^2}{4}\right)\exp\left(-\frac{\|\boldsymbol{\omega}\|^2 u^2}{4}\right)\exp\left(-\delta|u|^2\right) d\boldsymbol{\omega}$$

is a continuous spatio-temporal covariance function on $\mathbb{R}^d \times \mathbb{R}$. By operating in this expression and applying the appropriate Fourier transform, we obtain that

$$C(\mathbf{h}, u) = (2\sqrt{\pi})^d \frac{1}{\left(u^2 + c_0\right)^{d/2}} \exp\left(-\frac{\|\mathbf{h}\|^2}{u^2 + c_0}\right)\exp(-\delta|u|^2).$$

Note that the condition $\delta > 0$ is necessary to ensure that (i) is fulfilled when $\boldsymbol{\omega} = 0$. The resulting function taking limits when $\delta \to 0$ will also be a valid spatio-temporal covariance function, due to being the limit of a series of valid spatio-temporal covariance functions and will be expressed by:

$$C(\mathbf{h}, u) = \frac{\sigma^2}{\left(a^2u^2 + 1\right)^{d/2}} \exp\left(-\frac{b^2\|\mathbf{h}\|^2}{a^2u^2 + 1}\right), \tag{7.72}$$

which will be a parametric family of non-separable spatio-temporal covariance functions, with the vector of parameters $\boldsymbol{\theta} = (a, b, \sigma^2)'$, $a \geq 0$ being the time scaling parameter, $b \geq 0$ the space scaling parameter and $\sigma^2 = C(\mathbf{0}, 0) > 0$. Because of the redundancy in the parameters a, b and c_0, without loss of generality, in this example and the following ones, c_0 is set to 1. Due to the relationship that exists between a stationary covariance function and its corresponding semivariogram, we have that

$$\gamma(\mathbf{h}, u) = \sigma^2\left(1 - \frac{1}{\left(a^2u^2 + 1\right)^{d/2}} \exp\left(-\frac{b^2\|\mathbf{h}\|^2}{a^2u^2 + 1}\right)\right).$$

Example 7.9 (Cressie and Huang 1999. Example 2) *Consider the functions*

$$\rho(\boldsymbol{\omega}, u) = \exp\left(-\frac{\|\boldsymbol{\omega}\|^2 |u|}{4}\right)\exp\left(-\delta u^2\right), \qquad k(\boldsymbol{\omega}) = \exp\left(-\frac{c_0\|\boldsymbol{\omega}\|^2}{4}\right),$$

with $\delta > 0$ and $c_0 > 0$, which fulfill (i) and (ii). They give rise to the continuous spatio-temporal covariance model:

$$C(\mathbf{h}, u) \propto \frac{1}{\left(|u| + c_0\right)^{d/2}} \exp\left(-\frac{\|\mathbf{h}\|^2}{|u| + c_0}\right)\exp\left(-\delta u^2\right).$$

When $\delta \to 0$ we have the family of spatio-temporal covariance functions:

$$C(\mathbf{h}, u) = \frac{\sigma^2}{(a|u| + 1)^{d/2}} \exp\left(-\frac{b^2\|\mathbf{h}\|^2}{a|u| + 1}\right), \tag{7.73}$$

where $a \geq 0$ is the time scaling parameter, $b \geq 0$ is the space scaling parameter and $\sigma^2 = C(\mathbf{0}, 0) > 0$.

Example 7.10 (Cressie and Huang 1999. Example 3) *In this case consider*

$$\rho(\boldsymbol{\omega}, u) = \exp\left(-\|\boldsymbol{\omega}\|u^2\right) \exp\left(-\delta u^2\right), \qquad k(\boldsymbol{\omega}) = \exp\left(-c_0\|\boldsymbol{\omega}\|\right),$$

with $\delta > 0$ and $c_0 > 0$, which gives rise to the continuous spatio-temporal covariance model:

$$C(\mathbf{h}, u) \propto \frac{1}{\left(u^2 + c_0\right)^d} \left(1 + \frac{\|\mathbf{h}\|^2}{(u^2 + c_0)^2}\right)^{-(d+1)/2} \exp\left(-\delta u^2\right).$$

When $\delta \to 0$ we have the family of spatio-temporal covariance functions:

$$C(\mathbf{h}, u) = \frac{\sigma^2(a^2u^2 + 1)}{\left((a^2u^2 + 1)^2 + b^2\|\mathbf{h}\|^2\right)^{(d+1)/2}}, \tag{7.74}$$

where $a \geq 0$ is the time scaling parameter, $b \geq 0$ is the space scaling parameter and $\sigma^2 = C(\mathbf{0}, 0) > 0$.

Example 7.11 (Cressie and Huang 1999. Example 4) *Let*

$$\rho(\boldsymbol{\omega}, u) = \exp\left(-\|\boldsymbol{\omega}\||u|\right) \exp(-\delta u^2), \qquad k(\boldsymbol{\omega}) = \exp\left(-c_0\|\boldsymbol{\omega}\|\right),$$

with $\delta > 0$ and $c_0 > 0$, which gives rise to the continuous spatio-temporal covariance model:

$$C(\mathbf{h}, u) \propto \frac{1}{\left(|u| + c_0\right)^d} \left(1 + \frac{\|\mathbf{h}\|^2}{\left(|u| + c_0\right)^2}\right)^{-(d+1)/2} \exp\left(-\delta u^2\right).$$

If we consider $\delta \to 0$, we arrive at the spatio-temporal covariance model:

$$C(\mathbf{h}, u) = \frac{\sigma^2(a|u| + 1)}{\left((a|u| + 1)^2 + b^2\|\mathbf{h}\|^2\right)^{(d+1)/2}}, \tag{7.75}$$

where $a \geq 0$ is the time scaling parameter, $b \geq 0$ is the space scaling parameter and $\sigma^2 = C(\mathbf{0}, 0) > 0$.

Example 7.12 (Cressie and Huang 1999. Example 7) *In this example,*

$$\rho(\boldsymbol{\omega}, u) = \left(u^2 + 1 + (u^2 + c)\|\boldsymbol{\omega}\|^2\right)^{-(\nu+d)/2}\left(1 + c\|\boldsymbol{\omega}\|^2\right)^{(\nu+d)/2},$$
$$k(\boldsymbol{\omega}) = \left(1 + c\|\boldsymbol{\omega}\|^2\right)^{-(\nu+d)/2},$$

with $c > 0$ and $\nu > 0$, which fulfill conditions (i) and (ii) and which therefore give rise to a family of continuous spatio-temporal covariance functions on $\mathbb{R}^d \times \mathbb{R}$ given by:

$$C(\mathbf{h}, u) \propto \begin{cases} \frac{1}{(u^2+1)^{\nu}(u^2+c)^{d/2}}\left(\left(\frac{u^2+1}{u^2+c}\right)^{1/2}\|\mathbf{h}\|\right)^{\nu} \\ K_{\nu}\left(\left(\frac{u^2+1}{u^2+c}\right)^{1/2}\|\mathbf{h}\|\right), & \text{if } \|\mathbf{h}\| > 0, \\ \frac{1}{(u^2+1)^{\nu}(u^2+c)^{d/2}}, & \text{if } \|\mathbf{h}\| = 0, \end{cases}$$

where K_ν is the modified Bessel function of the second kind of order ν. Therefore, the function:

$$C(\mathbf{h}, u) = \begin{cases} \frac{\sigma^2(2c^{d/2})}{(a^2u^2+1)^{\nu}(a^2u^2+c)^{d/2}\Gamma(\nu)}\left(\frac{b}{2}\left(\frac{a^2u^2+1}{a^2u^2+c}\right)^{1/2}\|\mathbf{h}\|\right)^{\nu} \\ \times K_{\nu}\left(b\left(\frac{a^2u^2+1}{a^2u^2+c}\right)^{1/2}\|\mathbf{h}\|\right), & \text{if } \|\mathbf{h}\| > 0, \\ \frac{\sigma^2 c^{d/2}}{(a^2u^2+1)^{\nu}(a^2u^2+c)^{d/2}}, & \text{si } \|\mathbf{h}\| = 0, \end{cases} \tag{7.76}$$

is a family of non-separable spatio-temporal covariance functions which depends on the vector of parameters $\boldsymbol{\theta} = (a, b, c, \nu, \sigma^2)'$, where $a \geq 0$ is the time scaling parameter, $b \geq 0$ the space scaling parameter, $c > 0$ is a parameter of spatio-temporal interaction, $\nu > 0$ a smoothing parameter, and $\sigma^2 = C(\mathbf{0}, 0) > 0$. For $c = 1$ we have a separable covariance function.

Figure 7.9 shows different representations of the family of covariances (7.76) for different sets of possible values of the parameters that define it. The first example represents the covariance function obtained for $\boldsymbol{\theta} = (1, 1, 0.5, 0.5, 5)'$, the second that obtained for $\boldsymbol{\theta} = (1, 1, 1.5, 1, 5)'$, the third corresponds to $\boldsymbol{\theta} = (1, 5, 1.5, 0.5, 5)'$ and the fourth to $\boldsymbol{\theta} = (0.5, 1, 1.5, 0.5, 5)'$. Notice that for this model, the merely spatial covariance function $C(\mathbf{h}, 0)$ belongs to Matérn's model and is represented by a red dotted line, while the corresponding covariance $C(\mathbf{0}, u)$ is in green.

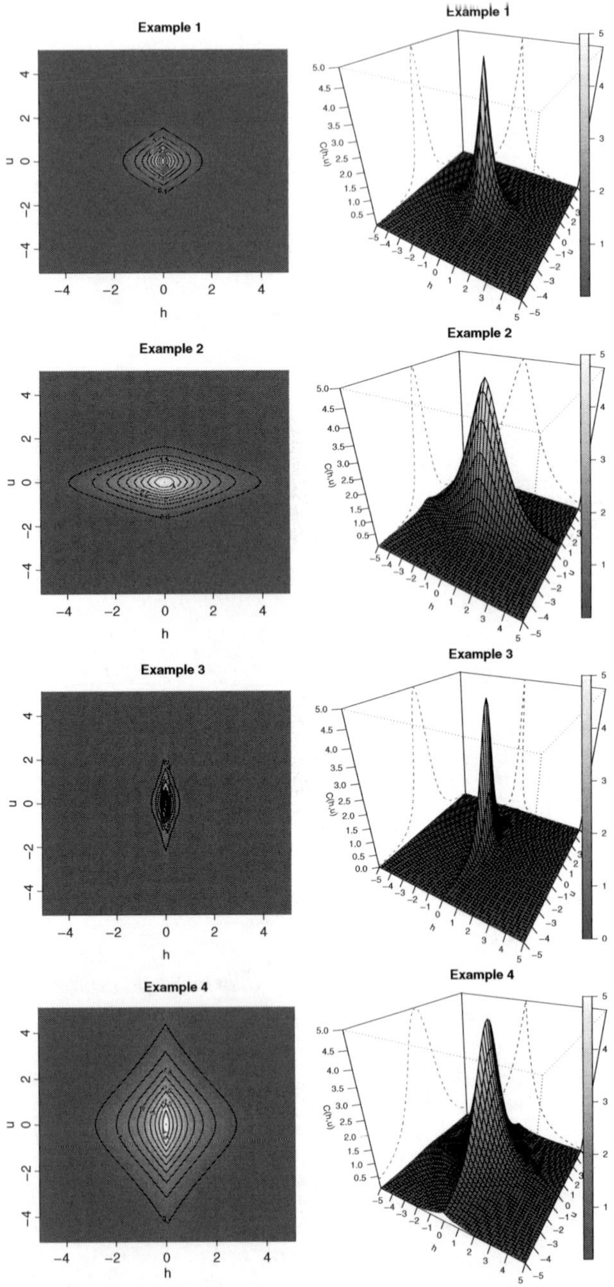

Figure 7.9 *2D and 3D different representations of the family of covariance functions (7.76). (See color figure in color plate section.)*

7.11 Models proposed by Gneiting

Gneiting (2002) recognizes that the approach taken by Cressie and Huang (1999) is novel and powerful, but considers that this procedure has a significant limitation: it is restricted to a small class of functions for which a closed-form solution to the d-variate Fourier integral is known. Therefore, it is no surprise that Gneiting uses the same approach as Cressie and Huang, but he avoids the limitation mentioned above and provides very general classes of valid spatio-temporal covariance models.

He does not make use of Fourier transforms (as in the procedure followed by Cressie and Huang), but rather allows for the construction of parametric spatio-temporal covariance functions directly in the spatio-temporal domain.

The class of models proposed by Gneiting involve completely monotone functions and Bernstein functions, both closely related to spatial statistics.

From the theorem of Bernstein (Feller 1966, p. 439), the general form of a completely monotone function is:

$$\varphi(t) = \int_0^{+\infty} e^{-rt} dF(r), \quad t > 0, \tag{7.77}$$

where F is non-decreasing. Therefore, Bernstein's theorem states that a completely monotone function is a mixture of exponential functions (a special case of this mixture is the expectation when $F(r) = f(r)dr$).

As can be appreciated, there is a relationship between completely monotone functions and Laplace transforms of finite and non-negative measures on $\mathbb{R}^+$. Some examples of completely monotone functions can be found in Table 7.1.

Isotropic covariance functions and completely monotone functions are closely related, as a continuous isotropic function $C(\mathbf{h}) = \varphi(\|\mathbf{h}\|^2)$ defined on $\mathbb{R}^d$ is a spatial covariance function for any given dimension d if, and only if, $\varphi(t), t \geq 0$ is completely monotone (Schoenberg 1938; Cressie 1993, p. 86).

It can be verified that a continuous and isotropic function, $\gamma(\mathbf{h}) = \psi(\|\mathbf{h}\|^2)$, defined on $\mathbb{R}^d$ is a semivariogram for any given dimension d if, and only if, $\psi(t), t \geq 0$, is a Bernstein function with $\psi(0) = 0$. Some Bernstein functions can be found in Table 7.2.

Table 7.1 Some completely monotone functions $\varphi(t), t > 0$.

Function	Parameters
$\varphi(t) = \exp(-ct^{\gamma})$	$c > 0, 0 < \gamma \leq 1$
$\varphi(t) = (2^{\upsilon-1}\Gamma(\upsilon))^{-1}\left(ct^{\frac{1}{2}}\right)^{\upsilon} K_{\upsilon}\left(ct^{\frac{1}{2}}\right)$	$c > 0, \upsilon > 0$
$\varphi(t) = (1 + ct^{\gamma})^{-\upsilon}$	$c > 0, 0 < \gamma \leq 1, \upsilon > 0$
$\varphi(t) = 2^{\upsilon}\left(\exp\left(ct^{\frac{1}{2}}\right) + \exp\left(-ct^{\frac{1}{2}}\right)\right)^{-\upsilon}$	$c > 0, \upsilon > 0$

Source: Gneiting (2002).

Table 7.2 Some Bernstein functions: positive-definite functions $\psi(t)$, $t > 0$, with completely monotonic derivatives (the functions have been standardized so that $\psi(0) = 1$).

Function	Parameters
$\psi(t) = (at^{\alpha} + 1)^{\beta}$	$a > 0, 0 < \alpha \leq 1, 0 \leq \beta \leq 1$
$\psi(t) = (at^{\alpha} + 1)$	$a > 0, 0 < \alpha \leq 1$
$\psi(t) = \ln\,(at^{\alpha} + b)\,/\ln\; b$	$a > 0, b > 1, 0 < \alpha \leq 1$
$\psi(t) = (at^{\alpha} + b)\,/(b\,(at^{\alpha} + 1))$	$a > 0, 0 < b \leq 1, 0 < \alpha \leq 1$

Source: Gneiting (2002).

On the grounds of the criteria 1 and 2 discussed in Feller (1978, p. 494), the following theorem can be formulated.

Theorem 7.11.1 *Let $\varphi(t)$, with $t > 0$, be a completely monotone function and $\psi(t)$, with $t \geq 0$, a Bernstein function, then the function*

$$C(\mathbf{h}, u) = \frac{\sigma^2}{\psi(|u|^2)^{d/2}}\varphi\left(\frac{||\mathbf{h}||^2}{\psi(|u|^2)}\right), \quad (\mathbf{h}, u) \in \mathbb{R}^d \times \mathbb{R} \tag{7.78}$$

is a valid spatio-temporal covariance function.

This theorem is demonstrated in Gneiting (2002). The covariance functions generated by (7.78) will be fully symmetric. The construction procedure that stems from Theorem 7.11.1 is very general and flexible.

All the examples of Cressie and Huang (1999) can be expressed in the form of (7.78), except for the case $c < 1$ in their Example 7, and their Examples 5 and 6 which are shown to be wrong. Even if the covariance function obtained is in general non-separable, the functions $\varphi(t)$ and $\psi(t)$ can be associated, respectively, with the spatial and temporal structure of the data.

Example 7.13 (Gneiting 2002. Example 1) *Consider the functions $\varphi(t) = \exp(-ct^{\gamma})$, with $t \geq 0$, and $\psi(t) = (at^{\alpha} + 1)^{\beta}$, with $t > 0$. We can see that for $c > 0$ and $0 < \gamma \leq 1, \varphi(t)$ is a completely monotone function and that for $a > 0, 0 < \alpha \leq 1$ and $0 \leq \beta \leq 1, \psi(t)$ is a positive function with a completely monotone derivative. Then, following Theorem 7.11.1, the function*

$$C(\mathbf{h}, u) = \frac{\sigma^2}{\left(a|u|^{2\alpha} + 1\right)^{\beta d/2}} \exp\left(-\frac{c||\mathbf{h}||^{2\gamma}}{\left(a|u|^{2\alpha} + 1\right)^{\beta\gamma}}\right), \quad (\mathbf{h}, u) \in \mathbb{R}^d \times \mathbb{R} \tag{7.79}$$

is a valid spatio-temporal covariance model on $\mathbb{R}^d \times \mathbb{R}$. Its product with the merely temporal covariance function $\left(a|u|^{2\alpha} + 1\right)^{-\delta}$ will give rise to the following valid *model of a spatio-temporal covariance function:*

$$C(\mathbf{h}, u) = \frac{\sigma^2}{\left(a|u|^{2\alpha} + 1\right)^{\delta+\beta d/2}} \exp\left(-\frac{c\|\mathbf{h}\|^{2\gamma}}{\left(a|u|^{2\alpha} + 1\right)^{\beta\gamma}} \right), \quad (\mathbf{h}, u) \in \mathbb{R}^d \times \mathbb{R}, \quad (7.80)$$

where $c > 0$ and $a > 0$ are two positive spatial and temporal scaling parameters, respectively, γ and α are two spatial and temporal smoothing parameters, respectively, which take values in $[0, 1]$, β is a spatio-temporal interaction parameter that takes values in $[0, 1]$, $\delta \geq 0$, and σ^2 is the variance of the spatio-temporal process. The merely spatial covariance function $C(\mathbf{h}, 0)$ belongs to the family of powered exponential covariances, whereas the merely temporal function $C(\mathbf{0}, u)$ belongs to the family of Cauchy covariances. We obtain a separable spatio-temporal covariance model for $\beta = 0$, whereas for $\beta = 1$ we obtain the most extreme case of the non-separable model of this family. As β increases, spatio-temporal interaction grows stronger.

Example 7.14 (Gneiting 2002. Example 2) *Consider the functions*

$$\varphi(t) = \left(2^{\nu-1}\Gamma(\nu)\right)^{-1} \left(ct^{1/2}\right)^{\nu} K_{\nu}\left(ct^{1/2}\right),$$

with $t \geq 0$, and $\psi(t) = (at^{\alpha} + 1)^{\beta}$, with $t \geq 0$. We can see that for $c > 0$ and $\nu > 0$, $\varphi(t)$ is a completely monotone function and that for $a > 0, 0 < \alpha \leq 1$ and $0 \leq \beta \leq 1, \psi(t)$ is a positive function with a completely monotone derivative. Then, according to Theorem 7.11.1, the function

$$C(\mathbf{h}, u) = \frac{\sigma^2}{2^{\nu-1}\Gamma(\nu)(a|u|^{2\alpha} + 1)^{\beta d/2}} \left(\frac{c\|\mathbf{h}\|}{(a|u|^{2\alpha} + 1)^{\beta/2}} \right)^{\nu} \times K_{\nu}\left(\frac{c\|\mathbf{h}\|}{(a|u|^{2\alpha} + 1)^{\beta/2}} \right)$$

is a valid spatio-temporal covariance model on $\mathbb{R}^d \times \mathbb{R}$.

Its product with the merely temporal covariance function $(a|u|^{2/\alpha} + 1)^{-\delta}$ gives rise to the following valid model of a spatio-temporal covariance function:

$$C(\mathbf{h}, u) = \frac{\sigma^2}{2^{\nu-1}\Gamma(\nu)(a|u|^{2\alpha} + 1)^{\delta+\beta d/2}} \left(\frac{c\|\mathbf{h}\|}{(a|u|^{2\alpha} + 1)^{\beta/2}} \right)^{\nu} \times K_{\nu}\left(\frac{c\|\mathbf{h}\|}{(a|u|^{2\alpha} + 1)^{\beta/2}} \right), \quad (7.81)$$

which depends on a vector of parameters $\boldsymbol{\theta} = (a, c, \alpha, \beta, \nu, \delta, \sigma^2)'$, where $c > 0$ and $a > 0$ are two positive spatial and temporal scaling parameters, respectively, $\nu > 0$ is a spatial smoothing parameter, $\alpha \in [0, 1]$ is a temporal smoothing parameter, $\beta \in [0, 1]$ is a parameter related to spatio-temporal interaction, δ is non-negative and $\sigma^2 > 0$ is the variance of the spatio-temporal process. As previously, we have a separable spatio-temporal covariance function for $\beta = 0$, whereas for $\beta = 1$ we obtain the least separable representative of this family of covariances.

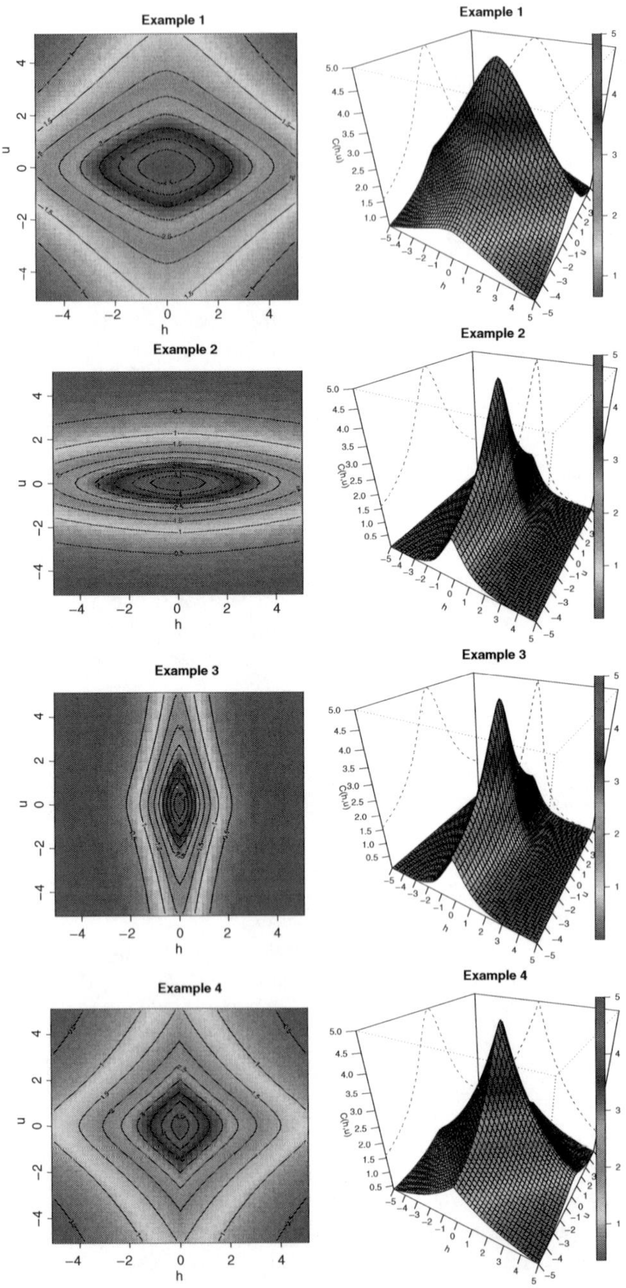

Figure 7.10 2D and 3D different representations of the family of covariance functions (7.81). *(See color figure in color plate section.)*

Figure 7.10 shows different representations of the family of covariances (7.81) for a set of possible values of the parameters that define it, in which we can see the effect of the variation of each of the above parameters. The first example represents the covariance function obtained for $\boldsymbol{\theta} = (0.5, 0.3, 0.9, 0, 1, 0.5, 5)'$, the second that obtained for $\boldsymbol{\theta} = (0.5, 0.3, 0.9, 0.9, 1, 0.5, 5)'$, the third corresponds to $\boldsymbol{\theta} = (0.5, 1.5, 0.9, 0, 1, 0.5, 5)'$ and the fourth that obtained with $\boldsymbol{\theta} = (0.5, 0.3, 0.9, 0, 0.5, 0.5, 5)'$. Notice that for this model, the merely spatial covariance function $C(\mathbf{h}, 0)$ belongs to Matérn's model and is represented by a dotted red line, while the corresponding covariance $C(\mathbf{0}, u)$, which is given by $\sigma^2/(a|u|^{2\alpha} + 1)^{\delta+\beta d/2}, u \in \mathbb{R}$, is in green.

7.12 Mixture models proposed by Ma

Ma, in the same line of research, and at the same time as Cressie and Huang and Gneiting, based his research on the shortage of models of non-separable covariance functions, a shortage that contrasts with the huge demand for this type of models on behalf of researchers in numerous fields of science. According to Ma, while separable models have many serious limitations when it comes to modeling space-time interaction, they are a useful and sound basis to generate non-separable spatio-temporal models using mixture procedures (Ma 2002).

The approach taken by Ma (2002, 2003b) to overcome the limitations of separable models is very simple and is based on mixtures of purely spatial and purely temporal covariance functions. More specifically, he proposes the scale mixture and the positive power mixture of separable covariance functions, for which the condition of validity can easily be verified and that, moreover, arise frequently in closed form. The approach considered by Ma applies both to continuous-time and discrete-time settings and does not require the random function under study to be stationary in space or in time (though we will assume stationarity in this section).

7.12.1 Covariance functions generated by scale mixtures

Let $C_s(\mathbf{h})$ be a stationary spatial covariance function on $\mathbb{R}^d$ and $C_t(u)$ a stationary temporal covariance function on T. It is obvious that for any given $a \in \mathbb{R}$ and $b \in T$, the functions $C_s(a\mathbf{h})$ and $C_t(bu)$ are valid stationary covariances on $\mathbb{R}^d$ and T, respectively. Therefore, the function generated by

$$C(\mathbf{h}, u) = \int_{\mathbb{R}\times T} C_s(a\mathbf{h})C_t(bu)d\mu(a, b), \tag{7.82}$$

is a stationary spatio-temporal covariance function on $\mathbb{R}^d \times T$, where $\mu(a, b)$ is a probability measure on $\mathbb{R} \times T$. As can be noted, $C(\mathbf{h}, u)$ reduces to a separable covariance function if $\mu(a, b)$ is separable, that is, if the rv's A and B are independent.

The following example illustrates the procedure of constructing covariance functions using scale mixtures.

Example 7.15 (Ma 2002. Example 5) *Let (A, B) be a random vector with bivariate binomial distribution, such that $p_{ij} = P(A = i, B = j)$, $i, j = 0, 1$. Then,*

$$\begin{aligned} C(\mathbf{h}, u) = {} & C_s(\mathbf{h})C_t(u)p_{11} + C_s(\mathbf{h})C_t(0)p_{10} \\ & + C_s(\mathbf{0})C_t(u)p_{01} + C_s(\mathbf{0})C_t(0)p_{00}, \ (\mathbf{h}, u), \in \mathbb{R}^d \times T \end{aligned} \tag{7.83}$$

is a stationary covariance function on $\mathbb{R}^d \times T$.

As can be appreciated, this type of modeling is closely related to the product-sum model.

7.12.2 Covariance functions generated by positive power mixtures

Let $C(\mathbf{h}, u)$ be the covariance function of a stationary rf in space and time. Let its correlation function be:

$$\rho(\mathbf{h}, u) = \frac{C(\mathbf{h}, u)}{C(\mathbf{0},0)}, \qquad (\mathbf{h}, u) \in \mathbb{R}^d \times T, \tag{7.84}$$

where $C(\mathbf{0},0) > 0$.

It is obvious that if $\rho(\mathbf{h}, u)$ is a correlation function on $\mathbb{R}^d \times T$, then its marginal functions $\rho(\mathbf{h}, 0)$ and $\rho(\mathbf{0},u)$ are purely spatial and temporal correlation functions on $\mathbb{R}^d$ and T, respectively. Now let us assume that $\rho_s(\mathbf{h})$ is a stationary spatial correlation function on $\mathbb{R}^d$ and $\rho_t(u)$ is a stationary temporal correlation function on T. Given $k \in \mathbb{N}$, $\rho_s^k(\mathbf{h})$ and $\rho_t^k(u)$ are also two stationary correlation functions, the former being spatial on $\mathbb{R}^d$ and the latter temporal on T. Using this as a basis, Ma generates spatio-temporal correlation functions that are mixtures of positive powers of $\rho_s(\mathbf{h})$ and $\rho_t(u)$. Mixtures of negative powers make no sense, as $\left(\rho_s(\mathbf{h})\right)^{-k}$ is only a valid correlation function if, and only if, $\left|\rho_s(\mathbf{h})\right| \equiv 1$.

The positive power mixture proposed by Ma is as follows:

$$\rho(\mathbf{h}, u) = \sum_{i=0}^{+\infty} \sum_{j=0}^{+\infty} \rho_s^i(\mathbf{h})\rho_t^j(u)p_{ij}, \qquad (\mathbf{h}, u) \in \mathbb{R}^d \times T, \tag{7.85}$$

where $\{p_{ij}, (i, j) \in \mathbb{N}_+^2\}$ are non-negative constants that meet the condition that $\sum_{i=0}^{+\infty} \sum_{j=0}^{+\infty} p_{ij} = 1$ (so that $\rho(\mathbf{0},0) = 1$). Therefore, we are really assuming that $p_{ij}, (i, j) \in \mathbb{N}_+^2$ is the probability function of a bivariate discrete non-negative random vector.

Let us verify that the above mixture is a stationary correlation function on $\mathbb{R}^d \times T$. For each $(i, j) \in \mathbb{N}_+^2$, $\rho_s^i(\mathbf{h})$ is a correlation function on $\mathbb{R}^d$, $\rho_t^j(u)$ is a correlation function on T, and, by Schur's theorem, $\rho_s^i(\mathbf{h})\rho_t^j(u)$ is a correlation function on $\mathbb{R}^d \times T$. So is $\rho_s^i(\mathbf{h})\rho_t^j(u)p_{ij}$. As a result, $\rho(\mathbf{h}, u) = \sum_{i=0}^{+\infty} \sum_{j=0}^{+\infty} \rho_s^i(\mathbf{h})\rho_t^j(u)p_{ij}$ is a correlation function on $\mathbb{R}^d \times T$. This can also be performed on an alternative approach: the conditional method. Assume that (A, B) is a non-negative, bivariate discrete random vector with a probability function given by $\{p_{ij}, (i,j) \in \mathbb{N}_+^2\}$, and for which $Z(\mathbf{s}, t)$ under the condition that $(A, B) = (i,j)$ has a separable correlation function $\rho_s^i(\mathbf{h})\rho_t^j(u)$. Then, the unconditional correlation function of $Z(\mathbf{s}, t)$ is given by $\rho(\mathbf{h}, u) = \sum_{i=0}^{+\infty}\sum_{j=0}^{+\infty} \rho_s^i(\mathbf{h})\rho_t^j(u)p_{ij}$.

Let $g(z_1, z_2) = E\left(z_1^A z_2^B\right)$, with $-1 \leq z_1, z_2 \leq 1$, be the probability-generating function of the random vector (A, B), which is always absolutely convergent in the unit rectangle $(-1, 1)^2$. Then (7.85) can be expressed as

$$\rho(\mathbf{h}, u) = g(\rho_s(\mathbf{h}), \rho_t(u)), \qquad (\mathbf{h}, u) \in \mathbb{R}^d \times T. \tag{7.86}$$

An important case of the mixture proposed by Ma arises when the entire mass of its discrete probability distribution $\{p_{ij}, (i,j) \in \mathbb{N}_+^2\}$ is confined to the line $i = j$, that is,

$$p_{ij} = \left\{\begin{matrix} p_k, & \text{if} & i = j \\ 0, & \text{if} & i \neq j \end{matrix}\right\} \tag{7.87}$$

where $p_k, k \in \mathbb{N}_+$, is a discrete probability function. This is really the equivalent of a weighted non-linear combination of separable correlation functions. The resulting spatio-temporal correlation function is:

$$\rho(\mathbf{h}, u) = \sum_{k=0}^{+\infty} (\rho_s(\mathbf{h})\rho_t(u))^k p_k, \quad (\mathbf{h}, u) \in \mathbb{R}^d \times T. \tag{7.88}$$

In order to illustrate this simple procedure, Ma (2002) provides a series of examples, including the following:

Example 7.16 (Ma 2002. Example 2. General case) *Let $g\left(z_1, z_2\right)$ be the probability generating function of a bivariate Poisson distribution (see, for example, Feller 1968, p. 279):*

$$g\left(z_1, z_2\right) = \exp\left(\lambda_1\left(z_1 - 1\right) + \lambda_2\left(z_2 - 1\right) + \lambda_{12}\left(z_1 z_2 - 1\right)\right), \quad \left(z_1, z_2\right) \in (-1, 1)^2 \tag{7.89}$$

λ_1, λ_2 and λ_{12} being non-negative constants, with $\lambda_{12} \leq \min\left(\lambda_1, \lambda_2\right)$. Then, using the expression $\rho(\mathbf{h}, u) = g(\rho_s(\mathbf{h}), \rho_t(u)), (\mathbf{h}, u) \in \mathbb{R}^d \times T$ as a basis and substituting $z_1 \left(= e^t\right)$ with $\rho_s(\mathbf{h})$ and $z_2 (= e^s)$ with $\rho_t(u)$ we have that:

$$\begin{aligned}\rho(\mathbf{h}, u) &= g\left(\rho_s(\mathbf{h}), \rho_t(u)\right) \\ &= \exp\left(\lambda_1\left(\rho_s(\mathbf{h}) - 1\right) + \lambda_2\left(\rho_t(u) - 1\right) + \lambda_{12}\left(\rho_s(\mathbf{h})\rho_t(u) - 1\right)\right)\end{aligned} \tag{7.90}$$

is a stationary correlation function on $\mathbb{R}^d \times T$, $\rho_s(\mathbf{h})$ and $\rho_t(u)$ being stationary correlation functions, the former being spatial and the latter temporal.

Example 7.17 (Ma 2002. Example 1) Case $p_{ij} = \left\{\begin{matrix} p_k, & \textit{if} & i = j \\ 0, & \textit{if} & i \neq j \end{matrix}\right\}$. *Consider the negative binomial distribution, the probability function and probability generating* *function of which are:*

$$p_k = \binom{\alpha + k - 1}{k} (1-\theta)^k \theta^\alpha, \; k \in \mathbb{R}_+ \tag{7.91}$$

$$g(z) = \left(\frac{1}{\theta - \theta(1-\theta)z} \right)^\alpha = \left(\frac{\theta}{1-(1-\theta)z} \right)^\alpha \quad 0 < \theta < 1, \; \alpha > 0 \tag{7.92}$$

Then, using the expression $\rho(\mathbf{h}, u) = g(\rho_s(\mathbf{h}), \rho_t(u)), (\mathbf{h}, u) \in \mathbb{R}^d \times T$ *as a basis and substituting* $z \left(= e^t\right)$ *with* $\rho_s(\mathbf{h})\rho_t(u)$ *we have that:*

$$\rho(\mathbf{h}, u) = g\left(\rho_s(\mathbf{h})\rho_t(u)\right) = \left(\frac{\theta}{1-(1-\theta)\,\rho_s(\mathbf{h})\rho_t(u)} \right)^\alpha \quad 0 < \theta < 1, \; \alpha > 0, \tag{7.93}$$

is a stationary correlation function on $\mathbb{R}^d \times T$, $\rho_s(\mathbf{h})$ *and* $\rho_t(u)$ *being stationary correlation functions, the former being spatial and the latter temporal.*

In the expression $\rho(\mathbf{h}, u) = \sum_{i=0}^{+\infty} \sum_{j=0}^{+\infty} \rho_s^i(\mathbf{h})\rho_t^j(u)p_{ij}$, $\rho_s(\mathbf{h})$ or $\rho_t(u)$ can be assumed to take both positive and negative values. However, as we can see in Example 7.16, the resulting spatio-temporal function may be positive throughout $(\mathbf{h}, u) \in \mathbb{R}^d \times T$. In order to construct a correlation function that takes positive and negative values, it is necessary to choose a probability generating function that takes negative values for $z_1(= e^t) \in (-1, 0)$ or $z_2(= e^s) \in (-1, 0)$. One example is as follows.

Example 7.18 (Ma 2002. Example 4) *Consider the bivariate logarithmic series distribution of Subrahmaniam (1966), which has the following probability generating function:*

$$g\left(z_1, z_2\right) = \frac{\log\left(1 - \theta_1 z_1 - \theta_2 z_2 - \theta_{12} z_1 z_2\right)}{\log \theta}, \tag{7.94}$$

where $\theta = 1 - \left(\theta_1 + \theta_2 + \theta_{12}\right), \theta_1, \theta_2, \theta_{12} \geq 0, \; 0 < \theta_1 + \theta_2 + \theta_{12} < 1$. *Then,*

$$\rho(\mathbf{h}, u) = \frac{\log\left(1 - \theta_1 \rho_s(\mathbf{h}) - \theta_2 \rho_t(u) - \theta_{12}\rho_s(\mathbf{h})\rho_t(u)\right)}{\log \theta}, \tag{7.95}$$

is a stationary autocorrelation function on $\mathbb{R}^d \times T$, $\rho_s(\mathbf{h})$ *and* $\rho_t(u)$ *being stationary correlation functions, the former being spatial and the latter temporal.*

Having said this, in light of some of the examples above and having confirmed the highly important role that the probability function $\{p_{ij}, (i,j) \in \mathbb{N}_+^2\}$ plays in the construction of spatio-temporal correlation functions, the logical question that arises is the following: Can its domain be extended from $\mathbb{N}_+^2$ to $\mathbb{R}_+^2$? Can the afore-mentioned procedure be extended to a continuous probability distribution on $\mathbb{R}_+^2$?

In other words, is

$$\rho(\mathbf{h}, u) = \int_0^{+\infty} \int_0^{+\infty} \rho_s^a(\mathbf{h}), \rho_t^b(u) p(a, b) da db, \quad (\mathbf{h}, u) \in \mathbb{R}^d \times T \tag{7.96}$$

with $p(a, b)$ the joint density function of a non-negative bivariate random vector (A, B), a valid (stationary) spatio-temporal correlation function? The answer is that the proposed extension has several definition problems:

- If $\rho_s(\mathbf{h})$ or $\rho_t(u)$ are negative, then the correlation function $\rho(\mathbf{h}, u)$ would not be properly defined, as it would take complex values.
- In general, $\rho_s^a(\mathbf{h})$ or $\rho_t^b(u)$ may not be a spatial and temporal correlation function, respectively, throughout $a > 0$ or $b > 0$. By way of example, $\rho_t(u) = cos^2(u)$ is a correlation function on $\mathbb{R}$, but for $a = \frac{1}{2}$, the power $\rho_t^{\frac{1}{2}}(u) = |cos(u)|$ is not a temporal correlation function.
- The expression (7.96) can be written as

$$\rho(\mathbf{h}, u) = \mathcal{L}(-\log(\rho_s(\mathbf{h})) - \log(\rho_t(u))), \quad (\mathbf{h}, u) \in \mathbb{R}^d \times T, \tag{7.97}$$

  $\mathcal{L}(\theta_1, \theta_2) = E(e^{-\theta_1 A - \theta_2 B})$ being the Laplace transform of $p(a, b)$. Unlike the probability generating function for a non-negative, bivariate discrete random vector, the Laplace transform of a non-negative, continuous, bivariate random vector does not always take finite values.

Ma (2003b) validates this result for $\rho_s(\mathbf{h}) = \exp(-\gamma_s(\mathbf{h}))$ and $\rho_t(u) = \exp(-\gamma_t(u))$, with $\gamma_s(\mathbf{h})$ being a spatial stationary semivariogram on $\mathbb{R}^d$ and $\gamma_t(u)$ a temporal stationary semivariogram on T. The proof of validity is based on the following theorem (in reality, Ma writes it in covariance function terms).

Theorem 7.12.1 (Ma 2003b) *Let $\mathcal{L}(\theta_1, \theta_2)$ be the Laplace transform of a non-negative bivariate random vector (A, B). Let $\gamma_s(\mathbf{h})$ be a spatial semivariogram on $\mathbb{R}^d$ and $\gamma_t(u)$ a temporal stationary semivariogram on* T. *Then,*

$$\rho(\mathbf{h}, u) = \mathcal{L}\left(\gamma_s(\mathbf{h}), \gamma_t(u)\right) \tag{7.98}$$

is a stationary spatio-temporal correlation function on $\mathbb{R}^d \times T$.

This result enables the author to state that a mixture of purely spatial and temporal semivariograms is a simple and effective way of constructing new spatio-temporal covariance models. Ma (2003b) provides several examples of how this theorem is applied.

7.13 Models generated by linear combinations proposed by Ma

It is a well-known fact that the difference or a linear combination of two spatio-temporal covariance functions (semivariograms) is not necessarily a valid covariance function (semivariogram). In general, there appears to be no simple condition that enables us to state that linear combinations are permissible models, unless their coefficients are non-negative. As a result, Ma (2005a, 2005b) explores the conditions of permissibility of the linear combination of two isotropic spatio-temporal covariance functions (or semivariograms) in space. There are several reasons for the author to investigate the difference or, in general, the linear combination of two covariance functions (semivariograms):

1. It is frequently desirable for the covariance function to be able to take negative values, or oscillate from positive to negative values when approaching zero as $\mathbf{h}$ tends to infinity.

2. When making inferences and kriging, it would be better to be able to determine the range of the parameter θ which makes the covariance function (7.99) permissible:

$$C((\mathbf{s}_1, t_1), (\mathbf{s}_2, t_2) = \theta C_1((\mathbf{s}_1, t_1), (\mathbf{s}_2, t_2)) + (1 - \theta)\, C_2((\mathbf{s}_1, t_1), (\mathbf{s}_2, t_2)). \quad (7.99)$$

3. It would be desirable to know how θ affects the domain of the covariance function (7.99), which, if one of its coefficients is negative, can only be valid in lower spatial dimensions.

4. There is strong demand for valid spatio-temporal models with a long-range dependence, as the literature provides evidence of far-reaching dependence in many spatial and temporal data sets.

From the above we can appreciate the important role played by the parameter θ in determining whether or not the linear combination yields a permissible model and in which dimension it is permissible. As anticipated, there are no simple conditions to determine this, except in the case of a convex linear combination where $\theta \in (0, 1)$.

After presenting this issue, Ma (2005a, 2005b) provides different stationary spatio-temporal covariance and semivariogram models generated by linear combinations of covariance functions and spatio-temporal semivariograms. The covariance functions and also the correlation functions and the semivariograms, form a convex cone on $\mathbb{R}^d \times T$. Therefore, given the non-negative constants a and b and two spatio-temporal covariance functions $C_1((\mathbf{s}_1, t_1), (\mathbf{s}_2, t_2))$ and $C_2((\mathbf{s}_1, t_1), (\mathbf{s}_2, t_2))$ on $\mathbb{R}^d \times T$, then $aC_1((\mathbf{s}_1, t_1), (\mathbf{s}_2, t_2)) + bC_2((\mathbf{s}_1, t_1), (\mathbf{s}_2, t_2))$ is also a covariance function on $\mathbb{R}^d \times T$, and the same can be said of the semivariograms or correlation functions.

Notwithstanding, such convex combinations, despite ensuring the validity of the resulting model, do not take negative values or oscillations from positive to negative values, which is necessary in order to be able to model certain natural processes. For this reason, general linear combinations are considered in the following form:

$$C((\mathbf{s}_1, t_1), (\mathbf{s}_2, t_2)) = aC_1((\mathbf{s}_1, t_1), (\mathbf{s}_2, t_2)) + bC_2((\mathbf{s}_1, t_1), (\mathbf{s}_2, t_2)), \quad (7.100)$$

where a, b are any two given constants, subject only to the constraint that $V(Z(\mathbf{s}, t)) = a\sigma_1^2 + b\sigma_2^2 \geq 0, \sigma_1^2$ and σ_2^2 being *a priori* variances of the covariance functions C_1 and C_2, respectively. Let us imagine the non-trivial case in which $V(Z(\mathbf{s}, t) > 0$. If

$a' = a\sigma_1^2$ and $b' = b\sigma_2^2$ are defined by $a' + b' > 0$, we have that:

$$C((\mathbf{s}_1, t_1), (\mathbf{s}_2, t_2)) = a'\rho_1((\mathbf{s}_1, t_1), (\mathbf{s}_2, t_2)) + b'\rho_2((\mathbf{s}_1, t_1), (\mathbf{s}_2, t_2)). \tag{7.101}$$

If this expression is divided by $V(Z(\mathbf{s}, t)) = a' + b' > 0$, we have that

$$\rho((\mathbf{s}_1, t_1), (\mathbf{s}_2, t_2)) = \theta\rho_1((\mathbf{s}_1, t_1), (\mathbf{s}_2, t_2)) + (1 - \theta)\rho_2((\mathbf{s}_1, t_1), (\mathbf{s}_2, t_2)), \tag{7.102}$$

whereby $\theta = a'/(a' + b')$ will be any given constant. Notice that the validity of the model (7.102) on $\mathbb{R}^d \times T$ will be guaranteed for any $\theta \in [0, 1]$, due to producing a convex combination. Ma (2005a) analyzes whether it is possible to increase the range of variation of θ in different specific situations under the condition of stationarity, which would give rise to combinations with negative values and achieve the practical advantages mentioned previously.

The following two theorems are extremely useful for generating spatio-temporal covariance functions via linear combinations that are formed using cosine transforms. More specifically, Theorem 7.13.1 provides spatio-temporal covariance functions with a power-law decay marginal temporal covariance. Theorem 7.13.2 is the same as Theorem 7.13.1, but with a power-law decay decreasing marginal spatial covariance.

Theorem 7.13.1 (Ma 2005a) *Let $d \in \mathbb{N}$ and let φ be a completely monotone function on $[0, \infty)$ with $\varphi(0) = 1$, then*

$$\begin{aligned} C(\mathbf{h}, u) = {} & \theta(1 + \alpha_1 |u|)^{-\frac{d}{2}} \varphi\left(\frac{\|\mathbf{h}\|^2}{1 + \alpha_1 |u|}\right) \\ & + (1 - \theta)(1 + \alpha_2 |u|)^{-\frac{d}{2}} \varphi\left(\frac{\|\mathbf{h}\|^2}{1 + \alpha_2 |u|}\right) \end{aligned} \tag{7.103}$$

is a stationary spatio-temporal covariance function on $\mathbb{R}^d \times \mathbb{R}$ if, and only if,

$$\theta \in \left[\left(1 - \frac{\alpha_2}{\alpha_1}\right)^{-1}, \left(1 - \frac{\alpha_1}{\alpha_2}\right)^{-1}\right]. \tag{7.104}$$

Likewise:

$$\begin{aligned} C(\mathbf{h}, u) = {} & \theta(1 + \alpha_1 u^2)^{-\frac{d}{2}} \varphi\left(\frac{\|\mathbf{h}\|^2}{1 + \alpha_1 u^2}\right) \\ & + (1 - \theta)(1 + \alpha_2 u^2)^{-\frac{d}{2}} \varphi\left(\frac{\|\mathbf{h}\|^2}{1 + \alpha_2 u^2}\right) \end{aligned} \tag{7.105}$$

is a stationary spatio-temporal covariance function on $\mathbb{R}^d \times \mathbb{R}$ if, and only if

$$\theta \in \left[\left(1 - \left(\frac{\alpha_2}{\alpha_1}\right)^{-1/2}\right)^{-1}, 1\right]. \tag{7.106}$$

Notice that both (7.103) and also (7.105) are spatio-temporal correlation functions on $\mathbb{R}^d \times \mathbb{R}$ under the conditions (7.104) and (7.106), respectively, regardless of the value of d, as the expressions (7.104) and (7.106) do not involve the spatial dimension parameter d.

The marginal temporal covariance of (7.103) and (7.105) is (remember that $\varphi(0) = 1$): $C(\mathbf{0},u) = \theta(1+\alpha_1|u|)^{-\frac{d}{2}} + (1-\theta)(1+\alpha_2|u|)^{-\frac{d}{2}}$ and $C(\mathbf{0},u) = \theta(1+\alpha_1 u^2)^{-\frac{d}{2}} + (1-\theta)(1+\alpha_2 u^2)^{-\frac{d}{2}}$ respectively, both of which have power-law decay.

The marginal spatial covariance in both cases is given by $C(\mathbf{h},0) = \varphi(\|\mathbf{h}\|^2)$, which is always non-negative throughout $\mathbf{h} \in \mathbb{R}^d$. However, the marginal temporal covariance will take negative values when θ is equal to the lowest value of the intervals $\left[\left(1-\frac{\alpha_2}{\alpha_1}\right)^{-1}, \left(1-\frac{\alpha_1}{\alpha_2}\right)^{-1}\right]$ and $\left[\left(1-\left(\frac{\alpha_2}{\alpha_1}\right)^{-1/2}\right)^{-1}, 1\right]$, respectively, which means (7.103) and (7.105) will take negative values in those cases. In order to illustrate how the theorem is applied, Ma (2005a) provides the following examples.

Example 7.19 (Ma 2005a. Example 4) *Let $\varphi(x) = (1+x)^{-\beta}, x \geq 0, \beta > 0$ a completely monotone function (third function in Table 7.1, where $c = \gamma = 1$). Then, by applying Theorem 7.13.1 we can state that*

$$C(\mathbf{h},u) = \theta(1+\alpha_1|u|)^{-\frac{d}{2}}\left(1+\frac{\|\mathbf{h}\|^2}{1+\alpha_1|u|}\right)^{-\beta} + (1-\theta)(1+\alpha_2|u|)^{-\frac{d}{2}}\left(1+\frac{\|\mathbf{h}\|^2}{1+\alpha_2|u|}\right)^{-\beta} \tag{7.107}$$

is a spatio-temporal stationary covariance function on $\mathbb{R}^d \times \mathbb{R}$, provided that

$$\theta \in \left[\left(1-\frac{\alpha_2}{\alpha_1}\right)^{-1}, \left(1-\frac{\alpha_1}{\alpha_2}\right)^{-1}\right].$$

The above covariance function has the power-law decay in time.

Example 7.20 (Ma 2005a) *Let $\varphi(x) = \left(-x^{\beta}\right) x \geq 0, 0 \leq \beta \leq 1$ be a completely monotone function. Then, if $\theta \in \left[\left(1-\left(\frac{\alpha_2}{\alpha_1}\right)^{-1/2}\right)^{-1}, 1\right]$, by applying Theorem 7.13.1 we can state that*

$$C(\mathbf{h},u) = \theta(1+\alpha_1 u^2)^{-\frac{d}{2}}\left(-\frac{\|\mathbf{h}\|^{2\beta}}{(1+\alpha_1 u^2)^{\beta}}\right) + (1-\theta)(1+\alpha_2 u^2)^{-\frac{d}{2}}\left(-\frac{\|\mathbf{h}\|^{2\beta}}{(1+\alpha_2 u^2)^{\beta}}\right) \tag{7.108}$$

is a spatio-temporal stationary covariance function on $\mathbb{R}^d \times \mathbb{R}$. Furthermore, this covariance function has a power-law decay temporal margin.

Theorem 7.13.2 (Ma 2005a) *Let $d \in \mathbb{N}$ and let φ be a completely monotone function with $\varphi(0) = 1$, then*

$$C(\mathbf{h}, u) = \theta(1 + \alpha_1 \|\mathbf{h}\|)^{-\frac{1}{2}} \varphi\left(\frac{u^2}{1 + \alpha_1 \|\mathbf{h}\|}\right) + (1 - \theta)(1 + \alpha_2 \|\mathbf{h}\|)^{-\frac{1}{2}} \varphi\left(\frac{u^2}{1 + \alpha_2 \|\mathbf{h}\|}\right), \quad (\mathbf{h}, u) \in \mathbb{R}^d \times \mathbb{R}, \tag{7.109}$$

is a spatio-temporal stationary covariance function on $\mathbb{R}^d \times \mathbb{R}$ if, and only if,

$$\theta \in \left[\left(1 - \left(\frac{\alpha_2}{\alpha_1}\right)^d\right)^{-1}, \left(1 - \frac{\alpha_1}{\alpha_2}\right)^{-1}\right]. \tag{7.110}$$

Likewise,

$$C(\mathbf{h}, u) = \theta(1 + \alpha_1 \|\mathbf{h}\|^2)^{-\frac{1}{2}} \varphi\left(\frac{u^2}{1 + \alpha_1 \|\mathbf{h}\|^2}\right) + (1 - \theta)(1 + \alpha_2 \|\mathbf{h}\|^2)^{-\frac{1}{2}} \varphi\left(\frac{u^2}{1 + \alpha_2 \|\mathbf{h}\|^2}\right), \tag{7.111}$$

is a spatio-temporal stationary covariance function on $\mathbb{R}^d \times \mathbb{R}$ if, and only if,

$$\theta \in \left[\left(1 - \left(\frac{\alpha_2}{\alpha_1}\right)^{d/2}\right)^{-1}, 1\right]. \tag{7.112}$$

In this case, the conditions of validity of (7.109) and (7.111) given by (7.110) and (7.112), respectively, depend on the value of d, unless θ takes values between 0 and $1 - \frac{\alpha_1}{\alpha_2}$, in which case the conditions of validity are independent of the dimension d.

As can be appreciated, the marginal temporal covariance coincides in both cases $\left(C(\mathbf{0},u) = \varphi\left(u^2\right)\right)$. The marginal spatial covariances decrease, in both cases, following a power-law decay as distance increases. As in the case of the previous theorem, when θ is equal to the lowest value of the intervals $\left[\left(1 - \left(\frac{\alpha_2}{\alpha_1}\right)^d\right)^{-1}, \left(1 - \frac{\alpha_1}{\alpha_2}\right)^{-1}\right]$ and $\left[\left(1 - \left(\frac{\alpha_2}{\alpha_1}\right)^{d/2}\right)^{-1}, 1\right]$, the spatial marginal covariance could take negative values and, as a result, (7.109) and (7.111) can generate both positive and negative values.

The examples below illustrate how this theorem is applied.

Example 7.21 (Ma 2005a) *Let $\varphi(x) = (1 + x)^{-\beta}, x > 0, \beta > 0$ be a completely monotone function (third function in Table 7.1, where $c = \gamma = 1$). Then, by applying*

Theorem 7.13.2, we can state that

$$C(\mathbf{h},u) = \theta(1+\alpha_1\|\mathbf{h}\|)^{-\frac{1}{2}}\left(1+\frac{u^2}{1+\alpha_1\|\mathbf{h}\|}\right)^{-\beta} + (1-\theta)(1+\alpha_2\|\mathbf{h}\|)^{-\frac{1}{2}}\left(\frac{u^2}{1+\alpha_2\|\mathbf{h}\|}\right)^{-\beta} \tag{7.113}$$

is a spatio-temporal stationary covariance function on $\mathbb{R}^d \times \mathbb{R}$*, provided that*

$$\theta \in \left[\left(1-\frac{\alpha_2}{\alpha_1}\right)^{-1}, \left(1-\frac{\alpha_1}{\alpha_2}\right)^{-1}\right].$$

Furthermore, this covariance function has a power-law decay marginal spatial covariance.

Example 7.22 (Ma 2005a) *Let* $\varphi(x) = \left(-x^{\beta}\right), x > 0, 0 \leq \beta \leq 1$ *be a completely monotone function. Then, if*

$$\theta \in \left[\left(1-\left(\frac{\alpha_2}{\alpha_1}\right)^{d/2}\right)^{-1}, 1\right],$$

by applying Theorem 7.13.2, we can state that

$$C(\mathbf{h},u) = \theta(1+\alpha_1\|\mathbf{h}\|^2)^{-\frac{1}{2}}\left(-\frac{|u|^{2\beta}}{\left(1+\alpha_1\|\mathbf{h}\|^2\right)^{\beta}}\right) + (1-\theta)(1+\alpha_2\|\mathbf{h}\|^2)^{-\frac{1}{2}}\left(-\frac{|u|^{2\beta}}{\left(1+\alpha_2\|\mathbf{h}\|^2\right)^{\beta}}\right) \tag{7.114}$$

is a spatio-temporal stationary covariance function on $\mathbb{R}^d \times \mathbb{R}$*. Furthermore, this covariance function has a power-law decay marginal spatial covariance.*

Ma (2005c) also analyzes the specific case of linear combinations of separable spatio-temporal covariance models. Following Ma (2005c), let us assume that α_k and $\beta_k (k = 1, 2)$ are positive constants with $\alpha_1 < \alpha_2$ and $\beta_1 < \beta_2$. Clearly, for $k = 1, 2$,

$$\exp(-\alpha_k\|\mathbf{h}\| - \beta_k|u|), \qquad (\mathbf{h},u) \in \mathbb{R}^d \times \mathbb{R}, \tag{7.115}$$

is a separable covariance function on $(\mathbf{h},u) \in \mathbb{R}^d \times \mathbb{R}$. Now the question is the following: When does their linear combination

$$C(\mathbf{h},u) = a_1\exp(-\alpha_1\|\mathbf{h}\| - \beta_1|u|) + a_2\exp(-\alpha_2\|\mathbf{h}\| - \beta_2|u|)) \ (\mathbf{h},u) \in \mathbb{R}^d \times \mathbb{R}, \tag{7.116}$$

where a_1 and a_2 are the constants subject to the obvious assumption $V(Z(\mathbf{h},u)) = a_1 + a_2 \geq 0$, define a valid spatio-temporal covariance function?

Let us ignore the trivial degenerate case in which $a_1 + a_2 = 0$ and consider function

$$\rho(\mathbf{h}, u) = \theta \exp(-\alpha_1 \|\mathbf{h}\| - \beta_1 |u|) + (1-\theta) \exp(-\alpha_2 \|\mathbf{h}\| - \beta_2 |u|)) \; (\mathbf{h}, u) \in \mathbb{R}^d \times \mathbb{R}, \tag{7.117}$$

with $\theta = \frac{a_1}{a_1+a_2}$.

It is clear that (7.117) is a spatio-temporal correlation function if $0 \leq \theta \leq 1$, as it is a convex combination of two separable correlation functions. For other values of θ it is the differences of two separable correlation functions.

A general form, in which stationary spatial and temporal covariances from the Matérn model are considered, and that includes (7.117) as a special case, is:

$$\begin{aligned} C(\mathbf{h}, u) &= \theta(\alpha_1 \|\mathbf{h}\|)^{\upsilon_1} K_{\upsilon_1}(\alpha_1 \|\mathbf{h}\|)(\beta_1 \, |u|)^{\upsilon_2} K_{\upsilon_2}(\beta_1 \, |u|) \\ &\quad + (1-\theta)(\alpha_2 \|\mathbf{h}\|)^{\upsilon_1} K_{\upsilon_1}(\alpha_2 \|\mathbf{h}\|)(\beta_2 \, |u|)^{\upsilon_2} K_{\upsilon_2}(\beta_2 \, |u|), \quad (\mathbf{h}, u) \in \mathbb{R}^d \times \mathbb{R}, \end{aligned} \tag{7.118}$$

where $\alpha_1, \alpha_2, \beta_1, \beta_2, \upsilon_1, \upsilon_2$ are positive constants, with $\alpha_1 < \alpha_2$ and $\beta_1 < \beta_2$, and K_υ is the modified Bessel function of the second kind of order v.

Ma (2005c), Theorem 3, demonstrates that (7.118) is a spatio-temporal covariance function on $\mathbb{R}^d \times \mathbb{R}$ if, and only if,

$$\theta \in \left[\left(1 - \left(\frac{\alpha_2}{\alpha_1}\right)^d \frac{\beta_2}{\beta_1} \right)^{-1}, \left(1 - \left(\frac{\alpha_1}{\alpha_2}\right)^{2\upsilon_1} \left(\frac{\beta_1}{\beta_2}\right)^{2\upsilon_2} \right) \right]. \tag{7.119}$$

In the case that $\alpha_1 = \alpha_2$ and $\beta_1 = \beta_2$, (7.119) is interpreted as for all real numbers θ. Notice that, while the upper bound of θ in (7.119) does not depend on the dimension d in space, the lower bound decreases as d increases and tends to zero as d approaches infinity.

Observe that if $\alpha_1 = \alpha_2$ or if $\beta_1 = \beta_2$, then the model (7.118) becomes the following separable spatio-temporal model:

$$C(\mathbf{h}, u) = (\alpha \|\mathbf{h}\|)^{\upsilon_1} K_{\upsilon_1}(\alpha \|\mathbf{h}\|)(\beta \, |u|)^{\upsilon_2} K_{\upsilon_2}(\beta \, |u|). \tag{7.120}$$

In the exponential case ($\upsilon_1 = \upsilon_2 = 1/2$) we have that:

$$C(\mathbf{h}, u) = \theta \exp(-\alpha_1 \|\mathbf{h}\|) \exp(-\beta_1 \, |u|) + (1-\theta) \exp(-\alpha_2 \|\mathbf{h}\|) \exp(-\beta_2 \, |u|), \tag{7.121}$$

where $\alpha_1, \alpha_2, \beta_1, \beta_2$ are positive constants, with $\alpha_1 \leq \alpha_2$ and $\beta_1 \leq \beta_2$, will be a spatio-temporal covariance function on $\mathbb{R}^d \times \mathbb{R}$ if, and only if:

$$\theta \in \left[\left(1 - \left(\frac{\alpha_2}{\alpha_1}\right)^d \frac{\beta_2}{\beta_1} \right)^{-1}, \left(1 - \left(\frac{\alpha_1}{\alpha_2}\right) \left(\frac{\beta_1}{\beta_2}\right) \right) \right]. \tag{7.122}$$

If $\alpha_1 = \alpha_2$ and $\beta_1 = \beta_2$, we obtain the following separable model:

$$C(\mathbf{h}, u) = \exp(-\alpha_1 \|\mathbf{h}\|) \exp(-\beta_1 \, |u|). \tag{7.123}$$

An important feature of the bounds in (7.119) and (7.122) is that they depend on the quotients $\frac{\alpha_1}{\alpha_2}$ and $\frac{\beta_1}{\beta_2}$ instead of the $\alpha's$ and $\beta's$ individually. This observation leads Ma to develop the following corollary, which is a special case of Theorem 7.12.1 when $\theta = 0$ or $\theta = 1$.

Corollary 1 (Special case of Theorem 7.12.1) *Assume that* $0 < \alpha_1 \leq \alpha_2, 0 < \beta_1 \leq \beta_2$ *and* θ *satisfies (7.122). If* $\mathcal{L}(a, b)$ *is the Laplace transform of a non-negative bivariate random vector* (A, B)*, then*

$$\rho(\mathbf{h}, u) = \theta\mathcal{L}(\alpha_1\|\mathbf{h}\|, \beta_1|u|) + (1-\theta)\mathcal{L}(\alpha_2\|\mathbf{h}\|, \beta_2|u|), \quad (\mathbf{h}, u) \in \mathbb{R}^d \times \mathbb{R}, \quad (7.124)$$

is a stationary correlation function on $\mathbb{R}^d \times \mathbb{R}$.

Theorem 7.13.3 (Ma 2005c) *Let* $\mathcal{L}(\theta_1, \theta_2)$ *be the Laplace transform of a non-negative random vector* (A, B)*. Then*

$$C(\mathbf{h}, u) = \theta\mathcal{L}(\alpha_1\|\mathbf{h}\|, \beta_1\,|u|) + (1-\theta)\mathcal{L}(\alpha_2\|\mathbf{h}\|, \beta_2\,|u|), \quad (7.125)$$

is a spatio-temporal stationary covariance function on $\mathbb{R}^d \times \mathbb{R}$.

7.14 Models proposed by Stein

Stein (2005a) provides a procedure that works in the spectral domain and generates a family of spectral densities that yield spatio-temporal stationary covariance functions that are generally speaking non-separable and analytically flexible. These densities are an extension of the Matérn model.

Separable covariance functions generally imply that small changes in the location of observations can result in significant changes in the correlation between certain linear combinations of observations. The source of this discontinuity can be explained by the fact that covariances defined in this way are no smoother far from the origin than at the origin. Stein's work shows that spatio-temporal covariance models that are smoother far from the origin yield optimum linear predictions with better properties. In Stein (2005a), the author reviews the different models presented previously, assessing the characteristics they display far from the origin and verifies this property. Stein proposes a model of spatio-temporal stationary covariance function that does have the property of being smoother far from the origin than at the origin. These covariances will be generated by way of the spectral density function:

$$f(\boldsymbol{\omega}, \tau) = (c_1(a_1^2 + \|\boldsymbol{\omega}\|^2)^{\alpha_1} + c_2(a_2^2 + |\tau|^2)^{\alpha_2})^{-\upsilon}, \quad (7.126)$$

with c_1 and c_2 being two positive constants, $a_1^2 + a_2^2 > 0, \alpha_1$ and α_2 two positive integers and $d_1/(\alpha_1\upsilon) + d_2/(\alpha_2\upsilon) < 2$. The covariance function associated to this spectral density will be given by:

$$C(\mathbf{h}, u) = \int\int e^{i(\boldsymbol{\omega}'\mathbf{h} + \tau u)} f(\boldsymbol{\omega}, \tau) d\boldsymbol{\omega} d\tau. \quad (7.127)$$

Applying one of the results presented in Stein (2005a), it is verified that under such conditions the spatio-temporal covariance function given by (7.127) has all the partial derivatives of all the orders for $(\mathbf{h}, u) \neq (\mathbf{0}, 0)$. Furthermore, it only depends on $||\mathbf{h}||$ and $|u|$. It also has an enormously useful practical quality insofar as we can ascertain the degree of temporal smoothing independently. Unfortunately, in very few cases do we have explicit expressions for the covariance functions (7.127) generated using this approach.

7.15 Construction of covariance functions using copulas and completely monotonic functions

This procedure involves spatial or temporal marginal semivariograms, $\gamma_s(\cdot), \gamma_t(\cdot)$, copulas, $K(\cdot,\cdot)$ and completely monotonic functions, $\varphi(\cdot)$. As can be seen in Porcu and Mateu (2007), there is a series of mixtures of the above expressions that lead to non-separable and stationary spatio-temporal covariance functions:

(i) $\int_0^1 \int_0^1 \exp(-w\gamma_s(\mathbf{h}) - v\gamma_t(u))dK(w, v)$

(ii) $\varphi\left(\gamma_s(\mathbf{h}) + \gamma_t(u)\right)$

(iii) $\varphi\left(||\mathbf{h}||^2 + \gamma_t(u)\right)$

(iv) $\varphi^{(2k)}(\gamma_s(\mathbf{h}))\varphi\left(||\mathbf{h}||^2 + \gamma_t(u)\right)$

(v) $\int_0^1 \int_0^1 \int_0^1 \varphi^{(2k)}(\gamma_s(\mathbf{h})w)\varphi\left(||\mathbf{h}||^2 v + \gamma_t(u)z\right) dK(w, v, z)$

It is necessary to clarify some points in case (i). It must be taken into account that the literature on the topic provides examples of bivariate copulas (see, for example, Nelsen 1999), making integration easy in the majority of cases. Furthermore, their closed form makes it possible to address a wide variety of practical situations. Finally, their structure is separable if, and only if, their distribution function is separable. Note that copulas allow for separability (the so-called product copulas), which makes it possible to tackle this case.

As regards (iv) and (v), take into account that in order to obtain a stationary structure, it is necessary for $\varphi^{(2k)}$ to be finite at the origin, otherwise the variance of the rf might not exist or be finite.

7.16 Generalized product-sum model

Let $\{C_{si} : i = 1, 2, \ldots, n\}$ and $\{C_{ti} : i = 1, 2, \ldots, n\}$ be two sets of spatial and temporal stationary covariance functions, respectively, continuous and integrable, for $n \in \mathbb{N}$. Let us consider the function

$$C(\mathbf{h}, u) = \sum_{i=1}^{n} k_i C_{si}(\mathbf{h}) C_{ti}(u), \quad (\mathbf{h}, u) \in \mathbb{R}^d \times \mathbb{R}. \tag{7.128}$$

As we saw in Section 7.13, the function (7.128) will be a valid spatio-temporal covariance function on $\mathbb{R}^d \times \mathbb{R}$ provided it is obtained using a convex linear combination, that is, the constants $k_1, \dots, k_n$ that define it must be non-negative and add up to unity.

Section (7.13) presents an initial approach to this problem, establishing a set of necessary and sufficient conditions for the case in which $n = 2$ and the spatial and temporal covariance functions have the same structure and belong to the Matérn model (or exponential model, which is a particular case). Gregori *et al.* (2008) generalize these results in order to cover a much broader spectrum of possibilities.

Therefore, this section addresses functions with that same structure (linear combinations), which is intuitively simple and easy to implement, but where the constants defining the linear combination are not subject to the constraint of being non-negative. Obviously, this generalization is of the utmost importance, as it will give rise to covariance functions that can take negative values, or which fluctuate between positive and negative, which has highly useful practical implications for modeling spatio-temporal dependence in certain natural and social processes. More specifically, Gregori *et al.* (2008) present a necessary condition and a sufficient condition for the model (7.128) to be valid in the general case whereby the constants $k_1, \dots, k_n$ are any given real values.

Proposition 1 (Gregori *et al.* 2008) *Let* $\{C_{si} : i = 1, 2, \dots, n\}$ *and* $\{C_{ti} : i = 1, 2, \dots, n\}$ *be two sets of n integrable and continuous, spatial and temporal covariance functions, respectively, with* $n \in \mathbb{N}$. *Let us consider the function:*

$$C(\mathbf{h}, u) = \sum_{i=1}^{n} k_i C_{si}(\mathbf{h}) C_{ti}(u), \quad (\mathbf{h}, u) \in \mathbb{R}^d \times \mathbb{R}, \tag{7.129}$$

with $k_1, \dots, k_n \in \mathbb{R}$. *Let* f_{si} *and* f_{ti}, *with* $i = 1, 2, \dots, n$ *be the Fourier transforms of the covariances* C_{si} *and* C_{ti}, *respectively. Let us assume that at least a pair of these functions* (f_{si}, f_{ti}) *are non-vanishing, which will be denoted, without losing generality, as* (f_{sn}, f_{tn}). *Let*

$$m_{ti} = \inf_{\tau \in \mathbb{R}} \frac{f_{ti}(\tau)}{f_{tn}(\tau)}, \quad M_{ti} = \sup_{\tau \in \mathbb{R}} \frac{f_{ti}(\tau)}{f_{tn}(\tau)}, \quad m_{si} = \inf_{\boldsymbol{\omega} \in \mathbb{R}^d} \frac{f_{si}(\boldsymbol{\omega})}{f_{sn}(\boldsymbol{\omega})}, \quad M_{si} = \sup_{\boldsymbol{\omega} \in \mathbb{R}^d} \frac{f_{si}(\boldsymbol{\omega})}{f_{sn}(\boldsymbol{\omega})}.$$

We can state that:

(i) If (7.129) is a spatio-temporal covariance function, then

$$k_n \geq -\sum_{i=1}^{n-1} k_i [M_{si} M_{ti} \mathbf{1}_{\{k_i \geq 0\}} + m_{si} m_{ti} \, \mathbf{1}_{\{k_i < 0\}}].$$

(ii) If $k_n \geq -\sum_{i=1}^{n-1} k_i [m_{si} m_{ti} \mathbf{1}_{\{k_i \geq 0\}} + M_{si} M_{ti} \mathbf{1}_{\{k_i < 0\}}]$, *then the function (7.129) is a spatio-temporal covariance function.*[7]

[7] See proof in Appendix D.

The above result will be particularized for the case in which $n = 2$, that is, for models generated as a linear combination of two separable spatio-temporal covariance models.

Proposition 2 (Proposition 1 for $n = 2$) *Let C_{s1} and C_{s2} be two spatial covariance functions defined on $\mathbb{R}^d$, and let C_{t1} and C_{t2} be two temporal covariance functions defined on $\mathbb{R}$, all of which are integrable and continuous. Let us consider the function*

$$C(\mathbf{h}, u) = k_1 C_{s1}(\mathbf{h})C_{t1}(u) + k_2 C_{s2}(\mathbf{h})C_{t2}(u), \quad (\mathbf{h}, u) \in \mathbb{R}^d \times \mathbb{R}, \tag{7.130}$$

with $k_1, k_2 \in \mathbb{R}$. Let f_{s1}, f_{s2}, f_{t1} and f_{t2} be the Fourier transforms of the covariance functions C_{s1}, C_{s2}, C_{t1} and C_{t2}, respectively where it will be assumed that f_{s2} and f_{t2} are non-vanishing functions. Let

$$m_{t1} = \inf_{\tau \in \mathbb{R}} \frac{f_{t1}(\tau)}{f_{t2}(\tau)}, \qquad M_{t1} = \sup_{\tau \in \mathbb{R}} \frac{f_{t1}(\tau)}{f_{t2}(\tau)},$$

$$m_{s1} = \inf_{\boldsymbol{\omega} \in \mathbb{R}^d} \frac{f_{s1}(\boldsymbol{\omega})}{f_{s2}(\boldsymbol{\omega})}, \qquad M_{s1} = \sup_{\boldsymbol{\omega} \in \mathbb{R}^d} \frac{f_{s1}(\boldsymbol{\omega})}{f_{s2}(\boldsymbol{\omega})}.$$

Then, (7.130) is a spatio-temporal covariance function if, and only if,[8]

$$k_2 \geq -k_1 \left(m_{s1} m_{t1} \, \mathbf{1}_{\{k_1 \geq 0\}} + M_{s1} M_{t1} \, \mathbf{1}_{\{k_1 < 0\}} \right).$$

If we now move on to consider the weighted sums of two separable spatio-temporal covariance functions such that the coefficients that define the linear combination, apart from being positive, add up to unity. Under this reparameterization, the model to be considered will be as follows:

$$C(\mathbf{h}, u) = \theta C_{s1}(\mathbf{h})C_{t1}(u) + (1 - \theta) C_{s2}(\mathbf{h})C_{t2}(u), \quad (\mathbf{h}, u) \in \mathbb{R}^d \times \mathbb{R} \tag{7.131}$$

with $\theta \in \mathbb{R}$, C_{s1} and C_{s2} being two spatial covariance functions defined on $\mathbb{R}^d$, and C_{t1} and C_{t2} being two temporal covariance functions defined on $\mathbb{R}$, all of which are integrable and continuous.

For the case of $n = 2$, the two following corollaries are of interest:

Corollary 1 *Let C_{s1} and C_{s2} be two spatial covariance functions defined on $\mathbb{R}^d$, and let C_{t1} and C_{t2} be two temporal covariance functions defined on $\mathbb{R}$, all of which will be integrable and continuous. Let us consider the function*

$$C(\mathbf{h}, u) = \theta C_{s1}(\mathbf{h})C_{t1}(u) + (1 - \theta) C_{s2}(\mathbf{h})C_{t2}(u), \quad (\mathbf{h}, u) \in \mathbb{R}^d \times \mathbb{R}, \tag{7.132}$$

[8] See proof in Appendix D.

with $\theta \in \mathbb{R}$. Let f_{s1}, f_{s2}, f_{t1} and f_{t2} be the Fourier transforms of the covariance functions C_{s1}, C_{s2}, C_{t1} and C_{t2}, respectively, where it will be assumed that f_{s2} and f_{t2} are not null-functions. Let

$$m_t = \inf_{\tau \in \mathbb{R}} \frac{f_{t1}(\tau)}{f_{t2}(\tau)}, \quad M_t = \sup_{\tau \in \mathbb{R}} \frac{f_{t1}(\tau)}{f_{t2}(\tau)},$$

$$m_s = \inf_{\boldsymbol{\omega} \in \mathbb{R}^d} \frac{f_{s1}(\boldsymbol{\omega})}{f_{s2}(\boldsymbol{\omega})}, \quad M_s = \sup_{\boldsymbol{\omega} \in \mathbb{R}^d} \frac{f_{s1}(\boldsymbol{\omega})}{f_{s2}(\boldsymbol{\omega})}.$$

Then (7.132) is a spatio-temporal covariance function if, and only if:

$$\left(1 - \max(1, M_s M_t)\right)^{-1} \leq \theta \leq \left(1 - \min(1, m_s m_t)\right)^{-1} \tag{7.133}$$

where $0^{-1} = -\infty$ and $(-\infty)^{-1} = 0$ in the left-hand side of the inequality, and $0^{-1} = +\infty$ in the right-hand side.

Notice that this result also makes it possible to define purely spatial covariance functions as linear combinations (not necessarily convexes) of purely spatial covariance functions, as shown below. This can be easily proved by applying all the above results to the merely spatial case.

Corollary 2 *Let C_{s1} and C_{s2} be two integrable and continuous spatial covariance functions defined on $\mathbb{R}^d$. Let us consider the function:*

$$C(\mathbf{h}) = \theta C_{s1}(\mathbf{h}) + (1 - \theta) C_{s2}(\mathbf{h}), \quad (\mathbf{h}) \in \mathbb{R}^d, \tag{7.134}$$

with $\theta \in \mathbb{R}$. Let f_{s1} and f_{s2} be the Fourier transforms associated with the covariance functions C_{s1} and C_{s2}, respectively, where it is assumed that f_{s2} is a non-vanishing function. Let

$$m = \inf_{\boldsymbol{\omega} \in \mathbb{R}^d} \frac{f_{s1}(\boldsymbol{\omega})}{f_{s2}(\boldsymbol{\omega})}, \quad M = \sup_{\boldsymbol{\omega} \in \mathbb{R}^d} \frac{f_{s1}(\boldsymbol{\omega})}{f_{s2}(\boldsymbol{\omega})}.$$

Then (7.134) is a spatial covariance function if, and only if:

$$(1 - \max(1, M))^{-1} \leq \theta \leq (1 - \min(1, m))^{-1}, \tag{7.135}$$

where $0^{-1} = -\infty$ and $(-\infty)^{-1} = 0$ on the left-hand side of the inequality and $0^{-1} = +\infty$ on the right-hand side.

Next, following Gregori *et al.* (2008), we proceed to calculate the bounds that define the possible variation range of θ which enables valid spatio-temporal covariance functions of the type (7.132) to be generated using certain models. In particular, we show the cases in which inputs are either Gaussian covariance functions or belong to the

Matérn model for which the associated spectral density function has an analytical expression.

Case 1[9] *Let* $G1(\mathbf{h} \mid \sigma^2_{g1}, \alpha_{g1})$ *and* $G2(\mathbf{h} \mid \sigma^2_{g2}, \alpha_{g2})$ *be two covariance functions that belong to the Gaussian model on* $\mathbb{R}^k$. *Let* $f_{G1}(\omega)$ *and* $f_{G2}(\omega)$ *be their respective spectral density functions and let*

$$m_{G1,G2} = \inf_{\boldsymbol{\omega}\in\mathbb{R}^d} \frac{f_{G1}(\boldsymbol{\omega})}{f_{G2}(\boldsymbol{\omega})}, \quad M_{G1,G2} = \sup_{\boldsymbol{\omega}\in\mathbb{R}^d} \frac{f_{G1}(\boldsymbol{\omega})}{f_{G2}(\boldsymbol{\omega})},$$

For this purpose, let us take into account that the covariance function of the Gaussian model is given by the expression:

$$G\left(\mathbf{h} \mid \sigma_g^2, \alpha_g\right) = \sigma_g^2 \exp\left(-\alpha_g \|\mathbf{h}\|^2\right) \mathbf{h} \in \mathbb{R}^d,$$

$\sigma_g^2 > 0$ *being the variance of the process and* α_g *a scaling parameter such that* $\alpha_g^{-1} > 0$ *indicates the speed at which the covariance decreases.*

Let us also take into account that the spectral density function associated with the Gaussian model is given by:

$$f_G(\omega) = \sigma_g^2 \pi^{\frac{k}{2}} \alpha_g^{-\frac{k}{2}} \exp\left(-\frac{\|\omega\|^2}{4\alpha_g}\right) \quad \omega \in \mathbb{R}^d,$$

hence, the ratio of spectral densities is:

$$\frac{f_{G_1}(\omega)}{f_{G_2}(\omega)} = \frac{\sigma^2_{g_1}}{\sigma^2_{g_2}} \left(\frac{\alpha_{g_1}}{\alpha_{g_1}}\right)^{-\frac{k}{2}} \exp\left(-\frac{1}{4}\left(\frac{1}{\alpha_{g_1}} - \frac{1}{\alpha_{g_2}}\right) \|\omega\|^2\right).$$

Consequently:

(i) If $\alpha_{g1} < \alpha_{g2}$, *then*

$$m_{G1,G2} = 0, \quad M_{G1,G2} = \frac{\sigma^2_{g1}}{\sigma^2_{g2}} \left(\frac{\alpha_{g2}}{\alpha_{g1}}\right)^{k/2}.$$

(ii) If $\alpha_{g1} = \alpha_{g2}$, *then*

$$m_{G1,G2} = \frac{\sigma^2_{g1}}{\sigma^2_{g2}}, \quad M_{G1,G2} = \frac{\sigma^2_{g1}}{\sigma^2_{g2}}.$$

(iii) If $\alpha_{g1} > \alpha_{g2}$, *then*

$$m_{G1,G2} = \frac{\sigma^2_{g1}}{\sigma^2_{g2}} \left(\frac{\alpha_{g2}}{\alpha_{g1}}\right)^{k/2}, \quad M_{G1,G2} = +\infty.$$

[9] See proof in Appendix D. We follow Martínez (2008).

Case 2[10] *Let* $M1(\mathbf{h} \mid \sigma^2_{m1}, \alpha_{m1}, \upsilon_{m1})$ *and* $M2(\mathbf{h} \mid \sigma^2_{m2}, \alpha_{m2}, \upsilon_{m2})$ *be two covariance functions that belong to the Matérn model on* $\mathbb{R}^k$. *Let* $f_{M1}(\omega)$ *and* $f_{M2}(\omega)$ *be their respective spectral density functions and let*

$$m_{M_1,M_2} = \inf_{\omega \in \mathbb{R}^d} \frac{f_{M_1}(\boldsymbol{\omega})}{f_{M_2}(\boldsymbol{\omega})}, \quad M_{M_1,M_2} = \sup_{\omega \in \mathbb{R}^d} \frac{f_{M_1}(\boldsymbol{\omega})}{f_{M_2}(\boldsymbol{\omega})}.$$

For the purposes of the case at hand, the covariance function associated with the Matérn model is given by:

$$M(\mathbf{h}|\sigma^2_m, \alpha_m, \upsilon_m) = \sigma^2_m(2^{1-\upsilon_m}/\Gamma(\upsilon_m))\big(\alpha_m \|\mathbf{h}\|\big)^{\upsilon_m} K_{\upsilon_m}(\alpha_m \|\mathbf{h}\|), \quad \mathbf{h} \in \mathbb{R}^k,$$

where $\sigma^2_m > 0$ *is the variance of the process,* α_m *is a scaling parameter such that* $\alpha_m^{-1} > 0$ *indicates the speed at which the process decreases and* υ_m *measures the degree of smoothness of the associated process.*

The spectral density associated with this model is given by:

$$f_M(\boldsymbol{\omega}) = \sigma^2_m 2^{\upsilon_m - 1} \pi^{-k/2} \Gamma\left(\upsilon_m + \frac{k}{2}\right) \alpha_m^{2\upsilon_m} \left(\alpha^2_m \|\boldsymbol{\omega}\|^2\right)^{-\upsilon_m - \frac{k}{2}}, \quad \omega \in \mathbb{R}^k$$

and the ratio of its spectral density functions takes the expression:

$$\frac{f_{M_1}(\boldsymbol{\omega})}{f_{M_2}(\boldsymbol{\omega})} = \frac{\sigma^2_{m1}}{\sigma^2_{m2}} 2^{\upsilon_{m1} - \upsilon_{m2}} \frac{\Gamma(\upsilon_{m1} + \frac{k}{2})\, \alpha_{m1}^{2\upsilon_{m1}}}{\Gamma(\upsilon_{m2} + \frac{k}{2})\, \alpha_{m2}^{2\upsilon_m}} \frac{\left(\alpha^2_{m1} + \|\boldsymbol{\omega}\|^2\right)^{-\upsilon_{m1} - \frac{k}{2}}}{\left(\alpha^2_{m2} + \|\boldsymbol{\omega}\|^2\right)^{-\upsilon_{m2} - \frac{k}{2}}}, \tag{7.136}$$

hence:

(i) *If* $\upsilon_{m1} = \upsilon_{m2}$, *and* $\alpha_{m1} = \alpha_{m2}$, *then:*

$$m_{M1,M2} = \frac{\sigma^2_{m1}}{\sigma^2_{m2}}, \quad M_{M1,M2} = \frac{\sigma^2_{m1}}{\sigma^2_{m2}}.$$

(ii) *If* $\upsilon_{m1} = \upsilon_{m2}$, *and* $\alpha_{m1} > \alpha_{m2}$, *then:*

$$m_{M1,M2} = \frac{\sigma^2_{m1}}{\sigma^2_{m2}} \left(\frac{\alpha_{m2}}{\alpha_{m1}}\right)^k, \quad M_{M1,M2} = \frac{\sigma^2_{m1}}{\sigma^2_{m2}} \left(\frac{\alpha_{m1}}{\alpha_{m2}}\right)^{2\upsilon_{m1}}.$$

(iii) *If* $\upsilon_{m1} = \upsilon_{m2}$, *and* $\alpha_{m1} < \alpha_{m2}$, *then:*

$$m_{M1,M2} = \frac{\sigma^2_{m1}}{\sigma^2_{m2}} \left(\frac{\alpha_{m1}}{\alpha_{m2}}\right)^{2\upsilon_{m1}}, \quad M_{M1,M2} = \frac{\sigma^2_{m1}}{\sigma^2_{m2}} \left(\frac{\alpha_{m2}}{\alpha_{m1}}\right)^k.$$

---

[10] See proof in Appendix D. We follow Martínez (2008).

(iv) If $\upsilon_{m1}>\upsilon_{m2}$*, and* $\frac{\alpha_{m2}^2}{\alpha_{m1}^2} \geq \frac{\upsilon_{m2}+\frac{k}{2}}{\upsilon_{m1}+\frac{k}{2}}$*, then:*

$$m_{M1,M2} = 0, \quad M_{M1,M2} = \frac{\sigma_{m1}^2}{\sigma_{m2}^2} 2^{\upsilon_{m1}-\upsilon_{m2}} \frac{\Gamma\left(\upsilon_{m1} + \frac{k}{2}\right)}{\Gamma\left(\upsilon_{m2} + \frac{k}{2}\right)} \left(\frac{\alpha_{m2}}{\alpha_{m1}}\right)^k.$$

(v) If $\upsilon_{m1}>\upsilon_{m2}$*, and* $\frac{\alpha_{m2}^2}{\alpha_{m1}^2} < \frac{\upsilon_{m2}+\frac{k}{2}}{\upsilon_{m1}+\frac{k}{2}}$*, then:*

$$m_{M1,M2} = 0,$$

$$M_{M1,M2} = \frac{\sigma_{m1}^2}{\sigma_{m2}^2} \frac{\Gamma\left(\upsilon_{m1} + \frac{k}{2}\right)}{\Gamma\left(\upsilon_{m2} + \frac{k}{2}\right)} \left(\frac{\alpha_{m2}^2 - \alpha_{m1}^2}{2\left(\upsilon_{m2} - \upsilon_{m1}\right)}\right)^{\upsilon_{m2}-\upsilon_{m1}} \frac{\alpha_{m1}^{2\upsilon_{m1}}}{\alpha_{m2}^{2\upsilon_{m2}}} \times \frac{\left(\upsilon_{m2} + \frac{k}{2}\right)^{\upsilon_{m2}+\frac{k}{2}}}{\left(\upsilon_{m1} + \frac{k}{2}\right)^{\upsilon_{m1}+\frac{k}{2}}}.$$

(vi) If $\upsilon_{m1}<\upsilon_{m2}$*, and* $\frac{\alpha_{m2}^2}{\alpha_{m1}^2} \geq \frac{\upsilon_{m2}+\frac{k}{2}}{\upsilon_{m1}+\frac{k}{2}}$*, then:*

$$m_{M1,M2} = \frac{\sigma_{m1}^2}{\sigma_{m2}^2} \frac{\Gamma\left(\upsilon_{m1} + \frac{k}{2}\right)}{\Gamma\left(\upsilon_{m2} + \frac{k}{2}\right)} \left(\frac{\alpha_{m2}^2 - \alpha_{m1}^2}{2\left(\upsilon_{m2} - \upsilon_{m1}\right)}\right)^{\upsilon_{m2}-\upsilon_{m1}} \frac{\alpha_{m1}^{2\upsilon_{m1}}}{\alpha_{m2}^{2\upsilon_{m2}}} \times \frac{\left(\upsilon_{m2} + \frac{k}{2}\right)^{\upsilon_{m2}+\frac{k}{2}}}{\left(\upsilon_{m1} + \frac{k}{2}\right)^{\upsilon_{m1}+\frac{k}{2}}},$$

$$M_{M1,M2} = +\infty.$$

(vii) If $\upsilon_{m1}<\upsilon_{m2}$*, and* $\frac{\alpha_{m2}^2}{\alpha_{m1}^2} < \frac{\upsilon_{m2}+\frac{k}{2}}{\upsilon_{m1}+\frac{k}{2}}$*, then:*

$$m_{M1,M2} = \frac{\sigma_{m1}^2}{\sigma_{m2}^2} 2^{\upsilon_{m1}-\upsilon_{m2}} \frac{\Gamma\left(\upsilon_{m1} + \frac{k}{2}\right)}{\Gamma\left(\upsilon_{m2} + \frac{k}{2}\right)} \left(\frac{\alpha_{m2}}{\alpha_{m1}}\right)^k,$$

$$M_{M1,M2} = +\infty.$$

Case 3[11] *Let $G(\mathbf{h} \mid \sigma_g^2, \alpha_g)$ be a Gaussian covariance function and $M(\mathbf{h} \mid \sigma_m^2, \alpha_m, \upsilon_m)$ a covariance function that belongs to the Matérn model, both valid on $\mathbb{R}^k$. Let $f_G(\omega)$ and $f_M(\omega)$ be their respective spectral density functions and let*

$$m_{_{G,M}} = \inf_{\boldsymbol{\omega}\in\mathbb{R}^d} \frac{f_G(\boldsymbol{\omega})}{f_M(\boldsymbol{\omega})}, \quad M_{_{G,M}} = \sup_{\boldsymbol{\omega}\in\mathbb{R}^d} \frac{f_G(\boldsymbol{\omega})}{f_M(\boldsymbol{\omega})}.$$

In this case, the ratio of spectral densities of the covariance functions involved is:

$$\frac{f_G(\omega)}{f_M(\omega)} = \frac{\sigma_g^2}{\sigma_m^2}\pi^k \frac{2^{1-\nu_m}}{\Gamma\left(\nu_m + \frac{k}{2}\right)} \frac{\alpha_g^{-\frac{k}{2}}}{\alpha_m^{2\nu_m}} \exp\left(-\frac{\|\boldsymbol{\omega}\|^2}{4\alpha_g}\right)\left(\alpha_m^2 + \|\boldsymbol{\omega}\|^2\right)^{\upsilon_m+\frac{k}{2}} \tag{7.137}$$

whereby:

(i) If $\frac{\alpha_m^2}{\upsilon_m+\frac{k}{2}} > 4\alpha_g$, *then:*

$$m_{_{G,M}} = 0,$$

$$M_{_{G,M}} = \frac{\sigma_g^2}{\sigma_m^2} 2^{k+1}\pi^k \left(\frac{2\alpha_g}{\alpha_m^2}\right)^{\upsilon_m} \frac{\left(\upsilon_m + \frac{k}{2}\right)^{\upsilon_m+\frac{k}{2}}}{\Gamma\left(\upsilon_m + \frac{k}{2}\right)} \exp\left(\frac{\alpha_m^2}{4\alpha_g} - \left(\upsilon_m + \frac{k}{2}\right)\right).$$

(ii) If $\frac{\alpha_m^2}{\upsilon_m+\frac{k}{2}} \leq 4\alpha_g$, *then:*

$$m_{_{G,M}} = 0,$$

$$M_{_{G,M}} = \frac{\sigma_g^2}{\sigma_m^2} \frac{2^{1-\upsilon_m}}{\Gamma\left(\upsilon_m + \frac{k}{2}\right)} \left(\frac{\pi^2\alpha_m^2}{\alpha_g}\right)^{k/2}.$$

Table 7.3 presents the values of the constants obtained in Cases 1, 2 and 3, allowing us to obtain the range of variation of θ, which gives rise to spatio-temporal covariance functions from the generalized product-sum model (7.134).

More specifically, Table 7.3 provides the values m_s and M_s. Proceeding in the same way in the temporal case, we obtain m_t and M_t. Using these four values as a basis, we obtain the range of variation of θ by way of the expression:

$$\left(1 - \max(1, M_s M_t)\right)^{-1} \leq \theta \leq \left(1 - \min(1, m_s m_t)\right)^{-1}.$$

[11] See proof in the Appendix D. We follow Martínez (2008).

Table 7.3 Values m_t, M_t, m_s and M_s for $n = 2$ in the case in which Gaussian and Matérn functions have been combined in accordance with their parameters.

	Gaussian$_1$ / Gaussian$_2$	
Parameters	m_{G_1,G_2}	M_{G_1,G_2}
$\alpha_{g1} < \alpha_{g2}$	0	$\frac{\sigma^2_{g1}}{\sigma^2_{g2}}\left(\frac{\alpha_{g2}}{\alpha_{g1}}\right)^{k/2}$
$\alpha_{g1} = \alpha_{g2}$	$\frac{\sigma^2_{g1}}{\sigma^2_{g2}}$	$\frac{\sigma^2_{g1}}{\sigma^2_{g2}}$
$\alpha_{g1} > \alpha_{g2}$	$\frac{\sigma^2_{g1}}{\sigma^2_{g2}}\left(\frac{\alpha_{g2}}{\alpha_{g1}}\right)^{k/2}$	$+\infty$
	Matérn$_1$ / Matérn$_2$	
Parameters	m_{M_1,M_2}	M_{M_1,M_2}
$\upsilon_{m1} = \upsilon_{m2}$ and $\alpha_{m1} \geq \alpha_{m2}$	$\frac{\sigma^2_{m1}}{\sigma^2_{m2}}\left(\frac{\alpha_{m2}}{\alpha_{m1}}\right)^{k}$	$\frac{\sigma^2_{m1}}{\sigma^2_{m2}}\left(\frac{\alpha_{m1}}{\alpha_{m2}}\right)^{2\upsilon_1}$
$\upsilon_{m1} = \upsilon_{m2}$ and $\alpha_{m1} < \alpha_{m2}$	$\frac{\sigma^2_{m1}}{\sigma^2_{m2}}\left(\frac{\alpha_{m1}}{\alpha_{m2}}\right)^{2\upsilon_1}$	$\frac{\sigma^2_{m1}}{\sigma^2_{m2}}\left(\frac{\alpha_{m2}}{\alpha_{m1}}\right)^{k}$
$\upsilon_{m1} > \upsilon_{m2}$ and $\frac{\alpha^2_{m2}}{\alpha^2_{m1}} \geq \frac{\upsilon_{m2}+k/2}{\upsilon_{m1}+k/2}$	0	$\frac{\sigma^2_{m1}}{\sigma^2_{m2}} 2^{\upsilon_{m1}-\upsilon_{m2}} \frac{\Gamma(\upsilon_{m1}+k/2)}{\Gamma(\upsilon_{m2}+k/2)}\left(\frac{\alpha_{m2}}{\alpha_{m1}}\right)^{k}$
$\upsilon_{m1} > \upsilon_{m2}$ and $\frac{\alpha^2_{m2}}{\alpha^2_{m1}} < \frac{\upsilon_{m2}+k/2}{\upsilon_{m1}+k/2}$	0	$\frac{\sigma^2_{m1}}{\sigma^2_{m2}} \frac{\Gamma(\upsilon_{m1}+k/2)}{\Gamma(\upsilon_{m2}+k/2)}\left(\frac{\alpha^2_{m2}-\alpha^2_{m1}}{2(\upsilon_{m2}-\upsilon_{m1})}\right)^{\upsilon_{m2}-\upsilon_{m1}}$ $\frac{\alpha^{2\upsilon_{m1}}_{m1}}{\alpha^{2\upsilon_{m2}}_{m2}} \frac{(\upsilon_{m2}+k/2)^{\upsilon_{m2}+k/2}}{(\upsilon_{m1}+k/2)^{\upsilon_{m1}+k/2}}$

(continued overleaf)

Table 7.3 (*continued*).

	Gaussian / Matérn	
Parameters	$m_{G,M}$	$M_{G,M}$
$\upsilon_{m1} < \upsilon_{m2}$ and $\frac{\alpha_{m2}^2}{\alpha_{m1}^2} \geq \frac{\upsilon_{m2}+k/2}{\upsilon_{m1}+k/2}$	$\frac{\sigma_{m1}^2}{\sigma_{m2}^2} \frac{\Gamma(\upsilon_{m1}+k/2)}{\Gamma(\upsilon_{m2}+k/2)} \left(\frac{\alpha_{m2}^2-\alpha_{m1}^2}{2(\upsilon_{m2}-\upsilon_{m1})} \right)^{\upsilon_{m2}-\upsilon_{m1}} \frac{\alpha_{m1}^{2\upsilon_{m1}}}{\alpha_{m2}^{2\upsilon_{m2}}} \frac{(\upsilon_{m2}+k/2)^{\upsilon_{m2}+k/2}}{(\upsilon_{m1}+k/2)^{\upsilon_{m1}+k/2}}$	$+\infty$
$\upsilon_{m1} < \upsilon_{m2}$ and $\frac{\alpha_{m2}^2}{\alpha_{m1}^2} < \frac{\upsilon_{m2}+k/2}{\upsilon_{m1}+k/2}$	$\frac{\sigma_{m1}^2}{\sigma_{m2}^2} \cdot 2^{\upsilon_{m1}-\upsilon_{m2}} \cdot \frac{\Gamma(\upsilon_{m1}+k/2)}{\Gamma(\upsilon_{m2}+d/2)} \cdot \left(\frac{\alpha_{m2}}{\alpha_{m1}} \right)^k$	$+\infty$

	Gaussian / Matérn	
Parameters	$m_{G,M}$	$M_{G,M}$
$\frac{\alpha_m^2}{\upsilon_m+k/2} > 4\alpha_g$	0	$\frac{\sigma_g^2}{\sigma_m^2} 2^{k+1} \pi^k \left(\frac{2\alpha_g}{\alpha_m^2} \right)^{\upsilon_m} \frac{\left(\upsilon_m+\frac{k}{2}\right)^{\upsilon_m+\frac{k}{2}}}{\Gamma\left(\upsilon_m+\frac{k}{2}\right)} \exp\left(\frac{\alpha_m^2}{4\alpha_g} - \left(\upsilon_m + \frac{k}{2}\right)\right)$
$\frac{\alpha_m^2}{\upsilon_m+k/2} \leq 4\alpha_g$	0	$\frac{\sigma_g^2}{\sigma_m^2} \frac{2^{1-\upsilon_m}}{\Gamma(\upsilon_m+k/2)} \left(\frac{\pi^2\alpha_m^2}{\alpha_g} \right)^{k/2}$

The first part of Table 7.3 shows the values of m_s and M_s in the case when the spatial covariances are both Gaussian (or the values of m_t and M_t in the case in which the temporal covariances are both Gaussian), depending on the values of the parameters considered. The second part of the Table 7.3 shows these same constants in the case in which both covariance functions (spatial or temporal) belong to the Matérn model. Finally, the third part contains the values in the case in which the first spatial covariance function is Gaussian and the second Matérn (the same in the temporal case).

The values when the opposite is true (the first covariance functions belong to the Matérn model and the second are Gaussian) are obtained by taking m_s and M_s as the inverse of the corresponding values m_s and M_s respectively.

This table also allows us to obtain the constants θ and $(1-\theta)$ for the exponential case, as this is a particular case of the Matérn model ($\upsilon = \frac{1}{2}$).

Example 7.23 (Generalized product-sum model. Martínez 2008) *Assume that we have two spatial covariance functions belonging to the Matérn model with the same degree of smoothness υ and the same a priori variance σ_s^2, and two exponential covariance functions (a particular case of the Matérn model with $\upsilon = \frac{1}{2}$), also with the same a priori variance σ_t^2. Therefore, the model considered is:*

$$C(\mathbf{h}, u) = \theta C_{s1}(\mathbf{h})C_{t1}(u) + (1-\theta)C_{s2}(\mathbf{h})C_{t2}(u), \quad (\mathbf{h}, u) \in \mathbb{R}^d \times \mathbb{R}, \qquad (7.138)$$

with:

$$C_{s1}(\mathbf{h}) = M\left(\mathbf{h}|\sigma_s^2, \beta_1, \upsilon\right) = \sigma_s^2 \left(\frac{2^{1-\upsilon}}{\Gamma(\upsilon)}\right)\left(\beta_1 \|\mathbf{h}\|\right)^{\upsilon} K_{\upsilon}\left(\beta_1 \|\mathbf{h}\|\right)$$

$$C_{s2}(\mathbf{h}) = M\left(\mathbf{h}|\sigma_s^2, \beta_2, \upsilon\right) = \sigma_s^2 \left(\frac{2^{1-\upsilon}}{\Gamma(\upsilon)}\right)\left(\beta_2 \|\mathbf{h}\|\right)^{\upsilon} K_{\upsilon}\left(\beta_2 \|\mathbf{h}\|\right)$$

$$C_{t1}(\mathbf{h}) = M\left(\mathbf{h}|\sigma_t^2, \alpha_1, \upsilon = \frac{1}{2}\right) = \sigma_t^2 \exp\left(-\alpha_1 |u|\right)$$

$$C_{t2}(\mathbf{h}) = M\left(\mathbf{h}|\sigma_t^2, \alpha_2, \upsilon = \frac{1}{2}\right) = \sigma_t^2 \exp\left(-\alpha_2 |u|\right)$$

For example, from Table 7.3 we have that for $\beta_1 \geq \beta_2$ and $\alpha_1 \geq \alpha_2$:

$$m_s = \left(\frac{\beta_2}{\beta_1}\right)^d, \quad M_s = \left(\frac{\beta_1}{\beta_2}\right)^{2\upsilon}, \quad m_s = \frac{\alpha_2}{\alpha_1}, \quad M_t = \frac{\alpha_1}{\alpha_2}$$

then,

$$\max\left(1, M_s M_t\right) = \max\left(1, \left(\frac{\beta_1}{\beta_2}\right)^{2\upsilon} \frac{\alpha_1}{\alpha_2}\right) = \left(\frac{\beta_1}{\beta_2}\right)^{2\upsilon} \frac{\alpha_1}{\alpha_2}$$

$$\min\left(1, m_s m_t\right) = \min\left(1, \left(\frac{\beta_2}{\beta_1}\right)^{d} \frac{\alpha_2}{\alpha_1}\right) = \left(\frac{\beta_2}{\beta_1}\right)^{d} \frac{\alpha_2}{\alpha_1}$$

Hence, the spatio-temporal covariance function:

$$C(\mathbf{h}, u) = \theta C_{s1}(\mathbf{h})C_{t1}(u) + (1-\theta)C_{s2}(\mathbf{h})C_{t2}(u), \quad (\mathbf{h}, u) \in \mathbb{R}^d \times \mathbb{R},$$

is considered valid if, and only if:

$$\left(1 - \left(\frac{\beta_1}{\beta_2}\right)^{2\upsilon} \frac{\alpha_1}{\alpha_2}\right)^{-1} \leq \theta \leq \left(1 - \left(\frac{\beta_2}{\beta_1}\right)^{d} \frac{\alpha_2}{\alpha_1}\right)^{-1}.$$

Similarly:

- If $\beta_1 \geq \beta_2$ and $\alpha_1 < \alpha_2$, then the function (7.138) is a spatio-temporal covariance function if, and only if:

$$\left(1 - \left(\frac{\beta_2}{\beta_1}\right)^{d} \frac{\alpha_2}{\alpha_1}\right)^{-1} \leq \theta \leq \left(1 - \left(\frac{\beta_1}{\beta_2}\right)^{2\upsilon} \frac{\alpha_1}{\alpha_2}\right)^{-1}. \tag{7.139}$$

- If $\beta_1 < \beta_2$ and $\alpha_1 \geq \alpha_2$, then the function (7.138) is a spatio-temporal covariance function if, and only if:

$$\left(1 - \left(\frac{\beta_2}{\beta_1}\right)^{d} \frac{\alpha_1}{\alpha_2}\right)^{-1} \leq \theta \leq \left(1 - \left(\frac{\beta_1}{\beta_2}\right)^{2\upsilon} \frac{\alpha_2}{\alpha_1}\right)^{-1}. \tag{7.140}$$

- If $\beta_1 < \beta_2$ and $\alpha_1 < \alpha_2$, then the function (7.138) is a spatio-temporal covariance function if, and only if:

$$\left(1 - \left(\frac{\beta_2}{\beta_1}\right)^{d} \frac{\alpha_2}{\alpha_1}\right)^{-1} \leq \theta \leq \left(1 - \left(\frac{\beta_1}{\beta_2}\right)^{2\upsilon} \frac{\alpha_1}{\alpha_2}\right)^{-1}. \tag{7.141}$$

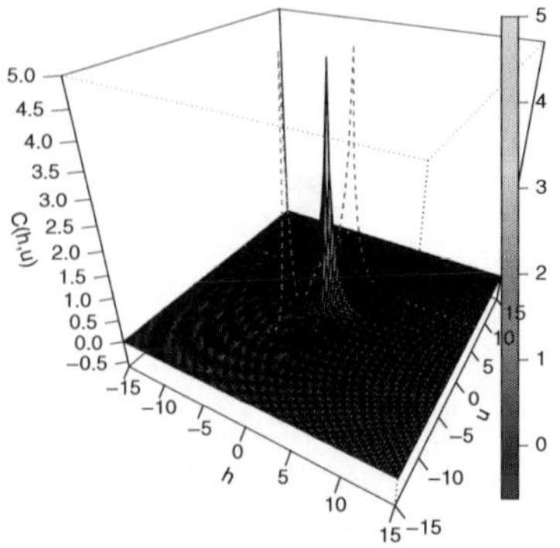

Figure 7.11 Representation of the generalized product-sum model (7.138) with negative covariances.

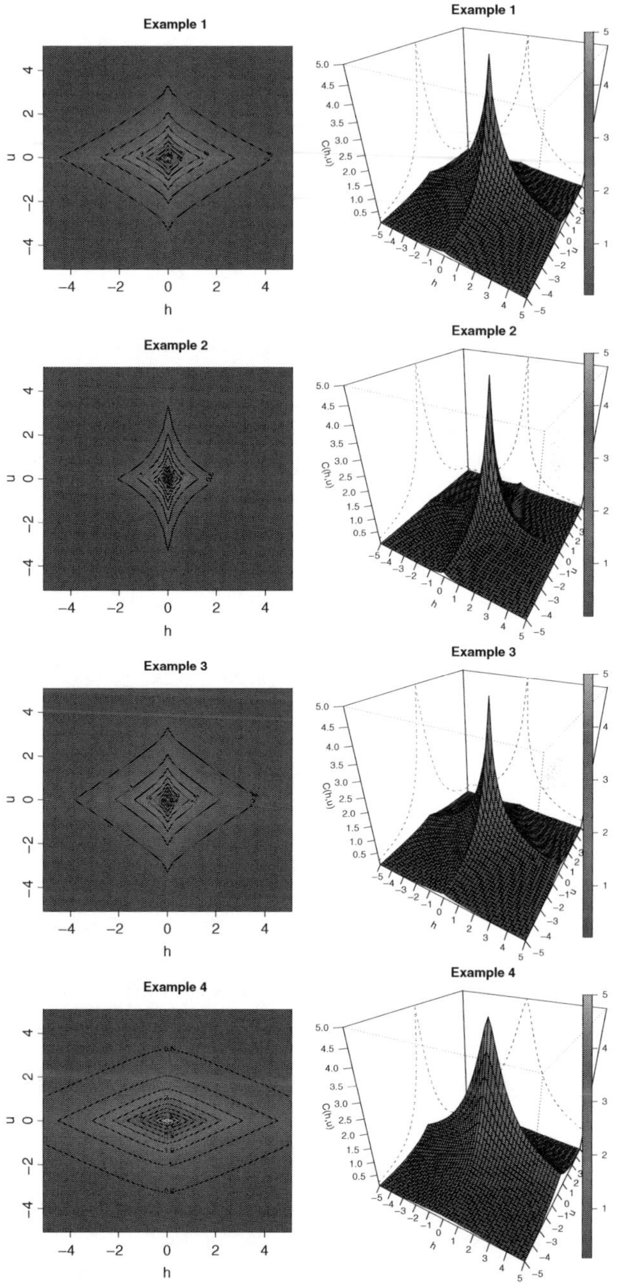

Figure 7.12 2D and 3D different representations of the generalized product-sum model (7.138). (See color figure in color plate section.)

As noted previously, one of the main advantages of the generalized product-sum model is that it allows for negative covariances or fluctuations between positive and negative values, which is of the utmost importance when modeling certain phenomena. For example, if in (7.138) we consider:

$$\alpha_1 = 0.5 < \alpha_2 = 1$$
$$\beta_1 = \beta_2 = 0.5$$
$$\nu = 0.3$$
$$\sigma^2 = \sigma_s^2\sigma_t^2 = 5$$

then, through (7.139) we have that the model is valid providing $-1 \leq \theta \leq 2$. If we consider $\theta = -1$, Figure 7.11 reveals that the covariance function can take negative values.

Figure 7.12 shows different representations of the family of covariance functions (7.138) for a set of possible values of the parameters that define it. More specifically, let $\boldsymbol{\theta} = (\alpha_1, \alpha_2, \beta_1, \beta_2, \nu_1, \nu_2, \theta, \sigma_{s1}^2, \sigma_{s2}^2, \sigma_{t1}^2, \sigma_{t2}^2)$ be the vector of parameters that defines the model (7.138). The covariance functions represented correspond to $\alpha_1 = 0.5, \alpha_2 = 1.5, \beta_1 = 0.3, \beta_2 = 0.6, \nu_1 = 0.3, \nu_2 = 0.3$ in Example 1, $\alpha_1 = 0.5, \alpha_2 = 1.5, \beta_1 = 1.5, \beta_2 = 0.6, \nu_1 = 0.3, \nu_2 = 0.3$ in Example 2, $\alpha_1 = 0.5, \alpha_2 = 1.5, \beta_1 = 0.3, \beta_2 = 1.2, \nu_1 = 0.3, \nu_2 = 0.3$ in Example 3, and $\alpha_1 = 0.5, \alpha_2 = 1.5, \beta_1 = 0.3, \beta_2 = 1.2, \nu_1 = 0.8, \nu_2 = 0.8$ in Example 4, where in all cases we have considered $\theta = 0.5$ and the same *a priori* spatial and temporal variance ($\sigma_{s1}^2 = \sigma_{s2}^2 = \sigma_s^2$ and $\sigma_{t1}^2 = \sigma_{t2}^2 = \sigma_t^2$) such that $\sigma^2 = \sigma_s^2\sigma_t^2 = 5$.

7.17 Models that are not fully symmetric

Certain spatio-temporal rf's are not fully symmetric due to their very nature. For example, as stated in Gneiting *et al.* (2007), environmental, atmospheric and geophysical processes are often influenced by prevailing winds or ocean currents, which are incompatible with the assumption of full symmetry. The covariance functions in the previous sections are fully symmetric, for which reason they cannot model these situations.

One way of tackling these problems is to generate stationary covariance functions of the form:

$$C(\mathbf{h}, u) = E_V[C_s(\mathbf{h} + Vu)], \quad (\mathbf{h}, u) \in \mathbb{R}^d \times \mathbb{R} \tag{7.142}$$

whereby $C_s(\mathbf{h})$ is the stationary spatial covariance function on $\mathbb{R}^d$ of a purely spatial random field, and it is supposed that the entire field moves time-forward with random velocity vector $V \in \mathbb{R}^d$. Similar constructions of stationary spatio-temporal covariance functions have been provided by Cox and Isham (1988) and Ma (2003a).

It is simple to verify the validity of covariance functions defined in (7.142), as:

$$\sum_{i=1}^{n}\sum_{j=1}^{n} a_i a_j C(\mathbf{s}_i - \mathbf{s}_j, t_i - t_j) = \sum_{i=1}^{n}\sum_{j=1}^{n} a_i a_j E_V\left(C_s(\mathbf{s}_i - \mathbf{s}_j) + V(t_i - t_j)\right)$$
$$= E_V\left(\sum_{i=1}^{n}\sum_{j=1}^{n} a_i a_j C_s(\mathbf{s}_i + Vt_i) - (\mathbf{s}_j + Vt_j)\right)$$
$$= E_V\left(\sum_{i=1}^{n}\sum_{j=1}^{n} a_i a_j C_s\left(\mathbf{r}_i - \mathbf{r}_j\right)\right) \geq 0,$$

because it is the expectation of a positive quantity, since $C_s(\mathbf{h})$ is a spatial covariance function on $\mathbb{R}^d$.

In order to specify the random velocity vector, Gneiting *et al.* (2007) provided various reasonable choices from a physical perspective. The simplest case is to select $V = v$, representing, for example, the mean of prevailing wind. Gupta and Waymire (1987) referred to this as the *frozen field* model.

Another possibility is that the velocity vector might attain a small number of distinct values (for example, representing wind regimens) or to identify its distribution with the empirical distribution of velocity vectors. The distribution of V can also be updated dynamically in accordance with the current state of the phenomenon, which will give rise to non-stationary covariance structures similar to those posited by Riishøjgaard (1998) and Huang and Hsu (2004), among others.

The literature provides several stationary spatio-temporal covariance functions that are not fully symmetric obtained through diffusion equations or stochastic equations of partial derivatives. For this purpose, Jones and Zhang (1997), Christakos (2000), Brown *et al.* (2000), Kolovos *et al.* (2004), Jun and Stein (2004) or Stein (2005a, 2005c) can be consulted, among others.

7.18 Mixture-based Bernstein zonally anisotropic covariance functions

Many of the covariance functions discussed previously are very flexible and easy to implement and estimate. However, they are limited as they are isotropic in space and fully symmetric. As regards the issue of zonal anisotropy in space, it is worth raising the following two questions: (i) Is it possible to construct covariance functions that are not necessarily $\mathcal{L}_p$ norms?, and (ii) Is it possible to obtain covariance functions with zonal anisotropy using mixtures? The answer to these questions is yes, according to the following result in Porcu et al. (2007).

Let $\mathcal{L}(.,.)$ be the bivariate Laplace transform of a non-negative random vector $\left(X_1, X_2\right)$. Let $\psi_i(\cdot), i = 1, \ldots, d$ and $\psi_T(\cdot)$ be positive-definite Bernstein functions on the positive real line (or continuous, increasing and concave functions on the positive real line). Then for $d = 1, 2 \ldots$

$$C(\mathbf{h}, u) = \mathcal{L}\left(\sum_{i=1}^{d} \psi_i\left(\|\mathbf{h}_i\|\right), \psi_T(u)\right), \tag{7.143}$$

is a non-separable and stationary spatio-temporal covariance function on $\mathbb{R}^d \times \mathbb{R}$.

The covariance functions obtained in this manner are not necessarily a $\mathcal{L}_p$ norm, and can be used in the case of zonal anisotropy in space, as proposed by Rouhani and Hall (1989). Note also that the class of Bernstein functions is not necessarily linked to the class of variograms. Even if in some cases the two classes overlap, it is relatively easy to demonstrate that a Bernstein function is not necessarily a variogram and vice versa. Another important advantage is that the composition $\psi \circ \gamma(\cdot)$ is a variogram (see Ma 2005a and the references included therein). Hence, using the expression (7.143), a broad class of spatio-temporal covariance functions with zonal anisotropy can be constructed.

The covariance functions given by (7.143) are a special case of a broader class of covariance functions proposed by Ma (2003b) in Theorem 7.18.1, which we now reproduce to homogenize the notation.

Theorem 7.18.1 (Ma 2003b) *Let $\mathcal{L}\left(\theta_1, \theta_2\right)$ be the Laplace transform of a non-negative random vector $\left(X_1, X_2\right)$. If $\gamma_1(\mathbf{h})$ is a purely spatial semivariogram on $\mathbb{R}^d$ and $\gamma_2(t)$ is a purely temporal semivariogram on $\mathbb{R}$, then:*[12]

$$C(\mathbf{h}, u) = \mathcal{L}\left(\gamma_1(\mathbf{h}), \gamma_2(t)\right), \tag{7.144}$$

is a spatio-temporal covariance function on $\mathbb{R}^d \times \mathbb{R}$.

That is, instead of using a spatial semivariogram, Porcu et al. (2007a) introduce a semivariogram that is a sum of semivariograms (and which is also a semivariogram), which allows to include different sills in each direction (with the same range).

By way of example, Porcu *et al.* (2007a) follow the line of work marked out by Shapiro and Botha (1991) and the extension to the spatio-temporal context performed by Fernández-Casal *et al.* (2003), and in order to tackle the anisotropy, they divide the spatial lag vector $\mathbf{h} \in \mathbb{R}^d$ into two sub-vectors: $\mathbf{h}_1$, of dimension d_1, and $\mathbf{h}_2$, of dimension $d_2 = d - d_1$, which makes it possible to obtain an anisotropic spatio-temporal covariance function from the result presented above. The example these authors propose and which illustrates how (7.143) is applied, is as follows (example also used in Ma 2003b).

Example 7.24 (Porcu *et al.* 2007a. Example 1) *For the Frechet-Hoeffding lower bound of bivariate copulas,*

$$F(u, v) = \max(u + v - 1, 0), \quad 0 \leq u, v \leq 1,$$

[12] See proof in Appendix D.

we have that for $\theta_1 = \theta_2 = \theta, \mathcal{L}(\theta,\theta) = \exp(-\theta)$ *and for* $\theta_1 \neq \theta_2$*:*

$$\begin{aligned}\mathcal{L}\left(\theta_1,\theta_2\right) &= \int_0^1 \int_0^1 \exp(-\theta_1 u)\exp(-\theta_2 v) dF(u,v) \\ &= \int_0^1 \exp(-\theta_1 u - \theta_2 u) du = \frac{\exp\left(-\theta_1\right) - \exp\left(-\theta_2\right)}{\theta_2 - \theta_1}.\end{aligned} \tag{7.145}$$

Let the following Bernstein functions be:

$$\begin{aligned}\psi_1(t) &= \left(a_1 t^{\alpha_1} + 1\right)^{\beta} \\ \psi_2(t) &= \frac{\left(a_2 t^{\alpha_2} + b\right)}{b\left(a_2 t^{\alpha_2} + 1\right)} \\ \psi_3(t) &= t^{\rho}\end{aligned} \tag{7.146}$$

where a_1 *and* a_2 *are positive scaling parameters,* $\alpha_1, \alpha_2 \in (0,2], \beta \in (0,1], b > 1$ *and* $\rho \in [0,1)$. *So, by setting: (i) In the Bernstein functions:* $t = \left\|\mathbf{h}_1\right\|$ *in* $\psi_1(t)$, $t = \left\|\mathbf{h}_2\right\|$ *in* $\psi_2(t)$ *and* $t = |u|$ *in* $\psi_3(t)$, *and then, (ii) in the Laplace transform:* $\theta_1 = \psi_1\left(\left\|\mathbf{h}_1\right\|\right) + \psi_2\left(\left\|\mathbf{h}_2\right\|\right)$ *and* $\theta_2 = \psi_3(t)$, *(7.145) is transformed as follows:*

$$\begin{aligned}\mathcal{L}\left(\theta_1,\theta_2\right) &= \frac{\exp\left(-\left(a_1\left\|\mathbf{h}_1\right\|^{\alpha_1}+1\right)^{\beta} - \frac{\left(a_2\left\|\mathbf{h}_2\right\|^{\alpha_2}+b\right)}{b\left(a_2\left\|\mathbf{h}_2\right\|^{\alpha_2}+1\right)}\right) - \exp\left(-|u|^{\rho}\right)}{|u|^{\rho} - \left(a_1\left\|\mathbf{h}_1\right\|^{\alpha_1}+1\right)^{\beta} - \frac{\left(a_2\left\|\mathbf{h}_2\right\|^{\alpha_2}+b\right)}{b\left(a_2\left\|\mathbf{h}_2\right\|^{\alpha_2}+1\right)}} \\ &= C_{\mathbf{s},t}\left(\mathbf{h}_1,\mathbf{h}_2,u\right),\end{aligned} \tag{7.147}$$

which is a permissible, non-separable spatio-temporal covariance function that, moreover, is spatially two-component anisotropic. The variance of the spatio-temporal rf, σ^2, *and a normalization constant* $k = \left(\frac{1}{2}\left(1-\exp(-2)\right)\right)^{-1}$, *can be included in the expression above in such a way that if* $\sigma^2 = 1$, *then* $C_{\mathbf{s},t}(\mathbf{0},\mathbf{0},0) = 1$, *thereby obtaining a spatio-temporal correlation function:*

$$C_{\mathbf{s},t}\left(\mathbf{h}_1,\mathbf{h}_2,u\right) = \sigma^2 k \frac{\exp\left(-\left(a_1\left\|\mathbf{h}_1\right\|^{\alpha_1}+1\right)^{\beta} - \frac{\left(a_2\left\|\mathbf{h}_2\right\|^{\alpha_2}+b\right)}{b\left(a_2\left\|\mathbf{h}_2\right\|^{\alpha_2}+1\right)}\right) - \exp\left(-|u|^{\rho}\right)}{|u|^{\rho} - \left(a_1\left\|\mathbf{h}_1\right\|^{\alpha_1}+1\right)^{\beta} - \frac{\left(a_2\left\|\mathbf{h}_2\right\|^{\alpha_2}+b\right)}{b\left(a_2\left\|\mathbf{h}_2\right\|^{\alpha_2}+1\right)}}.$$

The properties of this covariance function, in regard to its smoothness, can be assessed in view of its marginal covariances: $C_{\mathbf{s},t}\left(\mathbf{h}_1,\mathbf{0},0\right)$, $C_{\mathbf{s},t}\left(\mathbf{0},\mathbf{h}_2,0\right)$ *and* $C_{\mathbf{s},t}(\mathbf{0},\mathbf{0},u)$. *For example, it is not difficult to check that none of the three marginal covariances is differentiable at the origin.*

Finally, we reproduce some clarifications that Porcu *et al.* (2007a) consider of interest:

(i) The construction $C(\mathbf{h},u) = \mathcal{L}\left(\sum_{i=1}^{d} \psi_i\left(\|\mathbf{h}_i\|\right), \psi_t(u)\right)$ represents a special case of the general representation $Z(\mathbf{s},t) = Z_{d+1}\left(tW_{d+1}\right)\prod_{i=1}^{d} Z\left(\mathbf{s}_i, W_i\right)$, with $\mathbf{s} = \left(s_1, \ldots, s_d\right) \in \mathbb{R}^d$ and $t \in \mathbb{R}$, $\mathbf{W} = \left(W_1, \ldots, W_{d+1}\right)$ being a non-negative random vector with W_i independent of Z_i. It is relatively easy to see that the structure of the covariance associated with this random field is non-separable:

$$C(\mathbf{h},u) = \int_{\mathbb{R}_+^{d+1}} \exp\left(-\sum_{i=1}^{d} \psi_i\left(\|\mathbf{h}_i\|\right)\omega_i - \psi_t(|u|)\,\omega_{d+1}\right) dF(\omega),$$

where F is the $(d+1)$-variate distribution function associated with the random vector $\mathbf{W}$. If F is absolutely continuous with respect to the Lebesgue measure, then

$$C(\mathbf{h},u) = \int_{\mathbb{R}_+^{d+1}} \exp\left(-\sum_{i=1}^{d} \psi_i\left(|h_i|\right)\omega_i - \psi_t\left(\|\mathbf{h}_i\|\right)\omega_{d+1}\right) f(\omega)\,d\omega.$$

Notice that this construction allows for the separable case if, and only if, the random vector $\mathbf{W}$ is made up of independent rv's. In particular, the construction:

$$C(\mathbf{h},u) = \mathcal{L}\left(\sum_{i=1}^{d} \psi_i\left(\|\mathbf{h}_i\|\right), \psi_t(u)\right),$$

shows the existence of a stationary random field

$$Z(\mathbf{s},t) = Z_t\left(tW_2\right)\prod_{i=1}^{d} Z\left(\mathbf{s}_i, W_1\right)$$

where F is the probability distribution function associated with $\left(W_1, W_2\right)'$.

(ii) The Bernstein class can be used for both geometrically and zonally anisotropic models. Zonal anisotropy has been traditionally modeled using the sum model (Rouhani and Hall 1989), which is simple to construct and easy to implement, but considers dependence in different directions separately. Furthermore, this construction is based on semivariograms rather than covariograms. Dimitrakopoulos and Luo (1994) propose some models for geometrical anisotropy. Fernández-Casal *et al.* (2003) also propose anisotropic models, but they cannot be obtained in closed-form, the only possibility being numerical calculation. Therefore, the Bernstein class can shed some light on this issue. Indeed, we can obtain very simple closed-forms using very straightforward procedures.

(iii) Covariance functions belonging to Bernstein class are not necessarily based on L_p metrics.

(iv) The permissibility condition in (7.143) is much less restrictive than that which underlies Theorem 7.23 of Ma (2003b). It is easy to see that it is much simpler to check whether a function is continuous, increasing, and concave on the real line than to check whether it is conditionally negative-definite on $\mathbb{R}^d$.

(v) Some of the mathematical properties of the Bernstein class are linked to the use of linear operators, such as partial derivatives.

7.19 Non-stationary models

As in the case of full symmetry, adopting the hypothesis of stationarity is not a realistic approach when modeling certain natural and social phenomena. However, the assumption of stationarity has been overused on numerous occasions due to the fact that there is much less literature on non-stationary models, though fortunately this field of research has developed remarkably in recent years.

This section summarizes some of the main strategies used in the literature to construct non-stationary covariance functions. More specifically, we will analyze how the procedure to construct non-stationary spatial functions proposed by Fuentes and Smith (2001) is generalized to the spatio-temporal case by Chen *et al.* (2006) This procedure uses covariance functions with a structure of weighted sums of locally stationary covariances. We will also analyze the non-stationary models proposed in (Ma 2002, 2003c). Finally, we will describe the interesting non-stationary covariance functions proposed by Porcu and Mateu (2007).

7.19.1 Mixture of locally orthogonal stationary processes

Chen *et al.* (2006) propose a spatio-temporal generalization of the procedure to construct non-stationary spatial covariance functions using mixtures of locally orthogonal stationary processes posited by Fuentes and Smith (2001) and Fuentes (2002).

Let $Z(\mathbf{s}, t)$ be a zero mean non-stationary spatio-temporal process, with $(\mathbf{s}, t) \in D \times T, D \subseteq \mathbb{R}^d$ and $T \subseteq \mathbb{R}$. The procedure for constructing non-stationary spatio-temporal covariance functions proposed by Chen *et al.* (2006) consists in expressing the non-stationary and non-separable process $Z(\mathbf{s}, t)$ as a mixture of orthogonal locally stationary spatio-temporal processes:

$$Z(\mathbf{s}, t) = \sum_{i=1}^{k} K(\mathbf{s} - \mathbf{s}_i, t - t_i) Z_i(\mathbf{s}, t), \tag{7.148}$$

where $(\mathbf{s}_i, t_i)$ is the centroid of subregion $(S, T)_i$, $\{(S, T)_1, \ldots, (S, T)_k\}$ is a partition of the spatio-temporal domain of interest $D \times T$, $Z_i(\mathbf{s}, t)$ is a spatio-temporal process that is locally stationary in $(S, T)_i$ and which explains the spatio-temporal structure of the general process in that subregion, and $K(\mathbf{s} - \mathbf{s}_i, t - t_i)$ is a kernel function over space and time that assigns greater importance to the process $Z_i(\mathbf{s}, t)$ when $(\mathbf{s}, t)$ is near $(\mathbf{s}_i, t_i)$ and less importance when $(\mathbf{s}, t)$ is far from the centroid of the subregion, $(\mathbf{s}_i, t_i)$.

If the spatio-temporal process $Z(\mathbf{s}, t)$ is stationary in time but not in space, (7.148) can be written as:

$$Z(\mathbf{s}, t) = \sum_{i=1}^{k} K(\mathbf{s} - \mathbf{s}_i) Z_i(\mathbf{s}, t), \tag{7.149}$$

where in this case $Z_i(\mathbf{s}, t)$ is a spatio-temporal process that is locally stationary within subregion $D_i \times T$ and $K(\mathbf{s} - \mathbf{s}_i)$ is a weighting function centered at $\mathbf{s}_i$, $\{D_1 \times T, \dots, D_k \times T\}$ being a partition of the spatio-temporal domain $D \times T$ and $\mathbf{s}_i$ the centroid of D_i.

Similarly, if the spatio-temporal process $Z(\mathbf{s}, t)$ is stationary in space but not in time, (7.148) can be written as:

$$Z(\mathbf{s}, t) = \sum_{i=1}^{k} K(t - t_i) Z_i(\mathbf{s}, t), \tag{7.150}$$

where each item in the expression (7.150) is defined in the same way as the previous case.

7.19.2 Non-stationary models proposed by Ma

Ma acknowledges that in reality non-stationary processes are more common than stationary processes. For this reason it is crucial to develop procedures that allow the construction of non-stationary spatio-temporal covariance functions.

As we explained in Sections 7.12 and 7.13, in order to solve the problem of separability, Ma (2002) proposes scale mixtures and positive power mixtures of separable covariance functions for which the permissibility condition is easily verified and which, moreover, often appear in closed-form. In addition, the procedure proposed by Ma, which can be applied to both continuous and also discrete cases, does not require the process to be stationary in space or in time. Hence, in this section we will generalize the results set out in the sections mentioned above, which were confined to the stationary case.

We will also examine the technique described in Ma (2003c), which makes it possible to generate non-stationary spatio-temporal covariance functions and semi-variograms through linear combinations of stationary spatio-temporal covariance functions and intrinsically stationary variograms.

7.19.2.1 Construction of non-stationary spatio-temporal covariance functions using scale mixtures

As mentioned previously, here we generalize the procedure based on scale mixtures set out in Section 7.12.1 to the non-stationary case. Let $C_s(\mathbf{s}_1, \mathbf{s}_2)$ be a spatial covariance function on $\mathbb{R}^d$ and $C_t(t_1, t_2)$ a temporal covariance function on T. It is obvious that for any given $a \in \mathbb{R}$ and $b \in T$, the functions $C_s(a\mathbf{s}_1, a\mathbf{s}_2)$ and $C_t(bt_1, bt_2)$ are valid stationary covariance functions on $\mathbb{R}^d$ and T, respectively. Therefore, the function generated by:

$$C((\mathbf{s}_1,t_1),(\mathbf{s}_2,t_2)) = \int_{\mathbb{R}\times T} C_s(a\mathbf{s}_1,a\mathbf{s}_2)C_t(bt_1,bt_2)d\mu(a,b), \tag{7.151}$$

is a spatio-temporal stationary covariance function on $\mathbb{R}^d \times T$, where $\mu(a,b)$ is a probability measure on $\mathbb{R}\times T$.

Let us see how to construct a spatio-temporal process that possesses this covariance function. Let $\{Z_s(\mathbf{s}),\, \mathbf{s}\in\mathbb{R}^d\}$ be a spatial process with covariance $C_s(\mathbf{s}_1,\mathbf{s}_2)$ and $\{Z_t(t),\, t{\in}T\}$ a temporal process with covariance $C_t(t_1,t_2)$. Let us assume that both processes are independent. Let (A,B) be a bivariate random vector with the distribution function $\mu(a,b)$, independent of $\{Z_s(\mathbf{s}),\, Z_t(t),\, \mathbf{s}\in\mathbb{R}^d,\, t{\in}T\}$. Then the spatio-temporal process $\{Z(\mathbf{s},t),\, \mathbf{s}\in\mathbb{R}^d,\, t{\in}T\}$ with $Z(\mathbf{s},t) = Z_s(\mathbf{s}A)Z_t(tB)$, has (7.151) as a covariance function. If the rv's A and B are independent, then the distribution function $\mu(a,b)$ is separable and the function (7.151) is reduced to a separable covariance function.

As in the case of stationarity, we will now discuss some of the examples provided by Ma to illustrate how the procedure works in the case of non-stationarity.

Example 7.25 (Ma 2002. Example 6) *Let (A,B) be a non-negative bivariate random vector with the distribution function $\mu(a,b)$ and let $\mathcal{L}$ be the Laplace transform of $\mu(a,b)$. Consider the stationary spatial covariance function:*

$$C_s(\mathbf{s}_1,\mathbf{s}_2) = \exp\left(-\theta\|\mathbf{s}_1-\mathbf{s}_2\|\right),$$

with $\mathbf{s}_1,\mathbf{s}_2\in\mathbb{R}^d$ and $\theta>0$, and the non-stationary temporal covariance function:

$$C_s(t_1,t_2) = \exp\left(-\gamma\left|t_1-t_2\right|\right) - \exp\left(-\gamma t_1-\gamma t_2\right), \tag{7.152}$$

with $t_1,t_2\in\mathbb{R}_+$ and $\gamma>0$. Then,

$$C\left(\left(\mathbf{s}_1,t_1\right),\left(\mathbf{s}_2,t_2\right)\right) = \int_{\mathbb{R}_+^2} \exp\left(-\theta a\|\mathbf{s}_1-\mathbf{s}_2\|\right)\left(\exp\left(-\gamma b\left|t_1-t_2\right|\right) - \exp\left(-\gamma bt_1-\gamma bt_2\right)\right)d\mu(a,b), \tag{7.153}$$

is a spatio-temporal covariance function stationary in space.

Furthermore, by appropriately selecting θ,γ and $\mathcal{L}$, we can obtain different parametric families of covariance functions, because if $\mathcal{L}$ is the Laplace transform of the distribution function $\mu(a,b)$, then the foregoing covariance function can be expressed as:

$$C\left(\left(\mathbf{s}_1,t_1\right),\left(\mathbf{s}_2,t_2\right)\right) = \mathcal{L}\left(\theta\|\mathbf{s}_1-\mathbf{s}_2\|,\gamma\left|t_1-t_2\right|\right) - \mathcal{L}\left(\theta\|\mathbf{s}_1-\mathbf{s}_2\|,\gamma\left(t_1+t_2\right)\right).$$

For example, if we select the bivariate Cheriyan-Ramabhardran gamma distribution (cf. Kotz et al. *2000, p. 435), which has the Laplace transform:*

$$\mathcal{L}\left(z_1, z_2\right) = \left(1+z_1\right)^{-\alpha_1}\left(1+z_2\right)^{-\alpha_2}\left(1+z_1+z_2\right)^{-\alpha_{12}},$$
$$\alpha_1 \geq 0,\ \alpha 2 \text{ and } \alpha_{12} \geq 0,$$

then the resulting covariance function is:

$$C\left((\mathbf{s}_1,t_1),(\mathbf{s}_2,t_2)\right) =$$
$$\left(1+\theta\|\mathbf{s}_1-\mathbf{s}_2\|\right)^{-\alpha_1}\left(\left(1+\gamma\left|t_1-t_2\right|\right)^{-\alpha_2}\left(1+\theta\|\mathbf{s}_1-\mathbf{s}_2\|+\gamma\left|t_1-t_2\right|\right)^{-\alpha_{12}}\right.$$
$$\left.-\left(1+\gamma\left|t_1+t_2\right|\right)^{-\alpha_2}\left(1+\theta\|\mathbf{s}_1-\mathbf{s}_2\|+\gamma\left|t_1+t_2\right|\right)^{-\alpha_{12}}\right)). \tag{7.154}$$

Example 7.26 (Ma 2002. Example 7) *Let* (A,B) *be a non-negative bivariate random vector with the distribution function* $\mu(a,b)$ *and let* $\mathcal{L}$ *be the Laplace transform of* $\mu(a,b)$. *Let the stationary spatial covariance function be:*

$$C_s\left(\mathbf{s}_1,\mathbf{s}_2\right) = \exp\left(-\theta\|\mathbf{s}_1-\mathbf{s}_2\|\right),$$

with $\mathbf{s}_1,\mathbf{s}_2 \in \mathbb{R}^d$ *and* $\theta > 0$. *Consider the following the non-stationary temporal covariance function:*

$$C_s\left(t_1,t_2\right) = \exp\left(-\theta_1 t_1 - \theta_2 t_2 - \theta_{12}\max\left\{t_1,t_2\right\}\right),$$

with $t_1, t_2 \in \mathbb{R}_+$ *and* $\theta_1, \theta_2, \theta_{12} > 0$. *Then,*

$$C_{st}\left(\left(\mathbf{s}_1,t_1\right),\left(\mathbf{s}_2,t_2\right)\right) =$$
$$\int_{\mathbb{R}_+^2} \exp\left(-\theta\|\mathbf{s}_1-\mathbf{s}_2\|a\right)\exp\left(-\left(\theta_1 t_1+\theta_2 t_2+\theta_{12}\max\left\{t_1,t_2\right\}b\right)\right)d\mu(a,b), \tag{7.155}$$

is a spatio-temporal covariance function stationary in space. Let $\mathcal{L}$ *be the Laplace transform of the distribution function* $\mu(a,b)$. *Then the foregoing integral can be expressed as:*

$$C_{st}\left(\left(\mathbf{s}_1,t_1\right),\left(\mathbf{s}_2,t_2\right)\right) = \mathcal{L}(\theta\|\mathbf{s}_1-\mathbf{s}_2\|, \theta_1 t_1+\theta_2 t_2+\theta_{12}\max\left\{t_1,t_2\right\}), \tag{7.156}$$

which allows the generation of different parametric families of covariance functions by selecting the appropriate Laplace transform and values of the parameters θ_i.

7.19.2.2 Construction of non-stationary spatio-temporal covariance functions using positive power mixtures

In some cases, but not all, it is possible to extend the use of positive power mixtures, described in Section 7.12.2 to a general case. In this sense, applying the models (7.85) and (7.88) to the case of non-stationarity is possible only if the non-stationary spatial and temporal covariance functions considered are bounded (that is, if $|C_s(\mathbf{s}_1, \mathbf{s}_2)| < C_s$ and $|C_t(t_1, t_2)| < C_t$). In this case, the functions:

$$\rho((\mathbf{s}_1, t_1), (\mathbf{s}_2, t_2)) = \sum_{i=0}^{+\infty} \sum_{j=0}^{+\infty} \left(\frac{C_s(\mathbf{s}_1, \mathbf{s}_2)}{C_s} \right)^i \left(\frac{C_t(t_1, t_2)}{C_t} \right)^j p_{ij}, \tag{7.157}$$

$$\rho((\mathbf{s}_1, t_1), (\mathbf{s}_2, t_2)) = \sum_{k=0}^{+\infty} \left(\frac{C_s(\mathbf{s}_1, \mathbf{s}_2) C_t(t_1, t_2)}{C_s C_t} \right)^k p_k, \tag{7.158}$$

will be spatio-temporal correlation functions on $\mathbb{R}^d \times T$, with $\{p_{ij}(i,j) \in \mathbb{N}\}$ and $\{p_k(k) \in \mathbb{N}\}$ representing probability distributions. If the covariance functions C_s and C_t are not bounded, then (7.157) and (7.158) cannot be applied, as we know that:

$$g(\exp(t)) = E\left((\exp(t))^A\right)$$

is absolutely convergent in $(-1, 1)$, and

$$g(\exp(t+s)) = E\left((\exp(t))^A (\exp(s))^B\right)$$

in $(-1, 1)^2$. Of course, it will always be possible to define any given finite positive power mixture of non-stationary spatial and temporal covariance functions.

7.19.2.3 Construction of non-stationary spatio-temporal covariance functions using spatio-temporal stationary covariances and intrinsically stationary semivariograms

One interesting proposal made by Ma (2003c) was to construct non-stationary spatio-temporal covariance functions using stationary spatio-temporal covariance functions and intrinsically stationary semivariograms as a basis.

Generalizing the semivariogram analysis performed in Section 3.4, we find that if $\gamma(\mathbf{s}, t)$ is an intrinsically stationary semivariogram on $\mathbb{R}^d \times T$, then the function:[13]

$$\omega((\mathbf{s}_1, t_1), (\mathbf{s}_2, t_2)) = \gamma(\mathbf{s}_1, t_1) + \gamma(\mathbf{s}_2, t_2) - \gamma(\mathbf{s}_1 - \mathbf{s}_2, t_1 - t_2), \tag{7.159}$$

is a covariance function on $\mathbb{R}^d \times T$. This is the main argument used by Ma (2003c) to generate non-stationary spatio-temporal covariance functions. Other arguments employed by Ma include the functions:

$$\phi((\mathbf{s}_1, t_1), (\mathbf{s}_2, t_2)) = 2\left(\gamma(\mathbf{s}_1, t_1) + \gamma(\mathbf{s}_2, t_2)\right) - \gamma(\mathbf{s}_1 + \mathbf{s}_2, t_1 + t_2) - \gamma(\mathbf{s}_1 - \mathbf{s}_2, t_1 - t_2), \tag{7.160}$$

$$\kappa((\mathbf{s}_1, t_1), (\mathbf{s}_2, t_2)) = \gamma(\mathbf{s}_1 + \mathbf{s}_2, t_1 + t_2) - \gamma(\mathbf{s}_1 - \mathbf{s}_2, t_1 - t_2), \tag{7.161}$$

[13] See proof in Appendix D.

which verify the following relationship:

$$\omega((\mathbf{s}_1,t_1),(\mathbf{s}_2,t_2)) = \frac{1}{2}\phi((\mathbf{s}_1,t_1),(\mathbf{s}_2,t_2)) + \frac{1}{2}\kappa((\mathbf{s}_1,t_1),(\mathbf{s}_2,t_2)). \tag{7.162}$$

In particular, if $\phi((\mathbf{s}_1,t_1),(\mathbf{s}_2,t_2)) = 0$, then $\gamma(\mathbf{s},t)$ is a quadratic form and

$$\omega((\mathbf{s}_1,t_1),(\mathbf{s}_2,t_2) = \frac{1}{2}\kappa((\mathbf{s}_1,t_1),(\mathbf{s}_2,t_2))$$

is a covariance function on $\mathbb{R}^d \times T$.

Ma (2003c) demonstrates that if $\gamma(\mathbf{s},t)$ is an intrinsically stationary semivariogram, then the function $\phi((\mathbf{s}_1,t_1),(\mathbf{s}_2,t_2))$ defined by (7.160) is a non-negative covariance function on $\mathbb{R}^d \times T$, with the associated semivariogram $\frac{1}{2}\phi\left((\mathbf{s}_1+\mathbf{s}_2,t_1+t_2),(\mathbf{s}_1-\mathbf{s}_2,t_1-t_2)\right)$, which is non-negative. Therefore, $\phi((\mathbf{s}_1,t_1),(\mathbf{s}_2,t_2))$ is also non-negative as it can be rewritten as

$$\phi((\mathbf{s}_1,t_1),(\mathbf{s}_2,t_2)) = 2\gamma\left(\left(\frac{\mathbf{s}_1+\mathbf{s}_2}{2},\frac{t_1+t_2}{2}\right),\left(\frac{\mathbf{s}_1-\mathbf{s}_2}{2},\frac{t_1-t_2}{2}\right)\right).$$

If the semivariogram is stationary, we find that $\gamma(\mathbf{h},u) = C(\mathbf{0},0) - C(\mathbf{h},u)$. As a result, given a stationary spatio-temporal covariance function $C(\mathbf{h},u)$ on $\mathbb{R}^d \times T$, then the function:[14]

$$C(\mathbf{s}_1+\mathbf{s}_2,t_1+t_2) + C(\mathbf{s}_1-\mathbf{s}_2,t_1-t_2) - 2\left(C(\mathbf{s}_1,t_1) + C(\mathbf{s}_2,t_2) - C(\mathbf{0},0)\right) \tag{7.163}$$

is a non-negative and non-stationary spatio-temporal covariance function on $\mathbb{R}^d \times T$. It is also shown[15] that if $\gamma(\mathbf{s},t)$ is an intrinsically stationary semivariogram, then the function $\kappa((\mathbf{s}_1,t_1),(\mathbf{s}_2,t_2))$ defined by (7.161) is a covariance function on $\mathbb{R}^d \times T$. Consequently, given a stationary spatio-temporal covariance function $C(\mathbf{s},t)$ on $\mathbb{R}^d \times T$, the function:

$$C(\mathbf{s}_1-\mathbf{s}_2,t_1-t_2) - C(\mathbf{s}_1+\mathbf{s}_2,t_1+t_2) \tag{7.164}$$

is a non-stationary spatio-temporal covariance function on $\mathbb{R}^d \times T$. As can be appreciated, on the basis of these useful results, we can construct non-stationary covariance models using purely spatial and purely temporal intrinsically stationary semivariograms.

7.19.3 Non-stationary models proposed by Porcu and Mateu

Porcu and Mateu (2007) set out a procedure that allows for the construction of non-stationary models using completely monotone functions.

Let $\gamma(\mathbf{s},t)$ be an intrinsically stationary semivariogram with $\gamma(\mathbf{0},0) = 0$, and let $\varphi(t), t > 0$, be a completely monotone function with $\varphi(0) = 1$. Then, the function:

$$C((\mathbf{s}_1,t_1),(\mathbf{s}_2,t_2)) = \varphi\left(\psi\left(\mathbf{s}_1,\mathbf{s}_2,t_1,t_2\right)\right), \tag{7.165}$$

[14] See proof in Appendix D.

[15] See proof in Appendix D.

with

$$\psi\left(\mathbf{s}_1,\mathbf{s}_2,t_1,t_2\right) = \frac{1}{2}\left(\gamma(2\mathbf{s}_1,2t_1)+\gamma(2\mathbf{s}_2,2t_2)\right) - \left(\gamma(\mathbf{s}_1+\mathbf{s}_2,t_1+t_2)-\gamma(\mathbf{s}_1-\mathbf{s}_2,t_1-t_2)\right), \tag{7.166}$$

or

$$\psi\left(\mathbf{s}_1,\mathbf{s}_2,t_1,t_2\right) = 1+\frac{1}{2}\left(\gamma(2\mathbf{s}_1,2t_1)+\gamma(2\mathbf{s}_2,2t_2)\right) - \left(\frac{1+\gamma(\mathbf{s}_1+\mathbf{s}_2,t_1+t_2)}{1+\gamma(\mathbf{s}_1-\mathbf{s}_2,t_1-t_2)}\right), \tag{7.167}$$

is a valid non-stationary covariance function on $\mathbb{R}^d \times \mathbb{R}$.

Moreover, it is easy to verify that the functions:

$$C((\mathbf{s}_1,t_1),(\mathbf{s}_2,t_2)) = \varphi\left(\frac{\psi\left(\mathbf{s}_1,\mathbf{s}_2,t_1,t_2\right)}{1+\psi\left(\mathbf{s}_1,\mathbf{s}_2,t_1,t_2\right)}\right), \tag{7.168}$$

$$C((\mathbf{s}_1,t_1),(\mathbf{s}_2,t_2)) = \varphi\left(1-\frac{1-\psi\left(\mathbf{s}_1,\mathbf{s}_2,t_1,t_2\right)}{1+\psi\left(\mathbf{s}_1,\mathbf{s}_2,t_1,t_2\right)}\right), \tag{7.169}$$

$$C((\mathbf{s}_1,t_1),(\mathbf{s}_2,t_2)) = \varphi\left(\left(1+\psi\left(\mathbf{s}_1,\mathbf{s}_2,t_1,t_2\right)\right)^{\rho}-1\right), \tag{7.170}$$

ψ being the function given by (7.166) or (7.167) and $\rho \in [0,1]$, also generate valid non-stationary covariance functions on $\mathbb{R}^d \times \mathbb{R}$.

7.20 Anisotropic covariance functions by Porcu and Mateu

7.20.1 Constructing temporally symmetric and spatially anisotropic covariance functions

The assumption of isotropy is often unrealistic when modeling real phenomena. For this reason, it is therefore important to define models of covariance that are capable of capturing anisotropy in the spatial context.

There are two types of anisotropy (see Section 3.7). The first and simplest is called geometrical anisotropy whereby the variogram has the same sill but different ranges in different directions. The second is known as zonal anisotropy, and in this case the variogram has the same range in different directions, but different sills.

When modeling anisotropy using isotropic components it is assumed that there is not necessarily isotropy in the complete space, but separately in subspaces of complete dimension.

For example, on $\mathbb{R}^3$, if $\mathbf{h}=(h_1,h_2,h_3)$, then:

$$C_1(\mathbf{h}) := \tilde{C}_1\left(\|(\mathbf{h}_1,\mathbf{h}_2)\|,|\mathbf{h}_3|\right) \quad \text{or} \quad C_1(\mathbf{h}) := \tilde{C}_2\left(|\mathbf{h}_1|,|\mathbf{h}_2|,|\mathbf{h}_3|\right)$$

are two examples of covariance functions on $\mathbb{R}^3$ that are not isotropic but rather anisotropic functions modeled by isotropic components.

Porcu *et al.* (2006) model this type of covariance function using the following approach. The strategy they use is to create partitions in the vector of spatial lags $\mathbf{h} \in \mathbb{R}^d$ as follows: if $\mathbf{d} = (d_1, d_2, \ldots, d_n)$ and $\mathbf{h}, \mathbf{k} \in \mathbb{R}^d$ can always be written $\mathbf{h} = (\mathbf{h}_1, \mathbf{h}_2, \ldots, \mathbf{h}_n) \in \mathbb{R}^{d_1} \times \mathbb{R}^{d_2} \ldots \times \mathbb{R}^{d_n}$ (and the same can be said for $\mathbf{k}$), hence:

(i) $C(\mathbf{h}) = C(\mathbf{k})$ for any given $\mathbf{h}, \mathbf{k} \in \mathbb{R}^d$ if $\|\mathbf{h}_i\| = \|\mathbf{k}_i\|, \forall i = 1, 2, \ldots, n$.

(ii) The resulting covariance can be written as:

$$C(\mathbf{h}) = C\left(\|\mathbf{h}_1\|, \ldots, \|\mathbf{h}_n\|\right) \tag{7.171}$$

and it has been verified in Porcu *et al.* (2007b) that (7.171) admits the integral representation as a scale mixture of Bessel functions.

For the spatio-temporal domain, $\mathbf{d}$ always has cardinality $|\mathbf{d}| = d + 1$. Then, though there might be anisotropy in space, all the processes are considered symmetric in time.

7.20.2 Generalizing the class of spatio-temporal covariance functions proposed by Gneiting

As we noted in (7.11), Gneiting (2002) provides a flexible family of spatio-temporal covariance functions that are isotropic and temporally symmetric, that is, fully symmetric. In particular, the construction

$$C(\mathbf{h}, u) = \frac{\sigma^2}{\psi\left(|u|^2\right)^{d/2}} \varphi\left(\frac{\|\mathbf{h}\|^2}{\psi\left(|u|^2\right)}\right), \quad \mathbf{h} \in \mathbb{R}^d, u \in \mathbb{R}, \tag{7.172}$$

where φ is a completely monotone function and ψ a Bernstein function, is a spatio-temporal covariance function for any given $d = 1, 2, \ldots$.

Porcu *et al.* (2006), adapt Theorem 1 from Gneiting (2002) obtaining the following lemma as a result.

Lemma 7.20.1 (Porcu *et al.* 2006) *For $i = 1, 2, 3, 4$, let $d_i \in \mathbb{N} = \{1, 2, \ldots\}$. A continuous, bounded, symmetric and integrable function, $C : \mathbb{R}^{d_1} \times \mathbb{R}^{d_2} \times \mathbb{R}^{d_3} \times \mathbb{R}^{d_4} \to \mathbb{R}$, is a covariance function if, and only if, the function:*

$$C_{\omega_3\omega_4}(\mathbf{h}_1, \mathbf{h}_2) = \int\int e^{-i(\mathbf{h}_3'\omega_3 + \mathbf{h}_4'\omega_4)} C(\mathbf{h}_1, \mathbf{h}_2, \mathbf{h}_3, \mathbf{h}_4) d\mathbf{h}_3 d\mathbf{h}_4, \tag{7.173}$$

defined on $\mathbb{R}^{d_1} \times \mathbb{R}^{d_2}$ is a covariance function for almost all $\left(\boldsymbol{\omega}_3, \boldsymbol{\omega}_4\right) \in \mathbb{R}^{d_3} \times \mathbb{R}^{d_4}$.

Theorem 7.20.1 is the primary result of Porcu *et al.* (2006), which generalizes Theorem 2 in Gneiting (2002). The authors formulate this new result, first in general terms, before later using the subsequent results to build covariance functions with the spatial component being anisotropic and the temporal component being symmetric.

Theorem 7.20.1 *Let ψ_1 and ψ_2 be either (i) positive Bernstein functions or (ii) intrinsically stationary semivariograms non-null at the origin ($\psi \equiv \gamma$). Let $\mathcal{L}$ be the bivariate Laplace transform of a non-negative random vector (X_1, X_2) with distribution function F, that is, $\mathcal{L}(\theta_1, \theta_2) = \int_0^\infty \int_0^\infty e^{-r_1\theta_1 - r_2\theta_2} dF(r_1, r_2)$. Then:*

$$C(\mathbf{h}_1, \mathbf{h}_2, \mathbf{h}_3, \mathbf{h}_4) = \frac{\sigma^2}{\psi_1\left(\|\mathbf{h}_1\|^2\right)^{\frac{d_3}{2}} \psi_2\left(\|\mathbf{h}_2\|^2\right)^{\frac{d_4}{2}}} \mathcal{L}\left(\frac{\|\mathbf{h}_3\|^2}{\psi_1\left(\|\mathbf{h}_1\|^2\right)}, \frac{\|\mathbf{h}_4\|^2}{\psi_2\left(\|\mathbf{h}_2\|^2\right)}\right) \tag{7.174}$$

is a covariance function on $\mathbb{R}^{d_1} \times \mathbb{R}^{d_2} \times \mathbb{R}^{d_3} \times \mathbb{R}^{d_4}$.[16]

The following corollaries of the above theorem are of interest:

Corollary 1 ($d_1 = d_2 = 1$) *If $d_1 = d_2 = 1$, it is sufficient for ψ_1 and ψ_2 to be increasing and concave functions in $[0, \infty)$ to obtain that (7.174) is a stationary and non-separable covariance function*[17] *on $\mathbb{R}^{d_1} \times \mathbb{R}^{d_2} \times \mathbb{R}^{d_3} \times \mathbb{R}^{d_4}$.*

Observing the resulting from (7.174), we can see how simple it is to obtain a wide variety of closed-forms (see the corollaries below). The key ideas are: (i) the use of functions that can be represented as integrals of exponential functions (multivariate Laplace transforms); (ii) the separability of the vector of spatial lags in a given number of areas of isotropy, such that the Fourier transforms always involve separable Gaussian functions (and, therefore, we can always go back to multivariate Laplace transforms); and (iii) the use of criteria from Feller (1966), Berg and Forst (1975) and Schoenberg (1938).

Corollary 2 (Gneiting 2002) *Let φ be a completely monotone function and ψ a Bernstein function (or, equally, a semivariogram). Then the function:*

$$C(\mathbf{h}) = \frac{\sigma^2}{\psi\left(\|\mathbf{h}_2\|^2\right)^{\frac{d_1}{2}}} \varphi\left(\frac{\|\mathbf{h}_1\|^2}{\psi\left(\|\mathbf{h}_2\|^2\right)}\right) \tag{7.175}$$

is a non-separable stationary covariance function, where $\mathbf{h} \in \mathbb{R}^d = \mathbb{R}^{d_1} \times \mathbb{R}^{d_2}$ and $\mathbf{h} = (\mathbf{h}_1, \mathbf{h}_2)$ with $\mathbf{h}_i \in \mathbb{R}^{d_i}$ for $i = 1, 2$.

The proof is the trivial consequence of considering the marginal covariances of (7.174).

Corollary 3 (Another general form) *Let $d \in \mathbb{N}, d \geq 4$. An anisotropic covariance function can be constructed as follows:*

[16] See the proof of the theorem in Porcu *et al.* (2006)

[17] A proof of the Corollary 7.20.1 can also be found in Porcu *et al.* (2006).

1. *Choose a number of zones of isotropy, n (that is,* $\mathbb{R}^d = \mathbb{R}^{d_1} \times \ldots \times \mathbb{R}^{d_n}$*), such that each* $\mathbf{h} \in \mathbb{R}^d$ *can uniquely be expressed as* $\mathbf{h} = (\mathbf{h}_1, \ldots, \mathbf{h}_n)$ *with* $\mathbf{h}_i \in \mathbb{R}^{d_i}$*, for* $i = 1, \ldots, n$.
2. *Choose any* $k \in \mathbb{N}$ *such that* $n \geq 2k$.
3. *We choose a decomposition of the set* $\{1, 2, \ldots, n-k\} = \cup_{j=1}^{k} B_j$*, a disjoint union of non-empty sets.*
   Let $(\psi)_{i=1}^{n-k}$ *be a set of Bernstein functions (or semivariograms). Then, the function obtained through the* k*-variate Laplace transform:*

$$C(\mathbf{h}) = \frac{\sigma^2}{\prod_{j=1}^{k} \left(\sum_{i \in B_j} \psi_i \left(\|\mathbf{h}_i\|^2 \right) \right)^{\frac{d_{n-k+j}}{2}}} \times \mathcal{L} \left(\frac{\|\mathbf{h}_{n-k+1}\|^2}{\sum_{i \in B_j} \psi_i \left(\|\mathbf{h}_i\|^2 \right)}, \frac{\|\mathbf{h}_{n-k+2}\|^2}{\sum_{i \in B_j} \psi_i \left(\|\mathbf{h}_i\|^2 \right)}, \ldots, \frac{\|\mathbf{h}_k\|^2}{\sum_{i \in B_j} \psi_i \left(\|\mathbf{h}_i\|^2 \right)} \right) \tag{7.176}$$

is a covariance function on $\mathbb{R}^d$ *that is anisotropic on* $\mathbb{R}^{d_1} \times \ldots \times \mathbb{R}^{d_n}$.

Focusing on the spatio-temporal case, Corollary 4 shows how to build spatio-temporal covariance functions with the spatial component being anisotropic and the temporal one being symmetric.

Corollary 4 (Spatially component-wise anisotropic and temporally symmetric covariance functions) *Let* ψ_1, ψ_2 *be Bernstein functions, semivariograms or increasing and concave functions in the interval* $[0, \infty)$*. Then:*

$$C(h_1, h_2, h_3, u) = \frac{\sigma^2}{\psi_1 \left(|h_1|^2 \right)^{\frac{1}{2}} \psi_2 \left(|u|^2 \right)^{\frac{1}{2}}} \mathcal{L} \left(\frac{|h_2|^2}{\psi_1 \left(|h_1|^2 \right)}, \frac{|h_3|^2}{\psi_2 \left(|u|^2 \right)} \right), \tag{7.177}$$

where $h_i, u \in \mathbb{R}, i = 1, 2, 3$*, is a non-separable stationary spatio-temporal covariance function with spatially anisotropic components, defined on* $\mathbb{R}^3 \times \mathbb{R}$.

Notice that the arguments in (7.177) can be exchanged without affecting the validity of the resulting covariance function. Another interesting aspect is that this structure of covariance accepts four three-dimensional, six two-dimensional, and four one-dimensional marginal functions, which means we can build a wide range of spatial and spatio-temporal covariance functions using (7.177) as a basis. Furthermore, it is worth noting that two marginal bivariate functions are the type used in Gneiting (2002), as highlighted in the Corollary 1.

Finally, it is worth underlining that (7.177) has been presented in the product space $\mathbb{R}^3 \times \mathbb{R}$, but the extension to $\mathbb{R}^d \times \mathbb{R}, d > 3$, is direct, as it is sufficient to choose a suitable partition of the vector of spatial lags $\mathbf{h} \in \mathbb{R}^d$ to obtain the desired result.

Example 7.27 (Porcu *et al.* 2006. Example 2) *In order to illustrate the result (7.177), consider the bivariate Laplace transform of the Frechet lower bound for bivariate copulas:*[18]

$$\mathcal{L}(\theta_1, \theta_2) = \frac{e^{-\theta_1} - e^{-\theta_2}}{\theta_2 - \theta_1} \tag{7.178}$$

Therefore, using (7.177) as a basis, and setting $\theta_1 = \frac{|h_2|^2}{\psi_1(|h_1|^2)}$ *and* $\theta_2 = \frac{|h_1|^2}{\psi_2(|u|^2)}$, *we obtain:*

$$C(h_1, h_2, h_3, u) = \begin{cases} \dfrac{\sigma^2 \psi_1(|h_1|^2)^{1/2} \psi_2(|u|^2)^{1/2}}{|h_2|^3 \psi_1(|h_1|^2) - |h_2|^2 \psi_2(|u|^2)} \left(e^{-\frac{|h_2|^2}{\psi_1(|h_1|^2)}} - e^{-\frac{|h_3|^2}{\psi_2(|u|^2)}} \right) \\ \text{if } |h_3|^2 \psi_1\left(|h_1|^2\right) \neq |h_2|^2 \psi_2\left(|u|^2\right) \\ \dfrac{\sigma^2}{\psi_1(|h_1|^2)^{1/2} \psi_2\left(|u|^2\right)^{1/2}} e^{-\frac{|h_2|^2}{\psi_1\left(|h_1|^2\right)}} \\ \text{if } |h_3|^2 \psi_1\left(|h_1|^2\right) = |h_2|^2 \psi_2\left(|u|^2\right) \end{cases},$$

which is a non-separable component-wise anisotropic stationary spatio-temporal covariance function defined on $\mathbb{R}^3 \times \mathbb{R}$.

7.20.3 Differentiation and integration operators acting on classes of anisotropic covariance functions on the basis of isotropic components: 'La descente étendue'

Porcu *et al.* (2007b) provide the generalization of a procedure used in the setting of isotropic random process to the anisotropic scenario through isotropy within components.

In particular, the operator called *La descente* (related to the differentiation operator) acting in a special way on a covariance function and defined on a d-dimensional space, yields to new covariance functions defined on spaces with upper dimensions.

As pointed out in Porcu *et al.* (2007b), using differential operators yields some interesting mathematical characteristics of the component-wise isotropic covariance functions. This aspect is not new in the literature when considering permissible covariance functions that are isotropic and which represent the characteristic function of a spherical random vector that is symmetrically distributed. In colloquial terms, the main problem is detecting whether it is possible to preserve the condition of positive-definiteness, even not necessarily in the same Euclidean d-dimensional space, by applying integral or derivative operators to positive-definite radial functions

[18] See example in 7.24.

with support on $\mathbb{R}^d$. For this reason, in his work entitled *Les variables regionalisées et leur estimation: une application de la théorie des fonctions alétatoires aux sciences de la nature*, Matheron coined the term *Descente* to denote the differentiation of a positive-definite radial function.

The use of linear operators for the spatially anisotropic and spatio-temporal domain are extensively discussed in Porcu *et al.* (2007b). The results obtained therein are presented in order to illustrate that partial derivative operators, applied to the bivariate Laplace transform, and in particular to the Bernstein class, yield new functions that preserve permissibility in the spatio-temporal domain. This result is extremely interesting insofar as it implies a greater degree of flexibility in the proposed class. The results are obtained by means of a mixture of the spatial and temporal semivariograms γ_s and γ_t, respectively, with a bivariate Laplace transform, that is, $C(\mathbf{h}, u) = \mathcal{L}\left(\gamma_s(\mathbf{h}), \gamma_t(u)\right)$ (see Ma 2002), and the Bernstein class in Section 7.18.

On the one hand, if the measure W of the Laplace transform $\mathcal{L}(\cdot, \cdot)$ fulfills:

$$\int_0^\infty \int_0^\infty u_1^{k_1} u_2^{k_2} dW(u_1, u_2) < \infty$$

for a pair of integers $k_1, k_2 > 0$, and the spatial and temporal semivariograms vanish at the origin, then the function:

$$\rho(\mathbf{h}, u) := \frac{\frac{\partial^{k_1+k_2}\mathcal{L}(\gamma_s(\mathbf{h}),\gamma_t(u))}{\partial\theta_1^{k_1}\partial\theta_2^{k_2}}}{\frac{\partial^{k_1+k_2}\mathcal{L}(\mathbf{0},0)}{\partial\theta_1^{k_1}\partial\theta_2^{k_2}}}, \qquad \mathbf{h} \in \mathbb{R}^d, u \in \mathbb{R}, \tag{7.179}$$

results in the function φ_ρ (represented by $\varphi_\rho(s_1, s_2) = \rho(\mathbf{h}, u)$, given $\mathbf{h}$ and u, with $\|\mathbf{h}\| = s_1$ and $|u| = s_2$) remaining positive-definite for any given isotropic semivariograms.

On the other hand, and using the notation:

$$\mathcal{L}^*(s_1, \ldots, s_d, \tau) = \mathcal{L}\left(\sum_{i=1}^d \psi_i(s_i), \psi_t(\tau)\right), \qquad s_1, \ldots, s_d, \tau \in \mathbb{R} \tag{7.180}$$

and

$$D^{d+1}\mathcal{L}^*(s_1, \ldots, s_d, \tau) = \frac{\partial^{d+1}\mathcal{L}^*(s_1, \ldots, s_d, \tau)}{\partial s_1 \ldots \partial s_d \partial\tau}, \tag{7.181}$$

where:

- $\psi_1, \ldots, \psi_d, \psi_t$ are Bernstein functions such that $\psi_1'(0), \ldots, \psi_d'(0), \psi_t'(0)$ exist and are finite,
- considering $\mathcal{L}$ as in (7.145), its measure W fulfilling the condition:

$$\int_0^\infty \int_0^\infty u_1^d u_2 dW(u_1, u_2) < \infty, \tag{7.182}$$

- and, assuming that $\gamma_1, \ldots, \gamma_d, \gamma_t$ are univariate and intrinsically stationary semi-variograms defined on the real line, with $\gamma_1(0) = \ldots = \gamma_d(0) = \gamma_t(0) = 0$, the function:

$$\rho^*(s_1, \ldots, s_d, t) = \frac{D^{d+1}\mathcal{L}^*\left(\gamma_1(s_1), \ldots, \gamma_d(s_d), \gamma_T(t)\right)}{D^{d+1}\mathcal{L}^*(0, \ldots, 0, 0)} \tag{7.183}$$

is a non-separable stationary spatio-temporal correlation function that is valid on $\mathbb{R}^d \times \mathbb{R}$.

7.21 Spatio-temporal constructions based on quasi-arithmetic means of covariance functions

The theory of quasi-arithmetic means is a powerful tool for studying covariance functions in the spatio-temporal domain. Porcu *et al.* (2009b) use quasi-arithmetic functionals to make inferences about the permissibility of functions that, in general, are not permissible covariance functions. This is the case, for example, with two quite popular means, namely the geometric mean and the harmonic mean, for which the authors provide permissibility criteria. In addition, they show that the generator of the quasi-arithmetic means can be used as a link function to build non-separable spatio-temporal structures from the marginal spatial and temporal covariance functions. And finally, they use quasi-arithmetic functions to generalize existing results regarding the construction of non-stationary spatial covariance functions and discuss the application and limits of this process.

Definition 7.21.1 *The quasi-arithmetic mean M_n of the real variables $x_1, \ldots, x_n$ defined on the same real interval is any function of such variables satisfying:*

$$M_n\left(x_1, \ldots, x_n\right) = f^{-1}\left(\frac{1}{n}\sum_{i=1}^{n} f(x_i)\right), \tag{7.184}$$

where f is a continuous strictly monotonic real function.

Next we present some well-known examples of quasi-arithmetic mean. Table 7.4 lists an extended collection of examples.

Example 7.28 (The arithmetic mean)

$$f(x_i) = ax_i + b$$

$$\frac{1}{n}\sum_{i=1}^{n} f(x_i) = \frac{a(x_1 + \ldots + x_n)}{n} + b$$

$$\begin{aligned} M_n(x_1, \dots, x_n) &= f^{-1}\left(\frac{1}{n}\sum_{i=1}^{n} f(x_i)\right) \\ &= f^{-1}\left(\frac{a\sum_{i=1}^{n} x_i}{n} + b\right) = \frac{\sum_{i=1}^{n} x_i}{n}, \end{aligned}$$

which is defined for all real numbers.

Example 7.29 (The geometric mean)

$$\begin{aligned} f(x_i) &= \log x_i + b \\ \frac{1}{n}\sum_{i=1}^{n} f(x_i) &= \frac{(\log x_1 + \dots + \log x_n)}{n} + b \end{aligned}$$

$$\begin{aligned} M_n(x_1, \dots, x_n) &= f^{-1}\left(\frac{1}{n}\sum_{i=1}^{n} f(x_i)\right) \\ &= f^{-1}\left(\frac{\sum_{i=1}^{n} \log x_i}{n} + b\right) = \exp\left(\frac{\sum_{i=1}^{n} \log x_i}{n} + b - b\right) \\ &= \exp\left(\sum_{i=1}^{n} \log x_i^{\frac{1}{n}}\right) = \exp\left(\log\left(\prod_{i=1}^{n} x_i^{\frac{1}{n}}\right)\right) = \left(\prod_{i=1}^{n} x_i\right)^{\frac{1}{n}}, \end{aligned}$$

and is defined for strictly positive real numbers.

Table 7.4 Examples of quasi-arithmetic means.

Generator ($f(x)$)	$M_n\left(x_1, \dots, x_n\right)$	Name
$ax + b$	$\frac{1}{n}\sum_{i=1}^{n} ax_i + b$	Arithmetic
$ax^2 + b$	$\sqrt{\frac{1}{n}\sum_{i=1}^{n} ax_i^2 + b}$	Quadratic
$\log x + b$	$\sqrt[n]{\prod_{i=1}^{n} \log x_i + b}$	Geometric
$\frac{a}{x} + b$	$\frac{1}{\frac{1}{n}\sum_{i=1}^{n} \frac{1}{x_i}}$	Harmonic
$ax^{\alpha} + b \quad (\alpha \in \mathbb{R}_0)$	$\left(\frac{1}{n}ax^{\alpha} + b\right)^{\frac{1}{\alpha}}$	Power
$\exp(\alpha x) \quad (\alpha \in \mathbb{R}_0)$	$\frac{1}{\alpha}\log\left(\frac{1}{n}\sum_{i=1}^{n} \exp(\alpha x_i)\right)$	Exponential

Example 7.30 (The harmonic mean)

$$f(x_i) = \frac{a}{x_i} + b$$

$$\frac{1}{n}\sum_{i=1}^{n} f(x_i) = \frac{\frac{a}{x_1} + \ldots + \frac{1}{x_n}}{n} + b$$

$$M_n(x_1, \ldots, x_n) = f^{-1}\left(\frac{1}{n}\sum_{i=1}^{n} f(x_i)\right)$$

$$f^{-1}\left(\frac{a\sum_{i=1}^{n}\frac{1}{x_i}}{n} + b\right) = \frac{n}{\sum_{i=1}^{n}\frac{1}{x_i}}.$$

Obviously, all x_i must be non-null real numbers.

Porcu *et al.* (2009b) aim to answer, among others, the two following questions:

1. Let us consider an arbitrary number $n \in \mathbb{N}$ of covariance functions that are not necessarily defined in the same space. Their arithmetic mean and their product (that is, the geometric mean raised to the nth power) are valid covariance functions and, therefore, the kth power average of covariance functions (k being a positive natural number) also is. But, what can be said of other types of means? Insofar as quasi-arithmetic means constitute a general corpus that includes the arithmetic mean, the geometric mean, the power mean and the logarithmic mean as special cases, it seems natural to use quasi-arithmetic representations to obtain the conditions of permissibility for quasi-arithmetic averages of covariance functions.

2. Is it possible to find a class of "link functions" that generate valid non-separable covariance functions when applied to k covariance functions?

7.21.1 Multivariate quasi-arithmetic compositions

Having introducing the question, we use Porcu *et al.*'s (2009b) notation and define the quasi-arithmetic class of functionals.

Let Φ be the class of real-valued functions φ defined in a given domain $D(\varphi) \subset \mathbb{R}$, which admits inverse φ^{-1} defined in $D\left(\varphi^{-1}\right) \subset \mathbb{R}$, such that $\varphi\left(\varphi^{-1}(t)\right) = t$ for all $t \in D\left(\varphi^{-1}\right)$. In addition, let Φ_c and Φ_{cm} be the sub-classes of Φ, which are obtained by restricting φ to be convex or completely monotone, respectively, on the positive real line.

Porcu *et al.* (2009b) refer to the class below as the *quasi-arithmetic class of functionals*:

$$\mathfrak{Q} : = \left\{ \psi : D\left(\varphi^{-1}\right) \times \ldots \times D\left(\varphi^{-1}\right) \rightarrow \mathbb{R} : \psi(\mathbf{u}) \right.$$
$$\left. = \varphi\left(\sum_{i=1}^{n} \theta_i \varphi^{-1}(u_i) \right), \varphi \in \Phi \right\}, \tag{7.185}$$

where θ_i are non-negative weights and $\mathbf{u} = (u_1, \ldots, u_n)'$, for $n \geq 2$. Furthermore, the sub-classes of $\mathfrak{Q}$ that arise after forcing φ to belong, respectively, to Φ_c and Φ_c are called $\mathfrak{Q}_c$ and $\mathfrak{Q}_{cm}$.

If $\psi \in \mathfrak{Q}$, then φ_ψ can be written as a function that for any given non-negative vector $\mathbf{u}$, $\psi(\mathbf{u}) = \varphi_\psi\left(\sum_{i=1}^{n} \theta_i \varphi^{-1}(u_i)\right)$. In order to simplify the notation, we will write φ instead of φ_ψ providing this does not lead to confusion.

Definition 7.21.2 Quasi-arithmetic compositions *If $f_i : \mathbb{R}^{d_i} \rightarrow \mathbb{R}_+$, such that $\cup_i^n f_i(\mathbb{R}^{d_i}) \subset D\left(\varphi^{-1}\right)$ for a given $\varphi \in \Phi$, the composition of $f_1, f_2, \ldots, f_n$ with a generator function $\psi \in \mathfrak{Q}$ is defined as the functional:*

$$\mathcal{Q}_\psi\left(f_1, f_2, \ldots, f_n\right)(\mathbf{x}) = \psi\left(f(\mathbf{x}_1), f(\mathbf{x}_2), \ldots, f(\mathbf{x}_n)\right), \tag{7.186}$$

for $\mathbf{x} = \left(\mathbf{x}_1, \ldots, \mathbf{x}_n\right)'$, $\mathbf{x}_i \in \mathbb{R}^{d_i}$, $\mathbf{d} = \left(d_1, \ldots, d_n\right)$ and $|\mathbf{d}| = d$.

7.21.2 Permissibility criteria for quasi-arithmetic means of covariance functions on $\mathbb{R}^d$

As said previously, Porcu *et al.* (2009b) obtain permissibility criteria for quasi-arithmetic means of covariance functions on $\mathbb{R}^d$ by considering the following three cases:

(i) A general case in which the respective arguments of the covariance functions used for the quasi-arithmetic average have no restrictions.

(ii) The restriction to the component-wise isotropic case.

(iii) The further restriction to covariances that are isotropic and defined on the real line.

Proposition 1[19] (Proposition 2 in Porcu *et al.* 2009b)

(a) General case. Let $\varphi \in \Phi_{cm}$ and $C_i : \mathbb{R}^{d_i} \rightarrow \mathbb{R}, i = 1, \ldots, n$ be continuous stationary covariance functions such that $\cup_i^n C_i(\mathbb{R}^{d_i}) \subset D(\varphi^{-1})$ and $\mathbf{d} = \left(d_1, \ldots, d_n\right)'$, $|\mathbf{d}| = d$. If the functions $\mathbf{x}_i \rightarrow \varphi^{-1} \circ C_i\left(\mathbf{x}_i\right)$, are intrinsically stationary semivariograms on $\mathbb{R}^{d_i}$, then

$$\mathcal{Q}_\psi(C_1, \ldots, C_n)(\mathbf{x}_1, \ldots, \mathbf{x}_n) \tag{7.187}$$

is a stationary covariance function on $\mathbb{R}^d$.

[19] See proof in Appendix D.

(b) *Component-wise isotropy. Let* $\varphi \in \Phi_{cm}$ *and* $C_i : \mathbb{R}^{d_i} \to \mathbb{R}$, $i = 1, \ldots, n$ *be isotropic, continuous and stationary covariance functions such that* $\cup_i^n C_i(\mathbb{R}^{d_i}) \subset D(\varphi^{-1})$ *and* $\mathbf{d} = (d_1, \ldots, d_n)'$, $|\mathbf{d}| = d$. *If the functions* $\mathbf{x}_i \to \varphi^{-1} \circ C_i(\mathbf{x})$, *are Bernstein functions on the positive real line, then*

$$\mathcal{Q}_\psi(C_1, \ldots, C_n)(\|\mathbf{x}_1\|, \ldots, \|\mathbf{x}_n\|) \tag{7.188}$$

is a stationary and fully symmetric covariance function on $\mathbb{R}^d$.

(c) *One-dimensional covariances. Let* $\varphi \in \Phi_{cm}$ *and* $C_i : \mathbb{R}^{d_i} \to \mathbb{R}, i = 1, \ldots, n$, *be continuous and stationary covariance functions defined on the real line such that* $\cup_i^n C_i(\mathbb{R}) \subset D(\varphi^{-1})$ *and* $|\mathbf{d}| = n$. *If the functions* $\mathbf{x} \to \varphi^{-1} \circ C_i(\mathbf{x})$, *are continuous, not decreasing and concave on the positive real line, then*

$$\mathcal{Q}_\psi(C_1, \ldots, C_n)(|x_1|, \ldots, |x_n|) \tag{7.189}$$

is a stationary and fully symmetric covariance function on $\mathbb{R}^n$.

Note that (7.189) represents a permissibility condition of a covariance function that does not depend on the Euclidean metric, as it is a function of Manhattan or City Block distance (for a more in-depth discussion of the limitations of covariances that depend on the Euclidean norm, see Banerjee 2004).

The results contained in Proposition 1 are very important due to their implications for two classes of means: geometric means and harmonic means.

Corollary 1 (Geometric mean) *Let* $C_i : \mathbb{R}^{d_i} \to \mathbb{R}, i = 1, \ldots, n$ *be continuous permissible covariance functions. Let* θ_i $(i = 1, \ldots, n)$ *be non-negative weights that sum up to one. If the functions* $x \to -\ln\ (C_i(x)), x > 0$ *fulfill any relevant condition described in (a), (b) and (c) of Proposition 1, then:*

$$\mathcal{Q}_G(C_1, \ldots, C_n) = \prod_{i=1}^{n} C_i^{\theta_i} \tag{7.190}$$

is a covariance function.

Example 7.31 (Porcu *et al.* 2009b) *Let us consider the covariance function* $x \to (1 + x^\delta)^{-\epsilon}$, *x being a positive argument,* $\delta \in (0, 2]$ *and* ϵ *positive, also known as the generalized Cauchy class (Gneiting and Schlather 2004). It is possible to verify that the composition of this function with the natural logarithm is continuous, non-decreasing and concave on the positive real line for* $\delta \in (0, 1]$. *Another covariance function that complies with these requisites is* $x \to \exp(-x^\delta)$, *which is completely monotone for* $\delta \in (0, 1]$. *Note that these criteria are not applicable to covariance functions with compact support, such as the spherical model (Christakos 1992). Observe that the composition* $(1 - \ln\ x^\delta)^{-\epsilon} = (1 - \delta \ln\ x)^{-\epsilon} = (1 - \delta \ln\ C_i)^{-\delta}$ *is continuous, non-decreasing and concave on the positive real line for* $\delta \in (0, 1]$.

Observe that the composition $\exp - \left(-\ln\ x^{\delta}\right) = x^{\delta} = C_i^{\delta}$ *is a completely monotone function for* $\delta \in (0, 1]$*:*

- $x^{\delta} > 0,$
- $\frac{d^{(1)}x^{\delta}}{dx} = \delta x^{\delta-1} > 0,$
- $\frac{d^{(2)}x^{\delta}}{dx^2} = \delta\,(\delta - 1)\,x^{\delta-2} < 0,$
- $\frac{d^{(3)}x^{\delta}}{dx^3} = \delta\,(\delta - 1)\,(\delta - 2)\,x^{\delta-3} > 0,$
- ....
- $\frac{d^{(n)}x^{\delta}}{dx^n} = \delta\,(\delta - 1)\ldots(\delta - n + 1)\,x^{\delta-n},$

thereby revealing that $\frac{d^{(n)}x^{\delta}}{dx^n} = \delta\,(\delta - 1)\ldots(\delta - n + 1)\,x^{\delta-n}$ *is positive when n is an odd number and negative when* n *is an even number.*

Corollary 2 (Harmonic mean) *Let* $C_i : \mathbb{R}^{d_i} \rightarrow \mathbb{R}, i = 1, \ldots, n$ *be continuous permissible covariance functions. If the functions* $x \rightarrow C_i(x)^{-1}$, $x > 0$ *meet any given relevant condition described in (a), (b) or (c) of the Proposition 1, then for* $\theta_i \geq 0$ *such that* $\sum_{i=1}^{n} \theta_i = 1$,

$$\mathcal{Q}_H(C_1, \ldots, C_n) = \frac{1}{\sum_{i=1}^{n} \frac{\theta_i}{C_i}} \tag{7.191}$$

is a covariance function.

Example 7.32 (Porcu *et al.* 2009b) *Let us consider the covariance function* $x \rightarrow (1 + x^{\delta})^{-\epsilon}$ *again,* x *being a positive argument,* $\delta \in (0, 2]$ *and confining* ϵ *to the interval* $(0, 1]$. *It is easy to verify that this function meets the requisites of Corollary 4. Another function that meets these requisites is the so-called Dagum covariance function (Porcu* et al. *2007c), which is written* $x \rightarrow (1 + x^{-\delta})^{-\epsilon}$, *for* $\delta \in (0, 2]$ *and* $\epsilon \in (0, 1]$.

Observe that the composition $\left(1 + \frac{1}{x^{\delta}}\right)^{-\epsilon} = \left(1 + \frac{1}{C_i^{\delta}}\right)^{-\epsilon}$, *with* $\delta \in (0, 2]$ *and* ε *in the interval* $(0, 1]$ *meet the requirements of Corollary 2:*

- $\left(1 + \frac{1}{x^{\delta}}\right)^{-\epsilon} > 0.$
- $$\frac{d^{(1)}\left(1+\frac{1}{x^{\delta}}\right)^{-\epsilon}}{dx} = -\epsilon\left(1 + \frac{1}{x^{\delta}}\right)^{-\epsilon-1}\left(-\frac{\delta}{x^{(\delta+1)}}\right) = \epsilon\left(1 + \frac{1}{x^{\delta}}\right)^{-(\epsilon+1)}\left(\frac{\delta}{x^{(\delta+1)}}\right) > 0.$$
- $$\frac{d^{(2)}\left(1+\frac{1}{x^{\delta}}\right)^{-\epsilon}}{dx^2} = \delta\left(\begin{array}{l} -(\epsilon+1)\left(1 + \frac{1}{x^{\delta}}\right)^{-(\epsilon+2)}\left(\frac{\delta}{x^{(\delta+1)}}\right)\left(\frac{\delta}{x^{(\delta+1)}}\right) \\ +\left(1 + \frac{1}{x^{\delta}}\right)^{-(\epsilon+1)}\left(-\delta\,(\delta + 1)\,\frac{1}{x^{(\delta^2-\delta-1)}}\right) \end{array}\right) < 0.$$

By continuing the derivation we can see how the signs of the derivative alternate, making the Dagum covariance function completely monotone.

7.21.3 The use of quasi-arithmetic functionals to build non-separable, stationary, spatio-temporal covariance functions

The results presented previously for the spatial case can be very useful for building spatio-temporal covariance functions. However, we must take various aspects into account:

(i) The functional (7.185) should be adapted to be used in the spatio-temporal case, as it does not make much sense to consider a weighted mean of a spatial covariance with a temporal one.

(ii) In such a case, the use of weights prevents us from building non-separable models that allow for separability as a special case. Consequently, Porcu *et al.* (2009b) suggest reducing the functional in (7.185) to the case of trivial weights, for example $\theta_i = 1, \forall i$. It is relatively easy to demonstrate that the constraint relating to trivial weights does not affect the permissibility of the resulting covariance function, providing any of the restrictions imposed in (a), (b) or (c) of the Proposition 7.21.2 are complied with.

Bearing in mind the above considerations, it is possible to obtain new families of covariance functions whose analytical expressions are familiar in the field of probability modeling by means of copulas. All the families they propose, which we next reproduce, include models of separable covariance functions as special cases, depending on the parametric values of the generators $\varphi \in \Phi$.

Example 7.33 (Porcu *et al.* 2009b. Example 1. The Clayton family) *Let us consider the completely monotone function*

$$\varphi(x) = (1 + x)^{-1/\lambda_1}, \quad x > 0, \tag{7.192}$$

where λ_1 is a non-negative parameter, with inverse $\varphi^{-1}(y) = y^{-\lambda_1} - 1$. Observe that (7.192) is the function that generates the Clayton family of copulas (see Genest and MacKay 1986, for mathematical details about this class).

In addition, let us consider the covariance functions $C_s(\mathbf{h}) = (1 + \|\mathbf{h}\|)^{-1/\lambda_2}$ and $C_t = (1 + |u|)^{-1/\lambda_3}$, for λ_2 and λ_3 positive parameters. It is easy to verify that $\varphi^{-1}\left(C_i(y)\right) = (1 + y)^{\lambda_1/\lambda_i}, (i = 2, 3)$ is a positive function whose first derivative is completely monotone if, and only if $\lambda_1 < \lambda_i$. Under this constraint, and applying case (b) of Proposition 1, we have that: $\varphi(t) = (1 + t)^{-1/\lambda_1}$ and $\varphi^{-1}\left(f_i\left(x_i\right)\right) = f_i\left(x_i\right)^{-\lambda_1} - 1$.

Therefore, the quasi-arithmetic class of functionals is:

$$\psi\left(f_1\left(x_1\right), \dots, f_n\left(x_n\right)\right) = \varphi\left(\sum_{i=1}^{n} \theta_i \varphi^{-1}\left(f_i\left(x_i\right)\right)\right), \tag{7.193}$$

and setting $\theta_i = 1, \forall i$

$$\psi\left(f_1\left(x_1\right), \ldots, f_n\left(x_n\right)\right) = \varphi\left(\sum_{i=1}^{n} \theta_i \varphi^{-1}\left(f_i\left(x_i\right)\right)\right)$$
$$= \left(1 + \sum_{i=1}^{n} \varphi^{-1}\left(f_i\left(x_i\right)\right)\right)^{-1/\lambda_1}$$
$$= \left(1 + \sum_{i=1}^{n} \left(f_i\left(x_i\right)^{-\lambda_1} - 1\right)\right)^{-1/\lambda_1},$$

takes the form

$$\psi\left(f_1\left(x_1\right), \ldots, f_n\left(x_n\right)\right) = \left(1 + \left(f_1\left(x_1\right)^{-\lambda_1} - 1\right) + \left(f_2\left(x_2\right)^{-\lambda_1} - 1\right)\right)^{-1/\lambda_1}$$
$$= \left(f_1\left(x_1\right)^{-\lambda_1} + f_2\left(x_2\right)^{-\lambda_1} - 1\right)^{-1/\lambda_1}.$$

Now, writing

$$f_1\left(x_1\right) = C_s\left(\mathbf{h}\right) = \left(1 + \|\mathbf{h}\|\right)^{-\frac{1}{\lambda_2}}$$
$$f_2\left(x_2\right) = C_s\left(u\right) = \left(1 + |u|\right)^{-\frac{1}{\lambda_3}},$$

we have that:

$$C(\mathbf{h}, u) = \mathcal{Q}_\psi(C_s, C_t)(\mathbf{h}, u) = \left((1 + \|\mathbf{h}\|)^{\frac{\lambda_1}{\lambda_2}} + (1 + |u|)^{\frac{\lambda_1}{\lambda_3}} - 1\right)^{-1/\lambda_1}$$

or even better,

$$C(\mathbf{h}, u) = \mathcal{Q}_\psi(C_s, C_t)(\mathbf{h}, u) = \sigma^2((1 + \|\mathbf{h}\|)^{\rho_2} + (1 + |u|)^{\rho_3} - 1)^{-1/\lambda_1}, \tag{7.194}$$

is a valid non-separable and fully symmetric spatio-temporal covariance function, with $\rho_i = \lambda_1/\lambda_i, \lambda_i > 0, (i = 2, 3)$ *and* σ^2 *is a non-negative parameter that denotes the variance of the spatio-temporal process. It is interesting to note that the two marginal functions (spatial and temporal) belong to the generalized Cauchy class. This is desirable if we are interested in the local and global behaviour of the underlying random field.*

Another covariance function that preserves Cauchy margins can be obtained in the following way:

Let us consider the function $x \to x^{-\alpha}, t > 0$, *which belongs to* Φ_{cm} *for any given positive* α, *being the Laplace transform of the function* $x \to x^{-\alpha-1}/\Gamma(\alpha)$, *where* α *is a positive parameter and* Γ *the Euler Gamma function. Furthermore, consider the marginal spatial covariance function* $C_s(\mathbf{h}) = (1 + \|\mathbf{h}\|^{\delta})^{-\epsilon}$ *which belongs to* Φ_{cm} *if, and only if* $\delta \in (0, 2]$ *and* ϵ *is strictly positive. Finally, let us consider the marginal temporal covariance function* $C_t(u) = |u|^{-\rho}$, *which is a non-stationary covariance but respects the composition criteria, as* $\varphi^{-1} \circ C_t$ *is continuous, increasing and concave on the positive real line if, and only if* $\alpha < \rho$. *Then, it is simple to prove that* $\mathcal{Q}_{\psi}(C_s, C_t)(\mathbf{h}, u)$ *is a valid spatio-temporal function if, and only if* $\alpha < \epsilon$ *and* $\alpha < \rho$, *and that this covariance function has Cauchy-type marginal functions.*

Example 7.34 (Porcu *et al.* 2009b. Example 2. The Gumbel-Hougard family) *Let us consider the completely monotone function, also called the positive stable Laplace transform*

$$\varphi(x) = \exp\left(-x^{1/\lambda_1}\right), \quad x > 0, \tag{7.195}$$

where $\lambda_1 \geq 1$. *Equation (7.195) admits the inverse* $\varphi^{-1}(y) = (-\ln(y))^{\lambda_1}$. *This is the function that generates the Gumbel-Hougard family of copulas (the construction and mathematical details of which can be consulted in Nelsen 1999). Let us consider two covariance functions, one spatial and the other temporal, admitting the same analytical form, that is,* $C_s(\mathbf{h}) = \exp(-\|\mathbf{h}\|^{1/\lambda_2})$ *and* $C_t(u) = \exp\left(-|u|^{1/\lambda_3}\right)$. *Then, we can easily prove that* $\varphi^{-1}(C_s(y)) = y^{\lambda_1/\lambda_2}$, *and* $\varphi^{-1}\left(C_t(y)\right) = y^{\lambda_1/\lambda_3}$ *are always positive for* $y > 0$ *and possess completely monotone first derivatives if, and only if,* $\lambda_1 < \lambda_i, i = 2, 3$. *Therefore, the quasi-arithmetic class of functionals will be:*

$$\psi\left(f_1\left(x_1\right), \ldots, f_n\left(x_n\right)\right) = \varphi\left(\sum_{i=1}^{n} \theta_i \varphi^{-1}\left(f_i\left(x_i\right)\right)\right), \tag{7.196}$$

and setting $\theta_i = 1, \quad \forall i$

$$\begin{aligned}
\psi\left(f_1\left(x_1\right), \ldots, f_n\left(x_n\right)\right) &= \varphi\left(\sum_{i=1}^{n} \theta_i \varphi^{-1}\left(f_i\left(x_i\right)\right)\right) \\
&= \exp\left(-\sum_{i=1}^{n} \theta_i \left(-\ln \ \left(f_i\left(x_i\right)\right)\right)^{\lambda_1^{1/\lambda_1}}\right) \\
&= \exp\left(-\sum_{i=1}^{n} \theta_i \left(-\ln \left(f_i\left(x_i\right)\right)\right)\right) \\
&= \exp\left(\sum_{i=1}^{n} \ln \ \left(f_i\left(x_i\right)\right)\right) \\
&= \exp\left(\ln \ \prod_{i=1}^{n} f_i\left(x_i\right)\right) \\
&= \prod_{i=1}^{n} f_i\left(x_i\right),
\end{aligned}$$

hence, in terms of the foregoing covariances and including the term σ^2, a non-negative parameter that denotes the variance of the spatio-temporal process, we have that:

$$\begin{aligned}
C(\mathbf{h}, u) &= \mathcal{Q}_{\psi}(C_s, C_t)(\mathbf{h}, u) \\
&= \sigma^2 \exp\left(-\|\mathbf{h}\|^{1/\lambda_2}\right) \exp\left(-|u|^{1/\lambda_3}\right) \\
&= \sigma^2 \exp\left(-\left(\|\mathbf{h}\|^{\frac{\lambda_1}{\lambda_2}}\right)^{\frac{1}{\lambda_1}}\right) \exp\left(\left(-|u|^{\frac{\lambda_1}{\lambda_3}}\right)^{1/\lambda_1}\right) \\
&= \sigma^2 \exp\left(-\left(\|\mathbf{h}\|^{\frac{\lambda_1}{\lambda_2}} + |u|^{\frac{\lambda_1}{\lambda_3}}\right)^{\frac{1}{\lambda_1}}\right) \\
&= \sigma^2 \exp\left(-\left(\|\mathbf{h}\|^{\rho_2} + |u|^{\rho 3}\right)^{\frac{1}{\lambda_1}}\right) \qquad (7.197)
\end{aligned}$$

is a permissible non-separable fully symmetric spatio-temporal covariance function, with $\rho_i = \lambda_1/\lambda_i, i = 2, 3$.

Example 7.35 (Porcu *et al.* 2009b. Example 3. The power series family) *The so-called power series*

$$\varphi(x) = 1 - (1 - \exp(-x))^{1/\lambda_1}, \quad x > 0$$

with $\lambda_1 \geq 1$, allows the inverse $\varphi^{-1}(y) = -\ln\,(1 - (1 - y)^{\lambda_1})$. Let us consider a spatial covariance function and a temporal covariance function with the same analytical form, as in the example of the Gumbel-Hougard family. The composition $\varphi^{-1}(C_i(y)) = -\ln\left(1 - (1 - \exp(-y))^{\lambda_1/\lambda_i}\right), i = 2, 3$ is always positive for $y > 0$ and allows a first completely monotone derivative if, and only if $\lambda_1 < \lambda_i$. In this case, we have, in generic terms, the quasi-arithmetic class of functionals as:

$$\psi\left(f_1\left(x_1\right), \ldots, f_n\left(x_n\right)\right) = \varphi\left(\sum_{i=1}^{n} \theta_i \varphi^{-1}\left(f_i\left(x_i\right)\right)\right), \qquad (7.198)$$

and setting $\theta_i = 1, \forall i$

$$\begin{aligned}
\psi\left(f_1\left(x_1\right), \ldots, f_n\left(x_n\right)\right) &= \varphi\left(\sum_{i=1}^{n} \theta_i \varphi^{-1}\left(f_i\left(x_i\right)\right)\right) \\
&= 1 - \left(1 - \exp\left(-\left(-\sum_{i=1}^{n} \ln\left(1 - \left(1 - f_i\left(x_i\right)\right)^{\lambda_1}\right)\right)\right)\right)^{\frac{1}{\lambda_1}} \\
&= 1 - \left(1 - \exp\left(\sum_{i=1}^{n} \ln\left(1 - \left(1 - f_i\left(x_i\right)\right)^{\lambda_1}\right)\right)\right)^{\frac{1}{\lambda_1}}
\end{aligned}$$

$$= 1 - \left(1 - \exp\left(\ln \prod_{i=1}^{n} \left(1 - \left(1 - f_i\left(x_i\right)\right)^{\lambda_1}\right)\right)\right)^{\frac{1}{\lambda_1}}$$

$$= 1 - \left(1 - \prod_{i=1}^{n} \left(1 - \left(1 - f_i\left(x_i\right)\right)^{\lambda_1}\right)\right)^{\frac{1}{\lambda_1}},$$

hence, in terms of the foregoing covariances, we have that:

$$\begin{aligned} Q_{\psi}(C_s, C_t)(\mathbf{h}, u) &= 1 - (1 - \exp(-\|\mathbf{h}\|))^{\rho_1} - (1 - \exp(-|u|))^{\rho_2} \\ &\quad + (1 - \exp(-\|\mathbf{h}\|))^{\rho_1}(1 - \exp(-|u|))^{\rho_2}, \end{aligned} \tag{7.199}$$

is a permissible, non-separable, stationary, fully symmetric spatio-temporal covariance function, with $\rho_i = \lambda_1 / \lambda_i$, $(i = 2, 3)$.

Example 7.36 (Porcu *et al.* 2009b. Example 4. The Frank's semi-parametric family) *This example shows how a family of stationary covariance functions can be obtained using the proposed approach, even if the arguments of the quasi-arithmetic functional are not covariance functions. The Frank's family of copulas (Genest 1987) is generated by the function*

$$\varphi(x) = (1/\lambda) \ln\left(1 - (1 - \exp(-\lambda)) \exp(-x)\right),$$

and its inverse is given by $\varphi^{-1}(y) = -\ln\left((1 - \exp(-\lambda y)) / (1 - \exp(-\lambda))\right)$. *Nelsen (1999) shows that for* λ *positive,* φ *is the composition of an absolutely monotonic function with a completely monotonic function, that is, a completely monotonic function. Insofar as the inverse function is involved, it is easy to see that* $\varphi^{-1} \circ \gamma$ *(*γ *being an intrinsically stationary variogram) is negative-definite. In this case we have, in generic terms, the quasi-arithmetic class of functionals:*

$$\psi\left(f_1\left(x_1\right), \ldots, f_n\left(x_n\right)\right) = \varphi\left(\sum_{i=1}^{n} \theta_i \varphi^{-1}\left(f_i\left(x_i\right)\right)\right) \tag{7.200}$$

and setting $\theta_i = 1, \quad \forall i$

$$\psi(f_1\left(x_1\right), \ldots, f_n\left(x_n\right)) = \varphi\left(\sum_{i=1}^{n} \theta_i \varphi^{-1}\left(f_i\left(x_i\right)\right)\right)$$

$$= \frac{1}{\lambda}\left(\ln\left(1 - (1 - \exp(-\lambda)) \exp\left(\sum_{i=1}^{n} \ln \frac{1 - \exp\left(-\lambda f_i\left(x_i\right)\right)}{1 - \exp(-\lambda)}\right)\right)\right)$$

$$= \frac{1}{\lambda}\left(\ln\left(1 - (1 - \exp(-\lambda)) \exp\left(\ln \prod_{i=1}^{n} \frac{1 - \exp\left(-\lambda f_i\left(x_i\right)\right)}{1 - \exp(-\lambda)}\right)\right)\right)$$

$$= \frac{1}{\lambda} (\ln (1 - (1 - \exp(-\lambda))) \left(\prod_{i=1}^{n} \frac{1 - \exp\left(-\lambda f_i\left(x_i\right)\right)}{1 - \exp(-\lambda)} \right),$$

hence, using semivariograms, one spatial and other temporal, we have that:

$$\begin{aligned} C(\mathbf{h}, u) &= \mathcal{Q}_{\psi}(C_s, C_t)(\mathbf{h}, u) \\ &= \frac{1}{\lambda} \left(\ln \left(1 - \left(1 - \exp(-\lambda) \left(\frac{1 - \exp(-\lambda \gamma (\mathbf{h}))}{1 - \exp(-\lambda)} \right) \right. \right. \right. \\ &\qquad \times \left. \left. \left. \left(\frac{1 - \exp(-\lambda \gamma (u))}{1 - \exp(-\lambda)} \right) \right) \right) \right) \\ &= \frac{1}{\lambda} \left(\ln \left(1 - \left(\frac{(1 - \exp(-\lambda \gamma (\mathbf{h})))(1 - \exp(-\lambda \gamma (u)))}{1 - \exp(-\lambda)} \right) \right) \right), \end{aligned} \tag{7.201}$$

is a stationary spatio-temporal covariance function for the intrinsically stationary semivariograms γ_s and γ_t defined on $\mathbb{R}^d$ and $\mathbb{R}$, respectively.

7.21.4 Quasi-arithmeticity and non-stationarity in space

Porcu *et al.* (2009b) extend the approaches taken by Stein (2005b) and also by Paciorek and Schervish (2006) directly constructing spatial non-stationary covariances, using quasi-arithmetic functionals. More specifically, they propose the following theorem.

Theorem 7.21.1 (Porcu *et al.* 2009b) *Let Σ be a mapping of $\mathbb{R}^p \times \mathbb{R}^p$ to a series of positive-definite matrices $p \times p$, F a non-negative measure on $\mathbb{R}_+$, $\varphi_1, \varphi_2 \in \Phi_{cm}$ and g a non-negative function such that for a given $\mathbf{s}, \mathbf{s} \in \mathbb{R}^p$, $h_s = \varphi_2^{-1} \circ g(\cdot, \mathbf{s}) \in L^1(F)$. Let $\mathbf{\Sigma}(\mathbf{s}_1, \mathbf{s}_2) = 1/2 \left(\mathbf{\Sigma}(\mathbf{s}_1) + \mathbf{\Sigma}(\mathbf{s}_2)\right)$ and $\mathbf{Q}(\mathbf{s}_1, \mathbf{s}_2) = (\mathbf{s}_1, \mathbf{s}_2)'\Sigma(\mathbf{s}_1, \mathbf{s}_2)^{-1}(\mathbf{s}_1, \mathbf{s}_2)$. Then,*

$$C_s(\mathbf{s}_1, \mathbf{s}_2) = \frac{\left|\mathbf{\Sigma}(\mathbf{s}_1)\right|^{1/4} \left|\mathbf{\Sigma}(\mathbf{s}_2)\right|^{1/4}}{\left|\mathbf{\Sigma}(\mathbf{s}_1, \mathbf{s}_2)\right|^{1/2}} \int_0^\infty \varphi_1(\mathbf{Q}(\mathbf{s}_1, \mathbf{s}_2)\tau)\, \mathcal{Q}_{\psi_2}(\mathbf{g}_{\mathbf{s}_1}, \mathbf{g}_{\mathbf{s}_2})(\tau)\, dF(\tau), \tag{7.202}$$

with $\theta_i = 1, i = 1, 2$, is a non-stationary covariance function on $\mathbb{R}^d \times \mathbb{R}^d$.

Note that Stein's (2005b) result is a special case of (7.202), under the choice of $\varphi_1(t) = \exp(-t)$, t being positive and $\psi_2 := G$. So is the result in Paciorek and Schervish (2006).

The form proposed above has some drawbacks. The main problem is that it is very difficult to obtain a closed-form of the expression (7.202), unless one chooses $\psi_2 := G$.

Several examples of covariance functions (that can be integrated numerically) can be derived from the class (7.202). Below we show some closed-forms that Porcu *et al.* (2009b) obtained by letting $\psi_2 := G$.

Example 7.37 (Porcu *et al.* 2009b) *Let $dF(\tau) = d\tau$ and $\varphi_1(\tau) = \tau^{\lambda-1}$ which is a completely monotonic function for $\lambda \in (0,1), g(\tau, \alpha_i, \upsilon_i) = (1 + \alpha(\mathbf{s}_i)\tau)^{-\upsilon(s_i)}$, where α and υ are supposed to be strictly positive functions of $\mathbf{s}_i, (i = 1,2)$ and additionally $0 < \alpha(\mathbf{s}_i), \upsilon(\mathbf{s}_i) < \pi$. We can readily verify that all the integrability conditions in Theorem 7.21.1 are fulfilled. Taking*

$$k = \frac{|\boldsymbol{\Sigma}(\mathbf{s}_1)|^{1/4}|\boldsymbol{\Sigma}(\mathbf{s}_2)|^{1/4}}{|\boldsymbol{\Sigma}(\mathbf{s}_1, \mathbf{s}_2)|^{1/2}}\mathbf{Q}(\mathbf{s}_1, \mathbf{s}_2)^{\lambda-1}$$

and using [3.259.3] *in Gradshteyn and Ryzhik (1980), the following covariance function is obtained:*

$$C_s(\mathbf{s}_1, \mathbf{s}_2) = k\alpha\left(\mathbf{s}_1\right)^{-\lambda} B(\lambda, \upsilon\left(\mathbf{s}_1\right) + \upsilon\left(\mathbf{s}_2\right) - \lambda)$$
$$\times \quad {}_2F_1\left(\upsilon\left(\mathbf{s}_2\right), \lambda; \left(\mathbf{s}_1\right) + \upsilon\left(\mathbf{s}_2\right); 1 - \frac{\alpha\left(\mathbf{s}_2\right)}{\alpha\left(\mathbf{s}_1\right)}\right), \tag{7.203}$$

where $B(\cdot,\cdot)$ is the Beta function and ${}_2F_1$ is the Gauss hypergeometric function.

Example 7.38 (Porcu *et al.* 2009b) *Consider $F(d\tau) = \exp(-\tau)d\tau, \varphi_1(\tau) = \tau^{\upsilon-1}$, with $\upsilon \in (0,1), g(\tau; \mathbf{s}_i) = \exp\left(-\frac{\alpha(\mathbf{s}_i)}{2}\tau\right)$ with $\alpha\left(\mathbf{s}_i\right)$ being strictly positive $(i = 2,3)$ and using* [3.478.4] *in Gradshteyn and Ryzhik (1980), it is obtained:*

$$C_s(\mathbf{s}_1, \mathbf{s}_2) = 2k\left(\frac{\alpha\left(\mathbf{s}_1\right) + \alpha\left(\mathbf{s}_2\right)}{2}\right)^{-\upsilon/2} K_\upsilon\left(2\left(\frac{\alpha\left(\mathbf{s}_1\right) + \alpha\left(\mathbf{s}_2\right)}{2}\right)^{-1/2}\right),$$

is a valid non-stationary spatial covariance that allows for local geometric anisotropy, but which has a fixed smoothing parameter.

8

Spatio-temporal prediction and kriging

8.1 Spatio-temporal kriging

Once we know how to construct the empirical covariogram or semivariogram from the data (Chapter 6) and how to fit to it one of the theoretical models presented in Chapter 7, we are ready for spatio-temporal kriging prediction. The only tools we need are the optimal spatio-temporal predictor, the spatio-temporal kriging equations, which provide the weights of the spatio-temporal data of the regionalization observed in such a predictor, and the prediction variance, indicative of the accuracy of the predictions.

In this short chapter we present both the spatio-temporal kriging predictor and the spatio-temporal kriging equations accompanied by the prediction variance. However, both are merely generalizations of their spatial counterparts, and this is why in order to avoid redundancies and duplications, we only focus on their final expressions and some pertinent comments. We assume throughout the chapter that the observations and predictions are point-based. The reason for this is to minimize the length of the chapter, because (i) the point kriging equations can be immediately extended to block kriging equations; and (ii) as stated above, the spatio-temporal kriging equations, both point- and block-based, are straightforward generalizations of their spatial counterparts.

Point spatio-temporal kriging is aimed at predicting an unknown point value $Z(\mathbf{s}_0, t_0)$ at a non-observed point $(\mathbf{s}_0, t_0)$. In order to do so, all the information available about the regionalized variable is used, either at the points in the entire domain or in a subset of the domain called neighborhood.

Spatial and Spatio-Temporal Geostatistical Modeling and Kriging, First Edition.
José-María Montero, Gema Fernández-Avilés, and Jorge Mateu.

Companion Website: www.wiley.com/go/montero/spatial

Let $\{Z(\mathbf{s}, t)\colon \mathbf{s} \in D, t \in T\}$ with $D \subset \mathbb{R}^2$ and $T \subset \mathbb{R}$, be a spatio-temporal rf and let us assume throughout this chapter that the value of the rf has been observed on a set of n spatio-temporal locations $\{Z(\mathbf{s}_1, t_1), \ldots, Z(\mathbf{s}_n, t_n)\}$.

The objective will be to predict the value of the rf on a new spatio-temporal location $(\mathbf{s}_0, t_0)$, $Z(\mathbf{s}_0, t_0)$, where $\mathbf{s}_0$ will be a given spatial location on D and t_0 a moment in time on T. In order to predict the value of a spatio-temporal rf at a non-observed point $(\mathbf{s}_0, t_0)$, we will use the linear predictor

$$Z^*(\mathbf{s}_0, t_0) = \sum_{i=1}^{n} \lambda_i Z(\mathbf{s}_i, t_i), \tag{8.1}$$

which, as can be observed, is the straightforward generalization of the spatial kriging predictor. Obviously, as in the spatial case, we want the spatio-temporal predictor (8.1) to be the BLUP.

8.2 Spatio-temporal kriging equations

As in the spatial case, the spatio-temporal point kriging equations depend on the degree of stationarity attributed to the rf that supposedly generates the observed realization.

If $Z(\mathbf{s}, t)$ is a second-order stationary spatio-temporal rf with constant and known mean, μ, and known covariance function, $C(\mathbf{h}, u)$, the usual practice is to subtract the constant known mean from the values of the realization observed and then use the following spatio-temporal SK equations for prediction:

$$\sum_{j=1}^{n} \lambda_j C\left(\mathbf{s}_i - \mathbf{s}_j, t_i - t_j\right) = C\left(\mathbf{s}_i - \mathbf{s}_0, t_i - t_0\right), \quad \forall i = 1, \ldots, n, \tag{8.2}$$

which provides the predicted value of the spatio-temporal rf $Z(\mathbf{s}, t) - \mu$ at the location $(\mathbf{s}_0, t_0)$ where we are predicting. Thus, as a final step, we have to add μ to such a prediction.

The prediction error variance, which is not affected by a change of scale, is given by:

$$V\left(Z^*(\mathbf{s}_0, t_0) - Z(\mathbf{s}_0, t_0)\right) = C(\mathbf{0}, 0) - \sum_{i=1}^{n} \lambda_i C(\mathbf{s}_i - \mathbf{s}_0, t_i - t_0). \tag{8.3}$$

If $Z(\mathbf{s}, t)$ is (i) a second-order stationary spatio-temporal rf with constant mean μ, albeit unknown, and known covariance function $C(\mathbf{h}, u)$, or (ii) an intrinsic spatio-temporal rf, also with a constant but unknown mean μ, but with unbounded variance, spatio-temporal SK cannot be performed because the mean of the rf is unknown. We must therefore impose a condition of unbiasedness.

In these situations we use the spatio-temporal OK equations to obtain the optimal weights of the spatio-temporal kriging predictor. Such equations can be expressed, in the case (i), in terms of both the covariance function and the semivariogram (though, as in the purely spatial case, the use of the covariance function leads to a vicious circle), and in the case (ii) only in terms of the semivariogram, as the covariance is not defined at the origin.

The spatio-temporal OK kriging equations in terms of the covariance function are given by:

$$\begin{cases} \sum_{j=1}^{n} \lambda_j C(\mathbf{s}_i - \mathbf{s}_j, t_i - t_j) - \alpha = C(\mathbf{s}_i - \mathbf{s}_0, t_i - t_0), \forall i = 1, \ldots, n \\ \sum_{i=1}^{n} \lambda_i = 1 \end{cases}, \tag{8.4}$$

where α is the Lagrange multiplier associated with the condition of unbiasedness.

The corresponding prediction variance is as follows:

$$V\left(Z^*(\mathbf{s}_0, t_0) - Z(\mathbf{s}_0, t_0)\right) = C(\mathbf{0}, 0) - \sum_{i=1}^{n} \lambda_i C(\mathbf{s}_i - \mathbf{s}_0, t_i - t_0) + \alpha. \tag{8.5}$$

The spatio-temporal OK equations in semivariogram terms are:

$$\begin{cases} \sum_{j=1}^{n} \lambda_j \gamma\left(\mathbf{s}_i - \mathbf{s}_j, t_i - t_j\right) + \alpha = \gamma\left(\mathbf{s}_i - \mathbf{s}_0, t_i - t_0\right), \forall i = 1, \ldots, n \\ \sum_{i=1}^{n} \lambda_i = 1 \end{cases}, \tag{8.6}$$

with prediction variance given by:

$$V\left(Z^*(\mathbf{s}_0, t_0) - Z(\mathbf{s}_0, t_0)\right) = \sum_{i=1}^{n} \lambda_i \gamma(\mathbf{s}_i - \mathbf{s}_0, t_i - t_0) + \alpha. \tag{8.7}$$

Note that if the rf is second-order stationary the systems of kriging equations (8.4) and (8.6) are equivalent. One can check it by merely using the equivalence:

$$\gamma(\mathbf{h}, u) = C(\mathbf{0}, 0) - C(\mathbf{h}, u). \tag{8.8}$$

Should the rf be intrinsically stationary, we cannot run the system (8.4). However, the system (8.6) is operational as the variance of the rf does not appear in it.

Finally, if $Z(\mathbf{s}, t)$ is a spatio-temporal rf with drift, for which reason the mean of the rf is not constant but rather depends on the spatio-temporal locations $(\mathbf{s}, t)$, following the same reasoning as in the spatial case, the rf can be disaggregated into the following two components:

$$Z(\mathbf{s}, t) = \mu(\mathbf{s}, t) + e(\mathbf{s}, t), \tag{8.9}$$

with $\mu(\mathbf{s}, t)$ deterministic and $e(\mathbf{s}, t)$ stochastic and considered an intrinsically stationary spatio-temporal rf with zero expectation.

As in the spatial case, $\mu(\mathbf{s}, t)$, albeit unknown, can be expressed locally as follows:

$$\mu(\mathbf{s}, t) = \sum_{h=1}^{p} a_h f_h(\mathbf{s}, t), \tag{8.10}$$

where $\{f_h(\mathbf{s}, t),\ h = 1, \ldots, p\}$ are p known functions, a_h constant coefficients and p the number of terms employed in the approximation.

Having said that, the equations that should be used to obtain the optimal weights involved in the kriging predictor (the spatio-temporal UK equations) are given by:

$$\begin{cases} \sum_{j=1}^{n(\mathbf{s}_0,t_0)} \lambda_j \gamma_e((\mathbf{s}_i, t_i) - (\mathbf{s}_j, t_j)) + \sum_{k=1}^{p} \alpha_k f_k(\mathbf{s}_i, t_i) = \gamma_e((\mathbf{s}_i, t_i) - (\mathbf{s}_0, t_0)), \\ \qquad \forall i = 1, \ldots, n(\mathbf{s}_0, t_0) \\ \sum_{i=1}^{n(\mathbf{s}_0,t_0)} \lambda_i f_k(\mathbf{s}_i, t_i) = f_k(\mathbf{s}_0, t_0), \forall k = 1, \ldots, p \end{cases}, \tag{8.11}$$

the prediction variance being:

$$V\left(Z^*(\mathbf{s}_0, t_0) - Z(\mathbf{s}_0, t_0)\right) = \sum_{i=1}^{n(\mathbf{s}_0,t_0)} \lambda_i \gamma_e((\mathbf{s}_i, t_i) - (\mathbf{s}_0, t_0)) + \sum_{k=1}^{p} \alpha_k f_k(\mathbf{s}_0, t_0), \tag{8.12}$$

where α_k, $k = 1, \ldots, p$, are the Lagrange multipliers associated to the p conditions of unbiasedness.

Note that $i = 1, \ldots, n(\mathbf{s}_0, t_0)$ instead of $i = 1, \ldots, n$. This is due to the fact that the expression for the drift is only valid locally, hence $n(\mathbf{s}_0, t_0)$ represents the number of observations in a spatio-temporal region of $(\mathbf{s}_0, t_0)$ $\quad (n(\mathbf{s}_0, t_0) \leq n)$.

This method entails enormous practical difficulty. The semivariogram of $e(\mathbf{s}, t) = Z(\mathbf{s}, t) - \mu(\mathbf{s}, t)$ appears in the equations instead of the semivariogram of the rf. But due to not knowing the value of the drift at each point, we do not have the value of $e(\mathbf{s}_i, t_i)$ either and, therefore, we cannot construct the semivariogram $\gamma_e(\mathbf{h}, u)$. The possible solutions we consider to overcome this problem are extensions of the spatial procedures discussed at the end of Section 4.6.

Example 8.1 (Mapping with spatio-temporal SK) *In order to illustrate how spatio-temporal SK performs, let us consider a database consisting of* 140 *spatio-temporal data, simulated at the* 20 *locations shown in Figure 8.1 in seven instants of time, say, seven consecutive days. The simulated spatio-temporal database is shown in Table 8.1.*

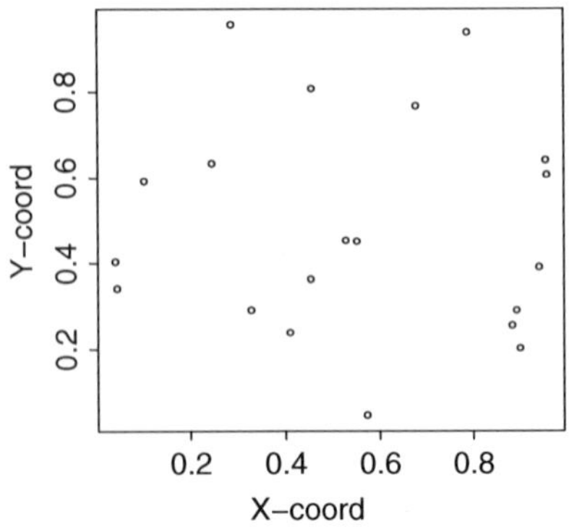

Figure 8.1 Location of the 20 sites observed.

Table 8.1 Simulated spatio-temporal database.

space \ time	1	2	3	4	5	6	7
1	0.08	−0.16	1.70	2.11	2.98	1.31	1.23
2	1.08	−0.61	0.06	−0.26	0.57	0.19	−0.55
3	−0.90	−1.81	−0.44	0.84	1.28	−0.31	−1.79
4	−0.35	−0.69	0.64	0.74	1.08	2.03	0.24
5	−0.23	−0.77	−0.26	0.52	1.34	1.26	−0.09
6	−0.65	−2.35	−1.49	−0.49	−0.59	−1.20	−1.28
7	−1.73	−1.83	−0.82	0.27	0.90	−1.40	−1.58
8	−0.39	−0.39	0.98	0.81	1.63	2.09	0.06
9	−1.63	−1.57	−0.94	0.00	1.02	−1.31	−1.53
10	−0.59	−1.72	−0.87	0.23	1.33	0.41	0.42
11	0.62	−0.67	−0.19	−0.62	0.42	0.93	1.03
12	−0.84	−1.65	−0.29	0.60	1.57	−0.55	−1.58
13	0.42	−0.91	−0.09	0.54	1.17	0.17	1.01
14	−1.11	−1.50	−0.53	0.12	0.03	−0.25	−0.40
15	0.33	−1.03	−0.08	0.97	1.94	0.02	0.58
16	−0.91	−0.99	0.10	0.28	1.21	2.24	−0.34
17	−0.22	−1.10	0.21	2.04	2.43	0.45	1.04
18	−0.65	−2.45	−1.51	0.01	0.00	−0.50	−0.63
19	−0.68	−1.64	−0.91	0.56	1.61	−0.01	−1.12
20	0.54	−0.60	−0.33	−0.64	0.31	0.99	0.72

The simulation was carried out using a mean-zero Gaussian process with doubly exponential covariance function (see Equation 7.26 and Figure 7.27) without the nugget effect, $C(\mathbf{0}, 0) = 2$, *and space and time scaling parameters* $a_s = 1$ *and* $a_t = 2$, *respectively. Our aim is to obtain the spatio-temporal SK prediction maps (together with the corresponding prediction variance maps) over the area represented*

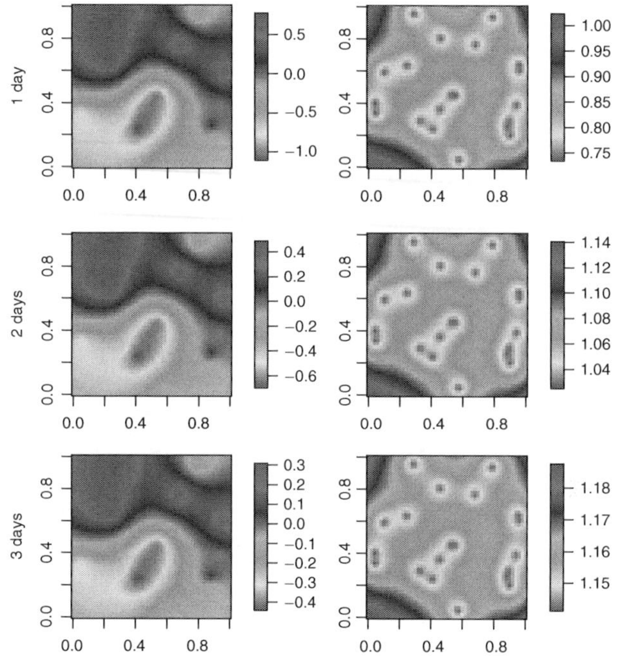

Figure 8.2 Spatio-temporal SK prediction maps (left panel) and prediction variance maps (right panel): 1-day, 2-day and 3-day time horizons. (See color figure in color plate section.)

in Figure 8.2 for 1-day, 2-day and 3-day time horizons, that is to say, for the next three days. Computations were made using `CompRandFld` *and* `RandomFields` *packages from R software.*

The maximization of the full standard likelihood (the dimension of this problem does not give rise to computational problems when using the standard ML method) resulted in the following estimated spatio-temporal covariance function (using REML the parameters are practically the same):

$$C(\mathbf{h}, u) = 1.2180 \exp\left(-\|\mathbf{h}\|/0.6534 - |u|/2.1832\right), \qquad (\mathbf{h}, u) \in \mathbb{R}^d \times \mathbb{R}, \quad (8.13)$$

which in turn led to the prediction maps depicted in Figure 8.2).

Although the three maps in the left panel of Figure 8.2 appear to be the same, the legends indicate that the predicted values tend to the mean of the process when the time horizon increases. The same can be said of the prediction variance-maps: they appear to be the same, but, as expected, the longer the prediction time horizon, the greater the prediction variance, its maximum being $C(\mathbf{0}, 0)$ in the covariance function used in the kriging equations (in our case, 1.2189).

In brief, as spatio-temporal kriging should not be used for predictions in locations very distant from the observed sites, no matter the time instant (because the *predictions will tend to the mean of the observed realization and the prediction variance to the sill of the fitted semivariogram), for the same reason it should not be used for predictions at time horizons other than the very short ones (because in such cases predictions will concentrate around the mean of the observations, the specific value depending on the spatial location, and the prediction variance around the sill of the semivariogram, or $C(\mathbf{0}, 0)$ in the covariance function, used for predictions). However, it is of great interest for "sandwich" predictions; that is, for predictions at times between the first and the last time in the realization observed and in locations surrounded by a large number of observed sites.*

Example 8.2 (Spatio-temporal geostatistical modeling of logCO* in Madrid 2008) *In Example 4.7 it was shown that the rf underlying the logCO* values in the city of Madrid exhibits a marked spatial dependence. In this example we extend the modeling of logCO* to the spatio-temporal framework by including the time argument and its interaction with space.*

More specifically, we focus on the values of logCO at 10 am, and use the data corresponding to the weeks 1 to 51 to carry out such a modeling. The data of week 52 are used to compare the predictions yielded by spatial and spatio-temporal SK. We had previously removed the seasonal effects, which were calculated using ANOVA.*

We use Gneiting's (2002) non-separable spatio-temporal covariance function (7.79):

$$C(\mathbf{h}, u, \boldsymbol{\theta}) = \sigma^2 \frac{1}{(a|u|^{2\alpha} + 1)} \exp\left(\frac{-c||\mathbf{h}||^{2\gamma}}{(a|u|^{2\alpha} + 1)^{\beta\gamma}}\right),$$

where c and a are positive spatial and temporal scaling parameters respectively, γ and α are spatial and temporal smoothing parameters, β is a spatio-temporal interaction parameter and σ^2 is the variance of the spatio-temporal process.

For $\beta = 0$ the above covariance function becomes separable, as β increases the spatio-temporal interaction grows stronger, and for $\beta = 1$ we obtain the most extreme case of non-separability of this family. Thus, this family of covariance functions compares the separable and non-separable case, in this way assessing the importance of considering the space-time interactions.

The parameters of the covariance function were estimated with `CompRandFld` *library in R software using the marginal likelihood method (the composite-likelihood is formed by marginal likelihoods) with the pairwise option (the composite-likelihood is defined by the pairwise likelihoods). The starting values were computing using the weighted moment estimator.*

The estimated spatio-temporal covariance functions obtained were:

$$C(\mathbf{h}, u, \boldsymbol{\theta}) = 0.1782 \frac{1}{(7.9948|u|^{2\cdot 0.2773} + 1)} \exp\left(-7.9275||\mathbf{h}||^{2\cdot 0.2456}\right)$$

for the separable spatio-temporal covariance, and

$$C(\mathbf{h}, u, \boldsymbol{\theta}) = 0.1783 \frac{1}{(0.4083|u|^{2\cdot 2} + 1)} \exp\left(\frac{-2.2481||\mathbf{h}||^{2\cdot 0.0668}}{(0.4036|u|^{2\cdot 2} + 1)^{0.98\cdot 0.6668}}\right)$$

for the non-separable spatio-temporal one.

Table 8.2 ME, MSE and MSDE.

	ME	MSE	MSDE
Separable	−0.229	0.117	16.185
Non-separable	−0.174	0.117	3.869

Table 8.2 lists the ME, MSE and MSDE statistics obtained for both the separable and non-separable covariance functions, when using them in the spatio-temporal SK equations to predict the values of logCO the last week of* 2008 *at the* 23 *monitoring stations operating in the city of Madrid. Although not ideal, they favor the non-separable hypothesis and point out the importance of considering space and time jointly.*

9

An introduction to functional geostatistics

9.1 Functional data analysis

Functional data analysis (FDA) statistically models rv's that take values in a space of functions, which is why they are called functional variables (fv's). In other words, FDA is about the analysis of information on curves or functions (it can be extended to surfaces or images).

As stated in Levitin *et al.* (2007), data collection technology has evolved over the last two decades so as to permit observations densely sampled over time, space, and other continua; these data usually reflect the influence of certain smooth functions that are assumed to underlie and to generate the observations. Thus, following Ullah and Finch (2013), the basic idea behind FDA is to express discrete observations arising from sampling over the above-mentioned continua (usually the continuum is time) in the form of a smooth function (to create functional data; convert the raw data into functional data is the first step in FDA) that represents the entire measured function as a single observation, and then to draw modeling and/or prediction information from a collection of functional data by applying statistical concepts using multivariate data analysis.

Some practical reasons for considering functional data include the following (see Ramsay 1982, 1988; Ramsay and Dalzell 1991; Müller 2011; Ferraty *et al.* 2007; and Mas and Pumo 2009, among others):

1. Smoothing and interpolation procedures can yield functional representations of a finite set of observations.
2. It is more natural to think through modeling problems in a functional form.

Spatial and Spatio-Temporal Geostatistical Modeling and Kriging, First Edition.
José-María Montero, Gema Fernández-Avilés, and Jorge Mateu.

Companion Website: www.wiley.com/go/montero/spatial

3. The objectives of an analysis can be functional in nature, as would be the case if finite data were used to estimate an entire function, its derivatives, or the values of other functions.

4. The timing intervals for data observations do not have to be equally spaced for all cases and can vary across cases.

5. FDA methods are not necessarily based on the assumption that the values observed at different times for a single subject are independent.

6. When dealing with huge databases, the data are usually interpreted as reflecting the influence of certain smooth functions that are assumed to underlie and to generate the observations. Classical multivariate statistical techniques do not take advantage of the additional information that could be implied by the smoothness of underlying functions. However, FDA methods do.

7. Because the FDA approach essentially treats the whole curve as a single entity, there is no concern about correlations between repeated measurements (this represents a change in philosophy towards the handling of time series and correlated data).

Thus, it is no surprise that (i) as stated in Levitin *et al.* (2007), FDA was designed to take advantage of replications, and in particular those produced by controlled experiments; and (ii) that since the pioneer research by Deville (1974) and above all since the work by Ramsay and Dalzell (1991), many techniques, not only univariate but also multivariate, have been considered from the functional perspective in the statistical literature: Descriptive and Exploratory Analysis, Linear Models, Generalized Linear Models, Analysis of Variance, Longitudinal Data Analysis, Principal Components Analysis, Factor Analysis, Canonical Correlation, Correspondence Analysis, Discriminant Analysis, Cluster Analysis, a variety of Non-Parametric methods, etc. A brief review of these procedures can be found in Ramsay and Silverman (1997, 2005) and Ferraty and Vieu (2006); Ullah and Finch (2013) carry out a systematic review of the applications of FDA.

Definition 9.1.1 (Ferraty and Vieu 2006) *A random variable, $\mathcal{X}$, is called a functional variable (fv) if it takes values in a infinite-dimensional space (or functional space). An observation of that variable, χ is called a functional observation.*

Thus, as stated in Levitin *et al.* (2007), a functional datum (or observation) is not a single observation but rather a set of measurements along a continuum that, taken together, are to be regarded as a single entity, curve or image. It could be said that the functional datum represents such a set of measurements (observations). To do this, a smooth curve is fitted to the discrete observations, which approximate the continuous underlying process. Then the discrete points are set aside and the functional objects retained for subsequent analyses.

At the risk of being repetitive (but this is an important issue), in terms of Ramsay and Silverman (2005), a functional datum for replication i arrives as a set of discrete measured values $y_{i1}, y_{i2}, \ldots, y_{iN}$. Then, such discrete values are converted to a function χ_i with values $\chi_i(t)$ computable for any desired value of the argument t (this is the first task in FDA). Finally, the discrete measured values are no longer used and FDA works with the functions χ_i, which are considered as single unitary entities. Note that, the symbol χ refers to a function, whereas $\chi(t)$ indicates the value of the function χ at the argument value t.

Is it appropriate to call a finite collection of observations related by means of a curve "functional"? From a semantic point of view, the answer is clearly no. However, as Ramsay and Silverman (2005, p. 38) indicate, the term "functional" in reference to observed data, refers to the intrinsic structure of the data rather than to their explicit form. Having a "functional observation" does not mean that we have the value of the curve at each point in time (if the continuum is time), because among other things this would imply observing an uncountable number of values. Instead, this implies assuming the existence of a function that gives rise to the finite observations available. In addition, such a function is assumed to be "smooth." A curve is said to be smooth when it is differentiable to a certain degree, implying that a number of derivatives can be derived or estimated from the data. As stated in Ramsay and Silverman (2005, p. 38), if this smooth property did not apply, there would be nothing much to be gained by treating the data as functional rather than just multivariate.

Definition 9.1.2 (Ferraty and Vieu 2006) *A functional dataset $\chi_1, \chi_2, \ldots, \chi_n$ is the observation of n functional variables $\mathcal{X}_1, \mathcal{X}_2, \ldots, \mathcal{X}_n$ identically distributed as $\mathcal{X}$.*

Since the functional data is artificially generated by the researcher using finite observations as a basis, now the question is the following: How do we go from a collection of discrete observations to a smooth function? The answer depends on whether the discrete observations are free or not of observational error. In the first case, the procedure to use would be interpolation; in the second, in order to "eliminate" the observational error, the conversion procedure would be smoothing. Therefore, FDA implies considering that the discrete observations available are not free of observation error.

For practical reasons, the procedure used to generate curves from the observed data is the linear combination of basis functions. The reason for this is that basis functions adapt well to storing information about functions and are a very good solution which combines flexibility and computational power. Furthermore, basis functions have the added advantage that their calculations are inserted into the context of traditional matrix algebra. B-spline basis functions (in the case of functional information that does not display marked cyclical variations) and Fourier basis functions (for periodic data) are the most used basis functions. Ramsay and Silverman (2005) and Ramsay *et al.* (2013) are two recommended references to learn about this topic.

Finally, the so-called basic statistics of classical analysis apply equally to the case of functional data. Once we have adjusted smooth curves $\chi_i, i = 1, 2, \ldots, n$, to the n sets of discrete measured values $y_{i1}, y_{i2}, \ldots, y_{iN}$ available (using interpolation if the data are error-free and smoothing otherwise), the basic functional descriptives are defined as follows (Ramsay and Silverman 2005):

- Mean function:

$$\bar{\chi}(t) = \frac{\sum_{i=1}^{n} \chi_i(t)}{n}.$$

- Variance function:

$$S_{\chi}^2(t) = \frac{\sum_{i=1}^{n} \left(\chi_i(t) - \bar{\chi}(t)\right)^2}{n}.$$

- Covariance function:

$$C\left(\chi(t_a), \chi(t_b)\right) = \frac{\sum_{i=1}^{n} \left(\chi_i(t_a) - \bar{\chi}(t_a)\right)\left(\chi_i(t_b) - \bar{\chi}(t_b)\right)}{n}, \quad \forall t_a, t_b \in T.$$

  It summarizes the linear dependence of the data across different values of the argument.

- Correlation function:

$$\rho\left(\chi(t_a), \chi(t_b)\right) = \frac{C\left(\chi(t_a), \chi(t_b)\right)}{\sqrt{S_{\chi}^2(t_a)\, S_{\chi}^2(t_b)}}, \quad \forall t_a, t_b \in T.$$

As can be appreciated, the covariance function and the correlation function are the functional analogies of the variance-covariance matrix and the correlation matrix, respectively in the statistical analysis of multivariate data.

- Cross-covariance function:

$$C\left(\chi(t_a), \mathcal{Y}(t_b)\right) = \frac{\sum_{i=1}^{n} \left(\chi_i(t_a) - \bar{\chi}(t_a)\right)\left(\mathcal{Y}_i(t_b) - \bar{\mathcal{Y}}(t_b)\right)}{n}, \quad \forall t_a, t_b \in T.$$

- Cross-correlation function:

$$\rho\left(\chi(t_a), \mathcal{Y}(t_b)\right) = \frac{C\left(\chi(t_a), \mathcal{Y}(t_b)\right)}{\sqrt{S_{\chi}^2(t_a)\, S_{\mathcal{Y}}^2(t_b)}}, \quad \forall t_a, t_b \in T.$$

The term "cross" implies measuring two characteristics over time.

Example 9.1 (Conversion of the raw data into functional data) *Let us consider the classical daily temperature data registered at* 35 *Canadian weather stations averaged over* 1960 *and* 1994. *The data were obtained from Ramsay and Silverman's home page (http://www.functionaldata.org/). The raw dataset consists of* 35 *sets of discrete measured values* $y_{i1}, y_{i2}, \ldots, y_{i365}$, *one for each monitoring station.*

Table 9.1 lists the raw data for the ten first days of the year and the five first monitoring stations; Figure 9.1 (left panel) depicts the sets of discrete measured values for the 35 *Canadian monitoring stations. As in most of the cases in reality, though the temperatures change slowly and continuously over time, we only have discrete observations sampled from this process.*

Table 9.1 Raw dataset: Canadian average annual weather cycle.

Day	Fredericton	Halifax	Sydney	Miramichi	Kentville	...
1	−14.10	−14.60	−8.60	−5.80	−24.90	...
2	−14.40	−14.70	−8.40	−5.60	−25.00	...
3	−15.00	−15.50	−9.20	−7.30	−26.50	...
4	−14.30	−14.30	−10.80	−7.00	−26.70	...
5	−16.20	−15.80	−12.00	−6.70	−26.70	...
6	−16.30	−16.90	−11.70	−8.90	−28.50	...
7	−14.60	−15.50	−12.10	−7.50	−28.80	...
8	−15.80	−16.60	−13.70	−6.90	−27.40	...
9	−16.80	−17.40	−13.20	−8.90	−27.00	...
10	−16.30	−17.70	−12.90	−8.80	−26.40	...
⋮	⋮	⋮	⋮	⋮	⋮	

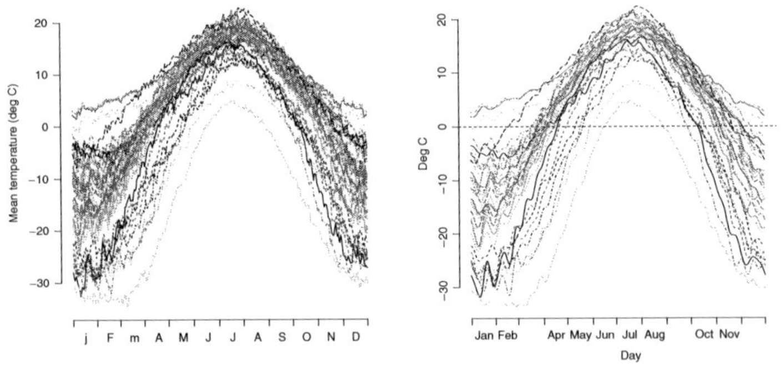

Figure 9.1 Raw daily temperature data (left panel) and their corresponding functional data (right panel) for 35 *Canadian monitoring stations (data are averaged over* 1960 *and* 1994*).*

Each of the 35 *sets of daily temperature data (composed of* 365 *values) is converted into a functional data (a smooth curve), which is supposed to represent the continuous temperature process at each weather station. Since data are periodic, a Fourier basis with an even number of basis functions is the most appropriate choice (Ramsay and Silverman 2005). This is why the* 35 *smooth fits shown in the right panel of the Figure 9.1 were obtained with* 65 *Fourier basis functions, the coefficients of such basis functions being obtained by WLS.*

Example 9.2 (Calculating basic statistics with functional data) *In this example the starting point is the* 35 *functional data obtained in Example 9.1 representing the average annual weather cycle at each of the weather stations where measurements were obtained. From these functional data, their mean, standard deviation (Figure 9.2, left panel and right panel, respectively) as well as the covariance*

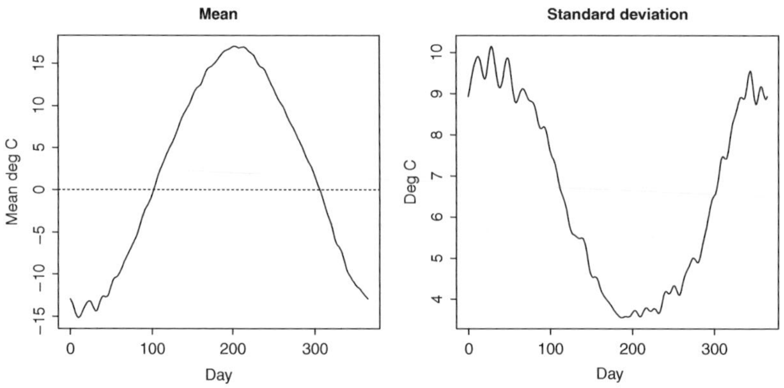

Figure 9.2 Mean and standard deviation function for functional data obtained in Example 9.1.

and the correlation functions across locations (Figures 9.3 and 9.4) have been calculated.

As said above, these functions are simple analogues of the classical measures and have similar interpretations. The mean function plot, for example, indicates that the summer months have the highest temperatures across Canada (around 15 *degrees Celsius). The standard deviation function plot suggests that the winter months have the greatest variability in recorded temperatures across the country – approximately* 9 *degrees Celsius, as compared to the summer months with a standard deviation of about* 4 *degrees Celsius. In the correlation function plot in Figure 9.4 it can be appreciated that the diagonal ridge running from the left to the right is exactly one, as is expected – it refers to the correlation of one day temperatures with the same day temperatures. Note also, as expected, that the correlation diminishes with the time lag, but when such a time lag is greater than* 180 *days, it starts increasing with it.*

9.2 Functional geostatistics: The parametric vs. the non-parametric approach

Just as standard statistical methods have been adapted to be used in the field of FDA, geostatistical procedures have recently been adapted to the functional context and the spatial prediction of curves is one more (sophisticated) geostatistical possibility. In the functional context the objective is the prediction of a functional data (a curve) at a non-observed location of interest as a linear combination of the functional data registered at the sites observed, the coefficients being real numbers – the closer the location observed to the non-observed location, the larger the coefficient of the curve at the site observed. The procedure described is known as kriging of functional data or, in line with the name of other statistical techniques when they are applied to functional data, functional kriging. And the only difference with merely spatial or

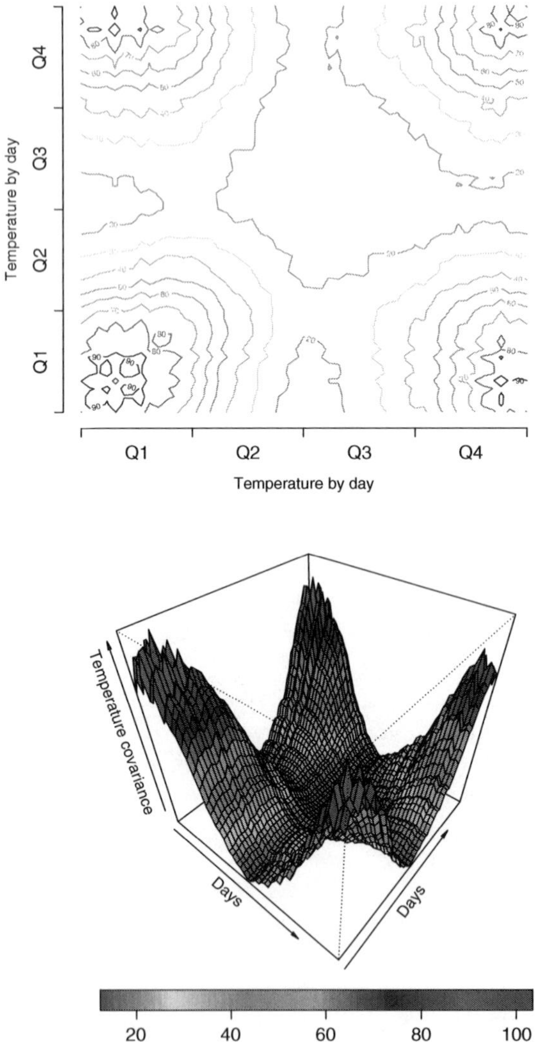

Figure 9.3 Covariance function across locations for functional data obtained in Example 9.1.

spatio-temporal kriging is that both the observation and the prediction support are curves instead of point or block data.

It can be challenging to represent a given set of functional data because representing functional data is troublesome *per se*. There are many ways of representing this type of data, depending on what the researcher intends to highlight, though the most common method (which does not imply it is the most informative) is to plot $\chi(t)$

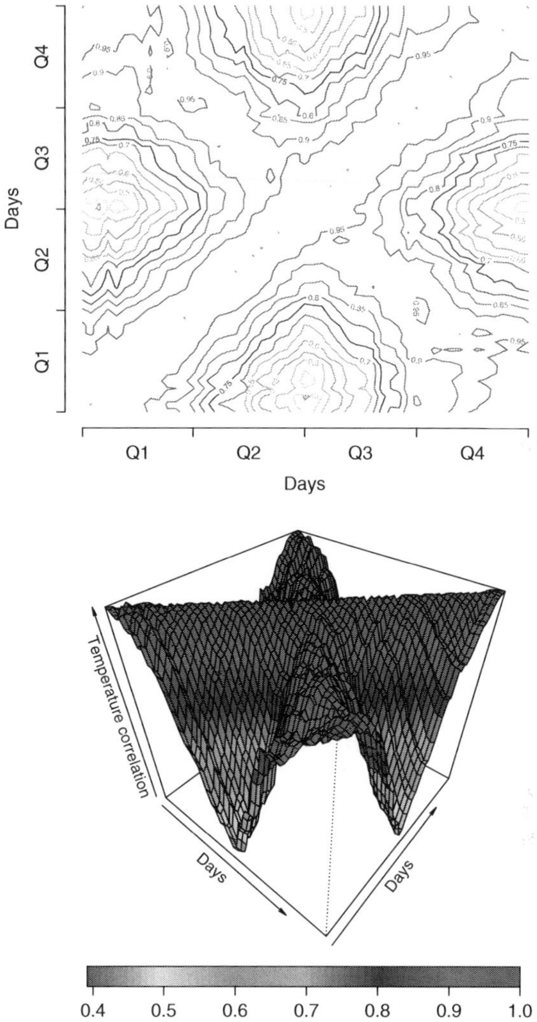

Figure 9.4 Correlation function across locations for functional data obtained in Example 9.1.

against t. However, in the case of functional data observed at a set of given locations, we prefer to represent them as in Figure 9.5, so that the figure includes not only the set of functional data but also the locations where they have been observed.

As stated in Section 9.1, Goulard and Voltz (1993) pioneered the study of functional kriging in a domain with spatial continuity, but they considered the curves known only at a few points and fitted a parametric model (assumed to be known and with a small number of parameters) to them in order to reconstruct the complete curve.

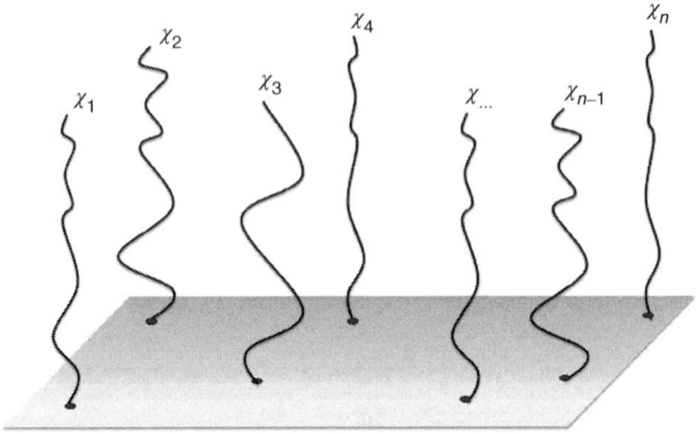

Figure 9.5 Functional data in a set of locations.

However, the shortfalls of the parametric approach by Goulard and Voltz (1993) have been largely overcome by the non-parametric approach by Giraldo (2009) and his co-authors (Giraldo *et al.* 2010; Delicado *et al.* 2010; Giraldo *et al.* 2011a, 2011b; Giraldo and Mateu 2012; Giraldo *et al.* 2012), which is the approach we follow in this book. Monestiez and Nerini (2008) and Nerini *et al.* (2010) are also part of the small body of literature produced on this topic after the pioneer article by Goulard and Voltz.

The approach by Giraldo (2009) is based on applying a non-parametric fitting pre-process to the observed data and is in line with the recent developments in FDA. Although the approach by Giraldo (2009) coincides formally with that of Goulard and Voltz (1993), the non-parametric approach of the former generates noticeable differences with the parametric perspective of the later. The cornerstone of the approach by Giraldo (2009), like the other non-parametric methods, is the choice of the smoothing parameters. However, as it will be seen later, this problem has been ingeniously solved by Giraldo by using a functional cross-validation process.

The approach being one or another, functional kriging is not only the way geostatistics predicts when data are functional in nature, but also a very interesting alternative to spatio-temporal kriging when dealing with massive datasets that imply a enormous computational burden.

In order to simplify the presentation, here we only focus on the functional version of the ordinary kriging (in the isotropic case), the ordinary kriging of functional data or, in short, functional ordinary kriging (FOK), but the method can be formalized for the non-stationary case and for situations where anisotropy is present.

For example, Reyes *et al.* (2012) have proposed a residual kriging approach for the non-stationary case. Caballero *et al.* (2013) have developed a universal kriging approach. Both methods are worked out specifically for functional data belonging to L^2. However, Menafoglio *et al.* (2013a, 2013b) establish a general and coherent theoretical framework for universal kriging prediction in any separable Hilbert space, not just L^2.

9.3 Functional ordinary kriging

9.3.1 Preliminaries

Let us consider a functional stochastic process of spatial nature $\{\mathcal{X}_{\mathbf{s}} : \mathbf{s} \in D \subseteq \mathbb{R}^d\}$, where normally $d = 2$, and let the functional rv associated with the point $\mathbf{s} \in D$ be called $\mathcal{X}_{\mathbf{s}}$. Let $\mathbf{s}_1, \ldots, \mathbf{s}_n$ be a set of arbitrary locations in D on which it is assumed we can observe a realization of the functional process, $\chi_{\mathbf{s}_1}, \ldots, \chi_{\mathbf{s}_n}$, and let t be a compact set in $\mathbb{R}$.

Let us also assume that $\{\mathcal{X}_{\mathbf{s}} : \mathbf{s} \in D \subseteq \mathbb{R}^d\}$ is second-order stationary and isotropic, that is, the mean and variance functions are constant and the covariance and semivariogram function only depend on the distance between the locations (that is, they only depend on $h = |\mathbf{h}| = \|\mathbf{h}\|, \mathbf{h}$ being the separation vector between two locations). In other words, we assume that for each $t \in T$ we have a second-order stationary and isotropic random process. In more formal terms, we assume that:

- $E\left(\mathcal{X}_{\mathbf{s}}(t)\right) = m(t), \quad \forall t \in T, \mathbf{s} \in D.$
- $V\left(\mathcal{X}_{\mathbf{s}}(t)\right) = \sigma^2(t), \quad \forall t \in T, \mathbf{s} \in D.$
- $C\left(\mathcal{X}_{\mathbf{s}_i}(t), \mathcal{X}_{\mathbf{s}_j}(t)\right) = C(\mathbf{h}, t) = C_{\mathbf{s}_i,\mathbf{s}_j}(t), \ \forall \mathbf{s}_{i,}\mathbf{s}_j \in D,\ t \in T,$ with $h = \|\mathbf{s}_i - \mathbf{s}_j\|$
- $\frac{1}{2}V\left(\mathcal{X}_{\mathbf{s}_i}(t) - \mathcal{X}_{\mathbf{s}_j}(t)\right) = \gamma(\mathbf{h}, t) = \gamma_{\mathbf{s}_i,\mathbf{s}_j}(t), \ \forall \mathbf{s}_{i,}\mathbf{s}_j \in D,\ t \in T,$ with $h = \|\mathbf{s}_i - \mathbf{s}_j\|$ where the function $\gamma(\mathbf{h}, t)$, as a function of h is called a semivariogram of $\mathcal{X}_{\mathbf{s}}(t)$.

A word of caution on the above notation: Unlike in the non-functional case, in the expressions of the mean and variance functions, "(t)" does not indicate that they depend on t (because they, m and σ^2, respectively, are constant across D for each $t \in T$) but their value for such instant t. The same can be said for the covariance and semivariogram functions; for each $t \in T$, they only depend on h.

For the prediction of a functional data $\chi_{\mathbf{s}_0}$ in an unsampled site $\mathbf{s}_0 \in D$, the OK linear predictor is adapted to the functional data case:

$$\chi^*_{\mathbf{s}_0}(t) = \sum_{i=1}^{n} \lambda_i \chi_{\mathbf{s}_i}(t), \quad \lambda_{1,\ldots,}\lambda_n \in \mathbb{R}. \tag{9.1}$$

Note that for each $t \in T$, the predictor (9.1) coincides with the point OK predictor (Equation 4.1).

As can be seen, the only difference is that the predictor for functional data is based on curves rather than on one-dimensional data.

As said in Section 9.2 the predictor assigns more weight to the observed curves surrounding the non-observed site where we are interested in predicting the functional data than to those more distant. However, these can be considered more flexible predictors which take into account correlations in the functional index (see Giraldo 2009).

9.3.2 Functional ordinary kriging equations

To find the best linear unbiased predictor (BLUP) $\mathcal{X}_{\mathbf{s}_0}$ we proceed as in the non-functional case. That is, the equations from which we obtain the weights λ_i, which in this case will be called the functional ordinary kriging (FOK) equations, are obtained by imposing on the prediction error the conditions of (i) zero expectation (unbiasedness); and (ii) minimum variance.

As in the non-functional case, as the mean function is constant across D, the fulfillment of the unbiased condition requires the sum of the weights to equal one, $\sum_{i=1}^{n} \lambda_i = 1$.

As regards minimizing the prediction error variance, functional geostatistics proceeds by extending the criterion given in Myers (1982) in the multivariate geostatistics setting. In such a setting (Myers 1982; Ver Hoef and Cressie 1993; Wackernagel 1995), in the homotopic case, the data consist of observations of a vector of m rf's at different spatial locations $\mathbf{Z}(\mathbf{s}) = \left(Z(\mathbf{s}_1), \ldots, Z(\mathbf{s}_n)\right)'$. In this context, the variance of the prediction error, $V\left(\mathbf{Z}^*(\mathbf{s}_0) - \mathbf{Z}(\mathbf{s}_0)\right)$ is a matrix, and the BLUP of the m variables for the unsampled location $\mathbf{s}_0$ is obtained by minimizing $\sigma^2_{\mathbf{s}_0} = \sum_{i=1}^{m} V\left(\mathbf{Z}^*(\mathbf{s}_0) - \mathbf{Z}(\mathbf{s}_0)\right)$ subject to the subsequent conditions of unbiasedness. Therefore, the BLUP is obtained by minimizing the trace of the mean-squared prediction error matrix subject to the conditions of unbiasedness mentioned in Myers (1982).

Having said that, functional geostatistics extends this criterion to the functional context by merely replacing the sum in the above expression with an integral. Then, the n weights of the kriging predictor of $\mathcal{X}_{\mathbf{s}_0}(t)$ are obtained by solving the following optimization problem (in what follows we follow Giraldo 2009):

$$\min_{\lambda_1,\ldots,\lambda_n} \int_T V\left(\mathcal{X}^*_{\mathbf{s}_0}(t) - \mathcal{X}_{\mathbf{s}_0}(t)\right) dt \text{ subject to } \sum_{i=1}^{n} \lambda_i = 1, \tag{9.2}$$

where $\int_T V\left(\mathcal{X}^*_{\mathbf{s}_0}(t) - \mathcal{X}_{\mathbf{s}_0}(t)\right) dt = \sigma^2_{FOK}(\mathbf{s}_0)$ is called the prediction trace-variance. Due to the condition of unbiasedness and Fubini's theorem, we have that:

$$\begin{aligned}\int_T V\left(\mathcal{X}^*_{\mathbf{s}_0}(t) - \mathcal{X}_{\mathbf{s}_0}(t)\right) dt &= \int_T E\left(\mathcal{X}^*_{\mathbf{s}_0}(t) - \mathcal{X}_{\mathbf{s}_0}(t)\right)^2 dt \\ &= E\int_T \left(\mathcal{X}^*_{\mathbf{s}_0}(t) - \mathcal{X}_{\mathbf{s}_0}(t)\right)^2 dt.\end{aligned} \tag{9.3}$$

The expression to be minimized will be:

$$\int_T V\left(\mathcal{X}^*_{\mathbf{s}_0}(t) - \mathcal{X}_{\mathbf{s}_0}(t)\right) dt + 2\mu\left(\sum_{i=1}^{n} \lambda_i - 1\right), \tag{9.4}$$

where the integral in (9.4) can be written as:

$$\sigma^2_{FOK}(\mathbf{s}_0) = \int_T V\left(\mathcal{X}^*_{\mathbf{s}_0}(t) - \mathcal{X}_{\mathbf{s}_0}(t)\right) dt$$

$$
\begin{aligned}
&= \int_T V\left(\mathcal{X}^*_{\mathbf{s}_0}(t)\right) dt + \int_T V\left(\mathcal{X}_{\mathbf{s}_0}(t)\right) dt \\
&\quad - 2\int_T C\left(\mathcal{X}^*_{\mathbf{s}_0}(t), \mathcal{X}_{\mathbf{s}_0}(t)\right) dt \\
&= \int_T V\left(\sum_{i=1}^{n} \lambda_i \mathcal{X}_{\mathbf{s}_i}(t)\right) dt + \int_T \sigma^2(t)\, dt \\
&\quad - 2\int_T C\left(\sum_{i=1}^{n} \lambda_i \mathcal{X}_{\mathbf{s}_i}(t), \mathcal{X}_{\mathbf{s}_0}(t)\right) dt \\
&= \int_T V\left(\sum_{i=1}^{n} \lambda_i \mathcal{X}_{\mathbf{s}_i}(t)\right) dt + \int_T \sigma^2(t)\, dt \\
&\quad - 2\int_T C\left(\sum_{i=1}^{n} \lambda_i \mathcal{X}_{\mathbf{s}_i}(t), \sum_{i=1}^{n} \lambda_i \mathcal{X}_{\mathbf{s}_0}(t)\right) dt \\
&= \int_T \sum_{i=1}^{n}\sum_{j=1}^{n} \lambda_i\lambda_j C\left(\mathcal{X}_{\mathbf{s}_i}(t), \mathcal{X}_{\mathbf{s}_j}(t)\right) dt + \int_T \sigma^2(t)\, dt \\
&\quad - 2\int_T \sum_{i=1}^{n} \lambda_i C\left(\mathcal{X}_{\mathbf{s}_i}(t), \mathcal{X}_{\mathbf{s}_0}(t)\right) dt \\
&= \int_T \sum_{i=1}^{n}\sum_{j=1}^{n} \lambda_i\lambda_j C\left(\mathcal{X}_{\mathbf{s}_i}(t), \mathcal{X}_{\mathbf{s}_j}(t)\right) dt + \int_T \sigma^2(t)\, dt \\
&\quad - 2\sum_{i=1}^{n} \lambda_i \int_T C\left(\mathcal{X}_{\mathbf{s}_i}(t), \mathcal{X}_{\mathbf{s}_0}(t)\right) dt \\
&= \int_T \sum_{i=1}^{n}\sum_{j=1}^{n} \lambda_i\lambda_j C_{\mathbf{s}_i,\mathbf{s}_j}(t)\, dt \\
&\quad + \int_T \sigma^2(t)\, dt - 2\sum_{i=1}^{n} \lambda_i \int_T C_{\mathbf{s}_i,\mathbf{s}_0}(t)\, dt,
\end{aligned} \tag{9.5}
$$

where, as we assume second-order stationarity and isotropy, $C_{\mathbf{s}_i,\mathbf{s}_j}(t)$ and $C_{\mathbf{s}_i,\mathbf{s}_0}(t)$ denote the covariogram values evaluated at $\mathbf{h} = \|\mathbf{s}_i - \mathbf{s}_j\|$ and $\mathbf{h} = \|\mathbf{s}_i - \mathbf{s}_0\|$, respectively.

Hence, the expression to be minimized is:

$$
\int_T \sum_{i=1}^{n}\sum_{j=1}^{n} \lambda_i\lambda_j C_{\mathbf{s}_i,\mathbf{s}_j}(t)\, dt + \int_T \sigma^2(t)\, dt - 2\sum_{i=1}^{n} \lambda_i \int_T C_{\mathbf{s}_i,\mathbf{s}_0}(t)\, dt + 2\mu\left(\sum_{i=1}^{n} \lambda_i - 1\right). \tag{9.6}
$$

By partially deriving with respect to $\lambda_{1,\ldots,}\lambda_n$ and μ, and setting these partial derivatives to zero, we obtain the following system of $n+1$ equations:

$$\sum_{j=1}^{n} \lambda_j \int_T C_{\mathbf{s}_i,\mathbf{s}_j}(t)\,dt + \mu = \int_T C_{\mathbf{s}_i,\mathbf{s}_0}(t)\,dt, \quad i = 1,2,\ldots,n$$

$$\sum_{j=1}^{n} \lambda_j = 1, \tag{9.7}$$

where $\int_T C_{\mathbf{s}_i,\mathbf{s}_j}(t)\,dt$ denotes the trace-covariogram function of the process evaluated in $\mathbf{h} = \|\mathbf{s}_i - \mathbf{s}_j\|$ (Menafoglio *et al.* 2013a).

The above equations, which are called FOK equations in terms of the trace-covariogram, can be expressed in matrix terms as follows:

$$\begin{pmatrix} \int_T C_{\mathbf{s}_1,\mathbf{s}_1}(t)\,dt & \cdots & \int_T C_{\mathbf{s}_1,\mathbf{s}_n}(t)\,dt & 1 \\ \vdots & \ddots & \vdots & \vdots \\ \int_T C_{\mathbf{s}_n,\mathbf{s}_1}(t)\,dt & \cdots & \int_T C_{\mathbf{s}_n,\mathbf{s}_n}(t)\,dt & 1 \\ 1 & \cdots & 1 & 0 \end{pmatrix} \begin{pmatrix} \lambda_1 \\ \vdots \\ \lambda_n \\ \mu \end{pmatrix} = \begin{pmatrix} \int_T C_{\mathbf{s}_1,\mathbf{s}_0}(t)\,dt \\ \vdots \\ \int_T C_{\mathbf{s}_n,\mathbf{s}_0}(t)\,dt \\ 1 \end{pmatrix}. \tag{9.8}$$

We can also express equations (9.7) and (9.8) in terms of the trace-semivariogram $\left(\int_T \gamma_{i,j}(t)\,dt\right)$, see Giraldo (2009). On the basis of assuming stationarity, we have that:

$$\begin{aligned} \gamma_{\mathbf{s}_i,\mathbf{s}_j}(t) &= \frac{1}{2} V\left(\mathcal{X}_{\mathbf{s}_i}(t) - \mathcal{X}_{\mathbf{s}_j}(t)\right) \\ &= \frac{1}{2} E\left(\mathcal{X}_{\mathbf{s}_i}(t) - \mathcal{X}_{\mathbf{s}_j}(t)\right)^2 \\ &= \sigma^2(t) - C_{\mathbf{s}_i,\mathbf{s}_j}(t), \end{aligned} \tag{9.9}$$

hence:

$$\int_T C_{\mathbf{s}_i,\mathbf{s}_j}(t)\,dt = \int_T \sigma^2(t)\,dt - \int_T \gamma_{\mathbf{s}_i,\mathbf{s}_j}(t)\,dt. \tag{9.10}$$

Using this relationship as a basis, the n first equations of (9.7) transform into the following:

$$\sum_{j=1}^{n} \lambda_j \int_T C_{\mathbf{s}_i,\mathbf{s}_j}(t)\,dt + \mu = \int_T C_{\mathbf{s}_i,\mathbf{s}_0}(t)\,dt,$$

$$i = 1,2,\ldots,n.$$

$$\Downarrow$$

$$\sum_{j=1}^{n} \lambda_j \left(\int_T \sigma^2(t)\,dt - \int_T \gamma_{\mathbf{s}_i,\mathbf{s}_j}(t)\,dt \right) + \mu = \int_T \sigma^2(t)\,dt - \int_T \gamma_{\mathbf{s}_i,\mathbf{s}_0}(t)\,dt,$$

$$i = 1,2,\ldots,n.$$

$$\Downarrow$$

$$\sum_{j=1}^{n} \lambda_j \quad \int_T \sigma^2(t)\,dt - \sum_{j=1}^{n} \lambda_j \int_T \gamma_{\mathbf{s}_i,\mathbf{s}_j}(t)\,dt + \mu = \int_T \sigma^2(t)\,dt - \int_T \gamma_{\mathbf{s}_i,\mathbf{s}_0}(t)\,dt,$$

$$i = 1, 2, \ldots, n.$$

$$\Downarrow$$

$$\int_T \sigma^2(t)\,dt - \sum_{j=1}^{n} \lambda_j \int_T \gamma_{\mathbf{s}_i,\mathbf{s}_j}(t)\,dt + \mu = \int_T \sigma^2(t)\,dt - \int_T \gamma_{\mathbf{s}_i,\mathbf{s}_0}(t)\,dt,$$

$$i = 1, 2, \ldots, n.$$

$$\Downarrow$$

$$-\sum_{j=1}^{n} \lambda_j \int_T \gamma_{\mathbf{s}_i,\mathbf{s}_j}(t)\,dt + \mu = -\int_T \gamma_{\mathbf{s}_i,\mathbf{s}_0}(t)\,dt,$$

$$i = 1, 2, \ldots, n.$$

$$\Downarrow$$

$$\sum_{j=1}^{n} \lambda_j \int_T \gamma_{\mathbf{s}_i,\mathbf{s}_j}(t)\,dt - \mu = \int_T \gamma_{\mathbf{s}_i,\mathbf{s}_0}(t)\,dt,$$

$$i = 1, 2, \ldots, n. \tag{9.11}$$

Consequently, the FOK equations (9.8) can be rewritten in terms of the trace-semivariogram as follows:

$$\begin{pmatrix} \int_T \gamma_{s_1,s_1}(t)\,dt & \cdots & \int_T \gamma_{s_1,s_n}(t)\,dt & 1 \\ \vdots & \ddots & \vdots & \vdots \\ \int_T \gamma_{s_n,s_1}(t)\,dt & \cdots & \int_T \gamma_{s_n,s_n}(t)\,dt & 1 \\ 1 & \cdots & 1 & 0 \end{pmatrix} \begin{pmatrix} \lambda_1 \\ \vdots \\ \lambda_n \\ -\mu \end{pmatrix} = \begin{pmatrix} \int_T \gamma_{s_1,s_0}(t)\,dt \\ \vdots \\ \int_T \gamma_{s_n,s_0}(t)\,dt \\ 1 \end{pmatrix}. \tag{9.12}$$

Finally, from the n first equations in (9.7) we have that:

$$\sum_{i=1}^{n} \sum_{j=1}^{n} \lambda_i \lambda_j \int_T C_{\mathbf{s}_i,\mathbf{s}_j}(t)\,dt = \sum_{i=1}^{n} \lambda_i \int_T C_{\mathbf{s}_i,\mathbf{s}_0}(t)\,dt - \sum_{i=1}^{n} \lambda_i \mu, \tag{9.13}$$

hence, by incorporating this relationship into (9.5), we obtain the following expression of the prediction trace-variance of FOK:

$$\begin{aligned} \sigma^2_{FOK}(\mathbf{s}_0) &= \int_T \sum_{i=1}^{n} \sum_{j=1}^{n} \lambda_i \lambda_j C_{\mathbf{s}_i,\mathbf{s}_j}(t)\,dt + \int_T \sigma^2(t)\,dt - 2\sum_{i=1}^{n} \lambda_i \int_T C_{\mathbf{s}_i,\mathbf{s}_0}(t)\,dt \\ &= \sum_{i=1}^{n} \lambda_i \int_T C_{\mathbf{s}_i,\mathbf{s}_0}(t)\,dt - \sum_{i=1}^{n} \lambda_i \mu + \int_T \sigma^2(t)\,dt - 2\sum_{i=1}^{n} \lambda_i \int_T C_{\mathbf{s}_i,\mathbf{s}_0}(t)\,dt \end{aligned}$$

$$= -\sum_{i=1}^{n} \lambda_i \mu + \int_T \sigma^2(t)\,dt - \sum_{i=1}^{n} \lambda_i \int_T C_{\mathbf{s}_i,\mathbf{s}_0}(t)\,dt$$

$$= \int_T \sigma^2(t)\,dt - \sum_{i=1}^{n} \lambda_i \int_T C_{\mathbf{s}_i,\mathbf{s}_0}(t)\,dt - \mu.$$

Moreover, if we consider the relationship in (9.10), the trace-variance of the FOK prediction based on the trace-semivariogram is given by:

$$\sigma^2_{FOK}(\mathbf{s}_0) = \int_T \sigma^2(t)\,dt - \sum_{i=1}^{n} \lambda_i \left(\int_T \sigma^2(t)\,dt - \int_T \gamma_{s_i,s_0}(t)\,dt \right) - \mu$$

$$= \sum_{i=1}^{n} \lambda_i \int_T \gamma_{s_i,s_0}(t)\,dt - \mu.$$

The prediction error trace-variance, $\sigma^2_{FOK}(\mathbf{s}_0)$, is considered a measure of global uncertainty, in the sense that it is an integral version of the classical individual OK prediction variance. If we have a trace-semivariogram model, this parameter can be estimated and used to identify the regions that display the most uncertainty in predictions.

9.3.3 Estimating the trace-semivariogram

In order to solve the system of FOK equations in semivariogram terms, we need to have a trace-semivariogram. In this sense, as mentioned previously, due to assuming that the functional stochastic process has a constant mean (function), m, throughout the domain D, we have that:

$$V\left(\mathcal{X}_{\mathbf{s}_i}(t) - \mathcal{X}_{\mathbf{s}_j}(t)\right) = E\left(\mathcal{X}_{\mathbf{s}_i}(t) - \mathcal{X}_{\mathbf{s}_j}(t)\right)^2,$$

and due to Fubini's theorem:

$$\gamma_{\mathbf{s}_i,\mathbf{s}_j}(h) = \frac{1}{2} E\left(\int_T \left(\mathcal{X}_{\mathbf{s}_i}(t) - \mathcal{X}_{\mathbf{s}_j}(t)\right)^2 dt \right), \qquad \forall \mathbf{s}_i, \mathbf{s}_j \in D, \tag{9.14}$$

with $h = \|\mathbf{s}_i - \mathbf{s}_j\|$.

Then, a generalization of the classical MoM provides the following estimator of the trace-semivariogram:

$$\hat{\gamma}(h) = \frac{1}{2\#N(h)} \sum_{i,j \in N(h)} \int_T \left(\mathcal{X}_{\mathbf{s}_i}(t) - \mathcal{X}_{\mathbf{s}_j}(t)\right)^2 dt, \tag{9.15}$$

where $N(h) = \left\{ \left(\mathbf{s}_i, \mathbf{s}_j\right) : h = \|\mathbf{s}_i - \mathbf{s}_j\| \right\}$, and $\#N(h)$ is the number of distinct elements in $N(h)$. In the case of data irregularly distributed in space, it is common for the number of observations separated by a distance h not to be sufficient to guarantee the empirical semivariogram is representative. As a result, we proceed to replace $N(h) = \left\{ \left(\mathbf{s}_i, \mathbf{s}_j\right) : h = \left\| \mathbf{s}_i - \mathbf{s}_j \right\| \right\}$ with $N(h) = \left\{ \left(\mathbf{s}_i, \mathbf{s}_j\right) : \left\| \mathbf{s}_i - \mathbf{s}_j \right\| \in (h - \varepsilon, h + \varepsilon) \right\}$, with $\varepsilon > 0$ being a small value.

After estimating the trace-semivariogram for a sequence of K lag distances h_k, we fit a parametric semivariogram model $\gamma_\alpha(h)$, chosen from those considered valid (see Section 3.4) to the semivariogram points $\left(h_k, \hat{\gamma}(h_k)\right), k = 1, \ldots, K$, as in classical geostatistics – note that the trace-semivariogram is a real-valued function which has to fulfill the same set of requirements as its finite-dimensional analogue. An alternative to this parametric fit is to apply smoothing techniques (splines or local linear regression, see, for example, Wasserman 2006) to the set of semivariogram points to try and assess, approximately, $\hat{\gamma}(h)$ for any given value $h \in \mathbb{R}^+$.

However, the estimator that this method yields, $\tilde{\gamma}(h)$, requires more attention *because it is* conditionally negative-definite.

9.3.4 Functional cross-validation

As in classical geostatistics, in the functional kriging process, a valid semivariogram model has to be chosen to appropriately represent the empirical trace-semivariogram, and eventually used, and some other assumptions have been made (usually about stationarity and the number of basis functions). However, we should validate these assumptions, especially those about the semivariogram. Otherwise, the results obtained from the functional kriging process could lead us to erroneous conclusions.

In the framework of spatially correlated functional observations, the cross-validation (CV) method is, as in the case of point spatially correlated observations, the most popular validation procedure. In order to include in the training set as many functional observations as possible (unfortunately, in most situations the empirical information available is not large enough to allow a validation set), the leave-one-out CV method is the most used cross-validation method, but (depending on the sample size) leave-more-than one-out methods can be applied. In the functional data setting, the leave-one-out CV method is known as leave-one-out functional CV (FCV) method, and consists of: (i) removing each of the functional observations from the database (one at a time), using the rest of functional observations to predict a smoothed function at the removed location (obviously, using the functional kriging predictor); (ii) calculating the n squared errors resulting from the functional kriging prediction; and, finally, (iii) calculating the sum of such squared errors:

$$SSE_{FCV} = \sum_{i=1}^{n} \sum_{j=1}^{m} \left(\chi_{\mathbf{s}_i}\left(t_j\right) - \hat{\chi}_{\mathbf{s}_i}^{(i)}\left(t_j\right) \right)^2, \tag{9.16}$$

where, $\hat{\chi}_{\mathbf{s}_i}^{(i)}\left(t_j\right)$ represents the functional data predicted at $\mathbf{s}_i$, evaluated at t_j, $j = 1, \ldots, m$, when we temporarily remove $\mathbf{s}_i$ from the dataset and use the rest of the functional observations to make a functional kriging prediction at such a site.

Obviously, the best set of assumptions about the semivariogram, type of stationarity, number of basis functions, etc. is that which results in the smallest SSE_{FCV}.

In the specific case of FOK, using B-splines as a smoothing method, Giraldo (2009) proposes the following steps to carry out the FCV method:

- Minimize in $L \in \left[L_{\min}; L_{\max}\right]$ and $\eta \in \left(\eta_{\min}; \eta_{\max}\right)$ the function $FCV(L, \eta)$, L and η being the number of interior knots and the smoothing parameter, respectively. $FCV(L, \eta)$ is computed for each fixed value (L, η) as follows:

  1. For $i = 1, \ldots, n$ repeat:

     (i) Leave the site $\mathbf{s}_i$ out of the sample.

     (ii) For $i' \neq i$ fit a cubic spline to the observations of the functional variable at $t_1, \ldots t_m,\ t \in T$, using the parameters. Let $\tilde{\chi}_{s_{i'}}$ be the smoothed function.

     (iii) Using (9.15), estimate the empirical trace-semivariogram from the functional dataset created in (ii) and fit a parametric valid semivariogram model to the empirical trace-semivariogram.

     (iv) Solve the system of FOK equations (9.8), created in (ii) and the valid semivariogram chosen in (iii), and obtain the predicted functional data $\tilde{\chi}^{(i)}_{\mathbf{s}_i}$ at $\mathbf{s}_0 = \mathbf{s}_i$.

     (v) Compute a measure of distance between $\chi_{\mathbf{s}_i}$ and $\tilde{\chi}_{s_i^{(i)}}$ at $t_1, \ldots, t_m$, $t \in T$:

$$SSE_{FCV}(L, \eta) = \sum_{j=1}^{m} \left(\chi_{\mathbf{s}_i}(t) - \hat{\chi}^{(i)}_{\mathbf{s}_i}(t_j)\right)^2.$$

  2. Define $FCV\,(L, \eta) = \sum_{i=1}^{n} SSE_{FVC}\,(i)$.

- Use the optimal values (L^*, η^*) to smooth the whole sample. Then fit a parametric semivariogram model to the empirical trace-semivariogram and use it to predict the functional data at non-observed sites using the FOK equations in (9.8).

For practical purposes, we must make a couple of clarifications:

(a) When $t_1, \ldots t_m,\ t \in T$, are equally spaced, that is, when the regularity of the time series is constant, $SSE_{FCV}(i)$ is (up to a multiplicative constant) an approximation of the quantity $\int_T \left(\chi_{\mathbf{s}_i}(t) - \tilde{\chi}^{(i)}_{\mathbf{s}_i}(t)\right)^2 dt$.

(b) Estimating the trace-semivariogram implies calculating integrals, which in the case of fitting cubic splines with a common basis of B-splines, can be simplified as follows:

$$\begin{aligned}\int_T \left(\tilde{\chi}_{\mathbf{s}_i}(t) - \tilde{\chi}_{\mathbf{s}_j}(t)\right)^2 dt &= \int_T \left(\mathbf{c}_i^T \mathbf{B}(t) - \mathbf{c}_j^T \mathbf{B}(t)\right)^2 dt \\ &= \int_T \left(\left(\mathbf{c}_i^T - \mathbf{c}_j^T\right) \mathbf{B}(t)\right)^2 dt\end{aligned}$$

$$= (\mathbf{c}_i - \mathbf{c}_j)^T \left(\int_T \mathbf{B}(t) \mathbf{B}^T(t)\, dt \right) (\mathbf{c}_i - \mathbf{c}_j)^T$$
$$= (\mathbf{c}_i - \mathbf{c}_j)^T \mathbf{W} (\mathbf{c}_i - \mathbf{c}_j)^T,$$

where $\mathbf{c}$ is the vector of coefficients c_k in the cubic spline, $\mathbf{B}(t)$ is a matrix containing the set of $(L+4)$ cubic B−splines $B_k(t)$ (Ramsay and Silverman 2005), and the matrix $\mathbf{W}$ has the advantage of depending only on the knots and, therefore, is the same for all the locations $\mathbf{s}_i$. Some other technical simplifications can be seen in Giraldo (2009).

Obviously, these two considerations are two good reasons for using a common value of L for all the functions $\tilde{\chi}_{\mathbf{s}_i}$. The parameter η could vary from one location to another, but this would imply traditional cross-validation in the fitting stage of any $\tilde{\chi}_{\mathbf{s}_i}$, significantly raising the computing cost. This is the main reason why in practice a common η at all the locations is used.

Example 9.3 (FOK of *logCO in Madrid)** *Let us consider the database containing the* 28704 *logCO** *data described in Section 1.3 (the logCO** *measurement for each of the* 24 *hours (of a "typical working day") of the* 52 *weeks of* 2008 *at each of the* 23 *monitoring stations operating in the city of Madrid).*

*As ecologists claim that the monitoring stations operating in the city of Madrid are not located in the most polluted locations of the city (they accuse the municipal government of moving the monitoring stations to less polluted districts of the city in a bid to post lower levels of pollution), we have selected four supposedly very polluted locations ((*A*) Plaza Cibeles, (*B*) Plaza de Callao, (*C*) Plaza Carlos V (Atocha), and (*D*) Puerta del Sol) to check whether the level of logCO** *at such four locations exceeds that of those registered at the sites where the monitoring stations are placed or not. For this purpose we will use FOK. Computations were made with* `fda` *and* `geoR` *packages in R software.*

In the central panel of Figure 9.6 are shown the locations of the prediction sites (note that the central panel depicts 27 *monitoring stations, but four of them, those*

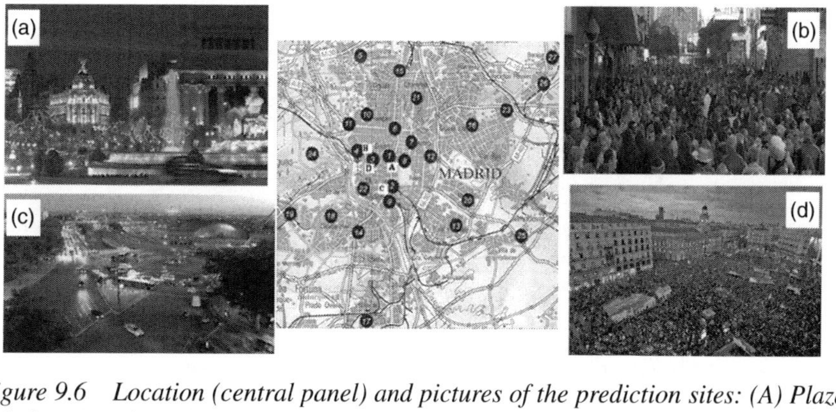

Figure 9.6 Location (central panel) and pictures of the prediction sites: (A) Plaza Cibeles, (B) Plaza de Callao, (C) Plaza Carlos V (Atocha), and (D) Puerta del Sol.

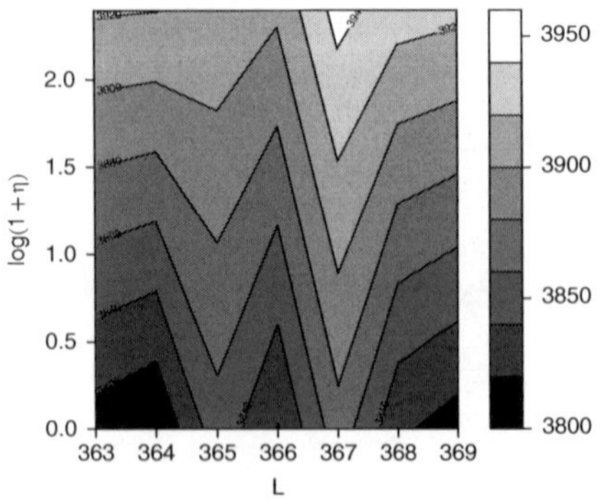

Figure 9.7 Contour plot for MSE_{FCV}.

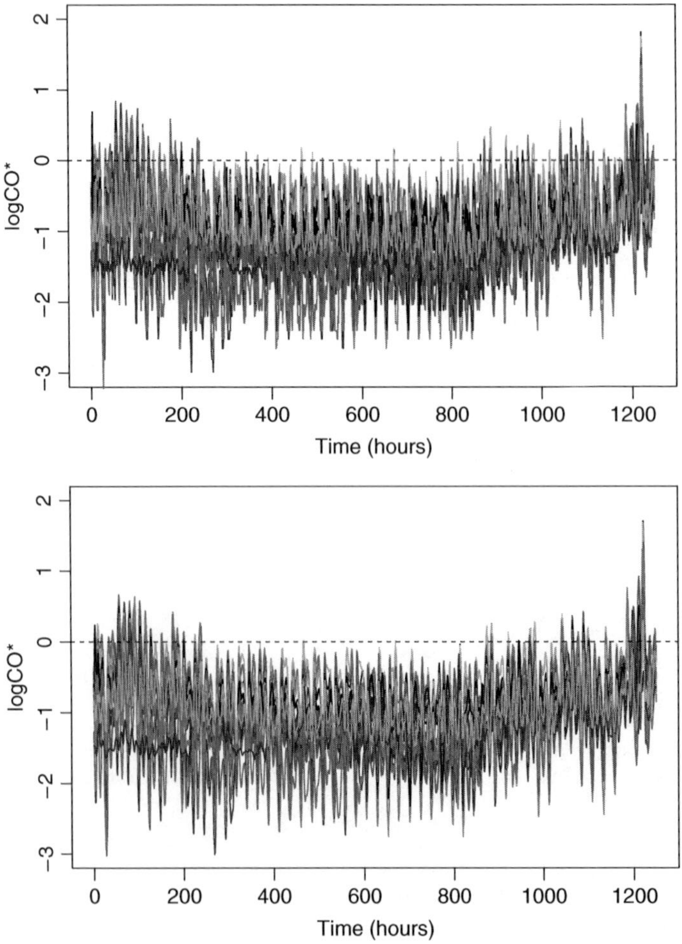

Figure 9.8 Original (top panel) and functional (bottom panel) data of logCO for the 23 monitoring stations operating in the city of Madrid in 2008. (See color figure in color plate section.)*

marked with numbers 2, 17, 26, *and* 27, *were not operative during* 2008*). The left and right panels show photos of such sites. As can be appreciated, the prediction sites are heavily transited areas located in the center of the city, visited daily by more than one million people.*

The first task to be done is to convert the 1248 *data we have for each monitoring station in a functional data entries (a single entity). This way, our functional database will consist of* 23 *functional data entries, one per monitoring station. For this purpose we use B-splines as basis functions to smooth the original series. The cross-validation process is used to find the optimal number of inner knots, L, and the value of the parameter of roughness penalty, η, which controls the trade-off between the fit of the observed data and the smoothness of the approximating spline. On the basis of a preliminary exploration of the data, the sets of possible values considered for L and η have been* $\{363, 364, \ldots, 369\}$ *and* $\{0, 1, 10, 10^2, 10^3\}$, *respectively. The values* $L^* = 366$ *and* $\eta^* = 0$ *were found to be those that minimize* $SSE_{FCV} = \sum_{i=1}^{n} \sum_{j=1}^{m} \left(\chi_{\mathbf{s}_i}(t_i) - \chi_{\mathbf{s}_i}^{(i)}(t_j) \right)^2$ *(the contour plot for the* MSE_{FCV} *is displayed in Figure 9.7, where it has been used as a logarithmic scale for η), and thus they were used to smooth the whole sample.*

Figure 9.8 shows the original data on the top panel, and the set of 23 *smoothed functions (functional data) on the bottom panel.*

Using WLS, an exponential semivariogram with a practical range of 4969 *m, a sill of* 233 *and no nugget effect has been fitted to the empirical trace-semivariogram which was used in the algorithm of the FCV function. Figure 9.9 shows the functional residuals derived from the cross-validation procedure of functional kriging.*

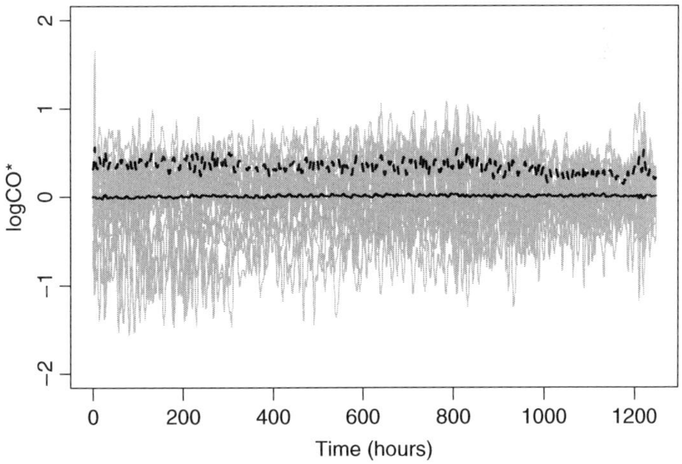

Figure 9.9 Functional residuals obtained through functional cross-validation. Mean function (continuous black line) and standard deviation function of residuals (dotted black line).

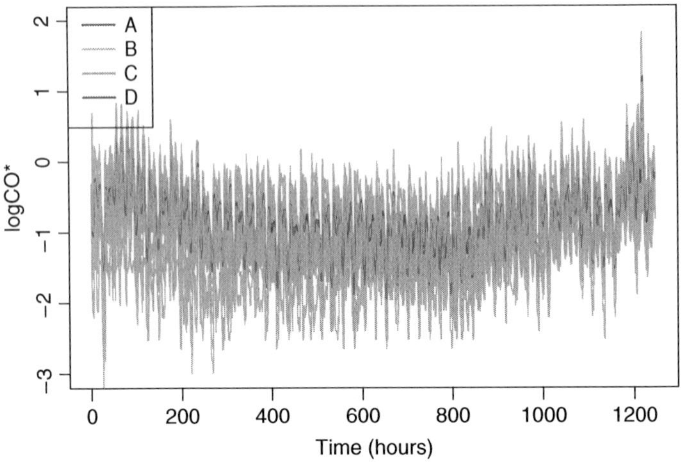

Figure 9.10 Predicted curves of logCO at (A) Plaza Cibeles, (B) Plaza de Callao, (C) Plaza Carlos V (Atocha), and (D) Puerta del Sol (bold lines) together with the observed functional data at the 23 monitoring stations.*

Let us say that, when predicting a specific curve in the FCV process, the predicted curve will be smoother than the corresponding observed one. In addition, when repeating the process for all the observed sites, the variability among the set of predicted curves is less than the variability among the observed ones. This is logical as kriging in itself is a smoothing method (leading to a decrease in variance).

Finally, using the FOK equations (9.8), and assuming that the above-mentioned exponential semivariogram with $range = 4969\,m$, $sill = 233$ *and* $nugget\ effect = 0$ *represents appropriately the existing spatial dependence, we obtain the prediction of the functional data at the four prediction sites.*

Figure 9.10 displays the curves representing the prediction of logCO at (A) Plaza Cibeles, (B) Plaza de Callao, (C) Plaza Carlos V (Atocha), and (D) Puerta del Sol, together with the curves observed at the sites where the monitoring stations are located. In Figure 9.10 it can be appreciated that the four predicted curves are in the top of the chart, which means that ecologists are right and the monitoring stations are not located in the most polluted points of the city. Note also that, as expected for the reasons given above, the predicted curves are smoother than the observed ones at the sites where the monitoring stations are placed.*

A

Spectral representations

Spectral methods are extremely useful when studying spatial (and spatio-temporal) rf's. In this Appendix, we briefly show some results relating to the spectral representation of continuous covariograms and semivariograms following Fernández-Casal (2004). A more detailed discussion on this topic can be found in Yaglom (1986) and Christakos (1992). Other classical references include Stein (1999), Chilès and Delfiner (1999), Schabenberger and Gotway (2005) and the references therein.

A.1 Spectral representation of the covariogram

It is a well-known fact that the class of covariograms that are valid in $\mathbb{R}^d$ is equivalent to the class of positive-definite functions in $\mathbb{R}^d$. Hence, it is possible to characterize functions of this type using their spectral representation, which makes obtaining valid covariogram models much easier.

Let us imagine a second-order stationary real-valued spatial rf in $\mathbb{R}^d$ having a covariogram $C(\mathbf{h})$ that is continuous at the origin.

According to Bochner's theorem (1955), as $C(\cdot)$ is a positive semi-definite function and continuous at the origin, it can be represented as follows:

$$C(\mathbf{h}) = \int_{\mathbb{R}^d} e^{i\boldsymbol{\omega}\mathbf{h}} dF(\boldsymbol{\omega}), \tag{A.1}$$

where dF is a positive finite measure, the converse also being true, and $\boldsymbol{\omega}$ represents the angular frequency.

If the function F is differentiable, Equation (A.1) can be expressed as:

$$C(\mathbf{h}) = \int_{\mathbb{R}^d} e^{i\boldsymbol{\omega}\mathbf{h}} f(\boldsymbol{\omega}) d(\boldsymbol{\omega}), \tag{A.2}$$

Spatial and Spatio-Temporal Geostatistical Modeling and Kriging, First Edition.
José-María Montero, Gema Fernández-Avilés, and Jorge Mateu.

Companion Website: www.wiley.com/go/montero/spatial

$f(\boldsymbol{\omega})$ being the spectral density function:

$$f(\boldsymbol{\omega}) = \frac{1}{(2\pi)^d}\int_{\mathbb{R}^d} e^{-i(\boldsymbol{\omega}\mathbf{h})}C(\mathbf{h})d(\mathbf{h}), \tag{A.3}$$

also verifying that $f(\omega) \geq 0$.

It is worth mentioning that $C(\mathbf{h})$ focuses on spatial dependency as a function of the separation vector $\mathbf{h}$, whereas $f(\boldsymbol{\omega}) \geq 0$ emphasizes the association of components of variability with frequencies. As stated in Schabenberger and Gotway (2005, p. 78), from the covariance function we glean the degree of continuity and the decay of spatial autocorrelation with increasing point separation. From the spectral density we glean periodicity in the process.

Due to the fact that in the case of a real-valued rf the covariogram is an even function, $C(\mathbf{h}) = C(-\mathbf{h})$, the exponential factors in the foregoing expressions can be replaced by cosines, obtaining:

$$C(\mathbf{h}) = \int_{\mathbb{R}^d} \cos(\boldsymbol{\omega}\mathbf{h})f(\boldsymbol{\omega})d\boldsymbol{\omega}, \tag{A.4}$$

and

$$f(\boldsymbol{\omega}) = \frac{1}{(2\pi)^d}\int_{\mathbb{R}^d} \cos(\boldsymbol{\omega}\mathbf{h})C(\mathbf{h})d(\mathbf{h}), \tag{A.5}$$

which is also an even function.

If the covariogram is isotropic, the Equations (A.4) and (A.5) can be transformed (in spherical coordinates, see Stein 1999, pp. 43–4, for example) into the corresponding isotropic expressions:

$$C(|\mathbf{h}|) = (2\pi)^{\frac{d}{2}}\int_0^\infty \frac{J_{(d-2)/2}(\lambda|\mathbf{h}|)}{(|\omega||\mathbf{h}|)^{(d-2)/2}} f(|\boldsymbol{\omega}|)|\boldsymbol{\omega}|^{(d-1)}d|\boldsymbol{\omega}|, \tag{A.6}$$

$$f(|\boldsymbol{\omega}|) = (2\pi)^{\frac{d}{2}}\int_0^\infty \frac{J_{(d-2)/2}(|\boldsymbol{\omega}||\mathbf{h}|)}{(|\boldsymbol{\omega}||\mathbf{h}|)^{(d-2)/2}} C(|\mathbf{h}|)|\mathbf{h}|^{(d-1)}d|\mathbf{h}|, \tag{A.7}$$

where $J_\upsilon(\cdot)$ is a Bessel function of order v, which can be expressed (see Abramowitz and Stegun 1965) as:

$$\begin{aligned} J_\upsilon(z) &= \frac{\left(\frac{1}{2}z\right)^\upsilon}{\pi^{1/2}\Gamma\left(\upsilon+\frac{1}{2}\right)}\int_o^\infty \cos(z\cos\theta)\sin^{2\upsilon}\theta d\theta = \\ &= \left(\frac{1}{2}z\right)^\upsilon \sum_{k=0}^\infty \frac{\left(-\frac{1}{4}z^2\right)^k}{k!\Gamma(\upsilon+k+1)}. \end{aligned} \tag{A.8}$$

Using the above as a basis, we can consider rewriting (A.6) so it is more manageable by defining:

$$\kappa_d(x) = \left(\frac{2}{x}\right)^{(d-2)/2}\Gamma\left(\frac{d}{2}\right)J_{(d-2)/2}(x), \tag{A.9}$$

which would yield:

$$C(|\mathbf{h}|) = \frac{2\pi^{d/2}}{\Gamma(\frac{d}{2})} \int_0^\infty \kappa_d(|\boldsymbol{\omega}||\mathbf{h}|) f(|\boldsymbol{\omega}|)|\boldsymbol{\omega}|^{d-1} d|\boldsymbol{\omega}| = \\ = \int_0^\infty \kappa_d(|\boldsymbol{\omega}||\mathbf{h}|) dG_d(|\boldsymbol{\omega}|), \tag{A.10}$$

where G_d is a non-decreasing bounded function in $(0, \infty)$ that takes the following form:

$$G_d(|\boldsymbol{\omega}^*|) = \int_{|\boldsymbol{\omega}|<|\boldsymbol{\omega}^*|} dF(\boldsymbol{\omega}), \tag{A.11}$$

dF being a symmetrical positive measure.

Furthermore, bearing in mind (A.8) and (A.9), we can check that $\kappa_d(0) = 1$, and therefore,

$$C(\mathbf{0}) = \int_0^\infty dG_d(|\boldsymbol{\omega}|). \tag{A.12}$$

Generally speaking, the expressions in Equation (A.10) include trigonometrical functions when d is odd, and Bessel functions of integer order when d is even. For example, simplifying the notation, in the particular cases of $d = 1, 2$ and 3, we would obtain:

- $d = 1$: $\kappa_1(x) = \cos(x)$, $C(|\mathbf{h}|) = \int_0^\infty \cos(|\boldsymbol{\omega}||\mathbf{h}|) dG_d(|\boldsymbol{\omega}|)$.
- $d = 2$: $\kappa_2(x) = J_0(x)$, $C(|\mathbf{h}|) = \int_0^\infty J_0(|\boldsymbol{\omega}||\mathbf{h}|) dG_d(|\boldsymbol{\omega}|)$.
- $d = 3$: $\kappa_3(x) = \frac{\sin(x)}{x}$, $C(|\mathbf{h}|) = \int_0^\infty \frac{sen(|\boldsymbol{\omega}||\mathbf{h}|)}{(|\boldsymbol{\omega}||\mathbf{h}|)} dG_d(|\boldsymbol{\omega}|)$.

In Section 3.2 we already saw that a valid isotropic covariogram in $\mathbb{R}^d$ is also a valid isotropic covariogram in $\mathbb{R}^m, \forall m \leq d$. Furthermore, it can be deduced from the above results (bearing in mind that $\kappa_d((2d)^{1/2}x) \to e^{-x^2}$ uniformly in x when $d \to \infty$), that a function is a valid covariogram in any dimension if, and only if, it can be represented as follows:

$$C(|\mathbf{h}|) = \int_0^\infty e^{-|\boldsymbol{\omega}|^2|\mathbf{h}|^2} dG(|\boldsymbol{\omega}|), \tag{A.13}$$

G being a non-decreasing bounded function in $[0, \infty)$. Hence:

$$\kappa_\infty(x) = e^{-x^2}. \tag{A.14}$$

In general, the covariogram may not be continuous at the origin (nugget effect); in this case, it can be expressed as:

$$C(\mathbf{h}) = c_0 I_{\{0\}}(\mathbf{h}) + C^0(\mathbf{h}), \tag{A.15}$$

where $c_0 \geq 0$ is the nugget effect, $I_{\{0\}}(\cdot)$ is the function that indicates the origin and $C^0(\cdot)$ is a continuous covariogram at the origin which accepts the corresponding spectral representation. Using (A.15) and (3.13) as a basis, we can deduce that the expression of the corresponding semivariogram is:

$$\gamma(\mathbf{h}) = C(\mathbf{0}) - C^0(\mathbf{h}) = c_0 + (C^0(\mathbf{0}) - C^0(\mathbf{h})), \quad \text{(A.16)}$$

for $\mathbf{h} \neq 0$ (remember that it is always verified that $\gamma(\mathbf{0}) = 0$).

A.2 Spectral representation of the semivariogram

The semivariogram of an intrinsically stationary rf is necessarily conditionally negative-definite. Therefore, in order to determine the conditions in which the converse is true, the following result is extremely useful:

If $g(\cdot)$ is a continuous function in $\mathbb{R}^d$ verifying $g(0) = 0$, then the following statements are equivalent (Schoenberg 1938; Neumann and Schoenberg 1941):

(i) $g(\mathbf{h})$ is conditionally negative semidefinite.

(ii) $e^{-tg(\mathbf{h})}$ is positive semidefinite (i.e. a covariogram), $\forall t > 0$.

(iii) $g(\cdot)$ takes the form:

$$g(\mathbf{h}) = \int_{\mathbb{R}^d} \frac{1 - \cos(\boldsymbol{\omega}\mathbf{h})}{|\boldsymbol{\omega}|^2} dF(\boldsymbol{\omega}) + Q(\mathbf{h}), \quad \text{(A.17)}$$

where $Q(\cdot) \geq 0$, is a quadratic form and dF is a positive, symmetrical and continuous measure at the origin which verifies:

$$\int_{\mathbb{R}^d} \frac{1}{1 + |\boldsymbol{\omega}|^2} dF(\boldsymbol{\omega}) < \infty. \quad \text{(A.18)}$$

Using this result as a basis, we can deduce (e.g. Cressie 1993, pp. 87–8) that if $\gamma(\cdot)$ is a conditionally negative semi-definite function with $Q(\cdot) = 0$ in the representation (A.17), then it is the variogram of an intrinsically stationary process in $\mathbb{R}^d$.

If the semivariogram is isotropic, (A.17) can be written as:

$$\gamma(|\mathbf{h}|) = \int_0^\infty \frac{1 - \kappa_d(|\boldsymbol{\omega}||\mathbf{h}|)}{|\boldsymbol{\omega}|^2} dG(|\boldsymbol{\omega}|), \quad \text{(A.19)}$$

where $\kappa_d(\cdot)$ is defined by (A.9) and G is a non-decreasing bounded function in $[0, \infty)$ which verifies:

$$\int_0^\infty \frac{1}{1 + |\boldsymbol{\omega}|^2} dG(|\boldsymbol{\omega}|) < \infty. \quad \text{(A.20)}$$

Furthermore, in the above conditions of the representation (A.19), we can see from Christakos (1984) that the verification of:

$$\lim_{|\mathbf{h}| \to \infty} \frac{\gamma(|\mathbf{h}|)}{|\mathbf{h}|^2} = 0, \quad \text{(A.21)}$$

is equivalent to the restriction (A.20). For the general case in which the semivariogram is not necessarily continuous at the origin, it is sufficient to express it, using the same notation as in (A.16), as:

$$\gamma(\mathbf{h}) = c_0 \left(1 - I_{\{0\}}(\mathbf{h})\right) + \gamma^0(\mathbf{h}), \tag{A.22}$$

$\gamma^0(\cdot)$ being a continuous semivariogram at the origin.

B

Probabilistic aspects of $U_{ij} = Z(\mathbf{s}_i) - Z(\mathbf{s}_j)$

Let $Z(\mathbf{s})$ be an intrinsically stationary Gaussian rf from which a set of values $\{Z(\mathbf{s}_1), \ldots, Z(\mathbf{s}_n)\}$ has been observed at the locations $\{\mathbf{s}_1, \ldots, \mathbf{s}_n\}$ of the area of interest. Let us consider all the possible differences $U_{ij} = Z(\mathbf{s}_i) - Z(\mathbf{s}_j)$, $i \neq j$. Assuming that the distribution of those differences is Gaussian, then $U_{ij} \sim G\left(0, 2\gamma_{ij}(\boldsymbol{\theta})\right)$, where $\gamma_{ij} = \gamma(\mathbf{s}_i - \mathbf{s}_j; \boldsymbol{\theta})$, a result that can be directly generalized to the spatio-temporal setting.

In effect, $2\hat{\gamma}(\mathbf{h})$ can be rewritten as a quadratic form in $\mathbf{Z} \equiv \left(Z\left(\mathbf{s}_1\right), \ldots, Z\left(\mathbf{s}_n\right)\right)'$:

$$2\hat{\gamma}(\mathbf{h}) = \frac{\mathbf{Z}'\left(Q(\mathbf{h})' Q(\mathbf{h})\right)\mathbf{Z}}{\#N(\mathbf{h})} \equiv \mathbf{Z}'\mathbf{A}(\mathbf{h})\mathbf{Z}, \tag{B.1}$$

with $Q(\mathbf{h})$ a matrix of order $(\#N(\mathbf{h}) \times n)$ the elements of which are $1, -1$, or 0. Therefore, $\mathbf{A}(\mathbf{h}) = \frac{Q(\mathbf{h})'Q(\mathbf{h})}{\#N(\mathbf{h})}$ is a symmetric matrix of order $(n \times n)$. If $\mu(\mathbf{s}) = E(Z(\mathbf{s}))$ is assumed to be constant on $\mathbf{s}$, as $\mathbf{1} \equiv (1, \ldots, 1)$ is an eigenvector of $\mathbf{A}(\mathbf{h})$, $E(\mathbf{Z}) = \mathbf{0}$ can be written without losing generality.

If $\mathbf{Z}(\mathbf{s})$ is $G(\boldsymbol{\mu}, \boldsymbol{\Sigma}(\boldsymbol{\theta}))$ it is easy to see that:

$$2\hat{\gamma}(\mathbf{h}) = \sum_{i=1}^{n} \lambda_i(\mathbf{h})\, \chi^2_{1,i}, \tag{B.2}$$

that is, $2\hat{\gamma}(\mathbf{h})$ is a linear combination of independent rv's with a Chi-squared distribution with one degree of freedom. The coefficients $\lambda_i(\mathbf{h})$, $i = 1, \ldots, n$, are the eigenvalues of the non negative-definite matrix $\mathbf{A}(\mathbf{h})\boldsymbol{\Sigma}(\boldsymbol{\theta})$.

Spatial and Spatio-Temporal Geostatistical Modeling and Kriging, First Edition.
José-María Montero, Gema Fernández-Avilés, and Jorge Mateu.

Companion Website: www.wiley.com/go/montero/spatial

The first two moments of $2\hat{\gamma}(\mathbf{h})$ are:

$$E(2\hat{\gamma}(\mathbf{h})) = tr(\mathbf{A}(\mathbf{h})\,\boldsymbol{\Sigma}(\boldsymbol{\theta})) \tag{B.3}$$

$$V(2\hat{\gamma}(\mathbf{h})) = 2tr(\mathbf{A}(\mathbf{h})\,\boldsymbol{\Sigma}(\boldsymbol{\theta})\mathbf{A}(\mathbf{h})\,\boldsymbol{\Sigma}(\boldsymbol{\theta})) \tag{B.4}$$

$$C\left(2\hat{\gamma}\left(\mathbf{h}_1\right), 2\hat{\gamma}\left(\mathbf{h}_2\right)\right) = 2tr\left(\mathbf{A}\left(\mathbf{h}_1\right)\boldsymbol{\Sigma}(\boldsymbol{\theta})\mathbf{A}\left(\mathbf{h}_2\right)\boldsymbol{\Sigma}(\boldsymbol{\theta})\right). \tag{B.5}$$

Of course, if $\mu(\mathbf{s})$ is not constant, $2\hat{\gamma}(\mathbf{h})$ is a poor estimator of the variogram and should not be used until the observations have been detrended.

C

Basic theory on restricted maximum likelihood

The idea underlying Restricted Maximum Likelihood (REML) is to estimate the variance components (in the context where it emerged), on the basis of the residuals resulting after adjusting by OLS the part of the model that contains the fixed effects.

Instead of directly using the vector of observations[1] $\mathbf{z}$, REML employs linear combinations of the observations such that they do not contain the fixed effects (in our case, they do not contain the mean of the process), whatever their value. These linear combinations are equivalent to the residuals obtained after adjusting for the fixed effects (which is why the procedure is also known as residual maximum likelihood).

If $\mathbf{z} = \mathbf{X}\boldsymbol{\beta} + \boldsymbol{\delta}$, the procedure starts with a set of linear combinations $\mathbf{k}'\mathbf{z} = \mathbf{k}'\mathbf{X}\boldsymbol{\beta} + \mathbf{k}'\boldsymbol{\delta}$, where the vectors $\mathbf{k}'\mathbf{z}$ are chosen such that they have no term in $\boldsymbol{\beta}$, which is achieved if $\mathbf{k}'\mathbf{X}\boldsymbol{\beta} = 0,\ \forall\boldsymbol{\beta}$, which also requires $\mathbf{k}'\mathbf{X} = \mathbf{0}'$.

Now the question is as follows: What form should $\mathbf{k}'$ take?

As $\mathbf{k}'\mathbf{X} = \mathbf{0}'$, then $\mathbf{X}'\mathbf{k} = \mathbf{0}$ and on the basis of the theory for solving linear equations (Searle 1982, Section 9.4b) $\mathbf{k} = \left(\mathbf{I} - \left(\mathbf{X}'\right)^{-}\mathbf{X}'\right)\mathbf{c}$, for any given vector $\mathbf{c}$ of the appropriate order. Therefore, as $\left(\mathbf{X}'\right)^{-}$ is a generalized inverse of $\mathbf{X}'$ we can write $\mathbf{k}' = \mathbf{c}'\left(\mathbf{I} - \mathbf{X}\mathbf{X}^{-}\right)$. Furthermore, as $\left(\mathbf{X}'\mathbf{X}\right)^{-}\mathbf{X}'$ is a generalized inverse of $\mathbf{X}$, another way of expressing $\mathbf{k}'$ is $\mathbf{k}' = \mathbf{c}'\left(\mathbf{I} - \mathbf{X}\left(\mathbf{X}'\mathbf{X}\right)^{-}\mathbf{X}'\right)$ or $\mathbf{k}' = \mathbf{c}'\left(\mathbf{I} - \mathbf{X}\mathbf{X}^{+}\right)$, being $\mathbf{X}\mathbf{X}^{+} = \mathbf{X}\left(\mathbf{X}'\mathbf{X}\right)^{-}\mathbf{X}'$.

Defining $\mathbf{M}$ as:

$$\mathbf{M} = \mathbf{I} - \mathbf{X}\mathbf{G}\mathbf{X}' = \mathbf{I} - \mathbf{X}\left(\mathbf{X}'\mathbf{X}\right)^{-}\mathbf{X}' = \mathbf{I} - \mathbf{X}\mathbf{X}^{+}, \tag{C.1}$$

[1] Since some of the equations in this Appendix include vectors and matrices with the same letter, in order to avoid confusion we denote vectors in bold lower case and matrices in bold upper case.

Spatial and Spatio-Temporal Geostatistical Modeling and Kriging, First Edition.
José-María Montero, Gema Fernández-Avilés, and Jorge Mateu.

Companion Website: www.wiley.com/go/montero/spatial

we have

$$\mathbf{k}' = \mathbf{c}'\mathbf{M}. \tag{C.2}$$

If $\mathbf{X}$ is a matrix of order $N \times p$ and rank r, there are only $N - r$ linearly independent vectors $\mathbf{k}'$ that fulfill $\mathbf{k}'\mathbf{X} = \mathbf{0}$ (see Searle 1982, Section 9.7.a). Using a set of such linearly independent vectors as rows in the matrix $\mathbf{K}'$ we have the following theorem, for $\mathbf{K}'\mathbf{X} = \mathbf{0}$, with $\mathbf{K}'$ having full row rank $N - r$ and $\mathbf{K}' = \mathbf{C}'\mathbf{M}$ for any given $\mathbf{C}$.

Theorem C.0.1 *(Searle et al. 1992, p. 451): If $\mathbf{K}'\mathbf{X} = \mathbf{0}'$, with $\mathbf{K}'$ having the full row rank $N - r$ and $\mathbf{V}$ being positive-definite, then,*

$$\mathbf{K}\left(\mathbf{K}'\mathbf{V}\mathbf{K}\right)^{-1}\mathbf{K}' = \mathbf{P} \quad \textit{for } \mathbf{P} \equiv \mathbf{V}^{-1} - \mathbf{V}^{-1}\mathbf{X}\left(\mathbf{X}'\mathbf{V}^{-1}\mathbf{X}\right)^{-}\mathbf{X}'\mathbf{V}^{-1} \tag{C.3}$$

Harville (1977) refers to $\mathbf{k}'\mathbf{z}$ for $\mathbf{k}'$ of this nature as being an "error contrast": its expected value is zero,

$$E\left(\mathbf{k}'\mathbf{z}\right) = \mathbf{k}'\mathbf{X}\boldsymbol{\beta} = \mathbf{0}. \tag{C.4}$$

The number of linearly independent error contrasts depends on $\mathbf{X}$. For $\mathbf{X}$ of order $N - r$ and rank r the equation

$$\mathbf{k}'\mathbf{X}\boldsymbol{\beta} = \mathbf{0} \;\; \forall\boldsymbol{\beta} \tag{C.5}$$

is fulfilled only by $N - r$ values of $\mathbf{k}'$. Therefore, when using a set of linearly independent vectors $\mathbf{k}'$ as rows in $\mathbf{K}'$ we must focus our attention on

$$\mathbf{K}'\mathbf{z} \;\; \text{for } \mathbf{K}' = \mathbf{T}\mathbf{M}, \tag{C.6}$$

where $\mathbf{K}'$ and $\mathbf{T}$ have a full row rank $N - r$. Obviously, it makes no sense to consider more than $N - r$ vectors $\mathbf{k}'$, because in such a case some of them will be linear combinations of others and, therefore they will be the values that correspond to $\mathbf{k}'\mathbf{z}$.

C.1 Restricted Maximum Likelihood equation

When $\mathbf{z} \sim G(\mathbf{X}\boldsymbol{\beta}, \boldsymbol{\Sigma})$, we have that, for $\mathbf{K}'\mathbf{X} = \mathbf{0}$,

$$\mathbf{K}'\mathbf{z} \sim G\left(\mathbf{0}, \mathbf{K}'\boldsymbol{\Sigma}K\right) \tag{C.7}$$

which, as can be appreciated, and this is the key point, does not depend on $\boldsymbol{\beta}$.

Then, the (restricted) likelihood function, its logarithm and the scalar transforms on its logarithm can easily be derived from the ML equations of the same name by performing the following substitutions: $\mathbf{z}$ for $\mathbf{K}'\mathbf{z}$, $\mathbf{X}$ for $\mathbf{K}'\mathbf{X}= \mathbf{0}$, and $\boldsymbol{\Sigma}$ for $\mathbf{K}'\boldsymbol{\Sigma}K$.

For example, since the ML estimating equation is

$$-2l(\boldsymbol{\beta}, \boldsymbol{\theta} \mid z) = (\mathbf{z} - \mathbf{X}\boldsymbol{\beta})\boldsymbol{\Sigma}_{\theta}^{-1}(\mathbf{z} - \mathbf{X}\boldsymbol{\beta}) + \log \boldsymbol{\Sigma}_{\theta} + n\log(2\pi), \tag{C.8}$$

the REML estimating equation is

$$-2l(\boldsymbol{\theta} \mid z) = (\mathbf{K}'\mathbf{z})\left(\mathbf{K}'\boldsymbol{\Sigma}_{\theta}\mathbf{K}\right)^{-1}(\mathbf{K}'\mathbf{z}) + \log \mathbf{K}'\boldsymbol{\Sigma}_{\theta}\mathbf{K} + n\log(2\pi). \tag{C.9}$$

D

Most relevant proofs (Chapter 7)

D.1 Product model: Peculiarity (ii) (Rodríguez-Iturbe and Mejia 1974; De Cesare *et al.* 1997)

The sum or product of a spatial semivariogram and a temporal semivariogram is not *necessarily* a valid model of semivariogram, whereas the combination of the two through the expression $\gamma(\mathbf{h}, u) = C_s(\mathbf{0})\gamma_t(u) + C_t(0)\gamma_s(\mathbf{h}) - \gamma_s(\mathbf{h})\gamma_t(u)$ is.

Indeed, in the context of the product model:

$$\begin{aligned}\gamma(\mathbf{h}, u) &= C(\mathbf{0}, 0) - C(\mathbf{h}, u) \\ &= C(\mathbf{0}, 0) - C_s(\mathbf{h})C_t(u) \\ &= C(\mathbf{0}, 0) - (C_s(\mathbf{0}) - \gamma_s(\mathbf{h}))(C_t(0) - \gamma_t(u)) \\ &= C_s(\mathbf{0})C_t(0) - C_s(\mathbf{0})C_t(0) + C_s(\mathbf{0})\gamma_t(u) \\ &\quad + C_t(0)\gamma_s(\mathbf{h}) - \gamma_s(\mathbf{h})\gamma_t(u) \\ &= C_s(\mathbf{0})\gamma_t(u) + C_t(0)\gamma_s(\mathbf{h}) - \gamma_s(\mathbf{h})\gamma_t(u).\end{aligned} \tag{D.1}$$

From the above expression we can deduce that:

(i) For the semivariogram associated with the separable covariance function to also be separable in terms of the product, the following would have to occur:

$$C_s(\mathbf{0}) = C_t(0) = 0, \tag{D.2}$$

or

$$C_s(\mathbf{0}) = \gamma_s(\mathbf{h}) = 0, \tag{D.3}$$

Spatial and Spatio-Temporal Geostatistical Modeling and Kriging, First Edition.
José-María Montero, Gema Fernández-Avilés, and Jorge Mateu.

Companion Website: www.wiley.com/go/montero/spatial

or

$$C_t(0) = \gamma_t(u) = 0, \tag{D.4}$$

or

$$\gamma_s(\mathbf{h}) = \gamma_t(u) = 0, \tag{D.5}$$

which are anomalous situations in the spatio-temporal case.

(ii) In order for the semivariogram associated with the separable covariance function to also be separable in terms of the sum, the following would have to occur: $C_s(\mathbf{0}) = C_t(0) = 1$, and, moreover, one of the two semivariograms (or both) would have to be null. But this last supposition is not compatible with the spatio-temporal scenario, because in this case,

$$\gamma(\mathbf{h}, u) = \gamma_t(u), \tag{D.6}$$

or

$$\gamma(\mathbf{h}, u) = \gamma_s(\mathbf{h}). \tag{D.7}$$

D.2 Product model: Peculiarity (iv) (Rodríguez-Iturbe and Mejia 1974; De Cesare *et al.* 1997)

Given that the product model, in the stationary case, is defined by:

$$\begin{aligned} C(\mathbf{h}, u) &= C_s(\mathbf{h})C_t(u) \\ C(\mathbf{h}, 0) &= C_s(\mathbf{h})C_t(0) \Rightarrow C_s(\mathbf{h}) = \frac{C(\mathbf{h}, 0)}{C_t(0)} \\ C(\mathbf{0}, u) &= C_s(\mathbf{0})C_t(u) \Rightarrow C_t(u) = \frac{C(\mathbf{0}, u)}{C_s(\mathbf{0})} \end{aligned} \tag{D.8}$$

then,

$$C(\mathbf{h}, u) = \frac{C(\mathbf{h}, 0)}{C_t(0)} \frac{C(\mathbf{0}, u)}{C_s(\mathbf{0})} = \frac{C(\mathbf{h}, 0)C(\mathbf{0}, u)}{C(\mathbf{0}, 0)}, \tag{D.9}$$

that is, the spatio-temporal covariance function can be expressed as the rescaled product of the *marginal covariance functions*, the rescaling factor being the variance of the rf (which coincides with the product of the variances of the merely spatial and merely temporal rf's).

D.3 Product-sum model: Semivariogram expression (7.29) (De Iaco *et al.* 2001)

When $Z(\mathbf{s}, t)$ is second-order stationary:

$$C_s(\mathbf{h}) = C_s(\mathbf{0}) - \gamma_s(\mathbf{h}) \tag{D.10}$$

and

$$C_t(u) = C_t(0) - \gamma_t(u). \tag{D.11}$$

Furthermore:

1. By definition $\gamma(\mathbf{0}, 0) = \gamma_s(\mathbf{0}) = \gamma_t(0) = 0$.
2. If $C(\mathbf{h}, u) = k_1 C_s(\mathbf{h})C_t(u) + k_2 C_s(\mathbf{h}) + k_3 C_t(u) \Rightarrow C(\mathbf{0}, 0) = k_1 C_s(\mathbf{0})C_t(0) + k_2 C_s(\mathbf{0}) + k_3 C_t(0)$.

Using the covariance function as a basis,

$$\begin{aligned}
C(\mathbf{h}, u) =& k_1(C_s(\mathbf{0}) - \gamma_s(\mathbf{h}))(C_t(0) - \gamma_t(u)) + k_2(C_s(\mathbf{0}) - \gamma_s(\mathbf{h})) \\
&+ k_3(C_t(0) - \gamma_t(u_t)) \\
=& k_1 C_s(\mathbf{0})C_t(0) - k_1 C_s(\mathbf{0})\gamma_t(u) - k_1 C_t(0)\gamma_s(\mathbf{h}) \\
&+ k_1\gamma_s(\mathbf{h})\gamma_t(u) + k_2 C_s(\mathbf{0}) - k_2\gamma_s(\mathbf{h}) + k_3 C_t(0) - k_3\gamma_t(u) \\
=& \gamma_s(\mathbf{h})(-k_1 C_t(0) - k_2) + \gamma_t(u)(-k_1 C_s(\mathbf{0}) - k_3) + k_1\gamma_s(\mathbf{h})\gamma_t(u) \\
&+ k_1 C_s(\mathbf{0})C_t(0) + k_2 C_s(\mathbf{0}) + k_3 C_t(0) \\
=& \gamma_s(\mathbf{h})(-k_1 C_t(0) - k_2) + \gamma_t(u)(-k_1 C_s(\mathbf{0}) - k_3) \\
&+ k_1\gamma_s(\mathbf{h})\gamma_t(u) + C(\mathbf{0}, 0) \\
=& \gamma_s(\mathbf{h})(-k_1 C_t(0) - k_2) + \gamma_t(u)(-k_1 C_s(\mathbf{0}) - k_3) \\
&+ k_1\gamma_s(\mathbf{h})\gamma_t(u) + C(\mathbf{h}, u) + \gamma(\mathbf{h}, u),
\end{aligned} \tag{D.12}$$

we arrive at:

$$\gamma(\mathbf{h}, u) = \gamma_s(\mathbf{h})(k_2 + k_1 C_t(0)) + \gamma_t(u)(k_3 + k_1 C_s(\mathbf{0})) - k_1\gamma_s(\mathbf{h})\gamma_t(u). \tag{D.13}$$

D.4 General product-sum model: Obtaining the constant *k* (De Iaco *et al.* 2001)

$$\begin{aligned}
k =& \frac{k_1}{k_s k_t} = \frac{\frac{k_s C_s(\mathbf{0}) + k_t C_t(0) - C(\mathbf{0},0)}{C_s(\mathbf{0})C_t(0)}}{(k_2 + k_1 C_t(0))(k_3 + k_1 C_s(0))} \\
=& \frac{k_s C_s(\mathbf{0}) + k_t C_t(0) - C(\mathbf{0}, 0)}{C_s(\mathbf{0})C_t(0)} \\
&\times \frac{1}{\left(\frac{C(\mathbf{0},0) - k_t C_t(0)}{C_s(\mathbf{0})} + \frac{k_s C_s(\mathbf{0}) + k_t C_t(0) - C(\mathbf{0},0)}{C_s(\mathbf{0})C_t(0)} C_t(0)\right)} \\
&\times \frac{1}{\left(\frac{C(\mathbf{0},0) - k_s C_s(\mathbf{0})}{C_t(0)} + \frac{k_s C_s(\mathbf{0}) + k_t C_t(0) - C(\mathbf{0},0)}{C_s(\mathbf{0})C_t(0)} C_s(\mathbf{0})\right)}
\end{aligned}$$

$$
\begin{aligned}
&= \frac{k_s C_s(\mathbf{0}) + k_t C_t(0) - C(\mathbf{0},0)}{C_s(\mathbf{0})C_t(0)} \\
&\quad \times \frac{1}{\left(\frac{C_t(0)(C(\mathbf{0},0)-k_t C_t(0)) + C_t(0)(k_s C_s(\mathbf{0}) + k_t C_t(0) - C(\mathbf{0},0))}{C_s(\mathbf{0})C_t(0)}\right)} \\
&\quad \times \frac{1}{\left(\frac{C_s(\mathbf{0})(C(\mathbf{0},0)-k_s C_s(\mathbf{0})) + C_s(0)(k_s C_s(\mathbf{0}) + k_t C_t(0) - C(\mathbf{0},0))}{C_s(\mathbf{0})C_t(0)}\right)} \\
&= \frac{k_s C_s(\mathbf{0}) + k_t C_t(0) - C(\mathbf{0},0)}{C_s(\mathbf{0})C_t(0)} \\
&\quad \times \frac{1}{\left(\frac{C_t(0)(C(\mathbf{0},0)-k_t C_t(0) + k_s C_s(\mathbf{0}) + k_t C_t(0) - C(\mathbf{0},0))}{C_s(\mathbf{0})C_t(0)}\right)} \\
&\quad \times \frac{1}{\left(\frac{C_s(\mathbf{0})(C(\mathbf{0},0)-k_s C_s(\mathbf{0}) + k_s C_s(\mathbf{0}) + k_t C_t(0) - C(\mathbf{0},0))}{C_s(\mathbf{0})C_t(0)}\right)} \\
&= \frac{k_s C_s(\mathbf{0}) + k_t C_t(0) - C(\mathbf{0},0)}{C_s(\mathbf{0})C_t(0)} \times \frac{1}{\left(\frac{C_t(0)(k_s C_s(\mathbf{0}))}{C_s(\mathbf{0})C_t(0)}\right)} \times \frac{1}{\left(\frac{C_s(\mathbf{0})(k_t C_t(0))}{C_s(\mathbf{0})C_t(0)}\right)}
\end{aligned} \tag{D.14}
$$

and by simplifying we arrive at:

$$
k = \frac{k_s C_s(\mathbf{0}) + k_t C_t(0) - C(\mathbf{0},0)}{(k_s C_s(\mathbf{0}))(k_t C_t(0))} \tag{D.15}
$$

D.5 General product-sum model: Theorem 7.8.1 (De Iaco *et al.* 2001)

In order to demonstrate that $\lim_{\mathbf{h}\to\infty}\lim_{u\to\infty}\gamma(\mathbf{h},u) = C(\mathbf{0},0)$, Equation (7.46), it is sufficient to recall the continuity of $C(\mathbf{h},u)$:

$$
\lim_{\mathbf{h}\to\infty}\lim_{u\to\infty} C(\mathbf{h},u) = 0. \tag{D.16}
$$

By substituting the above expression in $\gamma(\mathbf{h},u) = C(\mathbf{0},0) - C(\mathbf{h},u)$ we arrive at:

$$
\lim_{\mathbf{h}\to\infty}\lim_{u\to\infty} \gamma(\mathbf{h},u) = C(\mathbf{0},0). \tag{D.17}
$$

As for the demonstration of (7.47), using the following relationship:

$$
C(\mathbf{h},0) = (k_2 + k_1 C_t(0))C_s(\mathbf{h}) \ + k_3 C_t(0), \tag{D.18}
$$

implicit in the product-sum model, we can deduce that:

$$
\lim_{\mathbf{h}\to\infty} C(\mathbf{h},0) = \lim_{\mathbf{h}\to\infty} (k_2 + k_1 C_t(0))C_s(\mathbf{h}) \ + k_3 C_t(0)
$$

$$= (k_2 + k_1 C_t(0)) \times 0 + k_3 C_t(0) \qquad \text{(D.19)}$$
$$= k_3 C_t(0).$$

Finally, substituting the value of k_3 with its expression:

$$\begin{aligned}\lim_{\mathbf{h}\to\infty} C(\mathbf{h},0) &= k_3 C_t(0)\\ &= \left(\frac{C(\mathbf{0},0) - k_s C_s(\mathbf{0})}{C_t(0)}\right) C_t(0) \qquad \text{(D.20)}\\ &= C(\mathbf{0},0) - k_s C_s(\mathbf{0}).\end{aligned}$$

Consequently,

$$C(\mathbf{0},0) - \lim_{\mathbf{h}\to\infty} C(\mathbf{h},0) = k_s C_s(\mathbf{0}), \qquad \text{(D.21)}$$

that is,

$$\lim_{\mathbf{h}\to\infty} \gamma(\mathbf{h},0) = k_s C_s(\mathbf{0}). \qquad \text{(D.22)}$$

In the same way, using the following relationship as a basis:

$$C(\mathbf{0},u) = (k_1 C_s(\mathbf{0}) + k_3) C_t(u) + k_2 C_s(\mathbf{0}), \qquad \text{(D.23)}$$

and substituting $k_2 = \frac{C(\mathbf{0},0) - k_t C_t(0)}{C_s(\mathbf{0})}$ in the above expression, we can deduce that:

$$\lim_{u\to\infty} \gamma(\mathbf{0},u) = C(\mathbf{0},0) - \lim_{u\to\infty} C(\mathbf{0},u) = k_t C_t(0). \qquad \text{(D.24)}$$

D.6 General product-sum model: Theorem 7.8.2. (De Iaco *et al.* 2001)

If C is expressed as the product-sum of purely spatial and purely temporal covariances with coefficients $k_1 > 0,\ k_2 \geq 0$ and $k_3 \geq 0$ and

$$k = \frac{k_1}{k_s k_t} = \frac{k_s C_s(\mathbf{0}) + k_t C_t(0) - C(\mathbf{0},0)}{(k_s C_s(\mathbf{0}))(k_t C_t(0))}, \qquad \text{(D.25)}$$

where $k_s C_s(\mathbf{0}), k_t C_t(0)$ and $C(\mathbf{0},0)$ are the values of the sills of $\gamma(\mathbf{h},0), \gamma(\mathbf{0},u)$ and $\gamma(\mathbf{h}_s, h_t)$ as defined in Theorem 7.8.1, we can deduce that:

$$k_s = k_2 + k_1 C_t(0) > 0 \qquad \text{(D.26)}$$
$$k_t = k_3 + k_1 C_s(\mathbf{0}) > 0, \qquad \text{(D.27)}$$

hence $k > 0$. Moreover, the following inequalities are satisfied by assumption:

$$k_1 = \frac{k_s C_s(\mathbf{0}) + k_t C_t(0) - C(\mathbf{0},0)}{C_s(\mathbf{0}) C_t(0)} > 0 \Rightarrow C(\mathbf{0},0) < k_s C_s(\mathbf{0}) + k_t C_t(0), \qquad \text{(D.28)}$$

$$k_2 = \frac{C(\mathbf{0},0) - k_t C_t(0)}{C_s(0)} \geq 0 \Rightarrow C(\mathbf{0},0) \geq k_t C_t(0), \tag{D.29}$$

$$k_3 = \frac{C(\mathbf{0},0) - k_s C_s(\mathbf{0})}{C_t(0)} \geq 0 \Rightarrow C(\mathbf{0},0) \geq k_s C_s(\mathbf{0}), \tag{D.30}$$

The last two inequalities are fulfilled simultaneously when

$$C(\mathbf{0},0) \geq \max\{k_s C_s(\mathbf{0}), k_t C_t(0)\}. \tag{D.31}$$

For (D.28) and (D.31) to hold, k must be less than or equal to $\frac{1}{\max\{k_s C_s(\mathbf{0}), k_t C_t(0)\}}$, hence

$$k \in \left(0; \frac{1}{\max\left\{k_s C_s(\mathbf{0}), k_t C_t(0)\right\}}\right). \tag{D.32}$$

Consequently, if k fulfills

$$0 \leq k \leq \frac{1}{\max\{sill(\gamma(\mathbf{h},0)); sill(\gamma(\mathbf{0},u))\}}, \tag{D.33}$$

then

$$\gamma(\mathbf{h},u) = \gamma(\mathbf{h},0) + \gamma(\mathbf{0},u) - k\gamma(\mathbf{h},0)\gamma(\mathbf{0},u) \tag{D.34}$$

will be a valid spatio-temporal semivariogram, born out of a valid model of spatio-temporal covariance.

In fact, if

$$k = \frac{sill(\gamma(\mathbf{h},0)) + sill(\gamma(\mathbf{0},u)) - sill(\gamma(\mathbf{h},u))}{sill(\gamma(\mathbf{h},0))sill(\gamma(\mathbf{0},u))}, \tag{D.35}$$

then

$$\gamma(\mathbf{h},u) = \gamma(\mathbf{h},0) + \gamma(\mathbf{0},u) - k\gamma(\mathbf{h},0)\gamma(\mathbf{0},u) \tag{D.36}$$

is equivalent to

$$C(\mathbf{h},u) = k_1 C_s(\mathbf{h})C_t(u) + k_2 C_s(\mathbf{h}) + k_3 C_t(u). \tag{D.37}$$

The bounds on k and then the inequalities in (D.28) and (D.31) imply that $k_1 > 0, k_2 \geq 0$ and $k_3 \geq 0$.

D.7 Generalized product-sum model. Proposition 1[1] (Gregori *et al.* 2008)

(i) Let a function $C(\mathbf{h},u)$ be continuous and defined on $\mathbb{R}^d \times \mathbb{R}$ and with a spectral density function $f(\omega,\tau)$. Then, following Bochner's theorem, the fact that

[1] We follow Martínez (2008).

$C(\mathbf{h}, u)$ is a spatio-temporal covariance function is the same as if $f(\omega, \tau) \geq 0 \quad \forall(\boldsymbol{\omega}, \tau) \in \mathbb{R}^d \times \mathbb{R}$.
In this case, the spectral density function associated with the function $C(\mathbf{h}, u)$ defined by (7.128) is given by:

$$f(\boldsymbol{\omega}, \tau) = \sum_{i=1}^{n} k_i f_{si}(\boldsymbol{\omega}) f_{ti}(\tau), \quad (\boldsymbol{\omega}, \tau) \in \mathbb{R}^d \times \mathbb{R}. \tag{D.38}$$

Thus,

$$0 \leq f(\boldsymbol{\omega}, \tau) = \sum_{i=1}^{n} k_i f_{si}(\boldsymbol{\omega}) f_{ti}(\tau) = f_{sn}(\boldsymbol{\omega}) f_{tn}(\tau) \left(\sum_{i=1}^{n-1} k_i \frac{f_{si}(\boldsymbol{\omega})}{f_{sn}(\boldsymbol{\omega})} \frac{f_{ti}(\tau)}{f_{tn}(\tau)} + k_n \right). \tag{D.39}$$

As $C_{sn}(\mathbf{h})$ and $C_{tn}(u)$ are two covariance functions on $\mathbb{R}^d$ and $\mathbb{R}$, respectively, then $f_{sn}(\boldsymbol{\omega}) \geq 0$, $\forall \boldsymbol{\omega} \in \mathbb{R}^d$ and $f_{tn}(\tau) \geq 0$, $\forall \tau \in \mathbb{R}$. Therefore,

$$\begin{aligned} 0 \leq \sum_{i=1}^{n-1} k_i \frac{f_{si}(\boldsymbol{\omega})}{f_{sn}(\boldsymbol{\omega})} \frac{f_{ti}(\tau)}{f_{tn}(\tau)} + k_n \leq & \sum_{\substack{i=1 \\ k_i \geq 0}}^{n-1} k_i \frac{f_{si}(\boldsymbol{\omega})}{f_{sn}(\boldsymbol{\omega})} \frac{f_{ti}(\tau)}{f_{tn}(\tau)} \\ & + \sum_{\substack{i=1 \\ k_i < 0}}^{n-1} k_i \frac{f_{si}(\boldsymbol{\omega})}{f_{sn}(\boldsymbol{\omega})} \frac{f_{ti}(\tau)}{f_{tn}(\tau)} + k_n. \end{aligned} \tag{D.40}$$

By definition,

$$0 \leq m_{si} \leq \frac{f_{si}(\boldsymbol{\omega})}{f_{sn}(\boldsymbol{\omega})} \leq M_{si}, \ 0 \leq m_{ti} \leq \frac{f_{ti}(\tau)}{f_{tn}(\tau)} \leq M_{ti}, \ \forall(\boldsymbol{\omega}, \tau) \in \mathbb{R}^d \times \mathbb{R}. \tag{D.41}$$

Then,

$$0 \leq m_{si} m_{ti} \leq \frac{f_{si}(\boldsymbol{\omega})}{f_{sn}(\boldsymbol{\omega})} \frac{f_{ti}(\tau)}{f_{tn}(\tau)} \leq M_{si} M_{ti}, \ \forall(\boldsymbol{\omega}, \tau) \in \mathbb{R}^d \times \mathbb{R}. \tag{D.42}$$

Consequently,

$$\begin{aligned} 0 & \leq \sum_{\substack{i=1 \\ k_i \geq 0}}^{n-1} k_i \frac{f_{si}(\boldsymbol{\omega})}{f_{sn}(\boldsymbol{\omega})} \frac{f_{ti}(\tau)}{f_{tn}(\tau)} + \sum_{\substack{i=1 \\ k_i < 0}}^{n-1} k_i \frac{f_{si}(\boldsymbol{\omega})}{f_{sn}(\boldsymbol{\omega})} \frac{f_{ti}(\tau)}{f_{tn}(\tau)} + k_n \\ & \leq \sum_{\substack{i=1 \\ k_i \geq 0}}^{n-1} k_i M_{si} M_{ti} + \sum_{\substack{i=1 \\ k_i < 0}}^{n-1} k_i m_{si} m_{ti} + k_n, \end{aligned} \tag{D.43}$$

which implies that:

$$k_n \geq -\sum_{i=1}^{n-1} k_i (M_{si} M_{ti} \mathbf{1}_{\{k_i \geq 0\}} + m_{si} m_{ti} \mathbf{1}_{\{k_i < 0\}}). \tag{D.44}$$

(ii) Let us assume that

$$k_n \geq -\sum_{i=1}^{n-1} k_i(m_{si}m_{ti}\mathbf{1}_{\{k_i\geq 0\}} + M_{si}M_{ti}\mathbf{1}_{\{k_i<0\}}) \tag{D.45}$$

or, equivalently,

$$k_n + \sum_{i=1}^{n-1} k_i(m_{si}m_{ti}\mathbf{1}_{\{k_i\geq 0\}} + M_{si}M_{ti}\mathbf{1}_{\{k_i<0\}}) \geq 0 \tag{D.46}$$

Then, if this occurs, we proceed to demonstrate that the function $C(\mathbf{h}, u)$ defined by (7.128) is a spatio-temporal covariance function on $\mathbb{R}^d \times \mathbb{R}$, or in other words, that its spectral density function is non-negative, that is:

$$\begin{aligned} f(\boldsymbol{\omega}, \tau) &= \sum_{i=1}^{n} k_i f_{si}(\boldsymbol{\omega}) f_{ti}(\tau) \\ &= f_{sn}(\boldsymbol{\omega}) f_{tn}(\tau) \left(\sum_{i=1}^{n-1} k_i \frac{f_{si}(\boldsymbol{\omega})}{f_{sn}(\boldsymbol{\omega})} \frac{f_{ti}(\tau)}{f_{tn}(\tau)} + k_n \right) \geq 0, \forall (\boldsymbol{\omega}, \tau) \in \mathbb{R}^d \times \mathbb{R}. \end{aligned} \tag{D.47}$$

As $f_{sn}(\boldsymbol{\omega}) \geq 0$ and $f_{tn}(\tau) \geq 0$ due to being the spectral density functions associated with two covariance functions, then we have to prove that:

$$\begin{aligned} &\sum_{i=1}^{n-1} k_i \frac{f_{si}(\boldsymbol{\omega})}{f_{sn}(\boldsymbol{\omega})} \frac{f_{ti}(\tau)}{f_{tn}(\tau)} + k_n \\ &\quad = \sum_{\substack{i=1 \\ k_i\geq 0}}^{n-1} k_i \frac{f_{si}(\boldsymbol{\omega})}{f_{sn}(\boldsymbol{\omega})} \frac{f_{ti}(\tau)}{f_{tn}(\tau)} + \sum_{\substack{i=1 \\ k_i<0}}^{n-1} k_i \frac{f_{si}(\boldsymbol{\omega})}{f_{sn}(\boldsymbol{\omega})} \frac{f_{ti}(\tau)}{f_{tn}(\tau)} + k_n \geq 0. \end{aligned} \tag{D.48}$$

Using $0 \leq m_{si}m_{ti} \leq \frac{f_{si}(\boldsymbol{\omega})}{f_{sn}(\boldsymbol{\omega})} \frac{f_{ti}(\tau)}{f_{tn}(\tau)} \leq M_{si}M_{ti}, \forall (\boldsymbol{\omega}, \tau) \in \mathbb{R}^d \times \mathbb{R}$ as a basis, we have that:

$$\sum_{i=1}^{n-1} k_i \frac{f_{si}(\boldsymbol{\omega})}{f_{sn}(\boldsymbol{\omega})} \frac{f_{ti}(\tau)}{f_{tn}(\tau)} + k_n = \sum_{\substack{i=1 \\ k_i\geq 0}}^{n-1} k_i \frac{f_{si}(\boldsymbol{\omega})}{f_{sn}(\boldsymbol{\omega})} \frac{f_{ti}(\tau)}{f_{tn}(\tau)} + \sum_{\substack{i=1 \\ k_i<0}}^{n-1} k_i \frac{f_{si}(\boldsymbol{\omega})}{f_{sn}(\boldsymbol{\omega})} \frac{f_{ti}(\tau)}{f_{tn}(\tau)} + k_n \tag{D.49}$$

$$\geq \sum_{\substack{i=1 \\ k_i\geq 0}}^{n-1} k_i m_{si} m_{ti} + \sum_{\substack{i=1 \\ k_i<0}}^{n-1} k_i M_{si} M_{ti} + k_n, \tag{D.50}$$

which is non-negative.

D.8 Generalized product-sum model. Proposition 1[2] for $n = 2$ (Gregori *et al.* 2008)

Let us assume that $C(\mathbf{h}, u) = k_1 C_{s1}(\mathbf{h}) C_{t1}(u) + k_2 C_{s2}(\mathbf{h}) C_{t2}(u), (\mathbf{h}, u) \in \mathbb{R}^d \times \mathbb{R}$, is a spatio-temporal covariance function. Then its spectral density function will be non-negative, that is,

$$\begin{aligned} 0 \leq f(\boldsymbol{\omega}, \tau) &= k_1 f_{s1}(\boldsymbol{\omega}) f_{t1}(\tau) + k_2 f_{s2}(\boldsymbol{\omega}) f_{t2}(\tau) \\ &= f_{s2}(\boldsymbol{\omega}) f_{t2}(\tau) \left(k_1 \frac{f_{s1}(\boldsymbol{\omega})}{f_{s2}(\boldsymbol{\omega})} \frac{f_{t1}(\tau)}{f_{t2}(\tau)} + k_2 \right). \end{aligned} \tag{D.51}$$

As $C_{s2}(\mathbf{h})$ and $C_{t2}(u)$ are covariance functions, their associated spectral densities, $f_{s2}(\boldsymbol{\omega})$ and $f_{t2}(\tau)$, will be non-negative, hence:

$$0 \leq k_1 \frac{f_{s1}(\boldsymbol{\omega})}{f_{s2}(\boldsymbol{\omega})} \frac{f_{t1}(\tau)}{f_{t2}(\tau)} + k_2, \quad \forall (\boldsymbol{\omega}, \tau) \in \mathbb{R}^d \times \mathbb{R}, \tag{D.52}$$

which implies that:

$$-k_2 \leq k_1 \frac{f_{s1}(\boldsymbol{\omega})}{f_{s2}(\boldsymbol{\omega})} \frac{f_{t1}(\tau)}{f_{t2}(\tau)} \quad \forall (\boldsymbol{\omega}, \tau) \in \mathbb{R}^d \times \mathbb{R}. \tag{D.53}$$

If $k_1 \geq 0$, we have that:

$$-k_2 \leq k_1 \frac{f_{s1}(\boldsymbol{\omega})}{f_{s2}(\boldsymbol{\omega})} \frac{f_{t1}(\tau)}{f_{t2}(\tau)} \quad \forall (\boldsymbol{\omega}, \tau) \in \mathbb{R}^d \times \mathbb{R}, \tag{D.54}$$

and this inequality will be maintained for the infimum of both quotients, hence:

$$-k_2 \leq k_1 m_s m_t \quad \forall (\boldsymbol{\omega}, \tau) \in \mathbb{R}^d \times \mathbb{R}, \tag{D.55}$$

or:

$$k_2 \geq -k_1 m_s m_t \qquad \forall (\boldsymbol{\omega}, \tau) \in \mathbb{R}^d \times \mathbb{R}. \tag{D.56}$$

If $k_1 < 0$, we have that:

$$k_2 \geq -k_1 \frac{f_{s1}(\boldsymbol{\omega})}{f_{s2}(\boldsymbol{\omega})} \frac{f_{t1}(\tau)}{f_{t2}(\tau)} \quad \forall (\boldsymbol{\omega}, \tau) \in \mathbb{R}^d \times \mathbb{R}, \tag{D.57}$$

this inequality being satisfied in the supremum of both quotients:

$$k_2 \geq -k_1 M_s M_t \quad \forall (\boldsymbol{\omega}, \tau) \in \mathbb{R}^d \times \mathbb{R}. \tag{D.58}$$

Therefore, the following will be fulfilled:

$$k_2 \geq -k_1 (m_{s1} m_{t1} \mathbf{1}_{\{k_1 \geq 0\}} + M_{s1} M_{t1} \mathbf{1}_{\{k_1 < 0\}}). \tag{D.59}$$

The reciprocal function is immediate as a result of (ii) in Theorem 1.

[2] We follow Martínez (2008).

D.9 Generalized product-sum model. Corollary 1[3] of Proposition 2 (Gregori *et al.* 2008)

Let us consider $k_1 = \theta$ and $k_2 = 1 - \theta$. Following Theorem 1, (7.132) is a spatio-temporal covariance function if, and only if

$$1 - \theta \geq -\theta(m_s m_t \mathbf{1}_{\{\theta \geq 0\}} + M_s M_t \mathbf{1}_{\{\theta < 0\}}). \tag{D.60}$$

(i) Let us assume that $M_s M_t \geq m_s m_t \geq 1$ and that $\theta \geq 0$. If (7.132) is a spatio-temporal covariance function, by applying (D.60) we have that:

$$1 - \theta \geq -\theta(m_s m_t) \geq -\theta(M_s M_t),$$

which implies that

$$1 \geq \theta(1 - M_s M_t),$$

that is,

$$(1 - M_s M_t)^{-1} \leq \theta \leq +\infty,$$

hence $(1 - \max(1, M_s M_t))^{-1} \leq \theta \leq (1 - \min(1, m_s m_t))^{-1}$ is fulfilled. Reciprocally, if $(1 - \max(1, M_s M_t))^{-1} \leq \theta \leq (1 - \min(1, m_s m_t))^{-1}$, then $(1 - M_s M_t)^{-1} \leq \theta \leq +\infty$. Thus,

$$\theta \geq (1 - M_s M_t)^{-1} \geq (1 - m_s m_t)^{-1},$$

as $1 - M_s M_t \leq 1 - m_s m_t \leq 0$. Therefore, the condition (D.60) is satisfied and (7.132) is a spatio-temporal covariance function.

(ii) Let us assume that $M_s M_t \geq m_s m_t \geq 1$ and that $\theta < 0$. In this case, the equivalence of both expressions is immediate. The reason is that if (7.132) is a spatio-temporal covariance function, by applying (D.60) we have that:

$$1 - \theta \geq -\theta(M_s M_t) \Longleftrightarrow 1 \geq \theta(1 - M_s M_t)$$
$$\Longleftrightarrow (1 - M_s M_t)^{-1} \leq \theta \leq +\infty.$$

(iii) Let us assume that $M_s M_t \geq 1 \geq m_s m_t \geq 0$ and that $\theta \geq 0$. If (7.132) is a spatio-temporal covariance function, according to (D.60) we have that $1 - \theta \geq \theta m_s m_t$, whereby $-\theta m_s m_t \geq -\theta M_s M_t$. Therefore:

$$1 - \theta \geq -\theta(m_s m_t) \Rightarrow 1 \geq \theta(1 - m_s m_t) \Rightarrow (1 - m_s m_t)^{-1} \geq \theta$$
$$1 - \theta \geq -\theta(M_s M_t) \Rightarrow 1 \geq \theta(1 - M_s M_t) \Rightarrow (1 - M_s M_t)^{-1} \leq \theta,$$

[3] We follow Martínez (2008).

and hence the condition (4.72) is fulfilled. Reciprocally, if the condition (4.72) is fulfilled, then $(1 - M_s M_t)^{-1} \leq \theta \leq (1 - m_s m_t)^{-1}$. Then,

$$\theta \geq (1 - m_s m_t)^{-1} \Rightarrow 1 - \theta \geq -\theta(m_s m_t).$$

The condition (D.60) is therefore verified and (7.132) is a spatio-temporal covariance function.

(iv) Let us assume that $M_s M_t \geq 1 \geq m_s m_t \geq 0$ and that $\theta < 0$. If (7.132) is a spatio-temporal covariance function, (D.60) we have that $1 - \theta \geq -\theta M_s M_t$, whereby $-\theta M_s M_t \geq -\theta m_s m_t$. Therefore:

$$1 - \theta \geq -\theta(m_s m_t) \Rightarrow 1 \geq \theta(1 - m_s m_t) \Rightarrow (1 - m_s m_t)^{-1} \geq \theta$$

$$1 - \theta \geq -\theta(M_s M_t) \Rightarrow 1 \geq \theta(1 - M_s M_t) \Rightarrow (1 - M_s M_t)^{-1} \leq \theta,$$

thereby fulfilling the condition (7.133). Reciprocally, if the condition (7.133) is fulfilled, then $(1 - M_s M_t)^{-1} \leq \theta \leq (1 - m_s m_t)^{-1}$. Then,

$$\theta \geq (1 - M_s M_t)^{-1} \Rightarrow 1 - \theta \geq -\theta(M_s M_t).$$

Therefore, the condition (D.60) is verified and (7.132) is a spatio-temporal covariance function.

(v) Let us assume that $1 \geq M_s M_t \geq m_s m_t \geq 0$ and that $\theta \geq 0$. In this case, the equivalence between both expressions is immediate, as:

$$1 - \theta \geq -\theta(m_s m_t) \Longleftrightarrow 1 \geq \theta(1 - m_s m_t)$$
$$\Longleftrightarrow (1 - m_s m_t)^{-1} \geq \theta \geq -\infty.$$

(vi) Let us assume that $1 \geq M_s M_t \geq m_s m_t \geq 0$ and that $\theta < 0$. If (7.132) is a spatio-temporal covariance function, according to (D.60) we have that $1 - \theta \geq -\theta M_s M_t \geq -\theta m_s m_t$. Therefore,

$$1 \geq \theta(1 - m_s m_t) \Rightarrow (1 - m_s m_t)^{-1} \geq \theta \leq +\infty,$$

hence the condition (7.133) is fulfilled. Reciprocally, if the condition (7.133) is fulfilled, then $-\infty \leq \theta \leq (1 - m_s m_t)^{-1}$.
Then, $\theta \leq (1 - m_s m_t)^{-1} \leq (1 - M_s M_t)^{-1}$ as $0 \leq 1 - M_s M_t \leq 1 - m_s m_t \leq 1$. Therefore, the condition (D.60) is verified and (7.132) is a spatio-temporal covariance function.

D.10 Generalized product-sum model. Range of θ. Case 1: The Gaussian case[4] (Gregori *et al.* 2008)

The covariance function of the Gaussian model is given by:

$$G(\mathbf{h} \mid \sigma_g^2, \alpha_g) = \sigma_g^2 \exp\left(-\alpha_g \|\mathbf{h}\|^2\right), \ \mathbf{h} \in \mathbb{R}^d,$$

[4] We follow Martínez (2008).

$\sigma_g^2 > 0$ being the variance of the process and α_g a scaling parameter such that $\alpha_g^{-1} > 0$ indicates the speed at which the covariance decreases.

The spectral density function associated to the Gaussian model is given by:

$$f_G(\boldsymbol{\omega}) = \sigma_g^2 \pi^{\frac{k}{2}} \alpha_g^{-\frac{k}{2}} \exp\left(-\frac{\|\boldsymbol{\omega}\|^2}{4\alpha_g}\right), \quad \boldsymbol{\omega} \in \mathbb{R}^d.$$

Consequently, the quotient of spectral densities is given by:

$$\frac{f_{G_1}(\boldsymbol{\omega})}{f_{G_2}(\boldsymbol{\omega})} = \frac{\sigma_{g1}^2 \pi^{\frac{k}{2}} \alpha_{g1}^{-\frac{k}{2}} \exp\left(-\frac{\|\boldsymbol{\omega}\|^2}{4\alpha_{g1}}\right)}{\sigma_{g2}^2 \pi^{\frac{k}{2}} \alpha_{g_2}^{-\frac{k}{2}} \exp\left(-\frac{\|\boldsymbol{\omega}\|^2}{4\alpha_{g2}}\right)}$$

$$= \frac{\sigma_{g_1}^2}{\sigma_{g2}^2}\left(\frac{\alpha_{g_1}}{\alpha_{g_2}}\right)^{-\frac{k}{2}} \exp\left(-\frac{1}{4}\left(\frac{1}{\alpha_{g_1}} - \frac{1}{\alpha_{g_2}}\right)\|\boldsymbol{\omega}\|^2\right) = h(\|\boldsymbol{\omega}\|^2),$$

where $h(x) = a \exp(-bx)$, with $x > 0$, is a continuous function of a variable defined in $[0, +\infty[$ the performance of which depends on the sign of b. Of course, in $h(x)$ we have set $a = \frac{\sigma_{g1}^2}{\sigma_{g2}^2}\left(\frac{\alpha_{g1}}{\alpha_{g1}}\right)^{-\frac{k}{2}}$ and $b = \frac{1}{4}\left(\frac{1}{\alpha_{g_1}} - \frac{1}{\alpha_{g_2}}\right)$.

(i) If $\alpha_{g1} < \alpha_{g2}$, then $b > 0$, hence $h(x)$ will be decreasing. In this case, its highest value will coincide with its supremum, which is reached at $x = \|\boldsymbol{\omega}\|^2 = 0$ and will be $h(0) = a$, whereas its minimum will be reached at the limit and will be $\lim_{x\to+\infty} h(x) = 0$. Therefore,

$$m_{G_1G_2} = \inf_{\boldsymbol{\omega}\in\mathbb{R}^d} \frac{f_{G_1}(\boldsymbol{\omega})}{f_{G_2}(\boldsymbol{\omega})} = 0 \quad M_{G_1G_2} = \sup_{\boldsymbol{\omega}\in\mathbb{R}^d} \frac{f_{G_1}(\boldsymbol{\omega})}{f_{G_2}(\boldsymbol{\omega})} = \frac{\sigma_{g_1}^2}{\sigma_{g_2}^2}\left(\frac{\alpha_{g_1}}{\alpha_{g_2}}\right)^{-\frac{k}{2}}.$$

(ii) If $\alpha_{g1} = \alpha_{g2}$, then $b = 0$, hence $h(x)$ will be constant and equal in its entire domain. Therefore,

$$m_{G_1G_2} = \inf_{\boldsymbol{\omega}\in\mathbb{R}^d} \frac{f_{G_1}(\boldsymbol{\omega})}{f_{G_2}(\boldsymbol{\omega})} = \frac{\sigma_{g_1}^2}{\sigma_{g_2}^2} \quad M_{G_1G_2} = \sup_{\boldsymbol{\omega}\in\mathbb{R}^d} \frac{f_{G_1}(\boldsymbol{\omega})}{f_{G_2}(\boldsymbol{\omega})} = \frac{\sigma_{g_1}^2}{\sigma_{g_2}^2}.$$

(iii) If $\alpha_{g1} > \alpha_{g2}$, then $b < 0$, hence $h(x)$ will be increasing. In this case, its infimum will coincide with its minimum value, which is recorded at $x = \|\boldsymbol{\omega}\|^2 = 0$ and is worth $h(0) = a$, whereas its supremum is reached at the limit and will be $\lim_{x\to+\infty} h(x) = +\infty$. Therefore,

$$m_{G_1G_2} = \inf_{\boldsymbol{\omega}\in\mathbb{R}^d} \frac{f_{G_1}(\boldsymbol{\omega})}{f_{G_2}(\boldsymbol{\omega})} = \frac{\sigma_{g_1}^2}{\sigma_{g_2}^2}\left(\frac{\alpha_{g_1}}{\alpha_{g_2}}\right)^{-\frac{k}{2}} \qquad M_{G_1G_2} = \sup_{\boldsymbol{\omega}\in\mathbb{R}^d} \frac{f_{G_1}(\boldsymbol{\omega})}{f_{G_2}(\boldsymbol{\omega})} = +\infty.$$

D.11 Generalized product-sum model. Range of θ. Case 2: The Matérn case[5] (Gregori *et al.* 2008)

As noted in Section 7.16, the covariance function associated to the Matérn model is given by:

$$M(\mathbf{h}|\sigma_m^2, \alpha_m, \upsilon_m) = \sigma_m^2(2^{1-\upsilon_m}/\Gamma(\upsilon_m))(\alpha_m\|\mathbf{h}\|)^{\upsilon_m}K_{\upsilon_m}(\alpha_m\|\mathbf{h}\|), \quad \mathbf{h} \in \mathbb{R}^k,$$

where $\sigma_m^2 > 0$ is the variance of the process, α_m is a scaling parameter such that $\alpha_m^{-1} > 0$ indicates the speed at which the process decreases and υ_m measures the degree of smoothness of the process. The spectral density of this model is given by:

$$f_M(\boldsymbol{\omega}) = \sigma_m^2 2^{\upsilon_m - 1}\pi^{-k/2}\Gamma\left(\upsilon_m + \frac{k}{2}\right)\alpha_m^{2\upsilon_m}(\alpha_m^2 + \|\boldsymbol{\omega}\|^2)^{-\upsilon_m - \frac{k}{2}}, \quad \boldsymbol{\omega} \in \mathbb{R}^k.$$

Then, given two covariance functions belonging to this model, the quotient of their spectral density functions will be given by:

$$\frac{f_{M_1}(\boldsymbol{\omega})}{f_{M_2}(\boldsymbol{\omega})} = \frac{\sigma_{m1}^2 2^{\upsilon_{m1}-1}\pi^{-k/2}\Gamma\left(\upsilon_{m1} + \frac{k}{2}\right)\alpha_{m1}^{2\upsilon_{m1}}(\alpha_{m1}^2 + \|\boldsymbol{\omega}\|^2)^{-\upsilon_{m1} - \frac{k}{2}}}{\sigma_{m2}^2 2^{\upsilon_{m2}-1}\pi^{-k/2}\Gamma\left(\upsilon_{m2} + \frac{k}{2}\right)\alpha_{m2}^{2\upsilon_m}(\alpha_{m2}^2 + \|\boldsymbol{\omega}\|^2)^{-\upsilon_{m2} - \frac{k}{2}}}$$

$$= \frac{\sigma_{m1}^2}{\sigma_{m2}^2}2^{\upsilon_{m1}-\upsilon_{m2}}\frac{\Gamma\left(\upsilon_{m1} + \frac{k}{2}\right)}{\Gamma\left(\upsilon_{m2} + \frac{k}{2}\right)}\frac{\alpha_{m1}^{2\upsilon_{m1}}}{\alpha_{m2}^{2\upsilon_m}}\frac{(\alpha_{m1}^2 + \|\boldsymbol{\omega}\|^2)^{-\upsilon_{m1} - \frac{k}{2}}}{(\alpha_{m2}^2 + \|\boldsymbol{\omega}\|^2)^{-\upsilon_{m2} - \frac{k}{2}}},$$

where $h(x) = a(b+x)^e/(c+x)^f, x \geq 0$ is a continuous function of a variable defined on $[0, +\infty)$. Then,

$$a = \frac{\sigma_{m1}^2}{\sigma_{m2}^2}2^{\upsilon_{m1}-\upsilon_{m2}}\frac{\Gamma\left(\upsilon_{m1} + \frac{k}{2}\right)}{\Gamma\left(\upsilon_{m2} + \frac{k}{2}\right)}\frac{\alpha_{m1}^{2\upsilon_{m1}}}{\alpha_{m2}^{2\upsilon_m}},$$

$$b = \alpha_{m2}^2,$$

$$c = \alpha_{m1}^2,$$

$$e = \upsilon_{m2} + \frac{k}{2},$$

$$f = \upsilon_{m1} + \frac{k}{2}.$$

The supremum and infimum of the function $h(x)$ are reached on their bounds (at $x = 0$ or on their limit $x \to +\infty$) or at their singular points (which cancel out their first derivative). Their derivative is given by:

$$h(x) = a\frac{(b+x)^e}{(c+x)^f} \Rightarrow h'(x) = a\frac{(b+x)^{e-1}}{(c+x)^{f-1}}((e-f)x + (ec - fb)),$$

[5] We follow Martínez (2008).

which would have a singular point at $[0, +\infty)$ if, and only if $e - f \neq 0$, that is, if and only if $\upsilon_{m1} \neq \upsilon_{m2}$.

(i–iii) If $\upsilon_{m1} = \upsilon_{m2}$, then $e - f = 0$, hence the function $h(x)$ will reach its supremum and infimum at its extremes, and then:

$$h(0) = a\left(\frac{b}{c}\right)^e, \qquad \lim_{x\to+\infty} h(x) = \lim_{x\to+\infty} a\frac{(b+x)^e}{(c+x)} = a.$$

Then the infimum and the supremum of the function $h(x)$ will depend on the quotient b/c.

(i) If $\alpha_{m1} = \alpha_{m2}$, then $b = c$, hence $h(x) = a$ throughout its domain. In this case:

$$m_{M_1,M_2} = \inf_{\omega\in\mathbb{R}^k} \frac{f_{M_1}(\omega)}{f_{M_2}(\omega)} = a = \frac{\sigma^2_{m1}}{\sigma^2_{m2}},$$

$$M_{M_1,M_2} = \sup_{\omega\in\mathbb{R}^k} \frac{f_{M_1}(\omega)}{f_{M_2}(\omega)} = a = \frac{\sigma^2_{m1}}{\sigma^2_{m2}}.$$

(ii) If $\alpha_{m1} > \alpha_{m2}$, then $b < c$, hence the function $b/c < 1$. In this case:

$$\begin{aligned} m_{M_1,M_2} &= \inf_{\omega\in\mathbb{R}^k} \frac{f_{M_1}(\omega)}{f_{M_2}(\omega)} = a\left(\frac{b}{c}\right)^e \\ &= \frac{\sigma^2_{m1}}{\sigma^2_{m2}}\left(\frac{\alpha_{m1}}{\alpha_{m2}}\right)^{2\upsilon_{m1}}\left(\frac{\alpha_{m2}}{\alpha_{m1}}\right)^{2\upsilon_{m1}+k} = \frac{\sigma^2_{m1}}{\sigma^2_{m2}}\left(\frac{\alpha_{m2}}{\alpha_{m1}}\right)^k, \\ M_{M_1,M_2} &= \sup_{\omega\in\mathbb{R}^k} \frac{f_{M_1}(\omega)}{f_{M_2}(\omega)} = a = \frac{\sigma^2_{m1}}{\sigma^2_{m2}}\left(\frac{\alpha_{m1}}{\alpha_{m2}}\right)^{2\upsilon_{m1}}. \end{aligned}$$

(iii) If $\alpha_{m1} < \alpha_{m2}$, then $b > c$, which means the function $b/c > 1$. In this case:

$$m_{M_1,M_2} = \inf_{\omega\in\mathbb{R}^k} \frac{f_{M_1}(\omega)}{f_{M_2}(\omega)} = a = \frac{\sigma^2_{m1}}{\sigma^2_{m2}}\left(\frac{\alpha_{m1}}{\alpha_{m2}}\right)^{2\upsilon_{m1}},$$

$$M_{M_1,M_2} = \sup_{\omega\in\mathbb{R}^k} \frac{f_{M_1}(\omega)}{f_{M_2}(\omega)} = a\left(\frac{b}{c}\right)^e = \frac{\sigma^2_{m1}}{\sigma^2_{m2}}\left(\frac{\alpha_{m2}}{\alpha_{m1}}\right)^k.$$

(iv–vii) If $\upsilon_{m1} \neq \upsilon_{m2}$, then the function $h(x)$ has a singular point at $x^* = (fb - ec)/(e - f)$. At this point, the function $h(x)$ takes the value:

$$h(x^*) = a\frac{(b+x^*)^e}{(c+x^*)^f} = a\left(\frac{b-c}{e-f}\right)^{e-f}\frac{e^e}{f^f}.$$

Then the infimum and the supremum of the function $h(x)$ will be reached at this point or either of its two extremes,

$$h(0) = a\frac{b^e}{c^f}, \quad \lim_{x\to\infty} h(x) = \lim_{x\to\infty} a\frac{(b+x)^e}{(c+x)^f}.$$

(iv–v) If $\upsilon_{m1} > \upsilon_{m2}$, then $f > e$, hence:

$$\lim_{x\to\infty} h(x) = \lim_{x\to\infty} a\frac{(b+x)^e}{(c+x)^f} = 0.$$

Then, in this case $m = 0$. Now let us see where the supremum is reached, whether it is at $h(0)$ or $h(x^*)$. In order to do so, we calculate the quotient between the two quantities:

$$\frac{h(x^*)}{h(0)} = \frac{a\left(\frac{b-c}{e-f}\right)^{e-f}\frac{e^e}{f^f}}{a\frac{b^e}{c^f}} = \left(\frac{b-c}{e-f}\right)^{e-f}\left(\frac{e}{b}\right)^e\left(\frac{c}{f}\right)^f.$$

(iv) If $\frac{\alpha^2_{m2}}{\alpha^2_{m1}} \geq \frac{\upsilon_{m2}+k/2}{\upsilon_{m1}+k/2}$, then $c/f \leq b/e$, so

$$\begin{aligned}\frac{h(x^*)}{h(0)} &= \left(\frac{b-c}{e-f}\right)^{e-f}\left(\frac{e}{b}\right)^e\left(\frac{c}{f}\right)^f \\ &\leq \left(\frac{b-c}{e-f}\right)^{e-f}\left(\frac{e}{b}\right)^e\left(\frac{b}{e}\right)^f = \left(\frac{e(b-c)}{b(e-f)}\right)^{e-f} \leq 1.\end{aligned}$$

Then, in case $h(0) \geq h(x^*)$ the supremum is reached at $h(0)$ and takes the value:

$$\begin{aligned}M_{M_1,M_2} &= a\frac{b^e}{c^f} = \frac{\sigma^2_{m1}}{\sigma^2_{m2}}2^{\upsilon_{m1}-\upsilon_{m2}}\frac{\Gamma(\upsilon_{m1}+\frac{k}{2})\,\alpha_{m1}^{2\upsilon_{m1}}\alpha_{m2}^{2\upsilon_{m2}+k}}{\Gamma(\upsilon_{m2}+\frac{d}{2})\,\alpha_{m2}^{2\upsilon_{m2}}\alpha_{m1}^{2\upsilon_{m1}+k}} \\ &= \frac{\sigma^2_{m1}}{\sigma^2_{m2}}2^{\upsilon_{m1}-\upsilon_{m2}}\frac{\Gamma(\upsilon_{m1}+\frac{k}{2})}{\Gamma(\upsilon_{m2}+\frac{d}{2})}\left(\frac{\alpha_{m2}}{\alpha_{m1}}\right)^k.\end{aligned}$$

(v) If $\frac{\alpha^2_{m2}}{\alpha^2_{m1}} < \frac{\upsilon_{m2}+k/2}{\upsilon_{m1}+k/2}$, then $c/f > b/e$, hence:

$$\frac{h(x^*)}{h(0)} = \left(\frac{b-c}{e-f}\right)^{e-f}\left(\frac{e}{b}\right)^e\left(\frac{c}{f}\right)^f \geq \left(\frac{e(b-c)}{b(e-f)}\right)^{e-f} \geq 1.$$

Then, in case $h(0) \leq h(x^*)$ the supremum is reached at $h(x^*)$ and takes the value:

$$M_{M_1,M_2} = \left(\frac{b-c}{e-f}\right)^{e-f}\frac{e^e}{f^f}$$

$$= \frac{\sigma_{m1}^2}{\sigma_{m2}^2} 2^{\upsilon_{m1}-\upsilon_{m2}} \frac{\Gamma(\upsilon_{m1}+\frac{k}{2})}{\Gamma(\upsilon_{m2}+\frac{d}{2})} \frac{\alpha_{m1}^{2\upsilon_{m1}}}{\alpha_{m2}^{2\upsilon_{m2}}} \left(\frac{\alpha_{m2}^2-\alpha_{m1}^2}{\upsilon_{m2}-\upsilon_{m1}} \right)^{\upsilon_{m2}-\upsilon_{m1}}$$

$$\times \frac{(\upsilon_{m2}+\frac{k}{2})^{\upsilon_{m2}+\frac{k}{2}}}{(\upsilon_{m1}+\frac{k}{2})^{\upsilon_{m1}+\frac{k}{2}}}$$

$$= \frac{\sigma_{m1}^2}{\sigma_{m2}^2} \frac{\Gamma(\upsilon_{m1}+\frac{k}{2})}{\Gamma(\upsilon_{m2}+\frac{d}{2})} \left(\frac{\alpha_{m2}^2-\alpha_{m1}^2}{2\left(\upsilon_{m2}-\upsilon_{m1}\right)} \right)^{\upsilon_{m2}-\upsilon_{m1}} \frac{\alpha_{m1}^{2\upsilon_{m1}}}{\alpha_{m2}^{2\upsilon_{m2}}}$$

$$\times \frac{(\upsilon_{m2}+\frac{k}{2})^{\upsilon_{m2}+\frac{k}{2}}}{(\upsilon_{m1}+\frac{k}{2})^{\upsilon_{m1}+\frac{k}{2}}}.$$

(vi–viii) If $\upsilon_{m1} < \upsilon_{m2}$, then $f < e$, hence:

$$\lim_{x\to\infty} h(x) = \lim_{x\to\infty} a\frac{(b+x)^e}{(c+x)^f} = +\infty.$$

Then in this case the supremum is $M_{M_1,M_2} = +\infty$. Now let us see whether the infimum is reached at $h(0)$ or $h(x^*)$. In order to do so, we calculate the quotient between both quantities:

$$\frac{h(x^*)}{h(0)} = \left(\frac{b-c}{e-f}\right)^{e-f} \left(\frac{e}{b}\right)^e \left(\frac{c}{f}\right)^f.$$

(ix) If $\frac{\alpha_{m2}^2}{\alpha_{m1}^2} \geq \frac{\upsilon_{m2}+k/2}{\upsilon_{m1}+k/2}$, then proceeding in the same way as before, we arrive at:

$$m_{M_1,M_2} = \frac{\sigma_{m1}^2}{\sigma_{m2}^2} \frac{\Gamma(\upsilon_{m1}+\frac{k}{2})}{\Gamma(\upsilon_{m2}+\frac{d}{2})} \left(\frac{\alpha_{m2}^2-\alpha_{m1}^2}{2\left(\upsilon_{m2}-\upsilon_{m1}\right)} \right)^{\upsilon_{m2}-\upsilon_{m1}} \frac{\alpha_{m1}^{2\upsilon_{m1}}}{\alpha_{m2}^{2\upsilon_{m2}}}$$

$$\times \frac{(\upsilon_{m2}+\frac{k}{2})^{\upsilon_{m2}+\frac{k}{2}}}{(\upsilon_{m1}+\frac{k}{2})^{\upsilon_{m1}+\frac{k}{2}}}.$$

(x) If $\frac{\alpha_{m2}^2}{\alpha_{m1}^2} < \frac{\upsilon_{m2}+k/2}{\upsilon_{m1}+k/2}$, then proceeding as we did above, we arrive at:

$$m_{M_1,M_2} = \frac{\sigma_{m1}^2}{\sigma_{m2}^2} 2^{\upsilon_{m1}-\upsilon_{m2}} \frac{\Gamma(\upsilon_{m1}+\frac{k}{2})}{\Gamma(\upsilon_{m2}+\frac{d}{2})} \left(\frac{\alpha_{m2}}{\alpha_{m1}}\right)^{klk}.$$

D.12 Generalized product-sum model. Range of θ. Case 3: The Gaussian-Matérn case[6] (Gregori *et al.* 2008)

In this case, the quotient of the two spectral density functions is given by:

$$\frac{f_G(\omega)}{f_M(\omega)} = \frac{\sigma_g^2 \pi^{\frac{k}{2}} \alpha_g^{-\frac{k}{2}} \exp\left(-\frac{\|\omega\|^2}{4\alpha_g}\right)}{\sigma_m^2 2^{\nu_m - 1} \pi^{-\frac{k}{2}} \Gamma\left(\nu_m + \frac{k}{2}\right) \alpha_m^{2\nu_m} (\alpha_m^2 + \|\omega\|^2)^{-\upsilon_m - \frac{k}{2}}}$$

$$= \frac{\sigma_g^2}{\sigma_m^2} \pi^k \frac{2^{1-\nu_m}}{\Gamma\left(\nu_m + \frac{k}{2}\right)} \frac{\alpha_g^{-\frac{k}{2}}}{\alpha_m^{2\nu_m}} \exp\left(-\frac{\|\omega\|^2}{4\alpha_g}\right) (\alpha_m^2 + \|\omega\|^2)^{\upsilon_m + \frac{k}{2}},$$

where $h(x) = a \exp(-bx)(c+x)^e$ is a continuous function of a variable defined on $[0, +\infty[$, with:

$$a = \frac{\sigma_g^2}{\sigma_m^2} \pi^k \frac{2^{1-\nu_m}}{\Gamma\left(\nu_m + \frac{k}{2}\right)} \frac{\alpha_g^{-\frac{k}{2}}}{\alpha_m^{2\nu_m}}, \quad b = \frac{1}{4\alpha_g}, \quad c = \alpha_m^2, \quad e = \upsilon_m + \frac{k}{2}.$$

The supremum and infimum of the function $h(x)$ will be reached at its extremes (at $x = 0$ or $\lim_{x\to\infty} h(x)$) or at its singular points (those which cancel out their first derivative). In this case,

$$\lim_{x\to\infty} h(x) = 0,$$

hence the infimum of the foregoing quotient is $m = 0$.

The supremum on the other hand will be reached at $h(0)$ or at its singular points. In this case, $h'(x) = a(c+x)^{e-1} \exp(-bx)\left(-\frac{b}{c+x} + e\right)$, which is cancelled out at $x^* = \frac{e}{b} - c$, the value of the function at this point being $h(x^*) = a \exp(b(c-e))\left(\frac{e}{b}\right)^e$. Therefore, we must compare the values of $h(0)$ and $h(x^*)$ and determine which is the largest. The quotient of both is given by:

$$\frac{h(x^*)}{h(0)} = \frac{a \exp(b(c-e))\left(\frac{e}{b}\right)^e}{ac^e} = \exp(b(c-e))\left(\frac{e}{bc}\right)^e.$$

[6] We follow Martínez (2008).

(i) Let us assume that $\frac{\alpha_m^2}{\nu_m+\frac{k}{2}} > 4\alpha_g$. Then $bc > e$, which means that the foregoing quotient is always greater than or equal to unity. As a result:

$$M_{G,M} = a\exp\,(b(c-e))\left(\frac{e}{b}\right)^e$$
$$= \frac{\sigma_g^2}{\sigma_m^2}2^{k+1}\pi^k\left(\frac{2\alpha_g}{\alpha_m^2}\right)^{\nu_m}\frac{\left(\nu_m+\frac{k}{2}\right)^{\nu_m+\frac{k}{2}}}{\Gamma\left(\nu_m+\frac{k}{2}\right)}\exp\left(\frac{\alpha_m^2}{4\alpha_g}-\left(\nu_m+\frac{k}{2}\right)\right).$$

(ii) Let us assume that $\frac{\alpha_m^2}{\nu_m+\frac{k}{2}} \le 4\alpha_g$. Then $bc \le e$, which means that the foregoing quotient is always lesser than or equal to unity. As a result:

$$M_{G,M} = ac^e = \frac{\sigma_g^2}{\sigma_m^2}\frac{2^{1-\nu_m}}{\Gamma\left(\nu_m+\frac{k}{2}\right)}\left(\frac{\pi^2\alpha_m^2}{\alpha_g}\right)^{\frac{k}{2}}.$$

D.13 Mixture-based Bernstein zonally anisotropic covariance functions. Theorem 7.18.1 (Ma 2003b)

Let (X_1, X_2) be a non-negative random vector with probability distribution function $W(u, v)$. Then, its Laplace transform is given by:

$$\mathcal{L}(\theta_1,\theta_2) = \int_{\mathbb{R}_+^2}\exp\,(-u\theta_1 - v\theta_2)dW(u,v),\ \ \theta_1,\theta_2\in\mathbb{R}_+^2. \qquad \text{(D.61)}$$

For each $u \ge 0, \exp\,(-u\gamma_1(\mathbf{h}))$ is a spatial covariance function on $\mathbb{R}^d$ as $\gamma_1(\mathbf{h})$ is a spatial variogram on $\mathbb{R}^d$. Likewise, for each $v \ge 0, \exp\,(-u\gamma_1(t))$ is a temporal covariance function on $\mathbb{R}$. Therefore, for each $(u, v) \in \mathbb{R}_+^2, \exp\,(-u\gamma_1(\mathbf{h}) - u\gamma_1(t))$ is a spatio-temporal covariance function on $\mathbb{R}^d \times \mathbb{R}$. Consequently, $C(\mathbf{h}, u) = \mathcal{L}(\gamma_1(\mathbf{h}), \gamma_2(t))$ is a spatio-temporal covariance function on $\mathbb{R}^d \times \mathbb{R}$.

D.14 Construction of non-stationary spatio-temporal covariance functions using spatio-temporal stationary covariances and intrinsically stationary semivariograms. Equation (7.159) (Ma 2003c)

Let $\gamma((\mathbf{s}_1, t_1), (\mathbf{s}_2, t_2))$ be the semivariogram associated to the function:

$$\begin{aligned}\phi((\mathbf{s}_1,t_1),(\mathbf{s}_2,t_2)) =& 2(\gamma(\mathbf{s}_1,t_1)+\gamma(\mathbf{s}_2,t_2)) - \gamma(\mathbf{s}_1+\mathbf{s}_2,t_1+t_2)\\ &-\gamma(\mathbf{s}_1-\mathbf{s}_2,t_1-t_2)\end{aligned} \qquad \text{(D.62)}$$

Using the identity bellow as a basis (given the existence of the covariance):

$$\gamma((\mathbf{s}_1, t_1), (\mathbf{s}_2, t_2)) = \frac{1}{2}(C((\mathbf{s}_1, t_1), (\mathbf{s}_1, t_1)) + C((\mathbf{s}_2, t_2), (\mathbf{s}_2, t_2))) \\ - C((\mathbf{s}_1, t_1), (\mathbf{s}_2, t_2)), \tag{D.63}$$

which in the case of stationarity transforms into:

$$\gamma(\mathbf{h}, u) = C(\mathbf{0}, 0) - C(\mathbf{h}, u), \tag{D.64}$$

we have that (in our specific case):

$$\begin{aligned} C((\mathbf{s}_1, t_1), (\mathbf{s}_1, t_1)) =& 2(\gamma(\mathbf{s}_1, t_1) + \gamma(\mathbf{s}_1, t_1)) - \gamma(\mathbf{s}_1 + \mathbf{s}_1, t_1 + t_1) \\ & - \gamma(\mathbf{s}_1 - \mathbf{s}_1, t_1 - t_1) \\ C((\mathbf{s}_2, t_2), (\mathbf{s}_2, t_2)) =& 2(\gamma(\mathbf{s}_2, t_2) + \gamma(\mathbf{s}_2, t_2)) - \gamma(\mathbf{s}_2 + \mathbf{s}_2, t_2 + t_2) \\ & - \gamma(\mathbf{s}_2 - \mathbf{s}_2, t_2 - t_2) \\ C((\mathbf{s}_1, t_1), (\mathbf{s}_2, t_2)) =& 2(\gamma(\mathbf{s}_1, t_1) + \gamma(\mathbf{s}_2, t_2)) - \gamma(\mathbf{s}_1 + \mathbf{s}_2, t_1 + t_2) \\ & - \gamma(\mathbf{s}_1 - \mathbf{s}_2, t_1 - t_2), \end{aligned} \tag{D.65}$$

hence:

$$\begin{aligned} \gamma((\mathbf{s}_1, t_1), (\mathbf{s}_2, t_2)) &= \frac{1}{2}(C((\mathbf{s}_1, t_1), (\mathbf{s}_1, t_1)) + C((\mathbf{s}_2, t_2), (\mathbf{s}_2, t_2))) \\ &\quad - C((\mathbf{s}_1, t_1), (\mathbf{s}_2, t_2)) \\ &= \frac{1}{2}(2(\gamma(\mathbf{s}_1, t_1) + \gamma(\mathbf{s}_1, t_1)) - \gamma(\mathbf{s}_1 + \mathbf{s}_1, t_1 + t_1) \\ &\quad - \gamma(\mathbf{s}_1 - \mathbf{s}_1, t_1 - t_1)) + \frac{1}{2}((\gamma(\mathbf{s}_2, t_2) + \gamma(\mathbf{s}_2, t_2)) \\ &\quad - \gamma(\mathbf{s}_2 + \mathbf{s}_2, t_2 + t_2) - \gamma(\mathbf{s}_2 - \mathbf{s}_2, t_2 - t_2) \\ &\quad - 2(\gamma(\mathbf{s}_1, t_1) + \gamma(\mathbf{s}_2, t_2)) + \gamma(\mathbf{s}_1 + \mathbf{s}_2, t_1 + t_2) \\ &\quad + \gamma(\mathbf{s}_1 - \mathbf{s}_2, t_1 - t_2) \\ &= \frac{1}{2}(2(\gamma(\mathbf{s}_1, t_1) + \gamma(\mathbf{s}_1, t_1)) - \gamma(2\mathbf{s}_1, 2t_1) - \gamma(\mathbf{0}, 0)) \\ &\quad + \frac{1}{2}((\gamma(\mathbf{s}_2, t_2) + \gamma(\mathbf{s}_2, t_2)) - \gamma(2\mathbf{s}_2, 2t_2) - \gamma(\mathbf{0}, 0)) \\ &\quad - 2\gamma(\mathbf{s}_1, t_1) - 2\gamma(\mathbf{s}_2, t_2) + \gamma(\mathbf{s}_1 + \mathbf{s}_2, t_1 + t_2) \\ &+ \gamma(\mathbf{s}_1 - \mathbf{s}_2, t_1 - t_2) \\ &= \gamma(\mathbf{s}_1 + \mathbf{s}_2, t_1 + t_2) + \gamma(\mathbf{s}_1 - \mathbf{s}_2, t_1 - t_2) \end{aligned}$$

$$- \frac{1}{2}(\gamma(2\mathbf{s}_1, 2t_1) + \gamma(2\mathbf{s}_2, 2t_2))$$
$$= \phi((\mathbf{s}_1 + \mathbf{s}_2, t_1 + t_2), (\mathbf{s}_1 - \mathbf{s}_2, t_1 - t_2)). \tag{D.66}$$

In order to prove that $\phi((\mathbf{s}_1, t_1), (\mathbf{s}_2, t_2))$ is a valid covariance function, we only have to demonstrate that $\sum_{i=1}^{n} \sum_{j=1}^{n} a_i a_j C((\mathbf{s}_1, t_1), (\mathbf{s}_2, t_2)) \geq 0$ is verified $\forall n \in N, (\mathbf{s}_i, t_i) \in \mathbb{R}^d \times T, \{a_1 \dots a_n\} \subseteq \mathbb{R}$. In order to demonstrate the above, let us define, for $i = \pm 1 \dots \pm n$:

$$a_0^* = -2\sum_{i=1}^{n} a_i, \mathbf{s}_0^* = 0, t_0^* = 0, a_i^* = a_{|i|}, \mathbf{s}_i^* = sign(i)\mathbf{s}_{|i|}, t_i^* = sign(i)t_{|i|}. \tag{D.67}$$

Then,

$$\sum_{i=-n}^{n} a_i^* = a_1 + \dots + a_n - 2\sum_{i=1}^{n} a_i + a_1 + \dots + a_n = 0 \tag{D.68}$$

and

$$2\sum_{i=1}^{n}\sum_{j=1}^{n} a_i a_j \phi((\mathbf{s}_i, t_i), (\mathbf{s}_j, t_j))$$
$$= 2\sum_{i=1}^{n}\sum_{j=1}^{n} a_i a_j (2\gamma(\mathbf{s}_i, t_i) + 2\gamma(\mathbf{s}_j, t_j) - \gamma(\mathbf{s}_i + \mathbf{s}_j, t_i + t_j) - \gamma(\mathbf{s}_i - \mathbf{s}_j, t_i - t_j))$$
$$= 2\sum_{j=1}^{n} a_j \sum_{i=1}^{n} a_i 2\gamma(\mathbf{s}_i, t_i) + 2\sum_{i=1}^{n} a_i \sum_{j=1}^{n} a_j 2\gamma(\mathbf{s}_j, t_j)$$
$$- 2\sum_{i=1}^{n}\sum_{j=1}^{n} a_i a_j (\gamma(\mathbf{s}_i + \mathbf{s}_j, t_i + t_j) - \gamma(\mathbf{s}_i - \mathbf{s}_j, t_i - t_j))$$
$$= -2a_0^* \sum_{i=1}^{n} a_i^* \gamma(\mathbf{s}_i, t_i) - 2a_0^* \sum_{j=1}^{n} a_j^* \gamma(\mathbf{s}_j, t_j)$$
$$- 2\sum_{i=1}^{n}\sum_{j=1}^{n} a_i^* a_j^* (\gamma(\mathbf{s}_i + \mathbf{s}_j, t_i + t_j) - \gamma(\mathbf{s}_i - \mathbf{s}_j, t_i - t_j))$$
$$= -4a_0^* \sum_{i=1}^{n} a_i^* \gamma(\mathbf{s}_i, t_i)$$
$$- 2\sum_{i=1}^{n}\sum_{j=1}^{n} a_i^* a_j^* (\gamma(\mathbf{s}_i + \mathbf{s}_j, t_i + t_j) - \gamma(\mathbf{s}_i - \mathbf{s}_j, t_i - t_j))$$
$$= -2\sum_{i=-n}^{n} a_0^* a_i \gamma(\mathbf{s}_i, t_i) - \left(\sum_{i=1}^{n}\sum_{j=-n}^{-1} + \sum_{i=-n}^{1}\sum_{j=1}^{n} \right) a_{|i|}^* a_{|j|}^* \gamma(\mathbf{s}_{|i|} + \mathbf{s}_{|j|}, t_{|i|} + t_{|j|})$$

$$-\left(\sum_{i=1}^{n}\sum_{j=1}^{n}+\sum_{i=-n}^{-1}\sum_{j=-n}^{-1}\right)a^*_{|i|}a^*_{|j|}\gamma(\mathbf{s}_{|i|}-\mathbf{s}_{|j|},t_{|i|}-t_{|j|})$$

$$=-\sum_{i=-n}^{n}\sum_{j=-n}^{n}a^*_i a^*_j\gamma(\mathbf{s}^*_i-\mathbf{s}^*_j,t^*_i-t^*_j)\geq 0. \tag{D.69}$$

D.15 Construction of non-stationary spatio-temporal covariance functions using spatio-temporal stationary covariances and intrinsically stationary semivariograms. Equation (7.161) is a valid covariance function (Ma 2003c)

For $\forall n\in N, (\mathbf{s}_i,t_i)\in\mathbb{R}^d\times T, \{a_1\dots a_n\}\subseteq\mathbb{R}$, let us define:

$$a^*_0=-2\sum_{i=1}^{n}a_i, \mathbf{s}^*_0=0, t^*_0=0, a^*_i=a_{|i|}, \mathbf{s}^*_i=sign(i)\mathbf{s}_{|i|}, t^*_i=sign(i)t_{|i|}. \tag{D.70}$$

Then,

$$\sum_{i=-n}^{n}a^*_i=a_1+\dots+a_n-2\sum_{i=1}^{n}a_i+a_1+\dots+a_n=0 \tag{D.71}$$

Since $\gamma(\mathbf{s},t)$ is a semivariogram, we have:

$$2\sum_{i=1}^{n}\sum_{j=1}^{n}a_ia_j\kappa((\mathbf{s}_i,t_i),(\mathbf{s}_j,t_j))$$

$$=2\sum_{i=1}^{n}\sum_{j=1}^{n}a_ia_j(\gamma(\mathbf{s}_i+\mathbf{s}_j,t_i+t_j)-\gamma(\mathbf{s}_i-\mathbf{s}_j,t_i-t_j))$$

$$=\left(\sum_{i=1}^{n}\sum_{j=-n}^{-1}+\sum_{i=-n}^{-1}\sum_{j=1}^{n}a_{|i|}a_{|j|}\gamma(\mathbf{s}_{|i|}+\mathbf{s}_{|j|},t_{|i|}+t_{|j|})\right)$$

$$-\left(\sum_{i=1}^{n}\sum_{j=1}^{n}+\sum_{i=-n}^{-1}\sum_{j=-n}^{-1}a_{|i|}a_{|j|}\gamma(\mathbf{s}_{|i|}-\mathbf{s}_{|j|},t_{|i|}-t_{|j|})\right)$$

$$=\sum_{i=-n}^{n}\sum_{j=-n}^{n}a^*_i a^*_j\gamma(\mathbf{s}^*_i-\mathbf{s}^*_j,t^*_i-t^*_j)\geq 0. \tag{D.72}$$

D.16 Construction of non-stationary spatio-temporal covariance functions using spatio-temporal stationary covariances and intrinsically stationary semivariograms. Equation (7.163) Ma (2003c)

If $\gamma(\mathbf{s}, t)$ is an intrinsically stationary semivariogram, then

$$\begin{aligned}\phi((\mathbf{s}_1, t_1), (\mathbf{s}_2, t_2)) =& 2(\gamma(\mathbf{s}_1, t_1) + \gamma(\mathbf{s}_2, t_2)) - \gamma(\mathbf{s}_1 + \mathbf{s}_2, t_1 + t_2) \\ & - \gamma(\mathbf{s}_1 - \mathbf{s}_2, t_1 - t_2),\end{aligned} \tag{D.73}$$

which is a covariance function, transforms into:

$$\begin{aligned}\phi((\mathbf{s}_1, t_1), (\mathbf{s}_2, t_2)) &= 2(C(\mathbf{0}, 0) - C(\mathbf{s}_1, t_1) + C(\mathbf{0}, 0) - C(\mathbf{s}_2, t_2)) \\ &- (C(\mathbf{0}, 0) - C(\mathbf{s}_1 + \mathbf{s}_2, t_1 + t_2)) \\ &- (C(\mathbf{0}, 0) - C(\mathbf{s}_1 - \mathbf{s}_2, t_1 - t_2)) \\ &= C(\mathbf{s}_1 + \mathbf{s}_2, t_1 + t_2) + C(\mathbf{s}_1 - \mathbf{s}_2, t_1 - t_2) \\ &- 2(C(\mathbf{s}_1, t_1) + C(\mathbf{s}_2, t_2) - C(\mathbf{0}, 0)).\end{aligned} \tag{D.74}$$

D.17 Permissibility criteria for quasi-arithmetic means of covariance functions. Proposition 1 (Porcu *et al.* 2009b)

Due to Bernstein's theorem (Feller 1966, p. 439), $\varphi \in \Phi_{cm}$ if and only if:

$$\varphi(t) = \int_0^\infty \exp\ -rtdF(r), \tag{D.75}$$

F being a positive bounded measure. Then, in order to prove (a), consider that if $\varphi \in \Phi_{cm}$, the equation (7.187) can be rewritten as:

$$\varphi(t) = \int_0^\infty \exp\ -r\sum_{i=1}^{n} \theta_i \varphi^{-1}(C_i(\mathbf{x}))dF(r). \tag{D.76}$$

That said, if for each i, $\varphi^{-1} \circ C_i$ turns out to be a semivariogram, then $\sum_{i=1}^{n} \theta_i \varphi^{-1} \circ C_i$ also is. Therefore, according to Schoenberg's (1938) theorem, we have that for any given $r > 0$, the integral of the foregoing expression is a covariance function, thereby proving point (a).

As regards point (b), it is sufficient to note that, in accordance with Schoenberg's (1938) theorem, $\gamma: \mathbb{R}^d \rightarrow \mathbb{R}_+$ is the semivariogram associated with an isotropic and intrinsically stationary random field if and only if $\gamma(\mathbf{h}) = \psi(||\mathbf{h}||^2)$, ψ being a Bernstein function. Having said this, the rest of the proof is the same as in **(a)**.

Finally, in reference to point (c), as $\varphi^{-1} \circ C_i(\mathbf{x})$ are continuous, non-decreasing and concave functions on the positive real line, according to Pólya-type criteria (see Proposition 10.6 in Berg and Forst 1975) they are negative-definite on $\mathbb{R}$ and hence their sum also is. And as $\varphi \in \Phi_{cm}$, due to Schoenberg's theorem (cf. Berg and Forst 1975), is obtained, the result is presented.

Bibliography and further reading

Abramowitz, M. and Stegun, I. (1965) *Handbook of Mathematical Functions*. Dover Publications, New York.

Akaike, H. (2003) Toeplitz matrix inversion. *SIAM Journal of Applied Mathematics*, **14**, 234–41.

Armstrong, M. and Delfiner, P. (1980) Towards a more robust variogram: a case study on coal. Technical Report No. 671. Centre de Géostatistique, Fontainebleau, France.

Armstrong, M. and Wackernagel, H. (1988) The influence of the covariance function on the kriged estimator. *Sciences de la Terre, Scientifique, Informatique, Géologique*, **27**(II), 245–62.

Banerjee, S. (2004) On geodetic distance computations in spatial modelling. *Biometrics*, **61**, 617–25.

Banerjee, S., Carlin, B.P. and Gelfand, A.E. (2004) *Hierarchical Modeling and Analysis for Spatial Data*. Chapman & Hall/CRC Press, Boca Raton, FL.

Barry, J., Crowder, M. and Diggle, P. (1997) Parametric estimation of the variogram. Technical Report, Department of Maths and Stats, Lancaster University.

Barry, R. and Pace, K. (1997) Kriging with large data sets using sparse matrix techniques. *Communications in Statistics, Computation and Simulation*, **26**, 619–29.

Berg, C., Christensen, J. and Ressel, P. (1984) *Harmonic Analysis on Semigroups: Theory of Positive Definite and Related Functions*. Springer, New York.

Berg, C. and Forst, G. (1975) *Potential Theory on Locally Compact Abelian Groups*. Springer, New York.

Bevilacqua, M. (2008) Composite likelihood inference for space-time covariance models. Ph.D. thesis. University of Padua, Italy.

Spatial and Spatio-Temporal Geostatistical Modeling and Kriging, First Edition.
José-María Montero, Gema Fernández-Avilés, and Jorge Mateu.

Companion Website: www.wiley.com/go/montero/spatial

Bevilacqua, M., Gaetan, C., Mateu, J. and Porcu, P. (2012) Estimating space and space-time covariance functions for large data sets: a weighted composite likelihood approach. *Journal of the American Statistical Association*, **107**(497), 268–80.

Binsariti, A.A. (1980) Statistical analysis and stochastic modeling of the Cortaro aquifer in southern Arizona. Ph.D. dissertation. University of Arizona, Tucson, USA.

Bochner, S. (1933) Monotone funktionen, Stiltjes integrale und harmonische analyse. *Mathematische Annalen*, **108**, 378–410.

Brown, P.E., Karesen, K.F., Roberts, G.O. and Tonellato, S. (2000) Blur-generated non-separable space-time models. *Journal of the Royal Statistical Society Series B-Statistical Methodology*, **62**, 847–60.

Buhmann, M. (2000) Radial basis functions: the state-of-the-art and new results. *Acta Numerica*, **9**, 1–37.

Caballero, W., Giraldo, R. and Mateu, J. (2013) A universal kriging approach for spatial functional data. *Stochastic Environmental Research and Risk Assessment*, **27**, 1553–63.

Calder, C.A. and Cressie, N. (2007) Some topics in convolution-based spatial modeling. In *Proceedings of the 56th Session of the International Statistics Institute*, pp. 22–9. Lisbon, Portugal.

Caragea, P. and Smith, R. (2006) Approximate likelihoods for spatial processes. Technical Report. Department of Statistics, Iowa State University.

Carrera, J. and Samper, J. (1989) Hydrological modelling for the safety assessment of radioactive waste disposal, OCDE. Technical Report, OCDE.

Chen, L., Fuentes, M. and Davis, J.M. (2006) Spatial temporal statistical modelling and prediction of environmental processes. *Hierarchical Modelling for the Environmental Sciences*, pp. 121–144, Oxford University Press, Oxford.

Chilès, J.P. and Delfiner, P. (1999) *Geostatistics: Modeling Spatial Uncertainty*, John Wiley & Sons, Ltd, Chichester.

Christakos, G. (1984) On the problem of permissible covariance and variogram models. *Water Resources Research*, **20**(2), 251–65.

Christakos, G. (1992) A theory of spatiotemporal random fields and its application to space-time data processing. *IEEE Transactions on Systems, Man and Cybernetics*, **21**(4), 861–75.

Christakos, G. (2000) *Modern Spatiotemporal Geostatistics*. Oxford University Press, Oxford.

Christakos, G. and Hristopulos, D.T. (1998) *Spatiotemporal Environmental Health Modeling: A Tractatus Stochasticus*. Kluwer, Boston.

Cox, D.R. and Isham, V. (1988) A simple spatial-temporal model of rainfall. In *Proceedings of the Royal Society of London, A* **415**, 317–28.

Cressie, N. (1993) *Statistics for Spatial Data*. Wiley Series in Probability and Mathematical Statistics. John Wiley & Sons, Ltd, Chichester.

Cressie, N. and Hawkins, D.M. (1980) Robust estimation of the variogram. *Journal of the International Association for Mathematical Geology*, **12**(2), 115–25.

Cressie, N. and Huang, C. (1999) Classes of nonseparable, spatiotemporal stationary covariance functions. *Journal of the American Statistical Association*, **94**, 1330–40.

Cressie, N. and Majure, J.J. (1997) Spatio-temporal statistical modeling of livestock waste in streams. *Journal of Agricultural, Biological, and Environmental Statistics*, **2**(1), 24–47.

Curriero, F.C. and Lele, S. (1999) A composite likelihood approach to semivariogram estimation. *Journal of Agricultural, Biological, and Environmental Statistics*, **4**(1), 9–28.

De Cesare, L., Myers, D. and Posa, D. (1997) Spatial-temporal modeling of SO_2 in Milan district. In E.Y. Baafi and N.A. Schofield (eds), Geostatistics Wollongong'96, vol. 2, pp. 1031–42. Kluwer Academic Publishers, Dordrecht.

De Cesare, L., Myers, D.E. and Posa, D. (2001a) Estimating and modeling space-time correlation structures. *Statistics & Probability Letters*, **51**, 9–14.

De Cesare, L., Myers, D.E. and Posa, D. (2001b) Product-sum covariance for space-time modeling: an environmental application. *Environmetrics*, **12**, 11–23.

De Cesare, L., Myers, D.E. and Posa, D. (2002) FORTRAN programs for space-time modeling. *Computational Geosciences*, **28**, 205–12.

De Iaco, S., Myers, D.E. and Posa, D. (2001) Space-time analysis using a general product-sum model. *Statistics & Probability Letters*, **52**, 21–8.

De Iaco, S., Myers, D.E. and Posa, D. (2002a) Space-time variograms and a functional form for total air pollution measurements. *Computational Statistics & Data Analysis*, **41**, 311–28.

De Iaco, S., Myers, D.E. and Posa, D. (2002b) Nonseparable space-time covariance models: some parametric families. *Mathematical Geology*, **34**, 23–42.

De Iaco, S., Myers, D.E. and Posa, D. (2003) The linear coregionalization model and the product-sum space-time variogram. *Mathematical Geology*, **35**, 25–38.

Delicado, P., Giraldo, R., Comas, C. and Mateu, J. (2010) Statistics for spatial functional data. *Environmetrics*, **21**(3–4), 224–39.

Deville, J.C. (1974) Méthodes statistiques et numériques de l'analyse harmonique. *Annales INSEE*, **15**, 3–101.

Díaz-Viera, M. (2002) Geoestadadística Aplicada. Technical Report, Instituto de Geofísica, Universidad Nacional de Mexico.

Diggle, P.J. and Ribeiro, Jr. P.J. (2007) *Model Based Geostatistics*. Springer, New York.

Dimitrakopoulos, R. and Luo, X. (1994) Spatiotemporal modeling: covariances and ordinary kriging system. In R. Dimitrakopoulos (ed.), *Geostatistics for the Next Century*, pp. 88–93. Kluwer Academic Publisher, Dordrecht.

Emery, X. (2000) *Geoestadística lineal*. Departamento de Ingeniería de Minas. Facultad de Ciencias Físicas and Matemáticas. Universidad de Chile.

Feller, W. (1966) *An Introduction to Probability Theory and Its Applications*. vol. II. John Wiley & Sons, Ltd, Chichester.

Fennessy, P.J. and Neuman, S.P. (1982) Geostatistical analysis of aquifer test and water level data from the Madrid basin, Spain hydrology and water resources in Arizona and the Southwest. In *Proceedings of the 1982 Meetings of the Arizona Section-American Water Resources Association and the Hydrology Section-Arizona*. Nevada Academy of Science, Tempe, Arizona.

Fernández-Casal, R. (2004) *Geoestadística Espacio-Temporal: Modelos flexibles de variogramas anisotrópicos no separables*. Departamento de Estatística e Investigación Operativa. Universidade de Santiago de Compostela.

Fernández-Casal, R., González-Manteiga, W. and Febrero-Bande, M. (2003) Flexible spatio-temporal stationary variogram models. *Statistics and Computing*, **13**, 127–36.

Ferraty, F., Mas, A. and Vieu, P. (2007) Advances in nonparametric regression for functional variables. *Australian and New Zealand Journal of Statistics*, **49**, 1–20.

Ferraty, F. and Vieu, P. (2006) *Non Parametric Functional Data Analysis. Theory and Practice*. Springer, New York.

Fuentes, M. (2002) Spectral methods for nonstationary spatial processes. *Biometrika*, **89**, 197–210.

Fuentes, M. (2003) Testing for separability of spatial-temporal covariance functions. Technical Report No. 2533. North Carolina State University.

Fuentes, M. (2007) Approximate likelihood for large irregularly spaced spatial data. *Journal of the American Statistical Association*, **102**, 321–31.

Fuentes, M., Chen, L., Davis, J. and Lackmann, G. (2005) Modeling and predicting complex space-time structures and patterns of coastal wind fields. *Environmetrics*, **16**, 449–64.

Fuentes, M. and Smith, R. (2001) A new class of nonstationary spatial models. Research Report. Statistics Department, North Carolina State University.

Furrer, R., Genton, M. and Nychka, D. (2007) Covariance tapering for interpolation of large spatial datasets. *Journal of Computational and Graphical Statistics*, **15**, 502–23.

Gambolati, G. and Volpi, G. (1979) Groundwater contour mapping in Venice by stochastic interpolators. *Water Resources Research*, **15**(2), 281–90.

Genest, C. and MacKay, R.J. (1986) Copules archim 'ediennes et familles de lois bidimensionnelles dont les marges sont donn' ees. *Revue Canadienne de Statistique*, **14**(2), 145–9.

Genest, C. (1987) Frank's family of bivariate distributions. *Biometrika*, **74**, 549–55.

Genton, M. (1998) Variogram fitting by generalized least squares using an explicit formula for the covariance structure. *Mathematical Geology*, **30**, 323–45.

Giraldo, R. (2009) Geostatistical analysis of functional data. Ph.D. thesis. Universitat Politcnica de Catalunya.

Giraldo, R., Delicado, P. and Mateu, J. (2010) Continuous time-varying kriging for spatial prediction of functional data: an environmental application. *Journal of Agricultural, Biological, and Environmental Statistics*, **15**(1), 66–82.

Giraldo, R., Delicado, P. and Mateu, J. (2011a) Ordinary kriging for function-valued spatial data. *Environmental and Ecological Statistics*, **18**(3), 411–26.

Giraldo, R., Delicado, P. and Mateu, J. (2011b) Geostatistics with infinite dimensional data: a generalization of cokriging and multivariable spatial prediction. *Matemática: ICM-ESPOL*, **9**, 16–21.

Giraldo, R., Delicado, P. and Mateu, J. (2012) GEOFD: an R package for function-valued geostatistical prediction. *Revista Colombiana de Estadística*, **35**(3), 385–407.

Giraldo, R. and Mateu, J. (2012) Kriging for functional data. In A.H. El-Shaarawi and W.W. Piegorsch (eds), *Encyclopedia of Environmetrics*, second edition. John Wiley & Sons, Ltd, Chichester.

Godambe, V. (1991) *Estimating Functions*. Oxford University Press, New York.

Gomez, M. and Hazen, K. (1970) Evaluating sulfur and ash distribution in coal seams by statistical response surface regression analysis, US Bureau of Mines Report RI 7377.

Gneiting, T. (1999). Correlation functions for atmospheric data analysis. *Quarterly Journal of the Royal Meteorological Society*, **125**, 2449–64.

Gneiting, T. (2002) Stationary covariance functions for space-time data. *Journal of the American Statistical Association*, **97**, 590–600.

Gneiting, T. and Schlather, M. (2004) Stochastic models that separate fractal dimension and the Hurst effect. *Society for Industrial and Applied Mathematics Review*, **46**, 269–82.

Gneiting, T., Genton, M.G. and Guttorp, P. (2005) Geostatistical space-time models, stationarity, separability, and full symmetry. Technical Report No. 475. University of Washington.

Gneiting, T., Genton, M.G. and Guttorp, P. (2007) Geostatistical space-time models, stationarity, separability and full symmetry. In B. Finkenstadt, L. Held and V. Isham (eds) *Statistical Methods for Spatio-Temporal Systems*, pp. 151–75. Chapman and Hall/CRC, Boca Raton, FL.

Goulard, M. and Voltz, M. (1993) Geostatistical interpolation of curves: a case study in soil science. In A. Soares (ed.), *Geostatistics Tróia*, **92**, 2, 805–16. Kluwer Academic Press.

Gradshteyn, I.S. and Ryzhik, I.M. (1980) *Table of Integrals, Series, and Products*, sixth edition. Academic Press, San Diego.

Gregori, P., Porcu, E., Mateu, J. and Sasvári, Z. (2008) On potentially negative space time covariances obtained as sum of products of marginal ones. *Annals of the Institute of Statistical Mathematics*, **60**(4), 865–82.

Gunst, R.F. and Hartfield, M.I. (1997) Robust semivariogram estimation in the presence of influential spatial data values. In T.G. Gregoire, D.R. Brillinger, P.J. Diggle, E. Russek-Cohen, W.G. Warren and R.D. Wolfinger (eds), *Modeling Longitudinal and Spatially Correlated Data: Methods, Applications, and Future Directions, Lecture Notes in Statistics*, pp. 265–74. Springer, New York.

Gupta, V.K. and Waymire, E. (1987) On Taylor's hypothesis and dissipation in rainfall. *Journal of Geophysical Research*, **92**, 9657–60.

Hampel, F.R., Ronchetti, E.M., Rousseeuw, P.J. and Stahel, W.A. (1986) *Robust Statistics: The Approach Based on Influence Functions*. John Wiley & Sons, Ltd, Chichester.

Harville, D.A. (1977) Maximum likelihood approaches to variance component estimation and to related problems. *Journal of the American Statistical Association*, **72**(358), 320–38.

Heyde, C. (1997) *Quasi-Likelihood and Its Application: A General Approach to Optimal Parameter Estimation*. Springer, New York.

Higdon, D. (2002) Space and space-time modeling using process convolutions. In C. Anderson *et al.* (eds), *Quantitative Methods for Current Environmental Issues*, pp. 37–54. Springer, London.

Higdon, D., Swall, J. and Kern, J. (1999) Non-stationary spatial modeling. In J.M. Bernardo, J.O. Berger, A.P. Dawid, and A.F.M. Smith (eds), *Bayesian Statistics, 6*, pp. 761–8, Oxford University Press, Oxford.

Huang, H.C. and Hsu, N.J. (2004) Modeling transport effects on ground-level ozone using a non-stationary space-time model. *Environmetrics*, **15**, 251–68.

Jones, R.H. and Zhang, Y. (1997) Models for continuous stationary space-time processes. In *Modelling Longitudinal and Spatially Correlated Data*, pp. 289–98, Springer-Verlag, New York.

Journel, A.G. and Huijbregts, C.H.J. (1978) *Mining Geostatistics*. Academic Press, New York.

Jun, M. and Stein, M.L. (2004) An approach to producing space-time covariance functions on spheres, Technical Report No.18. Center for Integrating Statistical and Environmental Science, University of Chicago.

Jun, M. and Stein, M.L. (2008) Nonstationary covariance models for global data. *Annals of Applied Statistics*, **2**, 1271–89.

Kaufmann, C., Schervish, M. and Nychka, D. (2007) Covariance tapering for likelihood-based estimation in large spatial datasets. Technical Report, Statistical and Applied Mathematical Sciences Institute, Research Triangle Park, North Carolina.

Kitanidis, P.K. (1983) Statistical estimation of polynomial generalized covariance functions and hydrologic applications. *Water Resources Research*, **19**(4), 909–21.

Kolmogorov, A.N. (1930) Sur la notion de la moyenne. *Atti Accad. Naz. Lincei Mem. Cl. Sci. Fis. Mat. Natur. Sez.*, **12**(6), 388–91.

Kolovos, A., Christakos, G., Hristopulos, D.T. and Serre, M.L. (2004) Methods for generating non-separable spatiotemporal covariance models with potential environmental applications. *Advances in Water Resources*, **27**, 815–30.

Kotz, S., Balakrishnan, N. and Johnson, N.L. (2000) *Continuous Multivariate Distributions: Models and Applications*. John Wiley & Sons, Inc., New York.

Kyriakidis, P.C. and Journel, A.G. (2000) Geostatistical space-time models: a review. *Mathematical Geology*, **31**, 651–84, 1999 and **32**, 893, 2000.

Lahiri, S., Lee, Y. and Cressie, N. (2002) Efficiency of least squares estimators of spatial variogram parameters. *Journal of Statistical Planning and Inference*, **103**, 65–85.

Levitin, D.J., Nuzzo, R.L., Bradley, W.V. and Ramsay, J.O. (2007) Introduction to Functional Data Analysis. *Canadian Psychological*, **48**(3), 135–55.

Lindsay, B. (1988) Composite likelihood methods. *Contemporary Mathematics*, **80**, 221–39.

Lu, N. and Zimmerman, D.L. (2005) Testing for directional symmetry in spatial dependence using the periodogram. *Journal of Statistical Planning and Inference*, **129**, 369–85.

Luo, X. (1998) Spatiotemporal stochastic models for earth science and engineering applications. PhD thesis, McGill University, Montreal, Canada.

Ma, C. (2002) Spatio-temporal covariance functions generated by mixtures. *Mathematical Geology*, **34**, 965–75.

Ma, C. (2003a) Families of spatio-temporal stationary covariance models. *Journal of Statistical Planning and Inference*, **116**, 489–501,

Ma, C. (2003b) Spatio-temporal stationary covariance models. *Journal of Multivariate Analysis*, **86**, 97–107.

Ma, C. (2003c) Nonstationary covariance functions that model space-time interactions. *Statistics and Probability Letters*, **61**, 411–19.

Ma, C. (2005a) Linear combinations of space-time covariance functions and variograms. *IEEE Transactions on Signal Processing*, **53**, 857–64.

Ma, C. (2005b) Semiparametric spatio-temporal covariance models with the ARMA temporal margin. *Annals of the Institute of Mathematical Statistics*, **57**, 221–33.

Ma, C. (2005c) Spatio-temporal variograms and covariance models. *Advances in Applied Probability*, **37**, 706–25.

Mardia, K.V. and Marshall, R.J. (1984) Maximum likelihood estimation of models for residual covariance in spatial regression. *Biometrika*, **71**, 135–46.

Martínez, F. (2008) Modelización de la función de covarianza en procesos espacio-temporales: análisis y aplicaciones. Ph.D. dissertation. Department of Statistics. Universitat de València.

Mas, A. and Pumo, B. (2009) Functional linear regression with derivatives. *Journal of Nonparametric Statistics*, **21**, 19–40.

Matérn, B. (1960) Spatial variation: stochastic models and their application to some problems in forest surveys and other sampling investigations. *Meddelanden Fran Statens Skogsforskningsinstitut*, **49** 5, Stockholm.

Mateu, J. (2007) Computing limiting stochastic processes for spatial structure detection. *Journal of Numerical Analysis, Industrial and Applied Mathematics*, **2**(1–2), 79–102.

Mateu, J., Juan, P. and Porcu, E. (2007) Geostatistical analysis through spectral techniques: some words of caution. *Communications in Statistics: Computation and Simulation*, **36**(5), 1035–51.

Mateu, J., Fernández-Avilés, G. and Montero, J.M. (2013) On a class of non-stationary, compactly supported spatial covariance functions. *Stochastic Environmental Research and Risk Assessment*, **27**(2), 297–309.

Matheron, G. (1962) *Traité de Géostatistique appliquée*, vol. I. Éditions Technip, Paris.

Matheron, G. (1976) A simple substitute for conditional expectation: the disjunctive kriging. In *Advanced Geostatistics in the Mining Industry*, pp. 221–36. Reidel Publishing Company. Dordrecht.

Matheron, G. (1989) *Estimating and Choosing*. Springer, New York.

Menafoglio, A., Secchi, P. and Dalla Rosa, M. (2013a) A Universal Kriging predictor for spatially dependent functional data of a Hilbert Space. *Electronic Journal of Statistics*, **7**, 2209–40.

Menafoglio, A., Dalla Rosa, M. and Secchi, P. (2013b) Supplement to A universal kriging predictor for spatially dependent functional data of a Hilbert Space. *DOI:* 10.1214/00-EJS843SUPP.

Mitchell, M., Genton, M. and Gumpertz, M. (2005) Testing for separability of space-time covariances. *Environmetrics*, **16**, 819–31.

Mitchell, M., Genton, M. and Gumpertz, M. (2006) A likelihood ratio test for separability of covariances. *Journal of Multivariate Analysis*, **97**, 1025–43.

Montero, J.M. and Larraz, B. (2008) *Introducción a la Geoestadística Lineal*. Netbiblo, Spain.

Moreaux, G. (2008) Compactly supported radial covariance functions *Journal of Geodesy*, **82**, 431–43.

Moreaux, G., Tscherning, C. and Sanso, F. (1999) Approximation of harmonic covariance functions on the sphere by non-harmonic locally supported functions. *Journal of Geodesy*, **73**, 555–67.

Müller, H.G. (2011) Functional data analysis. In StatProb: *The Encyclopedia.* Sponsored by Statistics and Probability Societies.

Myers, D.E. (1982) Matrix formulation of co-kriging. *Journal of the International Association for Mathematical Geology*, **14**(3), 249–57.

Myers, D.E. and Journel, A.G. (1990) Variograms with zonal anisotropies and non-invertible kriging systems. *Mathematical Geology*, **22**, 779–85.

Myers, M.F., Rogers, D.J., Cox, J., Flahault, A. and Hay, S.I. (2000) Forecasting disease risk for increased epidemic preparedness in public health. In S.I. Hay, S.E. Randolph and D.J. Rogers (eds) *Remote Sensing and Geographic Information Systems in Epidemiology*, pp. 309–30. Academic Press, New York.

Nagumo, M. (1930) Äuber eine Klasse der Mittelwerte. *Japanese Journal of Mathematics*, **7**, 71–9.

Nelsen, R.B. (1999) *An Introduction to Copulas*. Springer, New York.

Nerini, D. and Monestiez, P. (2008) A cokriging method for spatial functional data with applications in oceanology. First International Workshop on Functional and Operational Statistics. Toulouse, France.

Nerini, D., Monestiez, P. and Manté, C. (2010) Cokriging for spatial functional data. *Journal of Multivariate Analysis*, **101**(2), 409–18.

Neumaier, A. and Groeneveld, E. (1998) Restricted maximum likelihood estimation of covariances in sparse linear models. *Genetics Selection Evolution*, **30**, 3–26.

Neuman, S.P. and Jacobsen, E.A. (1984) Analysis of nonintrinsic spatial variability by residual kriging with application to regional groundwater levels. *Mathematical Geology*, **16**(5), 499–521.

Neumann, J. von and Schoenberg, I.J. (1941) Fourier integrals and metric geometry. *Transactions of the American Mathematical Society*, **50**, 226–51.

Paciorek, C.J. and Schervish, M.J. (2006) Spatial modelling using a new class of nonstationary covariance functions. *Environmetrics*, **17**, 483–506.

Padoan, S. and Bevilacqua, M. (2013) CompRandFld: Composite-likelihood based Analysis of Random Fields. R package version 1.0.3. Available at: http://CRAN.R-project.org/package =CompRandFld

Pardo-Iguzquiza, E. (1998) Comparison of geostatistical methods for estimating the areal average climatological rainfall mean using data on precipitation and topography. *International Journal of Climatology*, **18**, 1031–47.

Patterson, A. and Thompson, R. (1971) Recovery of inter-block information when block sizes are unequal. *Biometrika*, **58**, 545–55.

Patterson, A. and Thompson, R. (1974) Maximum likelihood estimation of components of variance. In *Proceedings of Eighth International Biochemistry Conference*, pp. 197–207.

Pebesma, E.J. (2004) Multivariable geostatistics in S: the GSTAT package. *Computers & Geosciences*, **30**, 683–91.

Porcu, E., Gregori, P. and Mateu, J. (2006) Nonseparable stationary anisotropic space-time covariance functions. *Stochastic Environmental Research and Risk Assessment*, **21**, 113–22.

Porcu, E. and Mateu, J. (2007) Mixture-based modeling for space-time data. *Environmetrics*, **18**, 285–302.

Porcu, E., Mateu, J. and Bevilacqua, M. (2007a) Covariance functions which are stationary or nonstationary in space and stationary in time. *Statistica Neerlandica*, **61**(3), 358–82.

Porcu, E., Gregori, P. and Mateu, J. (2007b) La descente et la montée étendues: the spatially d-anisotropic and the spatiotemporal case. *Stochastic Environmental Research and Risk Assessment*, **21**(6), 683–93.

Porcu, E., Mateu, J., Zini, A. and Pini, R. (2007c) Modelling spatio-temporal data: a new variogram and covariance structure proposal. *Statistics and Probability Letters*, **77**, 83–9.

Porcu, E., Mateu, J. and Saura, F. (2008) New classes of covariance and spectral density functions for spatiotemporal modelling. *Stochastic Environmental Research and Risk Assessment*, **22**, 65–79.

Porcu, E., Gregori, P. and Mateu, J. (2009a) Archimedean spectral densities for nonstationary space-time geostatistics. *Statistica Sinica*, **19**(1), 273–86.

Porcu, E., Mateu, J. and Christakos, G. (2009b) Quasi-arithmetic means of covariance functions with potential applications to space-time data. *Journal of Multivariate Analysis*, **100**(8), 1830–44.

Porcu, E., Matkowski, J. and Mateu, J. (2010) On the non-reducibility of non-stationary correlation functions to stationary ones under a class of mean-operator transformations. *Stochastic Environmental Research and Risk Assessment*, **24**(5), 599–610.

Porcu, E. and Schilling, R. (2011) From Schoenberg to Pick-Nevanlinna: towards a complete picture of the variogram class. *Bernoulli*, **17**(1), 441–55.

R Core Team. (2014) R: A Language and Environment for Statistical Computing, R Foundation for Statistical Computing, Vienna, Austria. Available at: http://www.R-project.org/

Ramsay, J. (1982) When the data are functions? *Psychometrika*, **47**, 379–96.

Ramsay, J. (1988) Monotone regression splines in action. *Statistical Science*, **3**, 425–41.

Ramsay, J. and Dalzell, C. (1991) Some tools for functional data analysis. *Journal of the Royal Statistical Society*, **53**(3), 539–72.

Ramsay, J., Hooker, G. and Graves, S. (2013) *Functional Data Analysis with R and MATLAB*. Springer, New York.

Ramsay, J. and Silverman, B. (1997) *Functional Data Analysis*. Springer, New York.

Ramsay, J. and Silverman, B. (2005) *Functional Data Analysis*, second edition. Springer, New York.

Ramsay, J.O., Wickham, H., Graves, S. and Hooker, G. (2013) FDA: Functional Data Analysis. *R package version* 2.4.0. Available at: http://CRAN.R-project.org/package=fda

Reyes, A., Giraldo, R. and Mateu, J. (2012) Residual kriging for functional data: application to the spatial prediction of salinity curves. *Reporte interno de investigación*, **20**, Universidad Nacional de Colombia.

Ribeiro, Jr. P.J. and Diggle, P.J. (2001) geoR: a package for geostatistical analysis, *R-NEWS*, **1**(2): 15–18.

Riishøjgaard, L.P. (1998) A direct way of specifying flow-dependent background error correlations for meteorological analysis systems. *Tellus A*, **50**, 42–57.

Rivoirard, J. (1994) *Introduction to Disjunctive Kriging and Non-linear Geostatistics*. Oxford University Press, Oxford.

Rodríguez-Iturbe, I. and Mejia, J.M. (1974) The design of rainfall networks in time and space. *Water Resources Research*, **10**, 713–28.

Rouhani, S. and Hall, T.J. (1989) Space-time kriging of groundwater data. In M. Armstrong (ed.), *Geostatistics*, pp. 639–51. Kluwer Academic Publishers, Dordrecht.

Rouhani, S. and Myers, D.E. (1990) Problems in space-time kriging of hydrogeological data. *Math Geology*, **22**, 611–23.

Samper, F.J. (1986) Statistical methods for analyzing hydrological, hydrochemical and isotopic data aquifers. Ph.D. dissertation, Department of Hydrology and Water Resources. University of Arizona.

Samper, F.J. and Carrera, J. (1996) *Geoestadística. Aplicaciones a la hidrología subterránea.* CIMNE.

Schabenberger, O. and Gotway, C.A. (2005) *Statistical Methods for Spatial Data Analysis*. Chapman & Hall, London.

Schlather, M. (1999) Introduction to positive definite functions and to unconditional simulation of random fields Technical Report ST 99–10. Lancaster University.

Schlather, M., Menck, P., Singleton, R., Pfaff, B. and R Core team. (2013) RandomFields: Simulation and Analysis of Random Fields. R package version 2.0.66. Availavle at: http://CRAN.Rproject. org/package=RandomFields

Schoenberg, I.J. (1938) Metric spaces and completely monotone functions *Annals of Mathematics*, **39**, 811–41.

Searle, S.R. (1982) *Matrix Algebra Useful for Statistics*. John Wiley & Sons, Inc., New York.

Searle, S.R., Casella, G. and McCulloch, C.E. (1992) *Variance Components*., John Wiley & Sons, Inc., New York.

Shapiro, A. and Botha, J.D. (1991) Variogram fitting with a conditional class of conditionally nonnegative definite functions. *Computational Statistics and Data Analysis*, **11**, 87–96.

Staudte, R.G. and Sheather, S.J. (1990) *Robust Estimation and Testing*. John Wiley & Sons, Ltd., Chichester.

Stein, M.L. (1999) *Interpolation of Spatial Data. Some Theory of Kriging*. Springer-Verlag, Berlin.

Stein, M.L. (2004) Statistical methods for regular monitoring data. Technical Report, 15. Center for Integrating Statistical and Environmental Science. University of Chicago.

Stein, M.L., Chi, Z. and Welty, L.J. (2004) Approximating likelihoods for large spatial datasets, *Journal of the Royal Statistical Society*, *Ser. B*, **66**, 275–96.

Stein, M.L. (2005a) Space-time covariance functions. *Journal of the American Statistical Association*, **100**, 310–21.

Stein, M.L. (2005b) *Nonstationary spatial covariance functions*. Technical Report, University of Chicago.

Stein, M.L. (2005c) Statistical methods for regular monitoring data. *Journal of the Royal Statistical Society B*, **67**, 667–87.

Subrahmaniam, K. (1966) A test for ‘intrinsic correlation’ in the theory of accident proneness. *Journal of the Royal Statistical Society Series B-Statistical Methodology*, **28**, 180–9.

Taylor, G.I. (1938) The spectrum of turbulence. In *Proceedings of the Royal Society of London A*, **164**, 476–90.

Tobler, W. (1970 A computer movie simulating urban growth in the Detroit region. *Economic Geography*, **46**(2), 234–40.

Tukey, J.W. (1977) *Exploratory Data Analysis*. Addison-Wesley, Reading, MA.

Ullah, S. and Finch, C.F. (2013) Applications of functional data analysis: a systematic review. *BMC Medical Research Methodology*, **13**(43), 1–12.

Vecchia, A. (1998) Estimation and model identification for continuous spatial processes. *Journal of the Royal Statistical Society*, **50**, 297–312.

Ver Hoef, J.M. and Cressie, N. (1993) Multivariable spatial prediction. *Mathematical Geology*, **25**(2), 219–39.

Volpi, G. and Gambolati, G. (1978) On the use of a main trend for the kriging technique in hydrology. *Advances in Water Resources*, **1**(6), 345–9.

Wackernagel, H. (1995) *Multivariate Geostatistics: An Introduction with Applications*. Springer-Verlag, Berlin.

Wackernagel, H. (2003) *Multivariate Geostatistics*. Springer-Verlag, Berlin.

Wasserman, L. (2006) *All of Nonparametric Statistics*. Springer-Verlag, Berlin.

Webster, R. and Oliver, M.A. (2001) *Geostatistics for Environmental Scientists*. John Wiley & Sons, Ltd., Chichester.

Yaglom, A.M. (1986) *Correlation Theory of Stationary and Related Random Functions*, Vol. I: *Basic Results*. Springer-Verlag, Berlin.

Yates, S.R. and Warrick, A.W. (1986) Disjunctive kriging: overview of estimation and conditional probability. *Water Resources Research*, **22**(5), 615–21.

Zimmerman, D.L. and Zimmerman, M.B. (1991) A Monte Carlo comparison of spatial semivariogram estimators and corresponding ordinary kriging predictors. *Technometrics*, **33**, 77–91.

Index

Spatial and Spatio-Temporal Geostatistical Modeling and Kriging, First Edition.
José-María Montero, Gema Fernández-Avilés, and Jorge Mateu.

Companion Website: www.wiley.com/go/montero/spatial

WILEY SERIES IN PROBABILITY AND STATISTICS

ESTABLISHED BY WALTER A. SHEWHART AND SAMUEL S. WILKS

Editors: *David J. Balding, Noel A. C. Cressie, Garrett M. Fitzmaurice, Geof H. Givens, Harvey Goldstein, Geert Molenberghs, David W. Scott, Adrian F. M. Smith, Ruey S. Tsay, Sanford Weisberg*
Editors Emeriti: *J. Stuart Hunter, Iain M. Johnstone, Joseph B. Kadane, Jozef L. Teugels*

The ***Wiley Series in Probability and Statistics*** is well established and authoritative. It covers many topics of current research interest in both pure and applied statistics and probability theory. Written by leading statisticians and institutions, the titles span both state-of-the-art developments in the field and classical methods.

Reflecting the wide range of current research in statistics, the series encompasses applied, methodological and theoretical statistics, ranging from applications and new techniques made possible by advances in computerized practice to rigorous treatment of theoretical approaches.

This series provides essential and invaluable reading for all statisticians, whether in academia, industry, government, or research.

† ABRAHAM and LEDOLTER · Statistical Methods for Forecasting
AGRESTI · Analysis of Ordinal Categorical Data, *Second Edition*
AGRESTI · An Introduction to Categorical Data Analysis, *Second Edition*
AGRESTI · Categorical Data Analysis, *Third Edition*
ALSTON, MENGERSEN and PETTITT (editors) · Case Studies in Bayesian Statistical Modelling and Analysis
ALTMAN, GILL, and McDONALD · Numerical Issues in Statistical Computing for the Social Scientist
AMARATUNGA and CABRERA · Exploration and Analysis of DNA Microarray and Protein Array Data
AMARATUNGA, CABRERA, and SHKEDY · Exploration and Analysis of DNA Microarray and Other High-Dimensional Data, *Second Edition*
ANDĚL · Mathematics of Chance
ANDERSON · An Introduction to Multivariate Statistical Analysis, *Third Edition*
* ANDERSON · The Statistical Analysis of Time Series
ANDERSON, AUQUIER, HAUCK, OAKES, VANDAELE, and WEISBERG · Statistical Methods for Comparative Studies
ANDERSON and LOYNES · The Teaching of Practical Statistics
ARMITAGE and DAVID (editors) · Advances in Biometry
ARNOLD, BALAKRISHNAN, and NAGARAJA · Records
* ARTHANARI and DODGE · Mathematical Programming in Statistics
AUGUSTIN, COOLEN, DE COOMAN and TROFFAES (editors) · Introduction to Imprecise Probabilities
* BAILEY · The Elements of Stochastic Processes with Applications to the Natural Sciences

† Now available in a lower priced paperback edition in the Wiley–Interscience Paperback Series.
* Now available in a lower priced paperback edition in the Wiley Classics Library.

BAJORSKI · Statistics for Imaging, Optics, and Photonics

BALAKRISHNAN and KOUTRAS · Runs and Scans with Applications

BALAKRISHNAN and NG · Precedence-Type Tests and Applications

BARNETT · Comparative Statistical Inference, *Third Edition*

BARNETT · Environmental Statistics

BARNETT and LEWIS · Outliers in Statistical Data, *Third Edition*

BARTHOLOMEW, KNOTT, and MOUSTAKI · Latent Variable Models and Factor Analysis: A Unified Approach, *Third Edition*

BARTOSZYNSKI and NIEWIADOMSKA-BUGAJ · Probability and Statistical Inference, *Second Edition*

BASILEVSKY · Statistical Factor Analysis and Related Methods: Theory and Applications

BATES and WATTS · Nonlinear Regression Analysis and Its Applications

BECHHOFER, SANTNER, and GOLDSMAN · Design and Analysis of Experiments for Statistical Selection, Screening, and Multiple Comparisons

BEH and LOMBARDO · Correspondence Analysis: Theory, Practice and New Strategies

BEIRLANT, GOEGEBEUR, SEGERS, TEUGELS, and DE WAAL · Statistics of Extremes: Theory and Applications

BELSLEY · Conditioning Diagnostics: Collinearity and Weak Data in Regression

† BELSLEY, KUH, and WELSCH · Regression Diagnostics: Identifying Influential Data and Sources of Collinearity

BENDAT and PIERSOL · Random Data: Analysis and Measurement Procedures, *Fourth Edition*

BERNARDO and SMITH · Bayesian Theory

BHAT and MILLER · Elements of Applied Stochastic Processes, *Third Edition*

BHATTACHARYA and WAYMIRE · Stochastic Processes with Applications

BIEMER, GROVES, LYBERG, MATHIOWETZ, and SUDMAN · Measurement Errors in Surveys

BILLINGSLEY · Convergence of Probability Measures, *Second Edition*

BILLINGSLEY · Probability and Measure, *Anniversary Edition*

BIRKES and DODGE · Alternative Methods of Regression

BISGAARD and KULAHCI · Time Series Analysis and Forecasting by Example

BISWAS, DATTA, FINE, and SEGAL · Statistical Advances in the Biomedical Sciences: Clinical Trials, Epidemiology, Survival Analysis, and Bioinformatics

BLISCHKE and MURTHY (editors) · Case Studies in Reliability and Maintenance

BLISCHKE and MURTHY · Reliability: Modeling, Prediction, and Optimization

BLOOMFIELD · Fourier Analysis of Time Series: An Introduction, *Second Edition*

BOLLEN · Structural Equations with Latent Variables

BOLLEN and CURRAN · Latent Curve Models: A Structural Equation Perspective

BONNINI, CORAIN, MAROZZI and SALMASO · Nonparametric Hypothesis Testing: Rank and Permutation Methods with Applications in R

BOROVKOV · Ergodicity and Stability of Stochastic Processes

† Now available in a lower priced paperback edition in the Wiley–Interscience Paperback Series.

BOSQ and BLANKE · Inference and Prediction in Large Dimensions

BOULEAU · Numerical Methods for Stochastic Processes

* BOX and TIAO · Bayesian Inference in Statistical Analysis

BOX · Improving Almost Anything, *Revised Edition*

* BOX and DRAPER · Evolutionary Operation: A Statistical Method for Process Improvement

BOX and DRAPER · Response Surfaces, Mixtures, and Ridge Analyses, *Second Edition*

BOX, HUNTER, and HUNTER · Statistics for Experimenters: Design, Innovation, and Discovery, *Second Editon*

BOX, JENKINS, and REINSEL · Time Series Analysis: Forecasting and Control, *Fourth Edition*

BOX, LUCEÑO, and PANIAGUA-QUIÑONES · Statistical Control by Monitoring and Adjustment, *Second Edition*

* BROWN and HOLLANDER · Statistics: A Biomedical Introduction

CAIROLI and DALANG · Sequential Stochastic Optimization

CASTILLO, HADI, BALAKRISHNAN, and SARABIA · Extreme Value and Related Models with Applications in Engineering and Science

CHAN · Time Series: Applications to Finance with R and S-Plus®, *Second Edition*

CHARALAMBIDES · Combinatorial Methods in Discrete Distributions

CHATTERJEE and HADI · Regression Analysis by Example, *Fourth Edition*

CHATTERJEE and HADI · Sensitivity Analysis in Linear Regression

CHEN · The Fitness of Information: Quantitative Assessments of Critical Evidence

CHERNICK · Bootstrap Methods: A Guide for Practitioners and Researchers, *Second Edition*

CHERNICK and FRIIS · Introductory Biostatistics for the Health Sciences

CHILÈS and DELFINER · Geostatistics: Modeling Spatial Uncertainty, *Second Edition*

CHIU, STOYAN, KENDALL and MECKE · Stochastic Geometry and Its Applications, *Third Edition*

CHOW and LIU · Design and Analysis of Clinical Trials: Concepts and Methodologies, *Third Edition*

CLARKE · Linear Models: The Theory and Application of Analysis of Variance

CLARKE and DISNEY · Probability and Random Processes: A First Course with Applications, *Second Edition*

* COCHRAN and COX · Experimental Designs, *Second Edition*

COLLINS and LANZA · Latent Class and Latent Transition Analysis: With Applications in the Social, Behavioral, and Health Sciences

CONGDON · Applied Bayesian Modelling, *Second Edition*

CONGDON · Bayesian Models for Categorical Data

CONGDON · Bayesian Statistical Modelling, *Second Edition*

CONOVER · Practical Nonparametric Statistics, *Third Edition*

COOK · Regression Graphics

COOK and WEISBERG · An Introduction to Regression Graphics

* Now available in a lower priced paperback edition in the Wiley Classics Library.

COOK and WEISBERG · Applied Regression Including Computing and Graphics

CORNELL · A Primer on Experiments with Mixtures

CORNELL · Experiments with Mixtures, Designs, Models, and the Analysis of Mixture Data, *Third Edition*

COX · A Handbook of Introductory Statistical Methods

CRESSIE · Statistics for Spatial Data, *Revised Edition*

CRESSIE and WIKLE · Statistics for Spatio-Temporal Data

CSÖRGÖ and HORVÁTH · Limit Theorems in Change Point Analysis

DAGPUNAR · Simulation and Monte Carlo: With Applications in Finance and MCMC

DANIEL · Applications of Statistics to Industrial Experimentation

DANIEL · Biostatistics: A Foundation for Analysis in the Health Sciences, *Eighth Edition*

* DANIEL · Fitting Equations to Data: Computer Analysis of Multifactor Data, *Second Edition*

DASU and JOHNSON · Exploratory Data Mining and Data Cleaning

DAVID and NAGARAJA · Order Statistics, *Third Edition*

DAVINO, FURNO and VISTOCCO · Quantile Regression: Theory and Applications

* DEGROOT, FIENBERG, and KADANE · Statistics and the Law

DEL CASTILLO · Statistical Process Adjustment for Quality Control

DEMARIS · Regression with Social Data: Modeling Continuous and Limited Response Variables

DEMIDENKO · Mixed Models: Theory and Applications with R, *Second Edition*

DENISON, HOLMES, MALLICK and SMITH · Bayesian Methods for Nonlinear Classification and Regression

DETTE and STUDDEN · The Theory of Canonical Moments with Applications in Statistics, Probability, and Analysis

DEY and MUKERJEE · Fractional Factorial Plans

DILLON and GOLDSTEIN · Multivariate Analysis: Methods and Applications

* DODGE and ROMIG · Sampling Inspection Tables, *Second Edition*

* DOOB · Stochastic Processes

DOWDY, WEARDEN, and CHILKO · Statistics for Research, *Third Edition*

DRAPER and SMITH · Applied Regression Analysis, *Third Edition*

DRYDEN and MARDIA · Statistical Shape Analysis

DUDEWICZ and MISHRA · Modern Mathematical Statistics

DUNN and CLARK · Basic Statistics: A Primer for the Biomedical Sciences, *Fourth Edition*

DUPUIS and ELLIS · A Weak Convergence Approach to the Theory of Large Deviations

EDLER and KITSOS · Recent Advances in Quantitative Methods in Cancer and Human Health Risk Assessment

* ELANDT-JOHNSON and JOHNSON · Survival Models and Data Analysis

ENDERS · Applied Econometric Time Series, Third Edition

† ETHIER and KURTZ · Markov Processes: Characterization and Convergence

* Now available in a lower priced paperback edition in the Wiley Classics Library.

† Now available in a lower priced paperback edition in the Wiley–Interscience Paperback Series.

EVANS, HASTINGS, and PEACOCK · Statistical Distributions, *Third Edition*

EVERITT, LANDAU, LEESE, and STAHL · Cluster Analysis, *Fifth Edition*

FEDERER and KING · Variations on Split Plot and Split Block Experiment Designs

FELLER · An Introduction to Probability Theory and Its Applications, Volume I, *Third Edition,* Revised; Volume II, *Second Edition*

FITZMAURICE, LAIRD, and WARE · Applied Longitudinal Analysis, *Second Edition*

* FLEISS · The Design and Analysis of Clinical Experiments

FLEISS · Statistical Methods for Rates and Proportions, Third Edition

† FLEMING and HARRINGTON · Counting Processes and Survival Analysis

FUJIKOSHI, ULYANOV, and SHIMIZU · Multivariate Statistics: High-Dimensional and Large-Sample Approximations

FULLER · Introduction to Statistical Time Series, Second Edition

† FULLER · Measurement Error Models

GALLANT · Nonlinear Statistical Models

GEISSER · Modes of Parametric Statistical Inference

GELMAN and MENG · Applied Bayesian Modeling and Causal Inference from ncomplete-Data Perspectives

GEWEKE · Contemporary Bayesian Econometrics and Statistics

GHOSH, MUKHOPADHYAY, and SEN · Sequential Estimation

GIESBRECHT and GUMPERTZ · Planning, Construction, and Statistical Analysis of Comparative Experiments

GIFI · Nonlinear Multivariate Analysis

GIVENS and HOETING · Computational Statistics

GLASSERMAN and YAO · Monotone Structure in Discrete-Event Systems

GNANADESIKAN · Methods for Statistical Data Analysis of Multivariate Observations, *Second Edition*

GOLDSTEIN · Multilevel Statistical Models, *Fourth Edition*

GOLDSTEIN and LEWIS · Assessment: Problems, Development, and Statistical Issues

GOLDSTEIN and WOOFF · Bayes Linear Statistics

GRAHAM · Markov Chains: Analytic and Monte Carlo Computations

GREENWOOD and NIKULIN · A Guide to Chi-Squared Testing

GROSS, SHORTLE, THOMPSON, and HARRIS · Fundamentals of Queueing Theory, *Fourth Edition*

GROSS, SHORTLE, THOMPSON, and HARRIS · Solutions Manual to Accompany Fundamentals of Queueing Theory, *Fourth Edition*

* HAHN and SHAPIRO · Statistical Models in Engineering

HAHN and MEEKER · Statistical Intervals: A Guide for Practitioners

HALD · A History of Probability and Statistics and their Applications Before 1750

† HAMPEL · Robust Statistics: The Approach Based on Influence Functions

HARTUNG, KNAPP, and SINHA · Statistical Meta-Analysis with Applications

HEIBERGER · Computation for the Analysis of Designed Experiments

* Now available in a lower priced paperback edition in the Wiley Classics Library.
† Now available in a lower priced paperback edition in the Wiley–Interscience Paperback Series.

HEDAYAT and SINHA · Design and Inference in Finite Population Sampling

HEDEKER and GIBBONS · Longitudinal Data Analysis

HELLER · MACSYMA for Statisticians

HERITIER, CANTONI, COPT, and VICTORIA-FESER · Robust Methods in Biostatistics

HINKELMANN and KEMPTHORNE · Design and Analysis of Experiments, Volume 1: Introduction to Experimental Design, *Second Edition*

HINKELMANN and KEMPTHORNE · Design and Analysis of Experiments, Volume 2: Advanced Experimental Design

HINKELMANN (editor) · Design and Analysis of Experiments, Volume 3: Special Designs and Applications

HOAGLIN, MOSTELLER, and TUKEY · Fundamentals of Exploratory Analysis of Variance

* HOAGLIN, MOSTELLER, and TUKEY · Exploring Data Tables, Trends and Shapes

* HOAGLIN, MOSTELLER, and TUKEY · Understanding Robust and Exploratory Data Analysis

HOCHBERG and TAMHANE · Multiple Comparison Procedures

HOCKING · Methods and Applications of Linear Models: Regression and the Analysis of Variance, *Third Edition*

HOEL · Introduction to Mathematical Statistics, *Fifth Edition*

HOGG and KLUGMAN · Loss Distributions

HOLLANDER, WOLFE, and CHICKEN · Nonparametric Statistical Methods, *Third Edition*

HOSMER and LEMESHOW · Applied Logistic Regression, *Second Edition*

HOSMER, LEMESHOW, and MAY · Applied Survival Analysis: Regression Modeling of Time-to-Event Data, *Second Edition*

HUBER · Data Analysis: What Can Be Learned From the Past 50 Years

HUBER · Robust Statistics

† HUBER and RONCHETTI · Robust Statistics, *Second Edition*

HUBERTY · Applied Discriminant Analysis, *Second Edition*

HUBERTY and OLEJNIK · Applied MANOVA and Discriminant Analysis, *Second Edition*

HUITEMA · The Analysis of Covariance and Alternatives: Statistical Methods for Experiments, Quasi-Experiments, and Single-Case Studies, *Second Edition*

HUNT and KENNEDY · Financial Derivatives in Theory and Practice, *Revised Edition*

HURD and MIAMEE · Periodically Correlated Random Sequences: Spectral Theory and Practice

HUSKOVA, BERAN, and DUPAC · Collected Works of Jaroslav Hajek – with Commentary

HUZURBAZAR · Flowgraph Models for Multistate Time-to-Event Data

JACKMAN · Bayesian Analysis for the Social Sciences

† JACKSON · A User's Guide to Principle Components

* Now available in a lower priced paperback edition in the Wiley Classics Library.

† Now available in a lower priced paperback edition in the Wiley–Interscience Paperback Series.

JOHN · Statistical Methods in Engineering and Quality Assurance

JOHNSON · Multivariate Statistical Simulation

JOHNSON and BALAKRISHNAN · Advances in the Theory and Practice of Statistics: A Volume in Honor of Samuel Kotz

JOHNSON, KEMP, and KOTZ · Univariate Discrete Distributions, *Third Edition*

JOHNSON and KOTZ (editors) · Leading Personalities in Statistical Sciences: From the Seventeenth Century to the Present

JOHNSON, KOTZ, and BALAKRISHNAN · Continuous Univariate Distributions, Volume 1, *Second Edition*

JOHNSON, KOTZ, and BALAKRISHNAN · Continuous Univariate Distributions, Volume 2, *Second Edition*

JOHNSON, KOTZ, and BALAKRISHNAN · Discrete Multivariate Distributions

JUDGE, GRIFFITHS, HILL, LÜTKEPOHL, and LEE · The Theory and Practice of Econometrics, *Second Edition*

JUREK and MASON · Operator-Limit Distributions in Probability Theory

KADANE · Bayesian Methods and Ethics in a Clinical Trial Design

KADANE AND SCHUM · A Probabilistic Analysis of the Sacco and Vanzetti Evidence

KALBFLEISCH and PRENTICE · The Statistical Analysis of Failure Time Data, *Second Edition*

KARIYA and KURATA · Generalized Least Squares

KASS and VOS · Geometrical Foundations of Asymptotic Inference

† KAUFMAN and ROUSSEEUW · Finding Groups in Data: An Introduction to Cluster Analysis

KEDEM and FOKIANOS · Regression Models for Time Series Analysis

KENDALL, BARDEN, CARNE, and LE · Shape and Shape Theory

KHURI · Advanced Calculus with Applications in Statistics, *Second Edition*

KHURI, MATHEW, and SINHA · Statistical Tests for Mixed Linear Models

* KISH · Statistical Design for Research

KLEIBER and KOTZ · Statistical Size Distributions in Economics and Actuarial Sciences

KLEMELÄ · Smoothing of Multivariate Data: Density Estimation and Visualization

KLUGMAN, PANJER, and WILLMOT · Loss Models: From Data to Decisions, *Third Edition*

KLUGMAN, PANJER, and WILLMOT · Loss Models: Further Topics

KLUGMAN, PANJER, and WILLMOT · Solutions Manual to Accompany Loss Models: From Data to Decisions, *Third Edition*

KOSKI and NOBLE · Bayesian Networks: An Introduction

KOTZ, BALAKRISHNAN, and JOHNSON · Continuous Multivariate Distributions, Volume 1, *Second Edition*

KOTZ and JOHNSON (editors) · Encyclopedia of Statistical Sciences: Volumes 1 to 9 with Index

† Now available in a lower priced paperback edition in the Wiley–Interscience Paperback Series.

* Now available in a lower priced paperback edition in the Wiley Classics Library.

KOTZ and JOHNSON (editors) · Encyclopedia of Statistical Sciences: Supplement Volume

KOTZ, READ, and BANKS (editors) · Encyclopedia of Statistical Sciences: Update Volume 1

KOTZ, READ, and BANKS (editors) · Encyclopedia of Statistical Sciences: Update Volume 2

KOWALSKI and TU · Modern Applied U-Statistics

KRISHNAMOORTHY and MATHEW · Statistical Tolerance Regions: Theory, Applications, and Computation

KROESE, TAIMRE, and BOTEV · Handbook of Monte Carlo Methods

KROONENBERG · Applied Multiway Data Analysis

KULINSKAYA, MORGENTHALER, and STAUDTE · Meta Analysis: A Guide to Calibrating and Combining Statistical Evidence

KULKARNI and HARMAN · An Elementary Introduction to Statistical Learning Theory

KUROWICKA and COOKE · Uncertainty Analysis with High Dimensional Dependence Modelling

KVAM and VIDAKOVIC · Nonparametric Statistics with Applications to Science and Engineering

LACHIN · Biostatistical Methods: The Assessment of Relative Risks, *Second Edition*

LAD · Operational Subjective Statistical Methods: A Mathematical, Philosophical, and Historical Introduction

LAMPERTI · Probability: A Survey of the Mathematical Theory, *Second Edition*

LAWLESS · Statistical Models and Methods for Lifetime Data, *Second Edition*

LAWSON · Statistical Methods in Spatial Epidemiology, *Second Edition*

LE · Applied Categorical Data Analysis, *Second Edition*

LE · Applied Survival Analysis

LEE · Structural Equation Modeling: A Bayesian Approach

LEE and WANG · Statistical Methods for Survival Data Analysis, *Fourth Edition*

LePAGE and BILLARD · Exploring the Limits of Bootstrap

LESSLER and KALSBEEK · Nonsampling Errors in Surveys

LEYLAND and GOLDSTEIN (editors) · Multilevel Modelling of Health Statistics

LIAO · Statistical Group Comparison

LIN · Introductory Stochastic Analysis for Finance and Insurance

LINDLEY · Understanding Uncertainty, *Revised Edition*

LITTLE and RUBIN · Statistical Analysis with Missing Data, *Second Edition*

LLOYD · The Statistical Analysis of Categorical Data

LOWEN and TEICH · Fractal-Based Point Processes

MAGNUS and NEUDECKER · Matrix Differential Calculus with Applications in Statistics and Econometrics, *Revised Edition*

MALLER and ZHOU · Survival Analysis with Long Term Survivors

MARCHETTE · Random Graphs for Statistical Pattern Recognition

MARDIA and JUPP · Directional Statistics

MARKOVICH · Nonparametric Analysis of Univariate Heavy-Tailed Data: Research and Practice

MARONNA, MARTIN and YOHAI · Robust Statistics: Theory and Methods

MASON, GUNST, and HESS · Statistical Design and Analysis of Experiments with Applications to Engineering and Science, *Second Edition*

McCULLOCH, SEARLE, and NEUHAUS · Generalized, Linear, and Mixed Models, *Second Edition*

McFADDEN · Management of Data in Clinical Trials, *Second Edition*

* McLACHLAN · Discriminant Analysis and Statistical Pattern Recognition

McLACHLAN, DO, and AMBROISE · Analyzing Microarray Gene Expression Data

McLACHLAN and KRISHNAN · The EM Algorithm and Extensions, *Second Edition*

McLACHLAN and PEEL · Finite Mixture Models

McNEIL · Epidemiological Research Methods

MEEKER and ESCOBAR · Statistical Methods for Reliability Data

MEERSCHAERT and SCHEFFLER · Limit Distributions for Sums of Independent Random Vectors: Heavy Tails in Theory and Practice

MENGERSEN, ROBERT, and TITTERINGTON · Mixtures: Estimation and Applications

MICKEY, DUNN, and CLARK · Applied Statistics: Analysis of Variance and Regression, *Third Edition*

* MILLER · Survival Analysis, *Second Edition*

MONTGOMERY, JENNINGS, and KULAHCI · Introduction to Time Series Analysis and Forecasting

MONTGOMERY, PECK, and VINING · Introduction to Linear Regression Analysis, *Fifth Edition*

MORGENTHALER and TUKEY · Configural Polysampling: A Route to Practical Robustness

MUIRHEAD · Aspects of Multivariate Statistical Theory

MULLER and STOYAN · Comparison Methods for Stochastic Models and Risks

MURTHY, XIE, and JIANG · Weibull Models

MYERS, MONTGOMERY, and ANDERSON-COOK · Response Surface Methodology: Process and Product Optimization Using Designed Experiments, *Third Edition*

MYERS, MONTGOMERY, VINING, and ROBINSON · Generalized Linear Models. With Applications in Engineering and the Sciences, *Second Edition*

NATVIG · Multistate Systems Reliability Theory With Applications

† NELSON · Accelerated Testing, Statistical Models, Test Plans, and Data Analyses

† NELSON · Applied Life Data Analysis

NEWMAN · Biostatistical Methods in Epidemiology

NG, TAIN, and TANG · Dirichlet Theory: Theory, Methods and Applications

OKABE, BOOTS, SUGIHARA, and CHIU · Spatial Tesselations: Concepts and Applications of Voronoi Diagrams, *Second Edition*

OLIVER and SMITH · Influence Diagrams, Belief Nets and Decision Analysis

* Now available in a lower priced paperback edition in the Wiley Classics Library.

† Now available in a lower priced paperback edition in the Wiley–Interscience Paperback Series.

PALTA · Quantitative Methods in Population Health: Extensions of Ordinary Regressions
PANJER · Operational Risk: Modeling and Analytics
PANKRATZ · Forecasting with Dynamic Regression Models
PANKRATZ · Forecasting with Univariate Box-Jenkins Models: Concepts and Cases
PARDOUX · Markov Processes and Applications: Algorithms, Networks, Genome and Finance
PARMIGIANI and INOUE · Decision Theory: Principles and Approaches
* PARZEN · Modern Probability Theory and Its Applications
PEÑA, TIAO, and TSAY · A Course in Time Series Analysis
PESARIN and SALMASO · Permutation Tests for Complex Data: Applications and Software
PIANTADOSI · Clinical Trials: A Methodologic Perspective, *Second Edition*
POURAHMADI · Foundations of Time Series Analysis and Prediction Theory
POURAHMADI · High-Dimensional Covariance Estimation
POWELL · Approximate Dynamic Programming: Solving the Curses of Dimensionality, *Second Edition*
POWELL and RYZHOV · Optimal Learning
PRESS · Subjective and Objective Bayesian Statistics, *Second Edition*
PRESS and TANUR · The Subjectivity of Scientists and the Bayesian Approach
PURI, VILAPLANA, and WERTZ · New Perspectives in Theoretical and Applied Statistics
† PUTERMAN · Markov Decision Processes: Discrete Stochastic Dynamic Programming
QIU · Image Processing and Jump Regression Analysis
* RAO · Linear Statistical Inference and Its Applications, *Second Edition*
RAO · Statistical Inference for Fractional Diffusion Processes
RAUSAND and HØYLAND · System Reliability Theory: Models, Statistical Methods, and Applications, *Second Edition*
RAYNER, THAS, and BEST · Smooth Tests of Goodnes of Fit: Using R, *Second Edition*
RENCHER and SCHAALJE · Linear Models in Statistics, *Second Edition*
RENCHER and CHRISTENSEN · Methods of Multivariate Analysis, *Third Edition*
RENCHER · Multivariate Statistical Inference with Applications
RIGDON and BASU · Statistical Methods for the Reliability of Repairable Systems
* RIPLEY · Spatial Statistics
* RIPLEY · Stochastic Simulation
ROHATGI and SALEH · An Introduction to Probability and Statistics, *Second Edition*
ROLSKI, SCHMIDLI, SCHMIDT, and TEUGELS · Stochastic Processes for Insurance and Finance
ROSENBERGER and LACHIN · Randomization in Clinical Trials: Theory and Practice
ROSSI, ALLENBY, and MCCULLOCH · Bayesian Statistics and Marketing
† ROUSSEEUW and LEROY · Robust Regression and Outlier Detection

* Now available in a lower priced paperback edition in the Wiley Classics Library.
† Now available in a lower priced paperback edition in the Wiley–Interscience Paperback Series.

ROYSTON and SAUERBREI · Multivariate Model Building: A Pragmatic Approach to Regression Analysis Based on Fractional Polynomials for Modeling Continuous Variables

* RUBIN · Multiple Imputation for Nonresponse in Surveys

RUBINSTEIN and KROESE · Simulation and the Monte Carlo Method, *Second Edition*

RUBINSTEIN and MELAMED · Modern Simulation and Modeling

RUBINSTEIN, RIDDER, and VAISMAN · Fast Sequential Monte Carlo Methods for Counting and Optimization

RYAN · Modern Engineering Statistics

RYAN · Modern Experimental Design

RYAN · Modern Regression Methods, *Second Edition*

RYAN · Sample Size Determination and Power

RYAN · Statistical Methods for Quality Improvement, *Third Edition*

SALEH · Theory of Preliminary Test and Stein-Type Estimation with Applications

SALTELLI, CHAN, and SCOTT (editors) · Sensitivity Analysis

SCHERER · Batch Effects and Noise in Microarray Experiments: Sources and Solutions

* SCHEFFE · The Analysis of Variance

SCHIMEK · Smoothing and Regression: Approaches, Computation, and Application

SCHOTT · Matrix Analysis for Statistics, *Second Edition*

SCHOUTENS · Levy Processes in Finance: Pricing Financial Derivatives

SCOTT · Multivariate Density Estimation: Theory, Practice, and Visualization

* SEARLE · Linear Models

† SEARLE · Linear Models for Unbalanced Data

† SEARLE · Matrix Algebra Useful for Statistics

† SEARLE, CASELLA, and McCULLOCH · Variance Components

SEARLE and WILLETT · Matrix Algebra for Applied Economics

SEBER · A Matrix Handbook For Statisticians

† SEBER · Multivariate Observations

SEBER and LEE · Linear Regression Analysis, Second Edition

† SEBER and WILD · Nonlinear Regression

SENNOTT · Stochastic Dynamic Programming and the Control of Queueing Systems

* SERFLING · Approximation Theorems of Mathematical Statistics

SHAFER and VOVK · Probability and Finance: It's Only a Game!

SHERMAN · Spatial Statistics and Spatio-Temporal Data: Covariance Functions and Directional Properties

SILVAPULLE and SEN · Constrained Statistical Inference: Inequality, Order, and Shape Restrictions

SINGPURWALLA · Reliability and Risk: A Bayesian Perspective

SMALL and MCLEISH · Hilbert Space Methods in Probability and Statistical Inference

SRIVASTAVA · Methods of Multivariate Statistics

* Now available in a lower priced paperback edition in the Wiley Classics Library.

† Now available in a lower priced paperback edition in the Wiley–Interscience Paperback Series.

STAPLETON · Linear Statistical Models, *Second Edition*

STAPLETON · Models for Probability and Statistical Inference: Theory and Applications

STAUDTE and SHEATHER · Robust Estimation and Testing

STOYAN · Counterexamples in Probability, *Second Edition*

STOYAN and STOYAN · Fractals, Random Shapes and Point Fields: Methods of Geometrical Statistics

STREET and BURGESS · The Construction of Optimal Stated Choice Experiments: Theory and Methods

STYAN · The Collected Papers of T. W. Anderson: 1943–1985

SUTTON, ABRAMS, JONES, SHELDON, and SONG · Methods for Meta-Analysis in Medical Research

TAKEZAWA · Introduction to Nonparametric Regression

TAMHANE · Statistical Analysis of Designed Experiments: Theory and Applications

TANAKA · Time Series Analysis: Nonstationary and Noninvertible Distribution Theory

THOMPSON · Empirical Model Building: Data, Models, and Reality, *Second Edition*

THOMPSON · Sampling, *Third Edition*

THOMPSON · Simulation: A Modeler's Approach

THOMPSON and SEBER · Adaptive Sampling

THOMPSON, WILLIAMS, and FINDLAY · Models for Investors in Real World Markets

TIERNEY · LISP-STAT: An Object-Oriented Environment for Statistical Computing and Dynamic Graphics

TROFFAES and DE COOMAN · Lower Previsions

TSAY · Analysis of Financial Time Series, *Third Edition*

TSAY · An Introduction to Analysis of Financial Data with R

TSAY · Multivariate Time Series Analysis: With R and Financial Applications

UPTON and FINGLETON · Spatial Data Analysis by Example, Volume II: Categorical and Directional Data

† VAN BELLE · Statistical Rules of Thumb, *Second Edition*

VAN BELLE, FISHER, HEAGERTY, and LUMLEY · Biostatistics: A Methodology for the Health Sciences, *Second Edition*

VESTRUP · The Theory of Measures and Integration

VIDAKOVIC · Statistical Modeling by Wavelets

VIERTL · Statistical Methods for Fuzzy Data

VINOD and REAGLE · Preparing for the Worst: Incorporating Downside Risk in Stock Market Investments

WALLER and GOTWAY · Applied Spatial Statistics for Public Health Data

WEISBERG · Applied Linear Regression, *Fourth Edition*

WEISBERG · Bias and Causation: Models and Judgment for Valid Comparisons

WELSH · Aspects of Statistical Inference

WESTFALL and YOUNG · Resampling-Based Multiple Testing: Examples and Methods for *p*-Value Adjustment

† Now available in a lower priced paperback edition in the Wiley–Interscience Paperback Series.

* WHITTAKER · Graphical Models in Applied Multivariate Statistics

WINKER · Optimization Heuristics in Economics: Applications of Threshold Accepting

WOODWORTH · Biostatistics: A Bayesian Introduction

WOOLSON and CLARKE · Statistical Methods for the Analysis of Biomedical Data, *Second Edition*

WU and HAMADA · Experiments: Planning, Analysis, and Parameter Design Optimization, *Second Edition*

WU and ZHANG · Nonparametric Regression Methods for Longitudinal Data Analysis

YAKIR · Extremes in Random Fields

YIN · Clinical Trial Design: Bayesian and Frequentist Adaptive Methods

YOUNG, VALERO-MORA, and FRIENDLY · Visual Statistics: Seeing Data with Dynamic Interactive Graphics

ZACKS · Examples and Problems in Mathematical Statistics

ZACKS · Stage-Wise Adaptive Designs

* ZELLNER · An Introduction to Bayesian Inference in Econometrics

ZELTERMAN · Discrete Distributions – Applications in the Health Sciences

ZHOU, OBUCHOWSKI, and MCCLISH · Statistical Methods in Diagnostic Medicine, *Second Edition*

* Now available in a lower priced paperback edition in the Wiley Classics Library.